ECOSYSTEMS BIODIVERSITY

Edited by **Oscar Grillo**
and **Gianfranco Venora**

INTECHOPEN.COM

Ecosystems Biodiversity
Edited by Oscar Grillo and Gianfranco Venora

Published by InTech
Janeza Trdine 9, 51000 Rijeka, Croatia

Edition 2014

Publishing Process Manager Dragana Manestar

Technical Editor Teodora Smiljanic

Cover Designer InTech Design Team

Image Copyright Yellowj, 2010. Used under license from Shutterstock.com

Additional hard copies can be obtained from orders@intechopen.com

Ecosystems Biodiversity, Edited by Oscar Grillo and Gianfranco Venora
p. cm.
ISBN 978-953-307-417-7

INTECH

open science | open minds

Contents

Preface

The biosphere is an intricate net of carefully mixed life forms. Complex interactions of many macro and microscopic species of plants and animals, together with the rich arrays of symbiotic fungi and lichens, are the base of every ecosystem, from forests to coral reefs and freshwater to soils. All ecosystems are not only sources of rich biodiversity, but also extremely productive areas, offering a lot of benefits to mankind. A recent survey results showed that wetlands bring an estimated profit of trillions of dollars to mankind every year. Ecosystems, such as sea-grasses, tidal marshes and tropical forests are exceptionally important in removing CO_2 from the atmosphere, maintaining the water quality, and much more. Consequently, their progressive decline may accelerate the climate change, influencing flora and fauna composition and distribution and resulting in the loss of productivity.

Many ecological studies, environmental evaluations and monitoring have been conducted in the last decades, highlighting the current condition in which our planet is and focusing on future perspectives, and for this reason a lot of management strategies have been developed to try to solve the conservation problems.

Ecosystems Biodiversity presents comprehensive overviews and original studies focused on biological diversity and conservation of various ecosystems. This volume contains 19 chapters written by international experts, presenting thorough research results and critical reviews of the most relevant aspects and most ecologically interesting areas of the Earth. The book includes topics like preservation studies, hazards and conservation management, assessment of environmental variables affecting species diversity, richness and distribution; and also some consideration of the agro-ecosystems biodiversity and it's effects on the quality of human lifestyle.

Oscar Grillo
Stazione Sperimentale di Granicoltura per la Sicilia Caltagirone
Biodiversity Conservation Centre, University of Cagliari,
Italy

Gianfranco Venora
Stazione Sperimentale di Granicoltura per la Sicilia, Caltagirone
Italy

1

Biodiversity and Conservation of Coastal Lagoons

Josefina Garrido, Amaia Pérez-Bilbao
and Cesar João Benetti
University of Vigo
Spain

1. Introduction

Wetlands are sites of high biodiversity and productivity (Mitsch & Gosselink, 2000). They provide essential services, such as maintenance of atmosphere composition, key habitats for migratory species, and important nursery areas (Basset & Abbiati, 2004), but these ecosystems have suffered a serious decline worldwide due to human influence (Shine & Klemm, 1999; Solimini et al., 2008; Stenert & Maltchik, 2007). Declining water quality, drainage, eutrophication and catchment disturbances such as development, loss of natural vegetation and poor agricultural practices are changing the fundamental ecology of shallow lakes in much of the world (Drake et al., 2011). Different management strategies have been developed to solve these conservation problems, for instance the Europeans Natura 2000 network and Water Framework Directive (WFD).

Among these aquatic systems, coastal wetlands have been subject to massive environmental degradation and habitat destruction worldwide (Goudie, 1990). For example, more than 50% of the original area of coastal wetlands that existed in 1900 has been lost in most countries of Western Europe (Jones & Hughes, 1993). However, not until very recently have they become the focus of conservation interest per se (Abbiati & Basset, 2001; Barnes, 1999) with the declaration of Special Areas of Conservation as a consequence of their listing as a priority habitat type (1150*) on Annex I of the European Union Habitats Directive (Council Directive 92/43/EEC).

Coastal lagoons constitute a common coastal environment, occupying 13% of coastal areas worldwide (Kjerfve, 1994). The conservation of these habitats depends largely on the assessment of their natural characteristics, especially biodiversity, which is one of the main criteria used when elaborating wetland protection policies (Ramsar Convention Bureau, 2005). To assess the conservation status of wetlands correctly it is necessary to include studies of the invertebrate fauna. In this sense, the composition and abundance of benthic invertebrates is one of the most important criteria to be considered. Within this group of organisms are insects, especially Coleoptera and Hemiptera, which are two of the most common groups in these environments and also two of the most important groups in the freshwater food chain.

The purpose of this chapter is to study the importance of the invertebrate fauna in the conservation of coastal lagoons and to assess the effectiveness of protecting areas on the conservation of their biological values. The questions are: "Is the Natura 2000 network effective in protecting wildlife?" and "Are Coleoptera and Hemiptera assemblages good indicators of the environmental quality of coastal lagoons?" The chapter proceeds with some definitions of "lagoons", and the characterization of the ecology and description of its biological values. Next there is a brief summary of the conservation status of these assemblages within a case study area: coastal lagoons in the Autonomous Region of Galicia, in North-western Spain. The chapter presents results from studies conducted in two periods of time separated by 10 years: 1998 and 2008. The objective was to determine whether the invertebrate fauna has changed in that period and if so, what factors are responsible for that change. The three studied lagoons are catalogued as Special Areas of Conservation (SAC) under the European Union Habitats Directive and two of them are also protected by the Ramsar Agreement. To complement the biological data, several abiotic variables were recorded at the same time as fauna was sampled. The data corresponding to 1998 had already been published in Garrido & Munilla (2008).

The areas where these lagoons are located had suffered an increasing anthropogenic impact in recent years, mainly due to the expansion of tourist areas. Thus it is important to know if the protection figures are effective in the conservation of species and environments. The analysis of the data has allowed us to know the conservation state of the fauna and thus to assess the environmental health of these ecosystems.

2. Definition and characteristics of coastal lagoons

According to the *Interpretation Manual of European Union Habitats* (EUR April 25th, 2003), lagoons are expanses of shallow coastal salt water, of varying salinity and water volume, wholly or partially separated from the sea by sand banks or shingle, or, less frequently, by rocks. Salinity may vary from brackish water to hypersalinity depending on rainfall, evaporation and through the addition of fresh seawater from storms, temporary flooding of the sea in winter or tidal exchange.

Coastal lagoons are ecotones between terrestrial, freshwater and marine ecosystems (Basset & Abbiati, 2004). Sometimes they are mistaken for other coastal inland aquatic ecosystems, such as salt marshes and estuaries (Esteves et al., 2008). Kjerfve (1994) proposed a definition of coastal lagoons that differentiates them from other similar habitat types: an inland water body, usually oriented parallel to the coast, separated from the ocean by a barrier, connected to the ocean by one or more restricted inlets, and having depths which seldom exceed a couple of meters. A lagoon may or may not be subject to tidal mixing, and salinity can vary from that of a coastal fresh-water lake to a hypersaline lagoon, depending on the hydrologic balance. Lagoons formed as a result of rising sea level during the Holocene or Pleistocene and the building of coastal barriers by marine processes. A lagoon evolves from an estuary valley or shallow open embayment to a partially enclosed barrier-lagoon system, and then, with progressive infilling, to a marsh or deltaic-filled lagoon, ending the cycle with a depositional plain or with an eventual destruction by marine erosion (Nichols, 1989).

So, coastal lagoons can be distinguished according to several characteristics: they are close to the coastline, normally closer than one kilometre; they are not completely open to the sea; during the low tide they preserve part of the water isolated from the sea; the water body is separated from the sea and not completely surrounded by dunes; the vegetation does not cover all the wetland surface, leaving open water without submerged vegetation (Soria & Sahuquillo, 2009). Generally speaking, lagoons undergo some important temporal and spatial changes in their abiotic and biological characteristics. These abiotic gradients determine the structure of the biological assemblages (Kjerfve, 1994).

Typical lagoons are water bodies clearly separated from the sea by a sandbar, being their formation related to coastal dynamics. These ecosystems are very heterogeneous in physiography and hydrology (Ponti et al., 2011), due to the varying balance of water, salt, nutrients, particulate organic and inorganic matter (Orfanidis et al., 2008). They can receive freshwater from streams and brooks, runoff or groundwater, as well as seawater by the action of tides or waves. They are characterized by strong directional gradients of salinity, organic matter, nutrients and oxygen concentrations, which act as fine-mesh filters in selecting potential colonizer species (Basset, 2007).

Fig. 1. The Bodeira lagoon, located in the province of Pontevedra (Northwestern Spain).

The hydrological regime is determined by the communication with the open sea and the freshwater inputs. The influence of seawater can be (1) direct through channels, being influenced by tides, or (2) by the action of storms, which makes them unpredictable systems (Casado & Montes, 1995). In this sense, one of the most important characteristics of these ecosystems is the mixture of fresh and seawater, which produces a salinity gradient. According to Kjerfve (1994) coastal lagoons can span the range of salinities from hypersaline to completely fresh. In some cases, a positive vertical gradient of salinity can exist near the outlet if the water is deep enough. This factor is one of the main conditioners of the structure and the assemblages of the communities in these aquatic systems, as well as in the development of many species.

Their depth varies, although it does not exceed the 11 m. (Soria & Sahuquillo, 2009). They are generally shallow, with a high relation between surface and volume. Slope is usually little and the coastal zone is wide, so in many cases all the basin can be considered coastal.

Proximity to the sea makes wind exposure a greater factor than for other inland aquatic ecosystems, resulting in an important effect of vertical mixing and sediment mobilization (Kirk & Lauder, 2000). Primary production is evenly distributed throughout the lagoon without evident depth gradient, except in some special cases where there is a major presence of floating algae.

The organic matter of a lagoon comes from external inputs (runoff or pollution) and the degradation of the autochthonous organic matter. Mineralization of the organic matter can influence the oxygen balance negatively, and excessive inputs of nutrients can result in the eutrophication of the wetland. High levels of organic material cause a decrease in species diversity (Zaldívar et al., 2008).

Lagoons contain quaternary soils formed by the overflow of streams riverbeds. In many instances the action of the vegetation, especially helophytes, leads to the formation of peaty soils. Where the marine influence is greater than the continental input, sandy sediments carried by the wind are frequent. Their relatively low depth and the inputs make filling rates generally high, but these rates vary with the location, because each lagoon differs with respect to freshwater inflow, sediment input, tidal conditions and geomorphic characteristics (Nichols & Allen, 1981). Terrestrial inputs of sediments in the shallowest areas with small slopes reduce the depth and change the physiographical characteristics, which can affect the structure of biological assemblages. This phenomenon depends on the water recharge of the lagoon, the movements of the dunes or the accumulation of the vegetation.

Due to their special situation at the end of a basin, their transitional character (between continental and marine environments), and the interaction with the terrestrial ecosystem, coastal lagoons are high productivity areas (Basset, 2007; Basset et al., 2006a; Esteves et al., 2008; Kjerfve, 1994), but very unstable systems that tend to disappear due to the filling of the basin (Casado & Montes, 1995; Soria & Sahuquillo, 2009). On a geologic time scale, they are short-lived landscape features, in which the environmental transition is geologically rapid and can occur within decades to centuries (Kjerfve & Magill, 1989; Ward & Ashley, 1989).

3. Biodiversity of coastal lagoons

Coastal lagoons support a rich specific biodiversity, so we can find different groups of organisms, like plants, animals or microbes. They act as spawning grounds for marine fish and invertebrates, and behave as resting areas for many species of migratory birds (Aliaume et al., 2007).

Macrophytes can grow almost everywhere in the basin, due to its shallowness and the transparency of the water, except in more unstable or seasonally dry areas. Only nutrients determine the vegetation growth and the turbidity of the water by phytoplankton in cases of overconcentration. Lagoon vegetation constitutes a particular habitat for many species, especially birds and invertebrates, and is the substrate on which periphytic communities develop. They provide refuge, food and substrate for aquatic organisms (Diehl & Kornijów, 1998). Moreover, macrophytes create a micro climate with softened temperatures during the hot summer, produce oxygen and retain the upper horizon of the sediment helping to control the turbidity. In coastal lagoons the helophytes usually cover a more or less wide stretch in the shallowest areas, acting as a buffer zone for impacts on the open water. Submerged macrophytes can reduce eutrophication by stabilizing clear water states or outcompeting phytoplankton for nutrients and light (Søndergaard & Moss, 1998). But they can have a negative contribution to the productivity in most shallow lakes, because they influence water exchange with the offshore area, reducing water movement and increasing sedimentation rates (Arocena, 2007).

Fig. 2. Proliferation of the water lily (*Nymphaea alba*) in the Xuño lagoon (A Coruña, Northwestern Spain).

The microbial community is composed of common heterotrophic taxa of natural waters, the presence of coliforms in habitats suffering from human impact being of special interest (Soria & Sahuquillo, 2009). The bacterial community is basic in the energy flows, because it decomposes and remineralizates organic matter. The hydrology of the lagoon (freshwater or marine inputs) greatly influences the bacterial composition (Piccini & Conde, 2004). Changes in the ecosystem functioning can change the composition of the bacterial assemblages as an answer to the stress factor (Langenheder et al., 2003).

Phytoplankton in coastal lagoons is rich and diverse (Ramil et al., 2007), and as primary producers are an important part of the food web in freshwater environments. Phytoplankton is usually composed of diatoms, dinoflagellates, chlorophytes, cryptophytes and other microflagellates. In the coastal lagoons it is often responsible for the eutrophication phenomenon, with dominance of phytoplankton over other groups of organisms.

Zooplankton consists of heterotrophs that live suspended in the water column, and includes protists (flagellates and ciliates) and micro animals, mainly rotifers and microcrustaceans (cladocerans, copepods and ostracods). Overall, the plankton is consumed by fish, macroinvertebrates and some waterfowl, especially in brackish systems.

Benthos comprise the set of mobile or sessile aquatic organisms living in or on the substrate. They are very important in the recycling of the organic matter, facilitating the activity of the bacterial decomposers (Casado & Montes, 1995). With regard to photosynthetic organisms, besides being part of the plankton communities of microalgae (phytoplankton), many microscopic primary producers can grow attached to solid substrates, on rocks (epilithon), on the sediment (episammon) or on the vegetation (periphyton).

Zoobenthos are macroinvertebrates that live on the bottom (benthos)of water bodies, and on some macrophytes. In addition to the larvae of numerous species of insects (especially flies, dragonflies, beetles and bugs), zoobenthos includes flatworms, annelids, molluscs and crustaceans. In lentic ecosystems, zoobenthos is very important compared with the pelagic communities of animals, particularly in shallow systems and coastal areas. Surface area, water salinity, and outlet width and length, can actually be considered the key limiting, dimensions defining the environmental niche space for benthic macroinvertebrates in lagoon ecosystems (Basset et al., 2006b).

Fish are the main organisms of nekton, i. e. all the organisms that actively swim in the water. In coastal lagoons, the presence of fish is common and is related to the salinity of the system and its opening to the sea. Lentic ecosystems are an ideal habitat for waterfowl and amphibians. Nowadays amphibians are in decline worldwide, partly due to the decline of aquatic ecosystems where they live, largely coastal lagoons. On the other hand, waterfowl are important consumers in lagoons, but can also be sources of nutrients with a negative effect on eutrophication. Besides birds, amphibians and fishes, other vertebrates such as certain species of reptiles and mammals are typical of these ecosystems, although the diversity of these animals is more limited and only a few species are related to coastal lagoons.

4. Hazards and conservation management

Modification of extensive natural areas by human action, such as desiccation, urbanization, hydrologic modification and isolation of areas previously connected, has led to a reduction or disappearance of large areas of wetlands and coastal lagoons. Increasing of tourism pressure and the use of these ecosystems for different human activities such as aquaculture or fisheries has induced internal perturbations such as pollution or removal of indigenous species (Aliaume et al., 2007).

In coastal lagoons, species distribution is affected by heterogeneity of the environmental variable, such as water depth, hydrodynamic conditions, sediment characteristics, as well as by sources of anthropogenic disturbance that could interact with natural heterogeneity affecting patterns of species distribution, species diversity and ecosystem functioning (Ponti et al., 2011). Sometimes it is difficult to distinguish between natural heterogeneity and human disturbance.

Healthy lagoons produce food for birds and fish, and provide for several human extractive activities and significant food resources (aquaculture, fishing, hunting, etc.) (Sylaios & Theocharis, 2002). However, an increase in primary production by external inputs of nutrients, mainly anthropogenic, stresses the ecosystem favouring opportunistic species and reducing diversity. Unfortunately, coastal areas are highly impacted by eutrophication (Nixon, 1995).

These systems are very sensitive to potential climate change due to their dependence on continental and marine inputs (Álvarez-Cobelas et al., 2005). Recent climate changes are beginning to have effects on these environments (Soria & Sahuquillo, 2009). According to the Assessment Report of the Intergovernmental Panel on Climate Change (IPCC) (2007), many observed changes in freshwater biological systems are associated with rising water temperatures, as well as related changes in salinity, oxygen levels and circulation. Sea level rise and human development are together contributing to losses of coastal wetlands and increasing damage from coastal flooding in many areas. Climate change is likely to reduce flooded areas and increase eutrophication, increasing helophytes populations and grasslands and reducing submerged macrophytes.

One of the major consequences of the climate change is the spread of exotic species worldwide (Ricciardi, 2006), affecting the ecological functions of an ecosystem (Maezono & Miyashita, 2003; Ricciardi et al., 1997) and causing a loss of indigenous biodiversity (Rodríguez-Pérez et al., 2009; Witte et al., 2000).

It will also affect the biogeochemical cycles of coastal lagoons, especially in peaty environments, characterized by low concentrations of nutrients and organic matter accumulation. In this sense, the relative sedimentation ratio is extremely important in the assessment of lagoon behaviour under climate change scenarios, particularly regarding the prospects for accelerated sea level rise. Research into sedimentation rates is essential to formulating management plans for lagoons under present conditions of sea level rise, as it is to making an informed assessment of the hazards that might be posed by climate change (Kirk & Lauder, 2000).

5. Case study

5.1 Introduction

Major gaps still remain in aquatic conservation knowledge of lagoons and other coastal brackish water habitats in temperate regions, and much more emphasis needs to be placed on the investigation of the ecology of such systems to ensure their effective management (Barnes, 1999). Information about the relative biodiversity value of different water body types is a vital pre-requisite for many strategic conservation goals (Williams et al., 2003), including sustainable catchment management as required by the EC Water Framework Directive (2000/60/EC). In this sense, the composition and abundance of benthic invertebrates is one of the most important criteria to be considered.

Little is known of the communities of aquatic insects in coastal lagoons, especially when compared with freshwater ecosystems (Garrido & Munilla, 2008). Within this group of organisms are Coleoptera and Hemiptera, two of the most common groups in these environments. Water beetles are generally considered a suitable group to assess the environmental and conservation value of wetland sites and habitats (Bilton et al., 2006; Davis et al., 1987; Eyre & Rushton, 1989; Eyre et al., 1993; Foster, 1987, 1999; Foster et al., 1990; Oertli et al., 2005; Pérez-Bilbao & Garrido, 2009; Ribera & Foster, 1992; Sánchez-Fernández et al., 2006). The detailed study of the autoecology of aquatic insects, especially water beetles, has revealed the existence of many species with narrow ecological requirements including adaptations to saline environments (Foster, 2000; Greenwood & Wood, 2002). Aquatic Hemipteran assemblages are generally poorer in species than are water beetles, and seem to be more resilient to environmental change (Broering & Niedringhaus, 1988; Eyre & Foster, 1989; Roback, 1974; Savage, 1996; Tuly et al., 1991; Vierssen & Verhoeven, 1983). However, several families display differing tolerances to water pollution and are potential bioindicators of water quality (Jansson, 1977), so they can be used in terms of regional or global conservation planning of freshwater biodiversity (Polhemus & Polhemus, 2008).

The main objectives of this study were (a) to analyse the composition and structure of aquatic Coleoptera and Hemiptera assemblages in three protected coastal lagoons using data obtained in studies carried out in two periods of time separated by 10 years (1998 and 2008); (b) to assess the importance of the invertebrate fauna (Coleoptera and Hemiptera) in the conservation of these ecosystems; (c) and to assess the effectiveness of protecting areas (Natura 2000) on the conservation of their biological values studying the change in these assemblages after 10 years.

Fig. 3. The Vixán lagoon, located in the province of A Coruña (Northwestern Spain).

5.2 Study area

The study area comprised three coastal lagoons in the Autonomous Region of Galicia (North-western Spain): Bodeira, Xuño and Vixán (Figures 1-4). The three lagoons are located within Special Areas of Conservation (SAC) under the European Union Habitats Directive (Directive 92/43/EEC). Vixán and Bodeira are also protected by the Ramsar Agreement (Ramsar sites no. 598 and 452, respectively). Although each of the lagoons selected was completely separated from the sea by a barrier of sand dunes, seawater can enter by filtration or when sea storms break through the sand barrier. Freshwater is provided by small streams, rainfall and runoff. The extent of the standing water ranged from 0.8 Ha in Bodeira, to 11.2 Ha in Vixán and 2.3 Ha in Xuño. According to the thermal classification of lakes proposed by Lewis (1983), these lagoons are warm polymictic, without a stable thermal stratification. Bodeira and Xuño are fresh water systems, and Vixán is considered an oligohaline water body (Ramil et al., 2007).

Vixán has a relatively large reed (*Phragmites australis* (Cav.) Trin. Ex Steud) bed. Bodeira presents species such as *Polygonum amphibium* L., *Glyceria fluitans* (L.) R. Br., *Ranunculus* spp. L or *Myriophyllum* spp. Xuño has an important assemblage of the water lily (*Nymphaea alba*

L.) and also has different species of macrophytes including *Hydrocotyle vulgaris* L., *Ranunculus* spp. or *Veronica* spp.

The climate of the study area is mild and wet, with annual average temperatures around 14°C and annual rainfall between 1200 and 1800 mm. The landscape consists of a mosaic of farmland, heather (*Erica* spp.) and gorse (*Ulex* spp.) heathlands and pine (*Pinus pinaster* Ait.) forests in a densely populated rural setting (Garrido & Munilla, 2008).

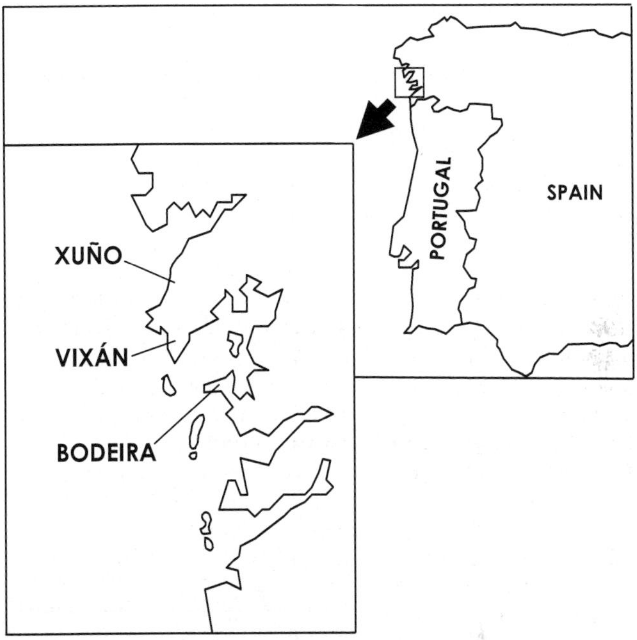

Fig. 4. Map of the study area showing the location of the lagoons under study (based on Garrido & Munilla, 2008).

5.3 Material and methods

5.3.1 Sampling methods and variables measured

The three lagoons were sampled in spring and summer in two periods, 1998 and 2008. The fauna was collected by sweeping an entomological water net (500 μm mesh, 30 cm diameter and 60 cm deep) across a 10m transect running parallel to the margin. This semi-quantitative method allows for direct comparisons across sites or time because sampling effort can be assumed equivalent. The material was preserved in 99% ethanol, and sorted out and identified to species level at the laboratory. After being studied, they were conserved in 70% ethanol and deposited in the scientific collection of the Entomology Laboratory at Vigo University. Water temperature, pH, conductivity, and dissolved oxygen were measured at each lagoon at the same time as the fauna was sampled. Salinity was only measured in 1998. As the salinity and conductivity values were highly correlated (r= 0.994) (Garrido & Munilla, 2008) and as this variable provides redundant information we decided not to measure it in 2008.

5.3.2 Data analysis

The structure of the assemblage was assessed using different diversity indices: Species richness (S); Rarefied species richness (ES); Abundance (N) and the Shannon-Wiener diversity index (H'). The values of rarefied richness were calculated for 100 individuals ES (100). ES and H' were calculated using the program PRIMER version 6. Analysis of variance (two-way ANOVA) was used to test for significant differences between the three lagoons and the two periods of time in both diversity indices and environmental variables. ANOVA was run using SPSS version 19.

Relationships between environmental variables with diversity indices and species were determined by a Pearson correlation test. Prior to this, the Kolmogorov-Smirnov test was used to verify the normal distribution of the data. Variables not showing normal distribution were logarithmically transformed ($\log_{10}$). These analyses were performed using SPSS version 19.

Complete linkage cluster analysis with Bray-Curtis coefficient was used to cluster the samples into groups and thus to be able to verify the changes in the Coleoptera and Hemiptera assemblages composition. This analysis was carried out with PRIMER version 6.

Canonical correspondence analysis (CCA) was used to analyze species-environment relationships in order to identify environmental factors potentially influencing water beetle assemblages. A Monte Carlo permutation test was performed to assess the significance of the relationships between environmental variables and species composition among study sites (Heino, 2000). CCA was carried out on global abundances, that is, total number of individuals collected at a site over the sampling period. Species with less than 10 individuals were removed from the analysis, which was performed using the CANOCO 4.5 program (Ter Braak & Šmilauer, 2002). The environmental factors used were pH, water temperature, dissolved oxygen and conductivity.

5.4 Results

In total, 71 species of 11 families of aquatic Coleoptera (55 species) and 8 families of aquatic Hemiptera (16 species) were collected (Table 1). The most representative families of beetles were Dytiscidae (20 species), Hydrophilidae (13 species), Haliplidae and Hydraenidae (5 species) and among bugs the most representative families were Corixidae (5 species) and Notonectidae (4 species).

Only 36 species (27 Coleoptera and 9 Hemiptera) were recorded in both periods, corresponding to approximately half the total species recorded (71). Besides, 20 species (15 Coleoptera and 5 Hemiptera) were only recorded in 1998 and 15 species (13 Coleoptera and 2 Hemiptera) were only recorded in 2008.

5.4.1 Biogeography

According to the classification proposed by Ribera et al. (1999) for Coleoptera and Millán et al. (2002) for Hemiptera, the species were classified in their respective chorological categories. In the three studied lagoons, species of four categories were captured: northern, trans-Iberian, southern and endemic (Figure 5). In the two years, most of the species belonged to the trans-Iberian category (84-88%), followed by northern (5-10%), endemics (4-5%) and southern (2%). It has to be noted that all the species of aquatic bugs are trans-Iberian, i. e. they are present in Europe north of Pyrenees, the Iberian Peninsula and north Africa (Table 1).

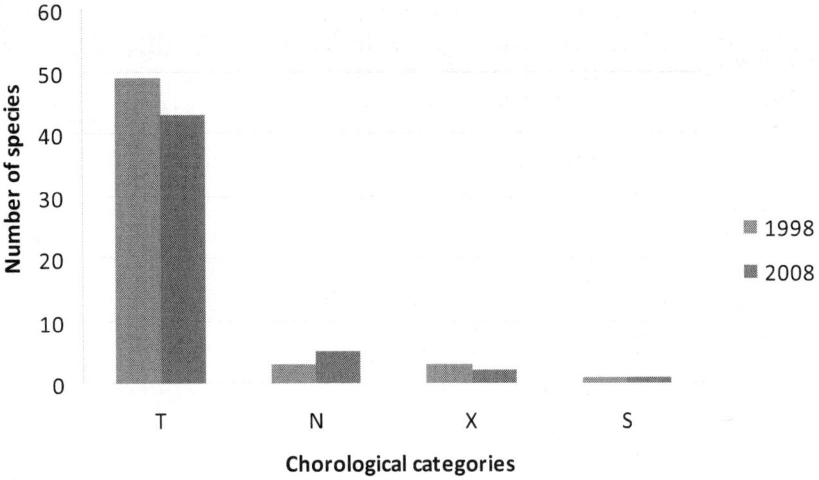

Fig. 5. Species composition of the chorological categories of aquatic Coleoptera and Hemiptera captured in the two periods of time. S: southern; X: endemic; N: northern; T: Trans-Iberian.

5.4.2 Abundance, species richness and diversity

The total abundance was of 10,166 individuals (4,395 Coleoptera and 3,904 Hemiptera). In 1998, 3,399 individuals were captured (1,005 beetles and 2,394 bugs), and in 2008 were 6,767 (3,390 Coleoptera and 3,377 Hemiptera). The greatest abundance was observed in the Bodeira lagoon in 2008, with 3,732 individuals collected (1,856 in spring and 1,876 in summer). On the contrary, the lowest value was obtained in Vixán in spring of 2008 with 206 individuals. In 1998 the hemipteran families with the highest number of individuals were Corixidae (2,032) and Pleidae (285). Among coleopterans, we found the highest abundances for Dytiscidae (381), Hydrophilidae (272) and Hydraenidae (180). On the contrary, in 2008 the hemipteran families with the highest number of individuals were Pleidae (2,533) and Corixidae (240). Among coleopterans, we found the highest abundances for Hydraenidae (2,045), followed by Hydrophilidae (670) and Dytiscidae (380).

In 1998 the most abundant species of aquatic bug found in the study area was *Hesperocorixa linnaei* (1,146), while in 2008 it was *Plea minutissima* (2,533). Within water beetles, the most abundant species in 1998 was *Hygrotus inaequalis* (202 individuals) and in 2008 it was *Ochthebius viridis fallaciosus* (1,812 individuals).

Only one species of Coleoptera (*Hygrotus inaequalis*) and one of Hemiptera (*Plea minutissima*) were recorded in all samples, in the three lagoons and in the two periods, which corresponds to only 1.42% of species. On the other hand, 10 species of water beetles (*Haliplus heydeni, H. guttatus, Hydroglyphus geminus, Hydroporus vespertinus, Graptodytes varius, Dytiscus marginalis, Helophorus flavipes, Hydrochus flavipennis, Berosus signaticollis* and *Dryops striatellus*) and 4 of bugs (*Nepa cinerea, Hesperocorixa sahlbergi, Notonecta glauca* and *Gerris gibbifer*) were recorded in just one of the samples, which corresponds to 19.72% of species.

Species	Bodeira 1998	Bodeira 2008	Vixán 1998	Vixán 2008	Xuño 1998	Xuño 2008	Distribution
COLEOPTERA							
Gyrinidae							
Gyrinus (Gyrinus) caspius Ménétriés, 1832		1	4		4	7	T
Haliplidae							
Peltodytes caesus (Duftschmid, 1805)	2		8	2	1	13	T
Peltodytes rotundatus (Aubé, 1836)		2	1				T
Haliplus (Neohaliplus) lineatocollis (Marshan, 1802)		2	4				T
Haliplus (Haliplus) heydeni Wehncke, 1875			1				T
Haliplus (Liaphlus) guttatus Aubé, 1836						2	T
Noteridae							
Noterus laevis Sturm, 1834	5	22	8	13		116	T
Paelobiidae							
Hygrobia hermanni (Fabricius, 1775)	6	5			51	3	T
Dytiscidae							
Liopterus haemorrhoidalis (Fabricius, 1787)		1	1				T
Laccophilus minutus (Linnaeus, 1758)	1	5	3	2	2	66	T
Hyphydrus aubei Ganglbauer, 1891	1	8	4	1	9	9	T
Hydrovatus clypealis Sharp, 1876		5		1		16	T
Bidessus goudoti (Laporte, 1835)	27	23	4		41	87	T
Hydroglyphus geminus (Fabricius, 1792)			1				T
Hygrotus inaequalis (Fabricius, 1777)	78	56	100	17	24	6	T
Hygrotus lagari (Fery, 1992)						2	S
Hydroporus vagepictus Fairmaire & Laboulbène, 1854					2	1	X
Hydroporus vespertinus Fery & Hendrich, 1988	1						X
Graptodytes aequalis Zimmermann, 1918			4				S
Graptodytes flavipes (Olivier, 1795)	1		14				T
Graptodytes ignotus (Mulsant, 1861)			2				T
Graptodytes varius (Aubé, 1838)						11	T
Stictonectes lepidus (Olivier, 1795)			3				T
Agabus bipustulatus (Linnaeus, 1767)		1	1		1		T
Rhantus (Rhantus) suturalis (McLeay, 1825)		2	47	8	4	1	T
Colymbetes fuscus (Linnaeus, 1758)		1					T
Dytiscus marginalis Linnaeus, 1758			1				T
Cybister (Scaphinectes) lateralimarginalis (De Geer, 1774)			3		1		T
Helophoridae							
Helophorus (Rhopalhelophorus) flavipes Fabricius, 1792				1			N
Helophorus (Rhopalhelophorus) minutus Fabricius, 1775	8	2	55	5	2	2	N
Hydrochidae							
Hydrochus angusi Valladares, 1988			1	7			X
Hydrochus angustatus Germar, 1824			1			26	T
Hydrochus flavipennis Küster, 1852						6	T
Hydrophilidae							
Berosus (Berosus) affinis Brullé, 1835	28		8		83	19	T
Berosus (Berosus) hispanicus Küster, 1847			57		2		T
Berosus (Berosus) signaticollis (Charpentier, 1825)						9	T
Paracymus scutellaris (Rosenhauer, 1856)		8		3		6	T
Anacaena lutescens (Stephens, 1829)			49	19	2	44	T

Table 1. Aquatic Coleoptera and Hemiptera species recorded in the lagoons Bodeira, Vixán and Xuño in the years 1998 and 2008, with their abundance and the biogeographical distribution.

Helochares (Helochares) lividus (Forster, 1771)	4				5		T
Helochares (Helochares) punctatus Sharp, 1869		149				79	N
Enochrus (Lumetus) fuscipennis (Thomson, 1884)		2		19		182	T
Enochrus (Lumetus) halophilus (Bedel, 1878)	1		7		1		T
Cymbiodyta marginella (Fabricius, 1792)			9	6		27	N
Hydrobius fuscipes (Linnaeus, 1758)		4				1	T
Limnoxenus niger (Gmelin, 1790)		25	12	29	1	39	N
Hydrophilus (Hydrophilus) pistaceus Laporte, 1840			2		1		T
Hydraenidae							
Hydraena testacea Curtis, 1830				4	1	14	T
Limnebius furcatus Baudi, 1872		26	27	6		182	T
Ochthebius (Asiobates) dilatatus Stephens, 1829			1	1			T
Ochthebius (Ochthebius) nanus Stephens, 1829	1		1				T
Ochthebius (Ochthebius) viridis fallaciosus Ganglbauer, 1901		851	147	25	2	936	T
Elmidae							
Oulimnius rivularis (Rosenhauer, 1856)		12	1	2			T
Dryopidae							
Dryops algiricus (Lucas, 1849)		78	4			8	T
Dryops luridus (Erichson, 1847)				2		6	T
Dryops striatellus (Fairmaire & Brisout, 1859)					5		T
HEMIPTERA							
Nepidae							
Nepa cinerea Linnaeus, 1758		1					T
Ranatra linearis (Linnaeus, 1758)	1			1			T
Corixidae							
Corixa panzeri (Fieber, 1848)	221	13	26	13	160		T
Hesperocorixa linnaei (Fieber, 1848)	385	91	29	50	732	73	T
Hesperocorixa sahlbergi (Fieber, 1837)			4				T
Sigara (Halicorixa) stagnalis stagnalis (Leach, 1817)	16		150		288		T
Sigara (Subsigara) scotti Douglas & Scott, 1868	20				1		T
Naucoridae							
Naucoris maculatus maculatus Fabricius, 1798	17	336	12	41	16	20	T
Notonectidae							
Anisops sardeus sardeus Herrich-Schäffer, 1849		153		8		1	T
Notonecta (Notonecta) glauca glauca Linnaeus, 1758			2				T
Notonecta (Notonecta) meridionalis Poisson, 1926			21			2	T
Notonecta (Notonecta) viridis Delcourt, 1909	1	4		2	4	1	T
Pleidae							
Plea minutissima minutissima Leach, 1817	238	1812	33	536	14	185	T
Hydrometridae							
Hydrometra stagnorum (Linnaeus, 1758)		12	1	1			T
Veliidae							
Microvelia (Microvelia) pygmaea (Dufour, 1833)		19	1	2			T
Gerridae							
Gerris (Gerris) gibbifer Schummel, 1832			1				T
Total abundance	1063	3732	876	827	1460	2208	

Table 1. (continued). Aquatic Coleoptera and Hemiptera species recorded in the lagoons Bodeira, Vixán and Xuño in the years 1998 and 2008, with their abundance and the biogeographical distribution.

In total, the species richness of the two periods of time (1998 and 2008) was of 71 species. In 1998, 56 species (42 Coleoptera and 14 Hemiptera) were captured, and 51 in 2008 (40 beetles and 11 bugs). If we analyse these results by lagoons, we can see that in 1998 Vixán presented the highest value of water beetle richness with 36 species, followed by Xuño (22) and Bodeira (14). On the other hand, ten years later it is Xuño that presents the highest richness with 31 species, followed by Bodeira (24) and Vixán (21). Aquatic bug richness did not vary too much between lagoons and in 1998 was again Vixán the richest one (11), followed by Bodeira (8) and Xuño (7). In 2008, both Bodeira and Vixán presented the same richness value with 9 species, and Xuño was still the poorest lagoon (6).

Rarefied species richness is the expected number of species for a given number of randomly sampled individuals and facilitates comparison of areas in which densities may differ (McCabe & Gotelli, 2000). The Bodeira lagoon had the lowest values in summer of 2008 (7.65) and in summer of 1998 (7.71). The highest values correspond to Vixán in spring of 1998 (23.75) and to Xuño in summer of 2008 (18.86).

Diversity indices	1998			2008		
	Mean ± SD	Minimum	Maximum	Mean ± SD	Minimum	Maximum
Richness S	22.17 ± 9.02	12	38	24.83 ± 3.97	20	30
Rarefied Richness ES (100)	12.57 ± 5.92	7.71	23.75	13.48 ± 3.93	7.65	18.86
Abundance N	566.50 ± 163.14	395	782	1127.83 ± 689.60	206	1876
Diversity H'(log2)	2.65 ± 0.83	1.95	4.18	2.46 ± 0.71	1.67	3.77

Table 2. Mean, minimum and maximum values of the diversity indices measured in 1998 and 2008.

The Shannon-Wiener index (H'(log2)) revealed that, in general, the studied lagoons presented high diversity values. The lowest diversity was recorded in Bodeira in summer of 2008 (1.67) and the highest in Vixán in spring of 1998 (4.18). The mean diversity values between the two periods of time were very similar. In 1998, the mean diversity was 2.65, while in 2008 it was 2.46.

Table 2 shows the mean, minimum and maximum values of the diversity indices for the coastal lagoons studied. There were no significant differences ($p < 0.05$) among periods (1998 and 2008) in any of the diversity indices, as evidenced by the ANOVA (Table 3).

Diversity indices and environmental variables	ANOVA Factor 1		ANOVA Factor 2	
	F	p	F	p
Richness S	0.439	0.522	1.078	0.38
Rarefied Richness ES (100)	0.099	0.759	2.469	0.14
Abundance N	3.765	0.081	0.981	0.412
Diversity H'(log2)	0.195	0.668	3.488	0.076
Water Temperature (°C)	0.006	0.941	0.64	0.55
pH	3.172	0.105	2.071	0.182
Conductivity (µS/cm)	0.989	0.343	1.166	0.355
Dissolved Oxygen (mg/l)	0.097	0.762	0.599	0.57

Table 3. Results of the ANOVA with years (1) and lagoons (2) as factors.

5.4.3 Environmental variables

Table 4 shows mean, minimum and maximum values of the environmental variables measured in the two years. The main result to highlight is the extreme value of conductivity measured in the summer of 1998 in the Vixán lagoon (34,816.60 µS/cm), much higher than in the other surveys. However, the ANOVA showed no significant differences ($p < 0.05$) among periods (1998 and 2008) in any of the environmental variables (Table 3).

Environmental variables	1998			2008		
	Mean ± SD	Minimum	Maximum	Mean ± SD	Minimum	Maximum
Water Temperature (°C)	21.45 ± 1.70	19.45	24.35	21.33 ± 3.33	16.40	25.60
pH	7.72 ± 0.92	6.70	8.95	7 ± 0.35	6.65	7.65
Conductivity (µS/cm)	6,296.55 ± 13,975.86	263.45	34,816.60	618.67 ± 443. 88	227	1,408
Dissolved Oxygen (mg/l)	8.75 ± 4.85	2.45	14.20	7.79 ± 5.85	2.03	16.50

Table 4. Mean, minimum and maximum values of the environmental variables measured in 1998 and 2008.

5.4.4 Correlations

The Pearson correlation test was performed to assess the relation between the environmental variables and the species and diversity indices. There were no significant correlations ($p<0.05$) between physicochemical variables and the ecological parameters. However, we found several significant and positive correlations between these variables and species. The species *Gyrinus caspius* was correlated with temperature ($r= 0.66$); *Corixa panzeri* ($r= 0.82$) and *Sigara scotti* ($r= 0.65$) were related to pH, while *Berosus hispanicus* ($r= 1.00$) and *Rhantus suturalis* ($r= 0.92$) presented high correlations with conductivity.

5.4.5 Influence of environmental variables on Coleoptera and Hemiptera assemblages

Figure 6 shows the results of the CCA. The eigenvalues for axes 1-4 were 0.466, 0.282, 0.226 and 0.076 respectively. Correlations for axes III and IV with environmental variables were low ($r < 0.5$), and only axes I and II were used for data interpretation. The cumulative percentage of variance for the species-environmental relation for these two axes was of 71.2%. The first two canonical axes were significant, as shown by the Monte Carlo permutation test ($p = 0.004$).

The first principal gradient is positively correlated with oxygen ($r = 0.453$) and temperature ($r = 0.442$), and negatively correlated with pH ($r = -0.585$). The second gradient is positively correlated with conductivity ($r = 0.621$) and negatively correlated with oxygen ($r = -0.779$), temperature ($r = -0.642$) and pH ($r = -0.447$).

According to the diagram, the species *Berosus hispanicus, Rhantus suturalis* and *Sigara stagnalis* are related to sites with high values of conductivity, while species such as *Sigara scotti* or *Corixa panzeri* prefer sites with high values of pH.

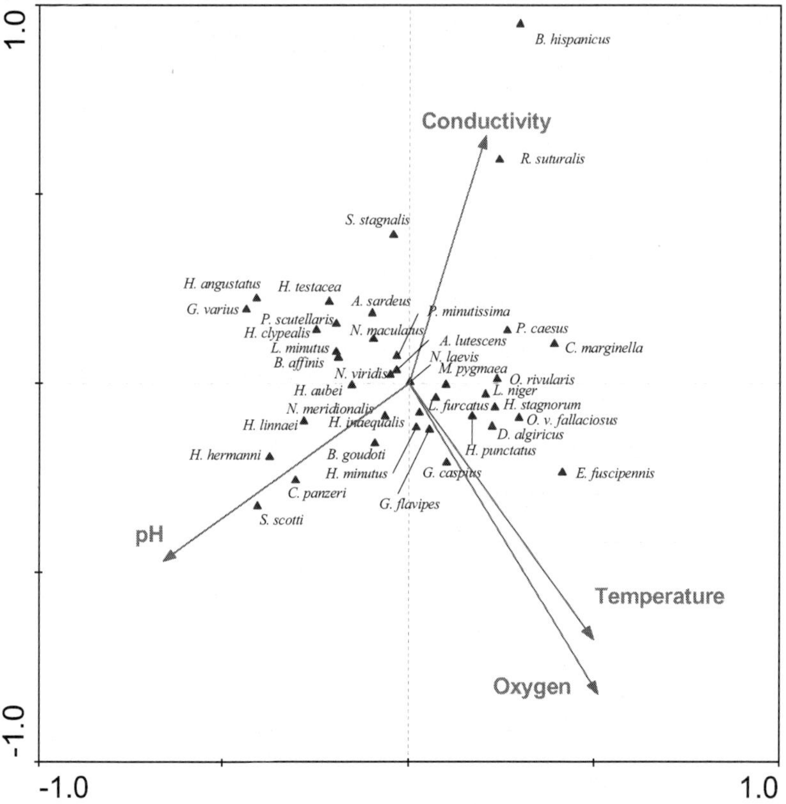

Fig. 6. Results of the canonical correspondence analysis (CCA) based on the aquatic Coleoptera and Hemiptera assemblages with respect to environmental variables. Arrows represent the environmental variables and triangles the species.

5.4.6 Assemblage composition

In 1998, 10 species were exclusive (not recorded in 2008) in Bodeira (7 Coleoptera and 3 Hemiptera), 26 in Vixán (21 Coleoptera and 5 Hemiptera) and 11 in Xuño (8 Coleoptera and 3 Hemiptera). On the other hand, in 2008, 22 species were exclusive (not recorded in 1998) in Bodeira (18 Coleoptera and 4 Hemiptera), 9 in Vixán (6 Coleoptera and 3 Hemiptera) and 18 in Xuño (16 Coleoptera and 2 Hemiptera).

The Bray-Curtis coefficient was used to calculate the affinity between samples, because it allows the comparison between the assemblages of different ecosystems or in the same ecosystem at different moments. The results ranged from 0 to 1. If the value obtained is close to 1, the populations will be more similar. In this study twelve samples were used, six in 1998 and six in 2008. The greatest degree of affinity was observed between spring and summer of 1998 in Bodeira (64.53%) and between Bodeira and Xuño in spring of 2008 (61.41%). If we compare the two periods of time, we observe a low degree of affinity in the three studied lagoons (Xuño: 10.79; Bodeira: 18.89; and Vixán: 23.84). In the figure 7, the temporal affinity is greater than the spatial one, i. e. the studied lagoons were more similar between them in the same year than the same lagoon in the two different periods of time. Two main groups corresponding to 1998 and 2008 are represented. In the "1998" group, it is important to note that Vixán is isolated because it had a low affinity with other lagoons (18%).

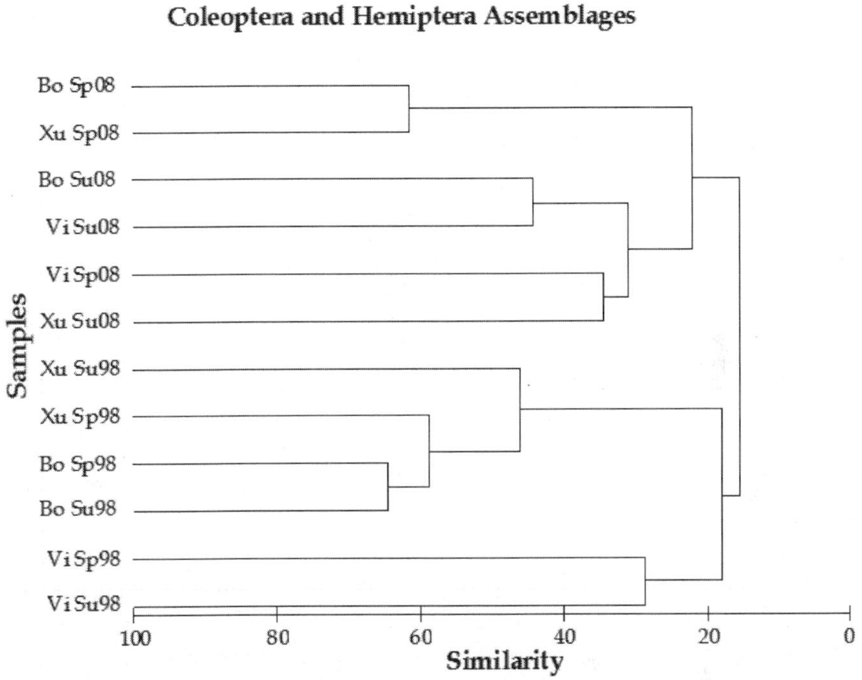

Fig. 7. Dendrogram from complete linkage clustering based on the Bray-Curtis coefficient.

6. Discussion

The aquatic Coleoptera and Hemiptera assemblages in the studied lagoons were predominantly composed of species that are widely distributed on the regional scale. Most of the species of water beetle collected had a wide Palearctic distribution and only three species could be considered endemic to Iberia: *Hydroporus vagepictus*, *Hydroporus vespertinus* and *Hydrochus angusi* (Ribera et al., 1999; Ribera & Vogler, 2004). The paucity of endemics among coastal water beetles has been attributed to the temporal dynamics of the lagoons and their recent origin in geological time, as most of the sites are 'temporary' on a scale of a few hundred years (Ribera et al., 1996) and this could also apply to aquatic Hemiptera. According to Ribera & Vogler (2000), stagnant water bodies present less endemic species than running waters, because stagnant species have a greater dispersal capability and thus a greater geographical distribution.

It should be noted that Vixán is one of the few localities where *Hydrochus angusi* (Pérez-Bilbao et al., 2009) and *Cymbiodyta marginella* (Pérez-Bilbao et al., 2010; Sáinz-Cantero & Garrido, 1996) have been recorded in the Iberian Peninsula. Regarding water bugs, we can highlight the species *Sigara scotti* and *Notonecta glauca glauca*, which are distributed in the Iberian Peninsula as a few scattered populations (Nieser & Montes, 1984; Nieser et al., 1994).
In the three lagoons, Coleoptera assemblages were less similar in species composition across sites than Hemiptera assemblages. Compared with other coastal stagnant water bodies in the Iberian Peninsula and the Balearic Islands, species richness of these coastal lagoons can be considered high. It is important to highlight that only three lagoons were sampled, when in other studies the number of sampling sites is much higher, e. g. 13 coastal wetlands in Ribera et al. (1996), 9 sites in Garrido et al. (1996, 1997), 22 ponds in Florencio et al. (2009) or 38 lagoons in Lucena-Moya et al. (2010). A richer Coleoptera assemblage seems to be the rule in aquatic insect communities where the proportion of Coleoptera to Hemiptera usually varies from 3:1 to 4:1 (Eyre & Foster, 1989; Lancaster & Scudder, 1987; Millán et al., 1997, 2001; Moreno et al., 1997). This lends further support to the idea that, for coastal brackish-water systems, Coleoptera assemblages are more valuable than Hemiptera assemblages for characterizing water bodies and providing information about their conservation status (Eyre & Foster, 1989; but see Hufnagel et al., 1999). In habitats with high conductivity values this proportion (3:1 to 4:1) can indicate a good conservation state (Sánchez-Fernández et al., 2007). On the other hand, fewer similarities in species composition among sites imply that a larger proportion of wetland sites will be needed to ensure the protection of entire assemblages (Margules et al., 1988).
According to Duigan et al. (1998), species richness is one of the most important parameters when assessing the conservation state of an ecosystem. The aquatic assemblages of the studied coastal lagoons seem to reflect adequate conditions for the conservation of biodiversity in terms of species richness. One aspect to consider in the conservation and management of aquatic ecosystems, apart from the danger of extinction, isolation and fragmentation, is the rarity of the species in terms of its distribution range, habitat specificity and abundance of their populations (Gaston, 1994). The existing information on habitat specificity of each species is not abundant in general and this is due to the lack of autoecological studies, because abundance data are often estimates, or simply indicate the presence/absence of species (Millán et al., 2002).
The observed values of physicochemical variables were within the range considered normal for these types of environments (fresh water and oligohaline water systems) in both periods (1998 and 2008). These values were similar to those obtained by other authors (Boix et al. 2008; Della Bella et al. 2008), except Vixán in summer 1998, probably due the entrance of sea water caused by a storm.
The results of the CCA corroborate the data obtained by the Pearson correlation tests for the relation between the environmental variables and the species. The response of beetles and bugs to the increase in conductivity seemed to define, at least, the species typical of sites with high values of this variable. Some species respond negatively to high conductivity (Garrido & Munilla, 2008), while others are adapted to sudden changes in this variable and can be considered tolerant. In this sense, ordination analysis identified a group of tolerant species (i. e. *Berosus hispanicus*, *Rhantus suturalis* and *Sigara stagnalis*), typical of habitats with high conductivity or salinity values (Arnold & Ormerod, 1997; Garrido & Munilla, 2008;

Hansen, 1987; Martinoy et al., 2006). Different authors have identified salinity as the most important factor regulating species composition (Britton & Johnson, 1987; Cognetti & Maltagliati, 2000; Martinoy et al., 2006; Timms, 1993). Thus, the two samples with higher salinities (Vixán in spring and summer of 1998) appear separated from the rest of the samples of 1998 (Figure 7).

In this study, *Gyrinus caspius* was positively correlated with temperature. In the literature there are no references concerning the correlation of this species with this factor. However, it is mainly found near the sea (Hansen, 1987), in lentic freshwater or subsalt systems (Sánchez-Fernández et al., 2007), and in high salt marshes (Bigot & Marazanof, 1966; Montes et al., 1982).

Some species, such as *Helochares puntactus* and *Hygrobia hermanni*, were not found in the oligohaline water body (Vixán). Both species were present only in the fresh water systems (Bodeira and Xuño), which could suggest their presence in waters with less conductivity. *H. puntactus* inhabits margins of ponds and lakes, generally eutrophic with grasses and other submerged macrophytes (Valladares, 1995) and pools in heathlands (Aquilina, 2010). On the other hand, Bigot & Marazanof (1966) and Montes et al. (1982) found *H. hermanni* in eutrophic and high salinity waters in the Guadalquivir Marshes (South of Spain).

The Bray-Curtis coefficient revealed significant differences in terms of faunal affinity between the two periods of time. The three lagoons studied presented significantly different assemblages 10 years later. The response of Coleoptera and Hemiptera assemblages to the passing of the time (10 years later) was a change in species composition, with the substitution of some species by others, not by changes in the richness values, as these were similar between the two periods of time. The change in species composition found in these ecosystems seems to be a consequence of the high temporal variability. The heterogeneity in physiography and hydrology of the coastal lagoons influences the faunal composition, and variations in the physicochemical characteristics can change the composition of the aquatic assemblages over time. These special conditions of the coastal lagoons were also mentioned by Martinoy et al. (2006)

In this sense, it is important to note that certain species occurred in only one of the years. Species such as *Enochrus halophilus* and *Sigara stagnalis* were very common in 1998 and were absent in 2008. Others, including *Hydrovatus clypealis*, *Paracymus scutellaris*, *Enochrus fuscipennis* and *Anisops sardeus sardeus*, were abundant in the three lagoons in 2008, but not found in 1998. *A. s. sardeus* is a trans-Iberian hemipteran, present in several environments, including moderately mineralised waters, and in artificial or newly created water bodies (Abellán et al., 2004), very common in the south of the Iberian Peninsula (Nieser & Montes, 1984).

In the same genus, certain species were substituted by another 10 years later. For example, in the genus *Graptodytes* three species were found in 1998 (*G. aequalis*, *G. flavipes*, *G. ignotus*), but only *G. varius* was collected in 2008. The same occurs with *Dryops*, which changes *D. striatellus* (1998) for *D. luridus* (2008), and *Enochrus*, in which *E. fuscipennis* substituted *E. halophilus*. In the Iberian Peninsula, *D. luridus* is much more common than *D. striatellus* (Montes & Soler, 1986). In fact, in England *D. striatellus* is a vulnerable species (Bilton et al., 2009). According to Hansen (1987) both *E. halophilus* and *E. fuscipennis* are found in temporary brackish pools above the high tide line or occasionally also in slightly saline water. *E. halophilus* is classified as "vulnerable" on the red list of Irish threatened species

(IUCN) (Foster et al., 2009). This species was found in wetlands in the Mediterranean coast of Spain in eutrophic or even in high salinity waters (Bigot & Marazanof, 1966; Ribera et al. 1996). *E. fuscipennis* has been found in stagnant water rich in organic matter (Valladares, 1995). According to Bilton et al. (2009) this species is widespread in ponds and other wetlands throughout the UK and north-west Europe.

The aquatic bug *Plea minutissima* is one of the most abundant species in this study. Other studies have also reported a high number of individuals of this insect in stagnant water bodies (Bloechl et al., 2010; González & Valladares, 1996). Another very abundant and common species of Hemiptera was *Hesperocorixa linnaei* (1,360 specimens recorded in 75% of the samples), which is different to the data obtained by Martinoy et al. (2006) in a study conducted in 84 water bodies in coastal ecosystems of the Empordà wetlands (NE Spain), where this species presented a very low abundance.

Some of the species found in these lagoons are rare and not common in other regions. For example, the case of *Ochthebius viridis fallaciosus*, which was the most abundant species of Coleoptera (1,961 individuals, recorded in 75% of the samples) should be noted. According to Foster et al. (2009), this subspecies is on the red list of threatened species (IUCN) in Ireland, classified as "near threatened", i.e. a species does not qualify for Critically Endangered, Endangered or Vulnerable now, but is close to qualifying for or is likely to qualify for a threatened category in the near future. In this country it is a sparsely distributed species that appears to be confined to brackish water.

This study demonstrates the importance of the conservation of these lagoons for maintaining aquatic biodiversity. In this regard, it should be said that the Natura 2000 network is effective in protecting wildlife, as far as these lagoons are concerned. If we compare data from 1998 to those obtained 10 years later, we note that there were no significant changes in the values of the diversity indices. In terms of faunal composition, it can be said that the observed changes are mainly due to instability, typical of the coastal lagoons. This can be highlighted by the specific changes in certain variables, such as the case of increased salinity in the summer of 1998.

Coleoptera and Hemiptera assemblages are good indicators of the environmental quality of coastal lagoons. Their responses to changes in environmental conditions differ. Some species are adapted to increasing in certain variables, for example, *Berosus hispanicus* and *Rhantus suturalis* correlated with high salinity. To conclude, it is important to note that the preservation of species is one of the main factors that should be taken into consideration in developing monitoring systems and assessing reference conditions, as stipulated by the Water Framework Directive (Council of the European Communities 2000).

7. References

Abbiati, M. & Basset, A. (2001). Ecological research and conservation of coastal ecosystems. *Aquatic Conservation: Marine and Freshwater Ecosystems*, 11, 233-234.

Abellán, P.; Sánchez-Fernández, D.; Barahona, J.; Velasco, J. & Millán, A. (2004). Coleópteros y hemípteros acuáticos con interés de conservación en el entorno de Calblanque. *Actas III Congreso de la Naturaleza de la Región de Murcia*, 89-97.

Aliaume, C.; Do Chi, T.; Viaroli, P. & Zaldívar, J.M. (2007). Coastal lagoons of Southern Europe: recent changes and future scenarios. *Transitional Waters Monographs,* 1, 1-12.

Álvarez-Cobelas, M.; Catalán, J.; García de Jalón, D.; Alonso, M.; Amores, V.; Armengol, J.; Artigas, J.; Barquín, J.; Bayo, M.; Boix, D.; Butturini, A.; Camacho, A.; Carrillo, P.; Casas, J.; Cirujano, S.; Comín, F.; Cruz-Pizarro, L.; Gaudes, A.; Lucena, J.; Menéndez, M.; Morales, R.; Moreno-Ostos, E.; Muñoz, I.; Pérez, C.; Pozo, J.; Quintana, X.; Reche, I.; Rodrigo, M.; Romaní, A.; Sabater, S.; Sánchez-Castillo, P.; Soria, J.M.; Suárez, M.L.; Bernal, S.; Sabater, F.; Toja, J.; Vega, J.C.; Vicente, I. & Vidal Abarca, M.R. (2005). Impactos sobre los ecosistemas acuáticos continentales, In: *Impactos del cambio climático en España,* pp. 113-146, Ministerio de Medio Ambiente, Madrid.

Aquilina, R. (2010). An assemblage of Coleoptera from ponds on Povington Heath, Dorset. *The Coleopterist,* 19(1), 25-29.

Arnold, S.L. & Ormerod, S.J. (1997). Aquatic macroinvertebrates and environmental gradients in Phragmites reedswamps: implications for conservation. *Aquatic Conservation: Marine and Freshwater Ecosystems,* 7, 153-163.

Arocena, R. (2007). Effects of Submerged Aquatic Vegetation on Macrozoobenthos in a Coastal Lagoon of the Southwestern Atlantic. *International Review of Hydrobiology,* 92(1), 33-47.

Barnes, R.S.K. (1999). The conservation of brackish-water systems: priorities for the 21st century. *Aquatic Conservation: Marine and Freshwater Ecosystems,* 9, 523-527.

Basset, A. (2007). Ecosystems and Society: do they really need to be bridged? *Aquatic Conservation: Marine and Freshwater Ecosystems,* 17, 551-553.

Basset, A. & Abbiati, M. (2004). Challenges to transitional water monitoring: ecological descriptors and scales. *Aquatic Conservation: Marine and Freshwater Ecosystems,* 14, S1-S3.

Basset, A.; Sabetta, L. & Carrada, G.C. (2006a). Conservation of transitional water ecosystems in the Mediterranean area: bridging basic ecological research and theories with requirements of application. *Aquatic Conservation: Marine and Freshwater Ecosystems,* 16, 439-440.

Basset, A.; Sabetta, L.; Fonnesu, A.; Mouillot, D.; Do Chi, T.; Viaroli, P.; Giordano, G.; Reizopoulou, S.; Abbiati, M. & Carrada, G.C. (2006b). Typology in Mediterranean transitional waters: new challenges and perspectives. *Aquatic Conservation: Marine and Freshwater Ecosystems,* 16, 441-455.

Bigot, L. & Marazanof, F. (1966). Notes sur l'ecologie des Coléoptères aquatiques des Marismas du Guadalquivir et premier inventaire des Coléoptères et Lépidoptères du Coto Doñana (Andalucía). *Annales de Limnologie,* 2, 491-502.

Bilton, D.T.; McAbendroth, L.C.; Bedford, A. & Ramsay, P.M. (2006). How wide to cast the net? Cross-taxon congruence of species richness, community similarity and indicator taxa in ponds. *Freshwater Biology,* 51, 578–590.

Bilton, D.T.; McAbendroth, L.C.; Nicolet, P.; Bedford, A.; Rundle, S.R. ; Foggo, A. & Ramsay, P.M. (2009). Ecology and conservation status of temporary and fluctuating ponds in two areas of southern England. *Aquatic Conservation: Marine and Freshwater Ecosystems,* 19, 134–146.

Bloechl, A.; Koenemann, S.; Philippi, B. & Melber, A. (2010). Abundance, diversity and succession of aquatic Coleoptera and Heteroptera in a cluster of artificial ponds in the North German Lowlands. *Limnologica*, 40, 215-225.

Boix, D.; Gascón, S.; Sala, J.; Badosa, A. Brucet, S. López-Flores, R.; Martinoy, M.; Gifre, J. & Quintana, X. D. (2008). Patterns of composition and species richness of crustaceans and aquatic insects along environmental gradients in Mediterranean water bodies. *Hydrobiologia* 597, 53–69.

Britton, R.H. & Johnson, A.R. (1987). An Ecological Account of a Mediterranean Salina: The Salin de Giraud, Camargue (S. France). *Biological Conservation*, 42, 185-230.

Broering, U. & Niedringhaus, R. (1988). On the ecology of aquatic Hemiptera (Hemiptera: Nepomorpha) in small ponds on the East Frisian Island of Norderney (FRG). *Fundamental and applied Limnology, Archiv für Hydrobiologie*, 111, 559-574.

Casado, S. & Montes, C. (1995). *Guía de los lagos y humedales de España*. J.M. Reyero, Madrid.

Cognetti, G. & Maltagliati, F. (2000). Biodiversity and Adaptative Mechanisms in Brackish Water Fauna. *Marine Pollution Bulletin*, 40, 7-14.

Davis, J.A.; Rolls, S.W. & Balla, S.A. (1987). The role of the Odonata and aquatic Coleoptera as indicators of environmental quality in wetlands, In: *The Role of Invertebrates in Conservation and Biological Survey*, J.D. Majer (ed.), pp. 31-42, Western Australian Department of Conservation and Land Management, Perth.

Della Bella, V.; Bazzanti, M.; Dowgiallo, M.G. & Iberite, M. (2008). Macrophyte diversity and physico-chemical characteristics of Tyrrhenian coast ponds in central Italy: implications for conservation. *Hydrobiologia*, 597, 85–95.

Diehl, S. & Kornijów, R. (1998). Influence of submerged macrophytes on trophic interactions among fish and macroinvertebrates, In: *Structuring Role of Submerged Macrophytes in Lakes*, E. Jeppesen, M. Søndergaard, M. Søndergaard & K. Christoffersen (eds.), pp. 24-46, Springer-Verlag, New York.

Drake, D.C.; Kelly, D. & Schallenberg, M. (2011). Shallow coastal lakes in New Zealand: current conditions, catchment-scale human disturbance, and determination of ecological integrity. *Hydrobiologia*, 658, 87-101.

Duigan, C.A.; Allot, T.E.H.; Monteith, D.T.; Patrick, S.T.; Lancaster, J. & Seda, J.M. (1998). The ecology and conservation of Llyn Idwal and Llyn Cwellyn (Snowdonia National Park, North Wales, U.K.)- two lakes proposed as Special Areas of Conservation in Europe. *Aquatic Conservation: Marine and Freshwater Ecosystems*, 8, 325-360.

Esteves, F.A.; Caliman, A.; Santangelo, J.M.; Guariento, R.D.; Farjalla, V.F. & Bozelli, R.L. (2008). Neotropical coastal lagoons: An appraisal of their biodiversity, functioning, threats and conservation management. *Brazilian Journal of Biology*, 68(4), 967-981.

Eyre, M.D. & Foster, G.N. (1989). A comparison of aquatic Hemiptera and Coleoptera communities as a basis for environmental and conservation assessments in static water sites. *Journal of Applied Entomology*, 108, 355-362.

Eyre, M.D. & Rushton, S.P. (1989). Quantification of conservation criteria using invertebrates. *Journal of Applied Entomology*, 26, 159-171.

Eyre, M.D.; Foster, G.N. & Young, A.G. (1993). Relationships between water-beetle distributions and climate variables: a possible index for monitoring global climatic change. *Fundamental and applied Limnology, Archiv für Hydrobiologie*, 127, 437-450.

Florencio, M.; Serrano, L.; Gómez-Rodríguez, C.; Millán, A. & Díaz-Paniagua, C. (2009). Inter- and intra-annual variations of macroinvertebrate assemblages are related to the hydroperiod in Mediterranean temporary ponds. *Hydrobiologia*, 634, 167–183.

Foster, G.N. (1987). The use of Coleoptera records in assessing the conservation status of wetlands, In: *The Use of Invertebrates in the Site Assessment for Conservation*, M.L. Luff (ed.), pp. 8-18, University of Newcastle upon Tyne, UK.

Foster, G.N. (1999). Biodiversity action plans for British water beetles. *Latissimus*, 11, 1-13.

Foster, G.N. (2000). The aquatic Coleoptera of British saltmarshes: extremes of generalism and specialism, In: British Saltmarshes, B.R. Sherwood, B.G. Gardiner & T. Harris (eds.), pp. 223-233, Linnean Society, London.

Foster, G.N.; Foster, A.P.; Eyre, M.D. & Bilton, D.T. (1990).Classification of water beetles assemblages in arable fenland and ranking of sites in relation to conservation value. *Freshwater Biology*, 22, 343-354.

Foster, G.N.; Nelson, B.H. & O'Connor, A. (2009). *Ireland Red List No. 1 – Water beetles.* National Parks and Wildlife Service, Department of Environment, Heritage and Local Government, Dublin, Ireland.

Garrido, J.; Sáinz-Cantero, C.E. & Díaz, J.A. (1996). Fauna entomológica del Parque Nacional de Doñana (Huelva, España) I. (Coleoptera, Polyphaga). *Nouvelle Revue d' Entomologie*, 13, 57–71.

Garrido, J.; Sáinz-Cantero, C.E. & Régil-Cueto, J.A. (1997). Fauna entomológica del Parque Nacional de Doñana (Huelva, España) II. (Coleoptera, Adephaga). *Nouvelle Revue d' Entomologie*, 14(4), 365–377.

Garrido, J. & Munilla, I. (2008). Aquatic Coleoptera and Hemiptera assemblages in three coastal lagoons of the NW Iberian Peninsula: assessment of conservation value and response to environmental factors. *Aquatic Conservation: Marine and Freshwater Ecosystems*, 18, 557-569.

Gaston, K.J. (1994). *Rarity*. Chapman & Hall, London.

González, S.C. & Valladares, L.F. (1996). The community of Odonata and aquatic Heteroptera (Gerromorpha and Nepomorpha) in a rehabilitated wetland: the Laguna de la Nava (Palencia, Spain). *Fundamental and applied Limnology, Archiv für Hydrobiologie*, 136(1), 89-104.

Goudie, A. (1990). *The Human Impact on the Natural Environment*. Basil Blackwell, Oxford.

Greenwood, M.T. & Wood, P.J. (2002). Effects of seasonal variation in salinity on a population of *Enochrus bicolor* Fabricius, 1792 (Coleoptera: Hydrophilidae) and implications for other beetles of conservation interest. *Aquatic Conservation: Marine and Freshwater Ecosystems*, 13, 21-34.

Hansen, M. (1987). *The Hydrophiloidea (Coleoptera) of Fennoscandia and Denmark. Fauna Entomologica Escandinavica, Vol. 18*, Scandinavian Science Press, ISBN 87-87491-35-4, Copenhagen, Denmark

Heino, J. (2000). Lentic macroinvertebrate assemblage structure along gradients in spatial heterogeneity, habitat size and water chemistry. *Hydrobiologia*, 418, 229–242.

Hufnagel, L.; Bakonyi, G. & Vásárhelyi, T. (1999). New approach for habitat characterization based on species lists of aquatic and semiaquatic bugs. *Environmental Monitoring and Assessment*, 58, 305–316.

IPCC (2007). *Climate Change 2007: Synthesis Report. Contribution of Working Groups I, II and III to the Fourth Assessment Report of the Intergovernmental Panel on Climate Change*, Core Writing Team, R.K. Pachauri & A. Reisinger (eds.), IPCC, Geneva, Switzerland.

Jansson, A. (1977). Micronectae (Heteroptera, Corixidae) as indicators of water quality in two lakes in southern Finland. *Annales Zoologici Fennici*, 14, 118–124.

Jones, T.A. & Hughes, J.M.R. (1993). Wetland inventories and wetlands loss studies: a European perspective, In: *Waterfowl and Wetland Conservation in the 1990s: a Global Perspective*, M. Moser, R.C. Prentice & J. Van Vessem (eds.), pp. 164-169, IWRB Special Publication, Slimbridge, UK.

Kirk, R.M. & Lauder, G.A. (2000). *Significant coastal lagoon systems in the South Island, New Zealand: coastal processes and lagoon mouth closure*. N.Z. Department of Conservation, Wellington.

Kjerfve, B. (1994). Coastal Lagoons, In: *Coastal Lagoon Processes*, B. Kjerfve (ed.), pp. 1-8, Elsevier Science Publishers, Amsterdam.

Kjerfve, B. & Magill, K.E. (1989). Geographic and hydrodynamic characteristics of shallow coastal lagoons. *Marine Geology*, 88, 187-199.

Lancaster, J. & Scudder, G.G.E. (1987). Aquatic Coeloptera and Hemiptera in some Canadian saline lakes: patterns in community structure. *Canadian Journal of Zoology*, 65, 1383–1390.

Langenheder, S.; Kisand, V.; Wikner, J. & Tranvik, L.J. (2003). Salinity as a structuring factor for the composition and performance of bacterioplankton degrading riverine DOC. *FEMS Microbiology Ecology*, 45, 189-202.

Lewis, W.M. Jr. (1983). A revised classification of lakes based on mixing. *Canadian Journal of Fisheries and Aquatic Sciences*, 40, 1779-1787.

Lucena-Moya, P.; Abraín, R.; Pardo, I.; Hermida, B. & Domínguez, M. (2010). Invertebrate species list of coastal lagoons in the Balearic Islands. *Transitional Waters Bulletin*, 4(1), 1-11.

Maezono, Y. & Miyashita, T. (2003). Community level impacts induced by introduced largemouth bass and bluegill in farm ponds in Japan. *Biological Conservation*, 109(1), 111–121.

Margules, C.R.; Nicholls, A.O. & Pressey, R.L. (1988). Selecting networks of reserves to maximise biological diversity. *Biological Conservation*, 43, 63–76.

Martinoy, M.; Boix, D.; Sala, J.; Gascón, S.; Gifre, J.; Argerisch, A.; de la Barrera, R.; Brucet, S.; Badosa, A.; López-Flores, R.; Méndez, M.; Utgé, J.M. & Quintana, X.D. (2006). Crustacean and aquatic insect assemblages in the Mediterranean coastal ecosystems of Empordà wetlands (NE Iberian Peninsula). *Limnetica*, 25, 665-682.

McCabe, D.J. & Gotelli, N.J. (2000). Effects of disturbance frequency, intensity, and area on assemblages of stream macroinvertebrates. *Oecologia*, 124, 270-279.

Millán, A., Moreno, J.L. & Velasco, J. (1997). Coleópteros y Heterópteros acuáticos del complejo lagunar del río Arquillo (Albacete). *Separatas de al-basit. Revista de estudios albacetenses*, 40.

Millán, A.; Moreno, J.L. & Velasco, J. (2001). Estudio faunístico y ecológico de los coleópteros y heterópteros acuáticos de las lagunas de Albacete. *Separatas de Sabuco. Revista de estudios albacetenses*, 1, 43–94.

Millán, A.; Moreno, J.L. & Velasco, J. (2002). *Los Coleópteros y Heterópteros acuáticos y semiacuáticos de la provincia de Albacete. Catálogo faunístico y estudio ecológico*, Instituto de Estudios Albacetenses Don Juan Manuel de la Excma. Diputación Provincial de Albacete, Albacete.

Mitsch, W.J. & Gosselink, J.G. (2000). *Wetlands*, John Wiley and Sons, New York.

Montes, C. ; Ramirez, L. & Soler, A. (1982). Variación estacional de las taxocenosis de odonatos, coleópteros y heterópteros acuáticos en algunos ecosistemas del Bajo Guadalquivir (SW. España) durante un ciclo anual. *Anales de la Universidad de Murcia, Ciencias*, 38(1-4), 19-100.

Montes, C. & Soler, A.G. (1986). *Lista faunística y bibliográfica de los coleópteros acuáticos Dryopoidea (Dryopidae & Elmidae) de la Península Ibérica e Islas Baleares. Listas de la flora y fauna de las aguas continentales de la Península Ibérica.* Asociación Española de Limnología, Madrid.

Moreno, J.L.;Millán, A.; Suárez, M.L.; Vidal-Abarca, M.R. & Velasco, J. (1997). Aquatic Coleoptera and Hemiptera assemblages in water bodies from ephemeral coastal streams ('ramblas') of south-eastern Spain. *Fundamental and applied Limnology, Archiv für Hydrobiologie*, 141, 93–107.

Nichols, M.M. (1989). Sediment accumulation rates and relative sea-level rise in lagoons. *Marine Geology*, 88, 201-219.

Nichols, M. & Allen, G. (1981). Sedimentary processes in coastal lagoons. *UNESCO Technical Bulletin in Marine Science*, 33, 27-80.

Nieser, N. & Montes, C. (1984). *Lista faunística y bibliográfica de los Heterópteros acuáticos (Nepomorpha & Gerromorpha) de España y Portugal.* Asociación Española de Limnología, Listas de la flora y fauna de las aguas continentales de la Península Ibérica. Madrid.

Nieser, N.; Baena, M.; Martínez-Avilés, J. & Millán, A. (1994). *Clave para la identificación de los heterópteros acuáticos (Nepomorpha & Gerromorpha) de la Península Ibérica - Con notas sobre las especies de las islas Azores, Baleares, Canarias y Madeira. Claves de identificación de la Flora y Fauna de las aguas continentales de la Península Ibérica.* Asociación Española de Limnología, Madrid.

Nixon, S.W. (1995). Coastal eutrophication: a definition, social causes, and future concerns. *Ophelia*, 41, 199–220.

Oertli, B.; Joye, D.A.; Castella, E.; Juge, R.; Lehmann, A. & Lachavanne, J-B. (2005). PLOCH: a standardized method for sampling and assessing the biodiversity in ponds. *Aquatic Conservation: Marine and Freshwater Ecosystems*, 15, 665-679.

Orfanidis, S.; Reizopoulou, S. & Basset, A. (2008). Transitional states in transitional waters. *Aquatic Conservation: Marine and Freshwater Ecosystems*, 18, S1-S3.

Pérez-Bilbao, A. & Garrido, J. (2009). Evaluación del estado de conservación de una zona LIC (Gándaras de Budiño, Red Natura 2000) usando los coleópteros acuáticos como indicadores. *Limnetica*, 28(1), 11-22.

Pérez-Bilbao, A.; Benetti, C.J. & Garrido, J. (2009). Nuevas aportaciones al conocimiento de la familia Hydrochidae (Coleoptera) en Galicia (Noroeste de España). *Boletín de la Sociedad Entomológica Aragonesa*, 45, 241-244.

Pérez-Bilbao, A.; Benetti, C.J. & Garrido, J. (2010). Coleópteros acuáticos (Polyphaga, Hydrophilidae) en lagunas de la Red Natura 2000 de Galicia (Noroestes de España). *Boletín de la Sociedad Entomológica Aragonesa*, 46, 255-260.

Piccini, C. & Conde, D. (2004). Cambios drásticos en la comunidad bacteriana de la laguna de Rocha y sus posibles implicaciones ambientales. *Revista Agrociencia*, 9(1-2), 269-275.

Polhemus, J.T. & Polhemus, D.A. (2008). Global diversity of true bugs (Heteroptera; Insecta) in freshwater. *Hydrobiologia*, 595, 379-391.

Ponti, M.; Casselli, C. & Abbiati, M. (2011). Anthropogenic disturbance and spatial heterogeneity of macrobenthic invertebrate assemblages in coastal lagoons: the study case of Pialassa Baiona (northern Adriatic Sea). *Helgoland Marine Research*, 65, 25-42.

Ramil, P.; Cillero, C.; García, N.; Codesido, D.; Rubinos, M. & Ferreiro, J. (2007). *Evaluación del estado de conservación de las lagunas costeras de Galicia: Propuestas de conservación y uso sostenible.* Instituto de Biodiversidade Agraria e Desenvolvemento Rural (IBADER), Lugo, Spain.

Ribera, I. & Foster, G.N. (1992). Uso de coleópteros acuáticos como indicadores biológicos (Coleoptera). *Elytron*, 6, 61-75.

Ribera, I.; Bilton, D.T.; Aguilera, P. & Foster, G.N. (1996). A north African-European transition fauna: water beetles (Coleoptera) from the Ebro Delta and other Mediterranean coastal wetlands in the Iberian Peninsula. *Aquatic Conservation: Marine and Freshwater Ecosystems*, 6, 121-140.

Ribera, I.; Hernando, C. & Aguilera, P. (1999). An annotated checklist of the Iberian water beetles (Coleoptera). *Zapateri, Revista aragonesa de entomología*, 8, 43-111.

Ribera, I. & Vogler, A.P. (2000). Habitat type as a determinant of species range sizes: the example of lotic-lentic differences in aquatic Coleoptera. *Biological Journal of the Linnean Society*, 71, 33-52.

Ribera, I. & Vogler, A.P. (2004). Speciation of Iberian diving beetles in Pleistocene refugia (Coleoptera, Dytiscidae). *Molecular Ecology*, 13, 179–193.

Ricciardi, A. (2006). Are modern biological invasions an unprecendent form of global change? *Conservation Biology*, 21(2), 329–336.

Ricciardi, A.; Whoriskey, F.G. & Rassmussen, J.B. (1997). The role of the zebra mussel (Dreissena polymorpha) in structuring macroinvertebrate communities on hard substrata. *Canadian Journal of Fisheries and Aquatic Sciences*, 54(11), 2596–2608.

Roback, S.S. (1974). Insects (Arthropoda: Insecta), In: *Pollution Ecology of Freshwater Invertebrates*, C.W. Hart Jr., & S.L.H. Fuller (eds.), pp. 313-376, Academic Press, Inc., New York.

Rodríguez-Pérez, H.; Florencio, M.; Gómez-Rodríguez, C.; Green, A.J.; Díaz-Paniagua, C. & Serrano, L. (2009). Monitoring the invasion of the aquatic bug *Trichocorixa verticalis verticalis* (Hemiptera: Corixidae) in the wetlands of Doñana National Park (SW Spain). *Hydrobiologia*, 634, 209-217.

Sáinz-Cantero, C.E. & Garrido, J. (1996). Primera cita de *Cymbiodyta marginella* (Fabricius, 1792) en España (Col., Hydrophilidae). *Bulletin de la Societé entomologique de France*, 101, 508.

Sánchez-Fernández, D.; Abellán, P.; Mellado, A.; Velasco, J. & Millán, A. (2006). Are water beetles good indicators of biodiversity in Mediterranean aquatic ecosystems? The case of the Segura river basin (SE Spain). *Biodiversity and Conservation*, 15, 4507-4520.

Sánchez-Fernández, D.; Abellán, P.; Camarero, F.; Esteban, I.; Gutiérrez-Cánovas, C.; Ribera, I.; Velasco, J. & Millán, A. (2007). Los macroinvertebrados acuáticos de las salinas de Añana (Álava, España): biodiversidad, vulnerabilidad y especies indicadoras. *Boletín de la Sociedad Entomológica Aragonesa*, 40, 233-245.

Savage, A.A. (1996). Density dependent and density independent relationships during a twenty-seven year study of the population dynamics of the benthic macroinvertebrate community of a chemically unstable lake. *Hydrobiologia*, 335, 115-131.

Shine, C. & Klemm, C. (1999). *Wetlands, Water and the Law: Using law to Advance Wetland Conservation and Wise Use*, IUCN, Gland.

Solimini, A.G.; Bazzanti, M.; Ruggiero, A. & Carchini, G. (2008). Developing a multimetric index of ecological integrity based on macroinvertebrates of mountain ponds in central Italy. *Hydrobiologia*, 597, 109-123.

Søndergaard, M. & Moss, B. (1998). Impact of submerged macrophyteson phytoplankton in shallow freshwater lakes, In: *Structuring Role of Submerged Macrophytes in Lakes*, E. Jeppesen, M. Søndergaard, M. Søndergaard & K. Christoffersen (eds), pp. 115-133, Springer-Verlag, New York.

Soria, J.M. & Sahuquillo, M. (2009). 1150 Lagunas costeras (*), In: *Bases ecológicas preliminares para la conservación de los tipos de hábitats de interés comunitario en España*. Ministerio de Medio Ambiente y Medio Rural y Marino, Madrid, España.

Stenert, C. & Maltchik, L. (2007). Influence of area, altitude and hydroperiod on macroinvertebrate communities in southern Brazil wetlands. *Marine and Freshwater Research*, 58, 993-1001.

Sylaios, G. & Theocharis, V. (2002). Hydrology and Nutrient Enrichment at Two Coastal Lagoon Systems in Northern Greece. *Water Resources Management*, 16, 171-196.

Ter Braak, C.J.F. & Šmilauer, P. (2002). *CANOCO Reference manual and CanoDraw for Window's User's guide: Software for Canonical Community Ordination (version 4.5)*. Microcomputer Power, Ithaca.

Timms, B.V. (1993). Saline lakes of the Paroo, inland New South Wales, Australia. *Hydrobiologia*, 267, 269-289.

Tuly, O.; McCarthy, T.K. & O'Donell, D. (1991). The ecology of the Corixidae (Hemiptera: Hemiptera) in the Corrib catchment, Ireland. *Hydrobiologia*, 210, 161-169.

Valladares, L.F. (1995). Los Palpicornia acuáticos de la provincia de León III. Helophoridae, Hydrochidae e Hydrophilidae (Coleoptera). *Boletín de la Asociación española de Entomología*, 19(1-2), 281-308.

Vierssen, W. & Verhoeven, J.T.A. (1983). Plant and animal communities in brackish supralitoral pools ('dobben') in the northern part of the Netherlands. *Hydrobiologia*, 98, 203-211.

Ward, L.G. & Ashley, G.M. (1989). Introduction: coastal lagoonal systems. *Marine Geology*, 88, 181-185.

Williams, P.; Whitfield, M.; Biggs, J.; Bray, S.; Fox, G.; Nicolet, P. & Sear, D. (2003). Comparative biodiversity of rivers, streams, ditches and ponds in an agricultural landscape in Southern England. *Biological Conservation*, 115, 329-341.

Witte, F.; Msuku, B.S.; Wanink, J.H.; Seehausen, O.; Katunzi, E.F.B.; Goudswaard, P.C. & Goldsmidt, T. (2000). Recovery of cichild species in lake Victoria: an examination of factors leading differential extinction. *Reviews in Fish Biology and Fisheries,* 10, 233-241.

Zaldívar, J.M.; Cardoso, A.C.; Viaroli, P.; Newton, A.; de Wit, R.; Ibáñez, C.; Reizopoulou, S.; Somma, F.; Razinkovas, A.; Basset, A.; Holmer, M. & Murray, N. (2008). Eutrophication in transitional waters: an overview. *Transitional Waters Monographs,* 1, 1-78.

2

Biodiversity of Coastal Lagoon Ecosystems and Their Vulnerability to Global Change

Rutger de Wit
Ecologie des Systèmes marins côtiers (Ecosym), UMR 5119, Université Montpellier 2, CNRS, IRD, Ifremer & Université Montpellier 1; Place Eugène Bataillon Université Montpellier 2, Montpellier Cedex 05 France

1. Introduction

In this chapter we define coastal lagoons according the definition proposed by Kjerfve (1994) as "shallow water bodies separated from the ocean by a barrier, connected at least intermittently to the ocean by one or more restricted inlets, and usually oriented shore-parallel". This definition is given in an introduction chapter of a multiauthor book edited by Kjerfve (1994). Nevertheless, in the scientific literature the term coastal lagoon is not always used. For example, in the North American ecological literature the term estuary has often been used rather than coastal lagoon for systems including "Laguna Madre", "Lake Pontchartrain" and "Waquoit Bay". In contrast, in the socio-historical context of Southwestern Europe, lagoons are well recognised and distinguished from estuaries as is shown by the vernacular languages that contain specific words for the coastal lagoons, their barriers and inlets. Thus, coastal lagoons are named "laguna" in Spanish and Italian, "lagoa" in Portuguese and "lagune" in French. However, the Spanish word "laguna" also refers to a relatively shallow freshwater lake system, and coastal lagoons are referred to as "lagunas costeras". In French, the word "lagon" refers to a marine water body that is separated from the open sea by a coral reef. Therefore, one has to be aware of possible confusions and different perceptions of coastal lagoons by different communities. Lagoon science is relatively young and had some difficulty of getting from the ground as was earlier highlighted in the classical book by Barnes (1980) titled "Coastal Lagoons; the natural history of a neglected habitat". Actually still today coastal lagoons in the tropical world have been poorly studied so far (Esteves et al., 2008). In Europe, nowadays a thriving community of lagoon scientists exists that joins together in biannual meetings since 2003 (European Coastal Lagoons Symposium, Eurolag), the latest (2011) has taken place in Aveiro, Portugal.

Coastal lagoon ecosystems are a particular type of estuarine systems where seawater mixes with fresh water from their continental catchments. These systems occupy about 12 % of the world coastlines (see Fig. 1). Salinities have been used since 1958 to classify the lagoons and other brackish or estuarine waters according the Venice system. Accordingly, a lagoon with salinities below 5 ppt is oligohaline; mesohaline waters have salinities between 5 and 18 ppt and polyhaline between 18 and 30 ppt. Lagoons with salinities above 30 ppt but below that of seawater are termed mixoeuhaline. Connections with the sea can be temporary or permanent through tidal inlets (see definition above by Kjerfve). This implies that tides are propagated in the lagoon when the inlets are open and tidal ranges in lagoon may range from <10 cm to 3 to 4 m. Tidal range has been used to define adjectives for coastal ecosystems including lagoon, i.e. microtidal (< 2m), mesotidal (2-4 m) and macrotidal (4-6 m). Coastal lagoons are particularly abundant along the coastlines of enclosed seas like the Mediterranean and Baltic Seas, which are both microtidal. In some lagoons, evaporation temporarily or permanently exceeds fresh water inflow and these lagoons become more saline (hyperhaline according the Venice system) than the adjacent sea. In other cases, the salinity of lagoons can drop tremendously during periods when fresh water input is high and exchanges with the sea are restricted. Such conditions can vary drastically during an annual cycle and because most lagoons are rather shallow, the water column salinities and temperatures often vary to a high extent. The term poikilohalinity is sometimes used to characterise the highly fluctuating salinity conditions that are so characteristic for many lagoons.

= barrier/lagoonal coastalines

Fig. 1. World distribution of barrier and lagoonal coastlines. Reproduced from Barnes (1980), Coastal Lagoons, Fig. 1.1 on page 2, with kind permission of Cambridge University Press.

2. Lagoon formation and creation of ecotones, biodiversity patterns

On a geological scale coastal lagoons are ephemeral ecosystems that change form and dimensions in time and space due to natural processes. This dynamic behaviour of coastal lagoons is nicely expressed in the title "Shifting Sands" a book on the coastal lagoons in Maryland, USA (Dennison et al., 2009). Most lagoons have been formed during periods of sea-level rise, so-called transgression periods. A first type of lagoons develops when the rising sea invades a relatively flat coastal plane and when a barrier is formed from marine sediments that separates the shallow water body from the sea (see Fig. 2A). When the barrier is attached to firm land it is called a "lido" or a "spit". However, tidal inlets often interrupt the barrier and a stretch of barrier separated by two inlets on either side is called a barrier island. A second type of lagoon forms from a classical estuary, when a barrier is subject to longshore drift that deviates the estuary from its original course and creates a large area of lagoonal environment (see Fig. 2B). The classical example is the triangular shaped Bassin d'Arcachon (De Wit, 2008). A third type is formed in coastal deltas, like in the Camargue region the delta of the Rhone (e.g. Etang de Vaccares) or in the Ebro, Po and Nile deltas. A fourth type is formed in more accidentated regions, e.g. by invasion of the sea in glacier vallies (Fjord type) or by tectonic processes. These latter lagoons can be rather deep while most of the other types are relatively shallow. In the Languedoc-Roussillon region in S. France (along the Gulf of Lions), 12 of 18 lagoons have a mean depth < 1 m and four between 1-2 m depth. Only the Thau lagoon, which is also the largest in this region (75 km^2), has an average depth of 4.5 m with deeper basins with around 10 m depth.

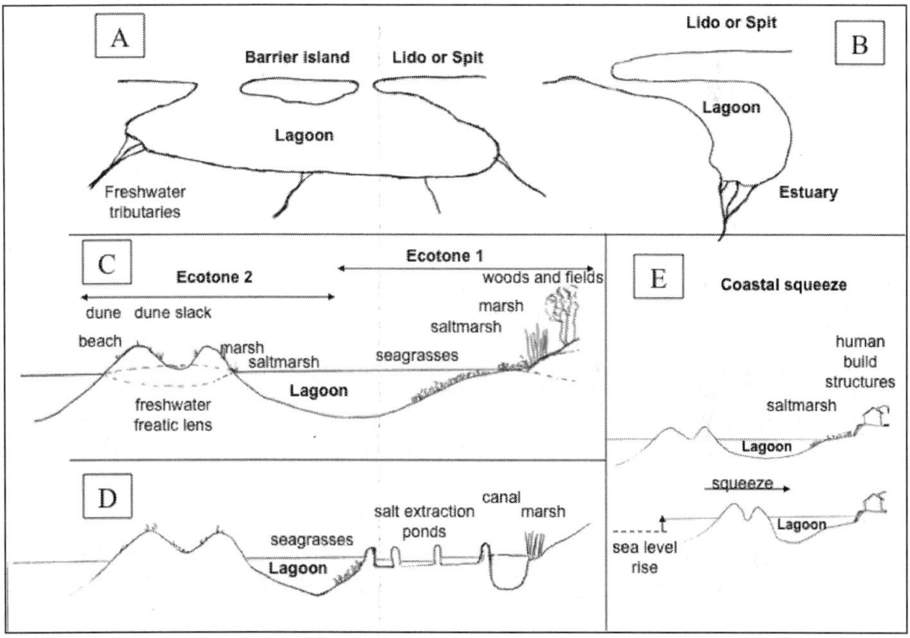

Fig. 2. A: Map of classical lagoon formed during sealevel rise by submersion of a coastal plain and formation of barrier (lido and barrier islands) from marine sediments. B: Bar built estuary evolved in a coastal lagoon. The longshore drift is oriented from right to left and has created a lido or spit that separates a lagoon from the sea. C: Ecotones in classical lagoons (cf. A). D: creation of salt extraction ponds in a coastal lagoon. The original ecotone (1) has been destroyed by creation of a canal to intercept freshwater runoff, while a new ecotone has been created. E: Effect of coastal squeeze in a lagoon where the inner shore is fixed by hard human structures. Sea level rise induces landward movement of the barrier and loss of the fringent salt marshes.

The lagoons existing nowadays are therefore relatively young ecosystems most of which have been formed during the Holocene transgression and particularly between 5,000 and 1,000 years BP. Barnes (1980) mentions that a typical lifetime can be 1,000 years for a coastal lagoon, but it seems that several lagoons have longer life spans. The Curonion lagoon on the Baltic Sea is the largest coastal lagoon in Europe (1540 km^2) and was formed since 5,000 years ago, while some Mediterranean lagoons existed since Roman times and other have been formed during the medieval period. For highly mobile species like fishes and birds, the lagoons systems are connected, because both easily migrate from one lagoon to the other through the sea and air, respectively. In contrast, for smaller size species with more restricted dispersion, a coastal lagoon can represent a geographically isolated environment. Hence, a major question is if the short geological lifetime of lagoons allows sufficient time for the evolution of novel species through geographic isolation? So far, this aspect remains poorly studied and perhaps endemic species specific to certain lagoons can be found in the future among micro-organisms and small benthic fauna like nematods (Esteves et al, 2008).

As highlighted before, the water body of coastal lagoons can be exposed to very strong fluctuations of salinities due to seasonal variations of precipitation and evaporation. In addition, in the shallow environments, the seasonal fluctuations of temperatures are often more pronounced than in the adjacent sea. This may present stressing conditions for many aquatic species and as a result, the biodiversity is often lower than can be found in comparable albeit more moderately fluctuating coastal environments. However, some species have clearly evolved from marine species to cope with the lagoonal fluctuations through radiative evolution. When the lagoon-adapted species and its evolutionary relative are morphologically very similar they are considered sibling species (Barnes, 1980). For example, *Cerastoderma glaucum*, the cockle of temperate coastal lagoons is the sibling species of *Cerastoderma edule*. Not surprisingly, *C. glaucum* has a higher tolerance range both for salinities and temperatures than *C. edule*. Three different species of *Hydrobia* snails exist, i.e. *H. ulvae*, *H. neglecta* and *H. ventrosa*. All three species are very similar as they are of similar size, attain the same age and feed on the same food and only differ by their salinity ranges (Barnes et al., 1980). Their distribution patterns corresponded to their salinity ranges of optimal growth, with *H. ulvae* occupying the open coastal environments and *H. neglecta* and *H. ventrosa* occupying the lagoonal environments of lower salinities in Danish coastal lagoons, which could be explained by a trade-off between salinity optima and resource competition (Fenchel and Kofoed, 1976). The zooplankton of the lagoon also comprise characteristic lagoon species that hardly occur in the adjacent sea including: *Acartia latisetosa, Acartia margalefi, Acartia tonsa, Calinipeda aquaedulcis*. Typical phytoplankton species in the lagoon are *Skeletonema costatum, Cyclotella* spp., *Amphora* sp. (a pennate benthic species), and typical foraminifera in the benthos include *Ammonia beccarii, Cribrononion granosum, Haynesina paucilocula* (Guerzoni and Tagliapietra, 2006).

Thus, while the large fluctuations in lagoons result in lower biodiversity of aquatic organisms in the water body and its associated sediments, at the landscape level the picture becomes a bit different. The lagoons, their barrier islands, coastal spits and their peripheral wetlands provide ecotones which creates highly diversified habitats that support a high level of biodiversity (see Fig. 2).

A first ecotone runs perpendicularly to the inner shoreline of the lagoon from terrestrial ecosystems like woods and fields- through freshwater marshes and swamps, brackish vegetation, salt marshes with halophytes and finally submerged aquatic vegetation in the lagoons (see Fig. 2C). The terrestrial aquatic interface is particularly important for

Productivity in coastal lagoons ranges from a poor 10 g C m^{-2} $year^{-1}$ to a very high value of 7,000 g C m^{-2} $year^{-1}$. The high productivity can be positive for sustaining biodiversity if there is an efficient trophic transfer. The fluctuating salinities are often a constraint and source of stress for a number of animal species. However, the relatively calm conditions and the high productivity of the lagoons are advantageous. As a result, many fish and invertebrate species have evolved a life cycle, which consists of spawning in the open sea where salinities are more constant and migration of juveniles into the lagoon where they can grow up and benefit from the calm conditions. The adults migrate back to the sea in autumn. Thus, lagoons can be considered as a nursery for many fish and invertebrate species and coastal fisheries clearly benefits from this nursery function. In general, coastal lagoons are rich in invertebrate species, e.g., the relatively small Thau lagoon (75 km^2) in S France is home for about 100 mollusc species of the 500 found in the Mediterranean Sea (Maxant and Quignard, 2005).

The primary productivity of the Magnoliophyte communities comprises both the plant material and the associated epiphytes. The epiphytes are often grazed by small invertebrates like snails *Gibbula* spp. and *Hydrobia* spp. (see above) and by small crustaceans and oligochaetes. Brent goose (see above) and other herbivorous waterfowl may directly feed on the leaves and stems of seagrass species. In general, lagoons can be very important for waterfowl, because it both sustains the herbivorous species as well as those that feed on small benthic invertebrates and on fishes. Thus, lagoons are often important for migrating waterfowl. Despite the fact, that waterfowl can feed on the plant material of the seagrasses it is generally believed that most of this material is not consumed as it is rather recalcitrant, may accumulate in sediments as refractory material and enters finally in a more slowly operating detritus food chain. However, in addition to the herbivorous bird species, seagrasses can be consumed by aquatic animals. A spectacular example are the Dugongs (*Dugong dugon*), large (upto 3 m length) marine herbivorous mammals (Mamifera) of the sirenian order that occur along the East coast of Africa, South Asia, the Indonesian archipelago and northern Australia. In temperate lagoons, seagrasses can be grazed by sea urchins (*Paracentrotus lividus*).

In conclusion, while brackish environments and particularly environments characterised by strongly fluctuating salinities are not prone to support very high biodiversity levels due to stress for many animal and plant species, it is remarkable that the lagoon environments are very important for biodiversity. But, it is important to consider the biodiversity at the landscape level. Hence, biodiversity is sustained by several factors: 1) the encounter of marine and continental flora and fauna, 2) the creation of many ecotones that run pronouncedly from dry to wet and from freshwater to (hyper) saline, 3) the presence of ecosystem engineers like the seagrasses that create specific habitats, 4) the mosaic structure of lagoons. The high productivity can be an asset for biodiversity if there is an efficient trophic transfer. However, increased eutrophication, and nutrient (N, P) overenrichment, which increases productivity, have often a negative impact on biodiversity as it has detrimental impacts on the seagrass meadows (e.g., De Wit et al., 2001, De Wit et al., 2005) and may disrupt food chains by creating dystrophic crisis in lagoons.

3. Some examples of interactions between humans and lagoons

Figure 3 is a photograph of a wall-painting representing the town of La Palma (Languedoc region, S. France) in the 13th century. It nicely shows the presence and activities of humans in the lagoon and its surrounding and gives a hint to the benefits people have obtained from lagoons. There is also substantial agricultural activity in the surrounding which certainly impacted the lagoon ecosystem, albeit to a much lesser extent than today (low fertilisation intensity and no pesticides used). The main activities in the lagoons were sailing, fishing and salt extraction. As can be seen on the photograph, salt ponds have been created as artificial structures. These have been managed to create and maintain a salinity gradient ranging from the salinity of the lagoon up to NaCl saturating conditions. Along this gradient interesting environments can be observed. The number of aquatic animal species decreases with increasing salinity and aquatic vegetation becomes simplified. From 70 g/L to 200 g/L the bottoms of the salt evaporation ponds are covered with so-called microbial mats, which are interesting stratified communities built by cyanobacteria that are very similar to stromatolites. At higher salinities an extremely eukaryotic halopilic algae *Dunaliella salina* and extremely halophilic archaea bloom in the water column colouring it orange and purple-red respectively (De Wit and Grimalt, 1992). In contrast, it was often necessary to protect these artificial pond system against intrusion of fresh water from continental runoff (see Fig. 2D). Therefore, in between the original shore of the lagoon and the salt extraction ponds, often a canal has been created to intercept the freshwater runoff and prevent it from flowing into the salt ponds. This way, the original ecotone across freshwater and salt marshes into the lagoon has been destroyed and a new ecotone has been created. While the high salinities result in lower biodiversity, it can be argued that the creation of the salt-evaporation ponds have resulted in habitat diversification allowing the original halophilic communities to coexist in the landscape with the interesting lagoonal biocoenoses, albeit at the expense of the earlier existing ecotone.

Fig. 3. Photograph of a wall painting in the Small town of La Palma, Languedoc, S. France (by Rutger de Wit).

Second, humans have tended to overexploit the natural resources in coastal lagoons. In general this has resulted in decreasing biodiversity that has become particularly dramatic since the 1950's. In many cases, the oysters where highly appreciated and have been overexploited and been depleted very early (Lotze et al., 2006) and often one or to centuries before the major dramatic biodiversity losses. In the 19th century this stimulated some entrepreneurial people to develop oyster farming in coastal lagoons as an alternative. Oysters were already farmed in the Gulf of Naples during Roman times. Initially the farmers used the autochthonous species, which in the Mediterranean and European Atlantic region was *Ostrea edulis*. However, in many cases diseases have decimated the stocks and this led to replacement by allochthonous species. The introduction of oyster farming raises the following questions: 1) Have the ecological roles of the natural populations of *Ostrea edulis* been replaced by the introduction and development of oyster farming?; 2) How should we evaluate the successive replacements by exotic species in oyster farming?; 3) Are oysters a sentinel for undesired environmental changes?; and 4) Does the socio-economic group of the oyster farmers positively contribute to the maintenance of biodiversity and ecosystem functions? Several of these questions are addressed in De Wit et al. (2011).
With the allochthonous oyster species humans also introduced undesired alien species. Currently, the Etang the Thau lagoon has about 20 % of alien macroalgal species, most originating from the Japanese coastal and marine environments, and therefore, the algal biodiversity in Thau has been characterised as a Japanese garden (Boudouresque et al., 2011). Most of them have been introduced with the foreign oysters. Some macroalgal species have become invasive and modify the local biodiversity patterns. Another animal alien species that causes many problems in mesohaline lagoons is *Ficopomatus enigmaticus* is a tube worm (Polychaeta) of 4-25 mm length that builds and lives within a carbonate tube (Grillas et al., 2006), which was first noticed in northern France in 1921. The accumulation of these carbonate tubing's results in recifs of several dm to a m height comparable to coral-like encrustation. The origin of this species is not clear and it has been suggested that it is originally an Australian species. It occurs in waters of variable salinity in temperate or warm temperate areas of both northern and southern hemispheres, and it was possibly introduced from Australia. Its preferred habitat is brackish water as found in estuaries and mesohaline lagoons. It is probable that it arrived while being transported on ships hulls or on commercial mollusc shells.

4. Lagoons and global change

Coastal lagoons are formed during periods of transgression under conditions of sea-level rise. However, the current rate of sealevel rise is much faster than observed in geological periods. This poses the question of the vulnerability of the lagoons. For example, rising sealevels and increasing erosion of lido's and barrier islands may result in complete disappearance of the lagoon at the land and seascape level, as the lagoon may be converted to an open bay. In other cases, the lido and the water body of the lagoon may move inward and a lagoonal setting remains possible in the future. Inward moving of lagoons is a natural phenomenon during periods of seawater level rise. However it depends a lot on the geomorphological and physical conditions in the land and seascape. Inward moving is often hampered by man-made structures to protect the land from erosion and and/or flooding and by urbanisation of the shoreline both at the seaside and interior shore of the lagoon. A nice example of this phenomenon is given in the book by Dennison et al. (2009) for Ocean city located on the barrier island Fenwick Island that separates the Isle of Wight Bay lagoon from the Atlantic ocean (Maryland, USA). Because of the heavy urbanisation the position of Fenwick Island has been fixed in space, but this is an uncomfortable situation as the area suffers from heavy erosion and occasionally from submersion by the sea during extreme high tides and storm events. In contrast, Assateague island, a non-urbanised barrier island located just to the south of Ocean city has moved about 600 m to the west (inland) since 1942. Assateague island and the lagoon Sinepuxent bay still comprise beautiful ecotones and all the natural aspects of lagoons although some modifications occur because Ocean city reduces the natural sand transport in the Northern part of Assateague island. This shows how natural systems have the capacity to adapt to relative sealevel rise which about 30 cm during the last 100 years and is expected to rise another 60-90 cm by 2100 (Dennison et al., 2009).

In general the interior shore of the lagoon also moves inland with relative sealevel rise. However, man-made structures can severely limit this inland migration (see Fig. 2E). As a first result, the lagoon is often losing many of the fringent saltmarshes and wetlands with a rising sea due to submersion and incorporation into the lagoon. Often, no new wetlands are created as humans often use the potentially remaining land for agriculture or housing. In addition, natural structures like cliffs may also limit the inward move of lagoons in some specific landscapes. Thus, the surfaces of coastal landscape surface in general and coastal lagoons in particular may become reduced by squeezing between an inward advancing shoreline and fixed boundary inland, a phenomenon referred to as coastal squeeze (Doody, 2004).

The shallow coastal lagoons may be particularly susceptible to increased temperatures and increased UV radiation, which are predicted to increase in 2100 by about 3 °C and 20%, respectively with respect to the current situation. This may have important consequences on the pelagic wood web (e.g., Vidussi et al., 2011). In addition, temperature changes may modify distribution patterns. For example, the seagrass *Zostera marina* has its southern distribution limit in the Mediterranean Sea. Thus, questions arises if *Zostera marina* will disappear from the Mediterranean lagoons and such a disappearance may have dramatic impacts on biodiversity, as this species is an important ecosystem engineer. Another aspect is that it is most important that lagoons have a high connectivity among them to allow species to migrate and accommodate to new geographic distributions that are compatible with the climate change. In contrast, very isolated lagoons are at higher risk as they may loose species and have low probabilities to be invaded by foreign species that are adapted to the new climatic conditions.

Finally an important aspect of global change in the marine environment is ocean acidification. Seawater used to have a pH of 8.2 and currently this has already decreased to 8.1 and a further decrease is expected to 7.9 by about 2011. While such a pH shift seems minor it may have important consequences in marine ecology, as this acidification is detrimental for calcifying organisms like some small algae, the coccolithophores, and shell-forming bivalves. However, shallow coastal lagoons with calcareous sediments like in the Languedoc region of S. France, will have the capacity to buffer the acidification.

$$CaCO_3 + H^+ \rightarrow HCO_3^- + Ca^{2+} \tag{1}$$

Equation 1 gives the chemical reaction for this dissolution. Bicarbonate is the predominant form of the carbonate equilibrium at pH around 8; hence the dissolution will consume protons and buffer the ocean acidification in shallow lagoons.

5. References

Barnes, R.S.K. (1980). Coastal lagoons; the natural history of a neglected habitat. 106 pp. Cambridge University Press, Cambridge

Boudouresque C.F., Klein J., Ruitton S. & Verlaque M. (2011). Biological Invasions: The Thau Lagoon, a Japanese biological island in the Mediterranean Sea. In: Ceccaldi, H.-J.; Dekeyser, I., Girault, M., Stora, G. (eds). Global Change: mankind-marine environment interactions - Part 3, pp 151-156. Springer Dordrecht, ISBN 978-90-481-8629-7

De Wit R., Mostajir B., Troussellier M. & Do Chi T. (2011). Environmental Management and Sustainable Use of Coastal lagon Ecosystems. Chapter 11. In: Friedman A.G., Lagoons: Biology, Management and Environmental Impact. Nova Science Publishers, Inc. New York, ISBN: 978-1-61761-738-6

De Wit, R. & Grimalt J.O. (1992) Microbial ecosystems in Spanish coastal salinas; an ecological and geochemical study of biomarkers. Limnetica, 8: 205-212.

De Wit, R. 2007. Biodiversity and Ecosystem Functioning in Transitional Waters; the point of view of a microbial ecologist. Transitional Waters Bulletin 1: 3-16

De Wit, R. 2008. Microbial diversity in the Bassin d'Arcachon coastal lagoon (SW France). Hydrobiologia 611, 5-15

De Wit, R. L.J. Stal, B.Aa. Lomstein, R.A. Herbert, H. Van Gemerden, P. Viaroli, V.-U. Cecherelli, F. Rodríguez-Valera, , B. Bartoli, G. Giordani, R. Azzoni, B. Schaub, D.T. Welsh, A. Donelly, A. Cifuentes, J. Antón, K. Finster, L.B. Nielsen, A.-G. Underlien Pedersen, A. Turi Neubeurer, M.A. Colangelo & S.K. Heijs, 2001. "ROBUST: The ROle of BUffering capacities in STabilising coastal lagoon ecosystems. Continental Shelf Research 21: 2021-2041.

De Wit, R., J. Leibreich, F. Vernier, F. Delmas, H. Beuffe, Ph. Maison, J.-C. Chossat, C. Laplace-Treyture, R. Laplana, V. Clavé, M. Torre, I. Auby, G. Trut, D. Maurer, P. Capdeville, 2005. Relationship between land-use in the agro-forestry system of les Landes, nitrogen loading to and risk of macro-algal blooming in the Bassin d'Arcachon coastal lagoon (SW France). Estuarine, Coastal and Shelf Science 62: 453-465

Dennison W.C., Thomas J.E., Cain C.J., Carruthers T.J.B., Hall M.R., Jesien R.V., Wazniak C.E., Wilson D.E. (2009) Shifting Sands: Environmental and cultural change in Maryland's Coastal Bays. IAN Press Cambridge MD, ISBN 978-0-9822305-0-3

Doody, J.P. (2004) 'Coastal squeeze' - an historical perspective. Journal of Coastal Conservation, 10/1-2, 129-138.

Duarte C.M. (2001) Seagrasses. In Encyclopedia of Biodiversity, vol. 5. (S.A. Levin Ed.) pp 255-268. Academic Press, San Diego

Esteves, FA., Caliman, A., Santangelo, JM., Guariento, RD., Farjalla, VF., Bozelli, RL. (2008) Neotropical coastal lagoons: an appraisal of their biodiversity, functioning, threats and conservation management. Brazilian Journal of Biology 68 no.4 suppl.0 São Carlos Nov. 2008 doi: 10.1590/S1519-69842008000500006

Fenchel T. & Kofoed, L.H. (1976). Evidence for exploitative interspecific competition in mud snails (Hydrobiidae). Oikos 27, 367-376

Grillas, P., Auby, I., Mesleard, F., Cramm, P., Morvan, R., Raveyrol, A., Dutrieux, E., Licari, M.-L., Crivelli, A. Guide méthodologigique de gestion des lagunes méditerranéennes. Tome 2 Les espèces. (241 pp). Région Languedoc-Roussillon, Montpellier (in French)

Guerzoni, S., Tagliapietra, D. (2006). Atlante della laguna venezia tra terra e mare. Marsilio Venice (in Italian) 241 pp ISBN 88-317-8764-0

Kjerfve, B. (1994). Coastal Lagoons, chapter 1. In: Kjerfve, B. (ed.) Coastal lagoon processes. pp 1-8. Elsevier Oceanography Series, Amsterdam

Lotze H.K., Lenihan H.S., Bourque B.J., Bradbury R.H., Cooke R.G., Kay M.C., Kidwell S.M., Kirby M.X., Peterson C.H., Jackson J.B.C. (2006) Depletion, degradation, and recovery potential of estuaries and coastal seas. Science 312:1806-1809

Maxant, F. & Quignard, J-P. (2005) Thau oasis de vie, Ecosystèmes, patrimoines et paysages sous-marins. Biotope Eds. 223 pp. ISBN 2914817010 (In French)

Vidussi F, Mostajir B, Fouilland E, Le Floc'h E, Nouguier J, Roques C, Got P, Thibault-Botha D, Bouvier T, Troussellier M (2011). Mediterranean plankton food web responses to experimental warming and ultraviolet B increases. Limnology and Oceanography 56 (1): 206-218

3

Distribution, Endemism and Conservation Status of Fishes in the Yangtze River Basin, China

Shaowen Ye, Zhongjie Li, Jiashou Liu,
Tanglin Zhang and Songguang Xie
State Key Laboratory of Freshwater Ecology and Biotechnology
Institute of Hydrobiology, Chinese Academy of Sciences
P.R. China

1. Introduction

The Yangtze River (Chang Jiang), with a length of 6300 km, is the longest river in Asia and the third-longest in the world. It originates from glaciers on the Geladandong Mountains of the Tibetan Plateau and follows a sinuous west to east route before emptying into the East China Sea at Shanghai. It drains one-fifth of China's land area and its river basin is home to one-third of China's population. Owing to a long geographical history, affluent water resources, an immense variety of lotic and lentic ecosystems, and the differences of physical environments in the upper, middle and lower reaches, the Yangtze River basin (a complex riverine-lacustrine network) is especially rich in fish fauna, representing high fish species richness and endemism in the Palearctic region. Being pregnant with plentiful fish resources, the basin is a cradle of inland fisheries in China. Its fishery yield accounts for about two-thirds of freshwater fishery production of the whole country (Liu & Cao, 1992). The Yangtze River Basin is therefore a globally significant area for preserving fish biodiversity and fishery resources.

However, the Yangtze River basin is also an area highly impacted by a long history of human use and environmental variation, and is further threatened by China's rapid economic development and the demands of over 400 million people living in the basin during the past few decades. Since the 1950s, loss of fish biodiversity in the Yangtze River basin have been accelerated by a series of direct and indirect effects of human activities and environmental changes, *e.g.* disappear, shrinkage and fragmentation of habitats for fish spawning, feeding and migration, overfishing, water pollution and invasion of exotic species. Documenting regional fish distribution and understanding major threats to fish biodiversity are necessary for protecting and recovering endangered fish species and natural communities. Thus, it is urgently needed to review the problems and threats facing fish resources in the Yangtze River basin, and to provide crucial information about which species are at risk and what factors threaten their existence for developing successful conservation strategies to slow the loss of fish biodiversity.

Since the 1930s, Chinese ichthyologists have investigated fish fauna and biodiversity in the Yangtze River basin, and a large amount of information has been published on taxonomic, biological and biogeographical aspects. In the present chapter, we collected and synthesized these scattered data from relevant literature including Chu (1955), Du (1962), Anonymous (1976, 1980), Yang (1987), Zeng (1990), Zhang (1991), Wu & Wu (1992), Ding (1994), Huang & Xie (1996), Chen (1998), Yue & Chen (1998), Li et al. (1998), Chu et al. (1999), Song et al. (1999), Yue (2000), Ni & Zhu (2005), Xie & Yang (2005), Zhuang et al. (2006), Ye et al. (2006; 2007), Ye (2007), Zhang & Li (2007), Wu & Zhong (2008), Zhang et al. (2008). Threatened status of fish species in the Yangtze River basin was compiled from China Species Red List (Wang & Xie, 2009) and China Red Data Book of Endangered Animals (Yue & Chen, 1998) into five threatened categories: extinct (EX), extinct in the wild (EW), critically endangered (CR), endangered (EN) and vulnerable (VU). Our main purposes were (1) to investigate large-scale distribution and endemic species composition of fishes in different reaches of the Yangtze River basin, (2) to rank the major threats to impaired or extinct fish species and quantify the relative contribution of intrinsic factors to fish endangerment, and (3) to provide recommendations for fish biodiversity conservation in the Yangtze River basin.

2. Fish distribution and endemism

2.1 Description of river reaches

The Yangtze River basin covers an area of about 1.8 million km^2 and lies across the three large topographic platforms of the Chinese mainland (Fig. 1). The riverhead is located on the Tibetan plateau, where the mean elevation is over 4500 m (Zeng, 1990). Upstream from Yichang (in Hubei Province) is the upper reach, with a narrow valley and a rocky channel of high gradient ratio. The length is more than 4300 km, and the drainage area is almost 100×10^4 km^2. The famous Three Gorges Dam is located in the upper mainstream. The range of the middle reach is from Yichang to Hukou (in Jiangxi Province), with a length of about 950 km and a drainage area of almost 68×10^4 km^2. Here, the river has a gentle gradient and takes a meandering course. It connects with shallow lakes of various sizes and numerous tributaries and forms an endemic Chinese compound ecosystem of inland water. From

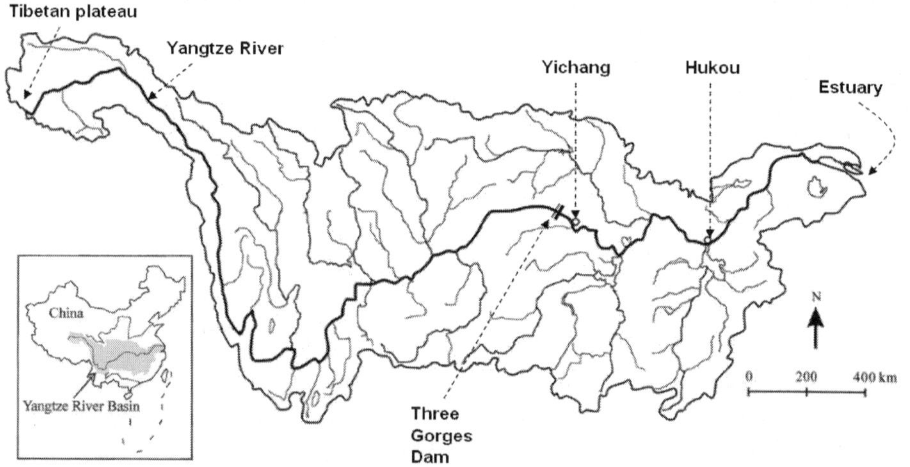

Fig. 1. Sketch map of the Yangtze River basin

Hukou downward to the mouth of the river is the lower reach, with a length of about 930 km. The drainage area is about $12×10^4$ km^2. This segment of the river wanders among plains and hills, and several large interior lakes, such as Lake Caohu and Lake Taihu in association with many tributaries, drain into the reach (Chen et al., 2000). In the estuary confluent with the East China Sea, the river forms a trumpet shaped delta.

2.2 Faunal composition

Our synthesis (see Appendix 1) showed that there are 416 fish species and subspecies (16 orders and 33 families) known from the entire Yangtze River basin, which amount to about 40% of the total freshwater fishes in China (Zhu, 1995) and far exceed the number of any other river systems in China. For example, the Pearl River has 294 and the Yellow River only has 150 fish species and subspecies (Cao, 1992). Of these fishes in the Yangtze River basin, 362 species and subspecies spend all their lives in fresh water (Table 1), and 11 species of migratory fishes include eight anadromous species (*Acipenser sinensis, Macrura reevesi, Coilia ectenes, C. mystus, Myxocyprinus asiaticus, Hemisalanx prognathus, Takifugu obscurus, T. flavidus*) and three catadromous species (*Anguilla japonica, A. marmorata, Trachidermus fasciatus*). The rest 43 species chiefly live in the brackish water of the estuary with a wide range of salt tolerance, moving regularly between coasts and the estuary. A significant feature of the fish fauna in the Yangtze River basin is the large amount of endemic fishes (Table 1). There are 178 endemic species that occupy 42.8% of the total number of fishes in the basin.

Order	Yangtze River basin			World [a]	
	Freshwater species	Species using freshwater [b]	Endemic species	Freshwater species	Species using freshwater [b]
Acipenseriformes	2	3	1	14	27
Anguilliformes	0	2	0	6	26
Clupeiformes	0	3	0	79	85
Cypriniformes	280	280	150	3268	3268
Siluriformes	40	40	20	2740	2750
Osmeriformes	6	8	1	82	86
Salmoniformes	2	2	1	45	66
Mugiliformes	0	4	0	1	7
Atheriniformes	1	1	0	210	240
Beloniformes	1	2	0	98	104
Cyprinodontiformes	3	3	0	996	1008
Synbranchiformes	1	1	0	96	99
Scorpaeniformes	0	1	0	60	62
Perciformes	26	50	4	2040	2335
Pleuronectiformes	0	4	0	10	20
Tetraodontiformes	0	12	1	14	22
Total	362	416	178	9759	10205

[a] Data are cited from Nelson (2006). [b] Freshwater species plus those species frequently occurring in freshwater that may otherwise be diadromous or simply entering fresh water in substantial numbers or in a substantial portion of their range.

Table 1. Species number of fishes in the Yangtze River basin and in the world

The most species-rich order in the Yangtze River basin is Cypriniformes (Table 1), for which the proportion of species number in the basin to those in the world is 8.6%. And there are four orders (Mugiliformes, Tetraodontiformes, Pleuronectiformes, Acipenseriformes) for which the proportion of number of species using freshwater in the basin to those in the world is more than 10%. This indicates that fish biodiversity in the Yangtze River basin plays an important role in the freshwater fish species bank of the world.

Cyprinidae, Cobitidae, Gobiidae, Bagridae and Homalopteridae are the most species- or endemic species-rich families (Fig. 2 and 3). Triplophysa (Cobitidae) and Schizothorax

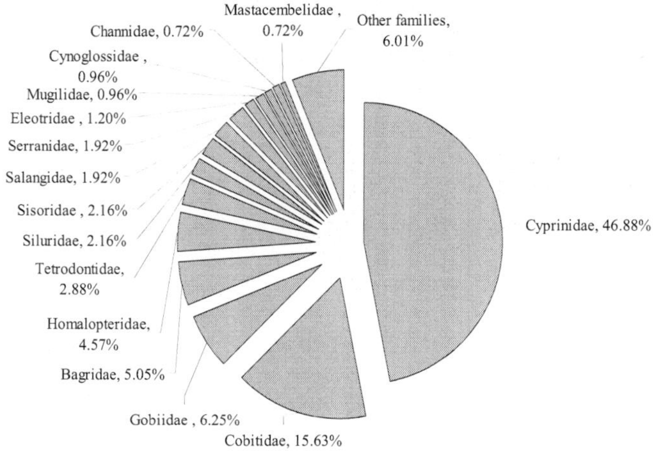

Fig. 2. Percentage of species number of the most species-rich families to the total species in the Yangtze River basin

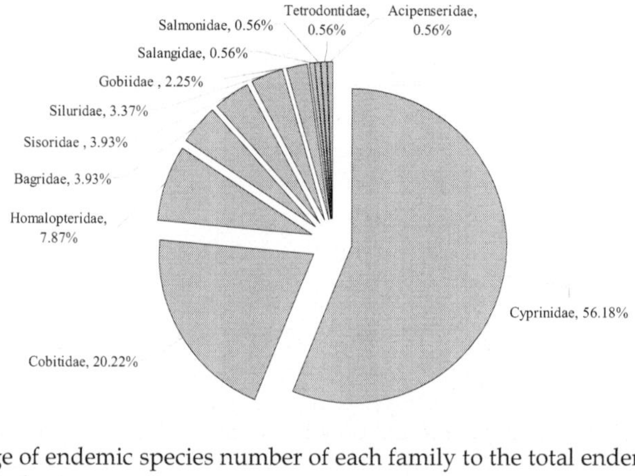

Fig. 3. Percentage of endemic species number of each family to the total endemic species in the Yangtze River basin

(Cyprinidae) are two genera with the highest species numbers in the basin, accounting for 6.5 and 4.1% of the total species, and 8.4 and 9.0% of the total endemic species, respectively (see Appendix 1). Tectonic activities, such as uplift of the Tibetan Plateau, were believed to be important and responsible for the higher diversity of the two genera (Cao et al., 1981). Tectonic activity has also been associated with the large numbers of endemic species in other river basins (Minckley et al., 1986).

2.3 Large-scale distribution

The fishes are not evenly distributed in the Yangtze River basin (Fig. 4). Among the 416 species, 206 species occupy only one of the five defined areas of the basin (riverhead, upper reach, middle reach, lower reach, and estuary). Common species that occur in two, three or four of the defined areas are all less than 25% of the total fish species number in the basin. Endemic fishes appear to have the similar distribution pattern. Only four endemic species (*Squalidus nitens, Saurogobio gymnocheilus, Parabotia bimaculata,* and *Pelteobagrus eupogon*) are common to the three reaches and estuary. Most of the 178 endemic fishes (116 species) occupy one of the define areas. In particular, 97 endemic species were found only in the upper reach (Table 2), indicating that the distribution of the endemic fishes intensively occurs there.

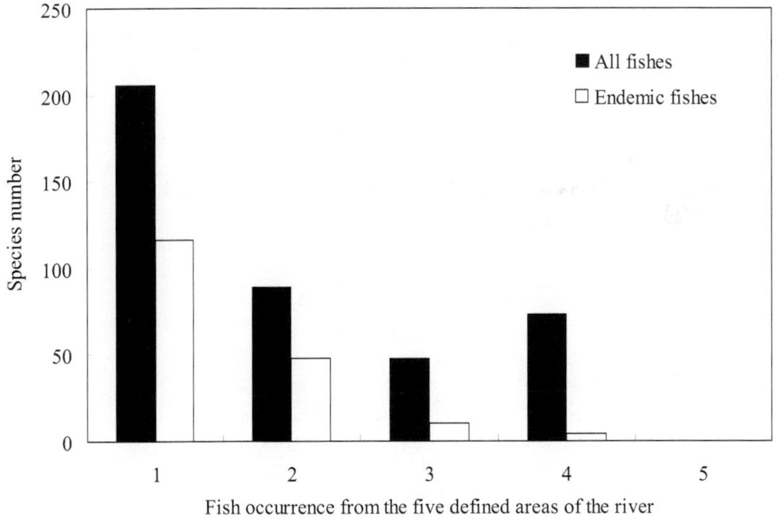

Fig. 4. Distribution range of fishes in the Yangtze River basin

In the riverhead waters, the number of fishes is very small, including five cyprinid fishes, seven species of genus Triplophysa, and two species of family Sisoridae (see Appendix 1). Here, the mean elevation is over 4500 m (Zeng, 1990) and only some specialists could live. Of these fishes, five species were found only in the riverhead (Table 1) and the other nine species were also reported in the upper reach. Nearly 60% species are endemic fishes, and one species (*Triplophysa tanggulaensis*) was recorded only in the riverhead (Table 2).

The upper reach has the highest species number, and more than half species are endemic (Table 2). There are 279 species and subspecies of fishes from a wide range of taxonomic categories. Taking food of adhering algae and invertebrates, most of them adapt to rapid current and inhabit the underlayer, while others (fishes of families Homalopteridae and Sisoridae) adhere to and climb on the stones of the riverbed. The Chinese paddlefish (*Psephurus gladius*) and Chinese sucker (*Myxocyprinus asiaticus*), which are distributed mainly in the Yangtze River and could be found in all of the three reaches and their main branches, at present occur chiefly in the upper reach. There were only two migratory fishes, but the Japanese eel (*Anguilla japonica*) disappeared after the Gezhouba Dam was constructed in the upper mainstream (Xie & Chen, 1999).
Fishes in the middle and lower reaches are also abundant, but with lower proportion of endemic fishes compared with the upper reach (Table 2). More migratory fishes (*Anguilla marmorata, Macrcura reevesi, Coilia ectenes, C. mystus, Myxocyprinus asiaticus, Hemisalanx prognathus, Trachidermus fasciatus, Takifugu obscurus, T. flavidus*) occur in the middle and lower reaches. Some euryhaline marine fishes inhabit the lower reach and the estuary of the river, such as *Neosalanx andersoni, Liza carinatus, Mugil cephalus, Terapon jarbua, Repomucenus olidus, Cynoglossus gracilis, and Takifugu spp.*

	All fishes (A)	Endemic fishes (E)	E/A (%)
Riverhead	14 (5)	8 (1)	57.1
Upper reach	279 (119)	147 (97)	52.7
Middle reach	227 (36)	70 (17)	30.8
Lower reach	158 (5)	23 (1)	14.6
Estuary	142 (41)	10 (0)	7.0

Note: The number in parentheses indicates the number of fish species found only in that area.

Table 2. Species number of fishes in different areas of the Yangtze River basin

3. Threats to fish biodiversity

3.1 Structural change of fish community

Historical variations in the fish community of Yangtze River's lakes can be summarized as follows: (1) decline in species richness and number of migratory species; (2) decrease in the abundance of piscivorous species in fish catches; and (3) diminution in age-specific body length of commercial fishes (Cao et al., 1991; Xie & Chen, 1999).
Here, four examples from Lake Donghu, Lake Honghu, Lake Liangzi and Lake Poyang illustrated the changes of species richness, ecological and trophic structures of fish community during the past half century (Table 3). Apparent reduction of species richness was observed in these lakes, especially a sharp decline in Lake Donghu that is a typical urban lake and therefore received more disturbances from human being. The number and percentage of migratory fishes decreased in all the four lakes, mainly due to river-lake blockage by the building of sluice gates. The comparison of fish catches indicated that the proportion of piscivorous fishes tended to decrease dramatically, mainly due to their relatively high vulnerability to over-fishing and habitat destruction or degradation.
Another five examples on four commercial fish species were used to demonstrate the size diminution of fishes in the Yangtze River's lakes. During the past decades, the age-specific

body lengths of crucian carp (*Carassius auratus auratus*), sneakhead fish (*Channa argus*), redfin culter (*Cultrichthys erythropterus*), and Chinese perch (*Siniperca chuatsi*) have shown a remarkable decline (Table 4). Extensive fishing might have removed more fast-growing individuals and thus left more stunted individuals (Cao et al., 1991).

Lake	Year	SR	Ecological group of fishes				Reference
			MS	RS	LS	PS	
Lake Donghu	before 1971	67	9 (13.4%)	9 (13.4%)	49 (73.2%)	5.22%	Huang & Xie, 1996
	1992-1994	39	5 (12.8%)	3 (7.7%)	31 (79.5%)	0.26%	Huang & Xie, 1996
Lake Honghu	1959	64	8 (12.5%)	12 (18.8%)	44 (68.7%)	10.0%	Song et al., 1999
	1992	57	4 (7.0%)	12 (21.1%)	41 (71.9%)	5.0%	Song et al., 1999
Lake Liangzi	Before 1974	72	10 (13.9%)	14 (19.4%)	48 (66.7%)	11.4%	Ye, 2007
	1997-1999	58	6 (10.3%)	4 (6.9%)	48 (82.8%)	9.4%	Xie & Yang, 2005
Lake Poyang	before 1980	117	14 (12.0%)	32 (27.4%)	71 (60.7%)	-	Zhang & Li, 2007
	1997-2000	101	7 (6.9%)	24 (23.8%)	70 (69.3%)	-	Zhang & Li, 2007

SR: Species richness; MS: Number and percentage of migratory species in fish community; RS: Number and percentage of riverine species in fish community; LS: Number and percentage of limnicolous species in fish community; PS: Percentage of piscivorous species in fish catches.

Table 3. Historical changes of fish community structure in Yangtze River's lakes

Fish species	Lake	Year	Age				Reference
			1	2	3	4	
Carassius auratus auratus	Lake Honghu	1960	61	118	184	231	Anonymous, 1976
		1988	38	63	83	99	Zhang, 1991
Channa argus	Lake Liangzi	1956	190	288	398	516	Du, 1962
		1998	198	305	366	443	Xie & Yang, 2005
Cultrichthys erythropterus	Lake Poyang	1958	115	159	218	235	Anonymous, 1976
		2006	131	148	176	191	Zhang et al., 2008
Siniperca chuatsi (Female)	Lake Poyang	1959	177	246	335	425	Anonymous, 1976
		1997	131	247	310	349	Li et al., 1998
Siniperca chuatsi (Male)	Lake Poyang	1959	173	220	324	-	Anonymous, 1976
		1997	127	228	289	328	Li et al., 1998

Table 4. Historical changes of age-specific body lengths (standard length, unit: mm) of fishes in lakes of the Yangtze River basin

3.2 Species endangerment

It was estimated that at least 20% of the world's freshwater fish species were seriously threatened or extinct, mainly as a result of habitat modification (competition for water, drainage, and pollution), species introduction, and commercial exploitation (Groombridge,

1992). In the Yangtze River basin, there are 65 threatened fish species included in the China Species Red List (Wang & Xie, 2009), belonging to 10 orders and 18 families (see Appendix 1). These fishes are classified into five threatened categories (Table 5), i.e., extinct (two species), extinct in the wild (two species), critically endangered (five species), endangered (36 species), and vulnerable (20 species). It should be noted that 69% of the threatened or extinct fishes are endemic species in the Yangtze River basin (Table 5).

As shown in Table 6, the two species extinct in the wild (*Anabarilius polylepis* and *Schizothorax parvus*) and one extinct species (*Anabarilius liui liui*) were known to be endemic only in the upper reach. Another extinct species (*Atrilinea macrolepis*) was endemic only in the middle reach. Of the five critically endangered species, Chinese paddlefish (*Psephurus gladius*) were recorded in the upper, middle, lower reaches and estuary, and *Euchiloglanis kishinouyei* were recorded in both riverhead and upper reach. The other three critically endangered species (*Megalobrama elongata, Schizothorax longibarbus*, and *Leiocassis longibarbus*) occurred only in the upper reach. Overall, there are much more threatened or extinct fish species distributed in the middle reach than in the other areas of the Yangtze River basin.

	All species (A)	Endemic species (E)	E/A (%)
Extinct (EX)	2	2	100
Extinct in the wild (EW)	2	2	100
Critically endangered (CR)	5	4	80
Endangered (EN)	36	27	75
Vulnerable (VU)	20	10	50
Total evaluated	65	45	69

Table 5. Species number of fishes in different threatened categories

	Threatened categories					Total evaluated
	Extinct	Extinct in the wild	Critically endangered	Endangered	Vulnerable	
Riverhead	0	0	1	2	1	4
Upper reach	1	2	5	28	13	48
Middle reach	1	0	1	9	10	20
Lower reach	0	0	1	8	8	9
Estuary	0	0	1	8	3	12

Table 6. Threatened status of fishes in different areas of the Yangtze River basin

3.3 Major threats

From the China Species Red List (Wang & Xie, 2009) and China Red Data Book of Endangered Animals (Yue & Chen, 1998), we extracted the recognized causes (major threats and intrinsic factors) of endangerment or extinction of the 65 species in the Yangtze River basin. We did not try to distinguish between ongoing and historical threats, because such information is often lacking, and the distinction itself is problematic in the case of habitat destruction. We tabulated the number of species threatened by four broad categories of threats: habitat loss/degradation, unreasonable harvesting, water pollution, and biological invasion. We then used the resulting database to determine the frequency of the threat categories and intrinsic factors (i.e. recruitment/reproduction, juvenile mortality, population density, growth rate, and dispersal range) for the threatened fish species. We further subdivided primary threat factors into finer categories, and determined the frequency of each of these causes of endangerment.

Fig. 5 presents a summary of the percentages of fish species in the Yangtze River basin imperilled by the major threats. Unreasonable harvesting is the most pervasive threat to the fishes in the Yangtze River basin, contributing to the endangerment of 86.2% of the listed species, followed by habitat loss/degradation (43.1%), water pollution (30.8%), and biological invasion (27.7%). In particular, overfishing (83.1%) was identifies as the most severe threat. The great frequency of overfishing stems mainly from China's large and poor population, along with widespread trade in fishery products. The use of devastating fishing methods, such as ever-finer meshed nets, poisons, explosives, and electric shocking, is also a contributing factor. Within the threats to habitat loss/degradation, damming and other hydraulic construction appeared to be the most significant threat. The construction of dams at cascades in the river and its tributaries has blocked the passage of migratory fish species, changed hydrological conditions, made their habitats fragmented, and brought about severe impacts on fish habitats and spawning environments. Water pollution influences fish resources not only through direct effects on fish survival, but also through negative impacts on various kinds of food organisms for fishes such as plankton and benthos, resulting in decreases in the abundance of natural fish stocks.

Invasion of non-indigenous species is also a serious threat to biodiversity in the basin. Fish compositions in some plateau lakes of the upper Yangtze River, such as Lake Dianchi and Lake Lugu, have been seriously affected by exotic species introductions (Xie et al., 2001). Exotic species compete with native fishes for food and other resources, leading to greatly reduced numbers of local species, and even the local extinction of some species (Chen et al., 1998). The number of native species of Lake Dianchi has declined from 25 in the 1940s, to 15 in 1978, and 8 in 1982. A recent fish fauna investigation conducted in the lake found only five native species, and the number of endemic species and subspecies had declined from nine in the 1950s to two (Xie & Chen, 1999; Xie et al., 2001).

Among the intrinsic factors we analyzed, limited dispersal and restricted range appear to have the highest contributions to the effects of the threats upon fishes in the river basin (Fig. 6). The other intrinsic factors lag behind these two primary factors in terms of their frequency for the imperilled and extinct fish species (Fig. 6). In comparison, the three factors (slow growth rate, population fluctuation, and high juvenile mortality) could be minor causes of endangerment in the Yangtze River basin.

The endemic fish distribution in the upper reach is represented by complicated assemblages, which are highly correlated to the topographic and geomorphic characteristics of the upper Yangtze River basin (He et al., 2011). Generally, these endemic species are well adapted to specific habitats with relatively small population size. For example, the fishes of Triplophysa and Schizothorax usually live in the tributaries and mountain streams with torrential flows and gravel riverbeds, whose origin, evolution and distribution are related to the uplift of the Tibetan Plateau (Cao et al., 1981). They are usually sensitive to environmental changes or disturbances, and more prone to extinction under the threats of overfishing and habitat loss, mainly due to their limited dispersal and restricted range.

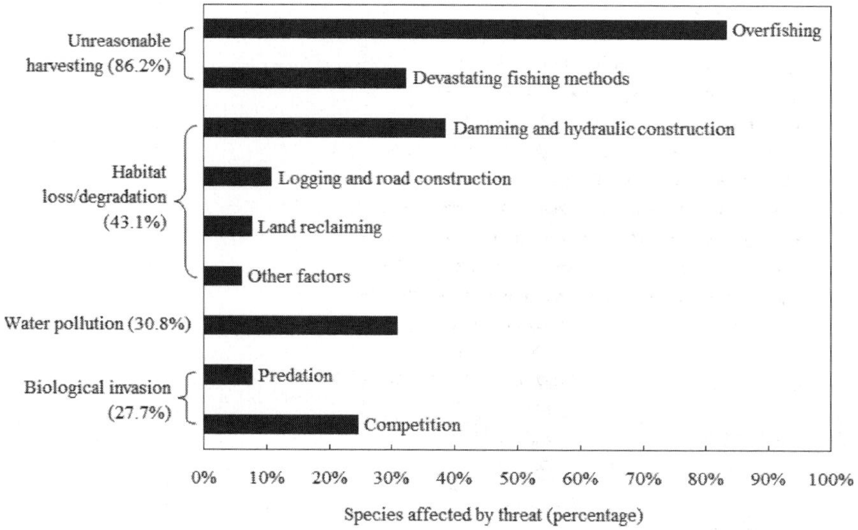

Fig. 5. Rank of the major threats to fish endangerment in the Yangtze River basin

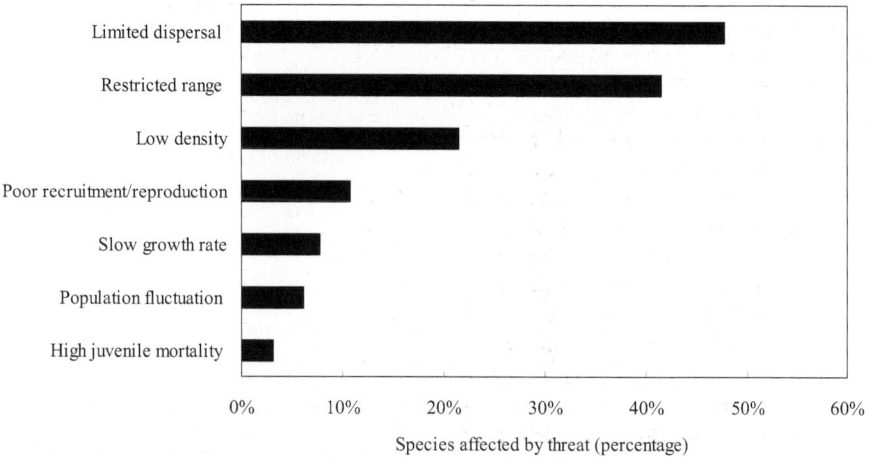

Fig. 6. Rank of the intrinsic factors for fish endangerment in the Yangtze River basin

Species that use distinct environments during different stages of their life histories (e.g. migratory fish) are at risk from elimination of crucial habitats or erection of structures that impede movement. For example, the Chinese paddlefish (*Psephurus gladius*) declined drastically in the Yangtze River after the Gezhouba Dam blocked access to its upstream spawning habitats, and the fish likely faces extinction (Wei et al., 1997). Furthermore, they are especially vulnerable to overfishing and other environmental hazards because the long maturation time limits rates of population recovery. On the other hand, fragmentation of resident populations also results from barriers, and may ultimately lead to species loss. During the late 1950s-1970s, sluice gates were constructed in almost all lakes of the middle and lower reaches except for Lake Dongting and Lake Poyang for water conservancy projects. The sluice gates blocked interchanges of fishes in the river-lakes ecosystem. Severance of fish exchange has brought about the decline in natural fish stocks in the river and its lakes, especially the decline in species richness and abundance of migratory fishes (Xie & Chen, 1999).

4. Conservation status and strategies

Restoration of the Yangtze River to its original state is impractical given the constraints prevailing in the region, but some degree of rehabilitation will be possible if relevant legislation and scientific information are promptly applied. Even though China lacks a truly comprehensive law on nature conservation, it did enact the China Wildlife Protection Law in 1989 to protect rare and endangered species. Although the law has yet to be completely enforced, China has established a legislative framework for biodiversity conservation and a number of action plans have been initiated (Xu et al., 1999). A fishery law dating from 1986 and revised in 2000 proscribes fishing of rare and precious aquatic animals. Within the last two decades, an interprovincial authority, the Administrative Commission of the Yangtze River Fisheries Resources, was established to control illegal fishing activities involving the use of electricity, explosives, and poisons, and to protect fish stocks in the Yangtze River. In February 2003, an annual fishing moratorium of three to six months was introduced to protect 8100 km of the river, including 4090 km of the mainstream and more than 4000 km of its tributaries. This measure builds on the well established and widespread practice of stocking Chinese rivers and lateral lakes with cultured fries of major carp species to maintain or enhance fishery yields (Fu et al., 2003). Such stocking cannot be regarded as a measure that contributes greatly to biodiversity conservation, because it has implications for the genetic variability of indigenous carp species. However, it may reduce harvesting pressure on endangered species that could benefit from the fishing moratorium.

Problems with protection of endemic fishes and implementation of wildlife protection legislation in the Yangtze River basin exist because the natural resources laws and regulations, especially for fisheries, have been formulated from the standpoint of economic value, which emphasizes utilization rather than protection. Thus the conservation of endangered species is mainly the responsibility of the Ministry of Agriculture, which has the historical remit of increasing or expanding the capture quotas of economically important species. Moreover, a separate authority, the Ministry of Water Resources, has an overlapping responsibility with the Ministry of Agriculture, but it is directed toward maintaining water supplies for human consumption and agriculture, whereas protection of water quality is under the jurisdiction of the State Environmental Protection Administration. The conflict between economic development and conservation plus the overlapping and sometimes contradictory responsibilities of government authorities have impeded actions that are needed to preserve biodiversity (Xu et al. 1999). Some progress has been made recently: in 2000, a 400-km section of the upper reach of the Yangtze River was designated as a reserve for rare fishes, but portions of it will be affected by rising water levels behind the Three Gorges Dam and by the dams planned for the Jinsha River tributary. Potential reserve sites for rare fishes elsewhere along the Yangtze River have been identified (Park et al., 2003), but they await official designation.

Since the Yangtze River basin is undergoing a very rapid deterioration as a consequence of human-induced changes, conservation strategies must be improved and expanded. The conservation of the fishes in the upper reach should be considered as a priority. Current knowledge of speciation theories indicates the importance of local processes in generating high endemicity of inland water fauna (Martens, 1997). The high endemicity of fish species in the upper reach indicates that the more immediate conservation need is to retain as many of the natural ecological processes and functions of the river as possible. A conservation recommendation, according to which the habitats rich in endemic species should be identified as the nature reserves, has been put forward (Liu & Cao, 1992). In view of the different endemic fish assemblages (with multiple habitats and different fish compositions) in the upper Yangtze River, different reserves aimed at different conservation objectives should be set up in order to preserve fish biodiversity (He et al., 2011). At present, there are three fish reserves, of which, one is for the protection of rare and unique species in the upper reach at the national level, and two are for the protection of Chinese sturgeon at the provincial level. The National Nature Reserve for Rare and Peculiar Fish Species in the upper Yangtze River is the only national protected zone across many provinces in China. Its management and protection tasks are very arduous. The effective realization of nature reserve functions can be pushed forward only through their tracing, monitoring and appraisal.

In the middle and lower reaches of the Yangtze River basin, the conservation strategies for fishes must take into account the complex life cycles and ontogenetic shifts in habitat requirements of migratory fishes. Restoring the stocks of the migratory fishes should be a priority in these reaches, because the severance between the Yangtze River and its connecting lakes is the most critical factor responsible for the decline of the abundance of the migratory fishes. In addition, fragmentation of the resident population resulting from the severance may ultimately lead to species loss. Ideally, restoration of natural fish stocks in the middle and lower reaches would include restoration of the free flow, so fishes can move freely between the Yangtze River and its lakes. Minimally, construction of a fish passage structure is necessary. Conventional fish ladders designed for salmonids may not be successful because most fishes do not jump. In the basin, there are more than 40 lake-river migratory fishes (species that migrate from lakes to breed in rivers; Liang et al., 1981). Most of them have been affected by severance (Xie & Chen, 1999). Compared to severed lakes, the river-connecting lakes are rich in river-lake migratory fishes (Liang et al., 1981; Liu & Wang, 2010). At present, considering the importance of river-connecting lakes for sustaining populations of the river–lake migratory fishes, river-connecting lakes, such as Lake Dongting and Lake Poyang, are also of high priority for conversation.

Conservation must be proactive. Our ability to ameliorate or mitigate the effects of human activities is predicated upon an adequate understanding of river ecology. But information on the prevailing situation with respect to habitat integrity and fish biodiversity in the Yangtze River basin is inadequate. Some published catalogs of fish faunal composition are based on dated literature or old museum collections instead of recent surveys. Information on fish conservation status is thus unreliable, and extinctions may not be evident until long after they have occurred. The basic task of compiling species inventories is one priority for biodiversity conservation, but it should be accompanied by assessments of population size and long-term viability (Dudgeon, 2000). These tasks will require monitoring strategies, such as those developed by Humphrey et al. (1995) in tropical Australia, and formulation of indices of river health that can be applied to individual habitats. Furthermore, explicit information on life histories of endangered species is highly necessary for successful conservation of the fishes.

Data sharing and collaboration between academic institutions and governmental agencies are particularly essential for the effectiveness of fish conservation. All research efforts are in vain if the resulting knowledge is not translated into social and/or political action. This may be the greatest challenge facing conservation ecologists. Ecological stations and networks should be established linking inland water biological laboratories from the universities and the institutions along the Yangtze River, in conjunction with museums, particularly regarding long-term ecological research. This will ensure that currently available information is used effectively. It is also necessary to establish and standardize databases of inland water organisms and associated environmental information.

5. Conclusion

The large amount and high endemicity of fish species have resulted in the importance of the Yangtze River basin in the freshwater fish species bank of the world. Fish biodiversity and freshwater ecosystems are seriously jeopardized by human activities in the basin. The decline of diversity and the loss of some fishes will have potential impacts on the global fish biodiversity, so more hotspots of fish biodiversity in the Yangtze River basin should be identified as nature reserves. There is an urgent need for integrated action and legislation to ensure that the endangered species are legally protected within their range. A mixture of strategies will be essential to preserve fish biodiversity in the long term. It must include reserves that protect key, biodiversity-rich water-bodies (e.g. tributaries in the upper reach and river-connecting lakes in the middle reach) and their catchments, as well as species- or habitat-centred plans that reconcile the protection of biodiversity and societal use of water resources. In parallel, conservation ecologists must more effectively communicate the importance and value of fish biodiversity to stakeholders and policy makers, so as to make certain that all available information on fish biodiversity is applied effectively to ensure its conservation.

6. Acknowledgment

The authors are grateful to the National Science Foundation of China (No. 30900182) and the "Special Fund for Agro-scientific Research in the Public Interest" (No. 200903048-04) for their support.

Order Family *Species*	Threat status	River-head	Upper reach	Middle reach	Lower reach	Estuary
I. Acipenseriformes						
Acipenseridae						
*1. Acipenser dabryanus**	EN		+	+		
2. A. sinensis	EN		+	+	+	+
Polyodontidae						
3. Psephurus gladius	CR		+	+	+	+
II. Anguilliformes						
Anguillidae						
4. Anguilla japonica			+	+	+	+
5. A. marmorata	EN				+	+
III. Clupeiformes						
Clupeidae						
6. Macrura reevesi	EN			+	+	+
Engraulidae						
7. Coilia ectenes				+	+	+
8. C. mystus					+	+
IV. Cypriniformes						
Catostomidae						
9. Myxocyprinus asiaticus	VU		+	+	+	+
Cyprinidae						
10. Zacco platypus			+	+	+	+
*11. Z. chengtui**	VU		+			
12. Opsariichthys bidens			+	+	+	+
13. Aphyocypris chinensis			+	+	+	+
*14. Gobiocypris rarus**	EN		+			
15. Phoxinus oxycephalus			+	+	+	
16. Atrilinea roulei	VU				+	
*17. A. macrolepis**	EX			+		
18. Mylopharyngodon piceus			+	+	+	+
19. Ctenopharyngodon idellus			+	+	+	+
20. Squaliobarbus curriculus			+	+	+	+
21. Luciobrama macrocephalus	EN		+	+	+	+
22. Ochetobius elongatus			+	+	+	+
23. Elopichthys bambusa			+	+	+	+
*24. Anabarilius liui liui**	EX		+			
*25. A. l. chenghaiensis**			+			
*26. A. brevianalis**			+			
*27. A. songmingensis**			+			
*28. A. qionghaiensis**	EN		+			
*29. A. xundianensis**			+			
*30. A. polylepis**	EW		+			
*31. A. alburnops**	EN		+			
32. Sinibrama wui			+	+	+	
33. S. macrops			+	+	+	
*34. S. taeniatus**			+			
*35. Ancherythroculter wangi**			+			
*36. A. kurematsui**			+			
*37. A. nigrocauda**			+			

Order **Family** *Species*	Threat status	River-head	Upper reach	Middle reach	Lower reach	Estuary
38. *Pseudoleubuca sinensis*			+	+	+	+
39. *P. engraulis*			+	+	+	+
40. *Toxabramis swinhonis*			+	+	+	+
41. *Hemiculter leucisculus*			+	+	+	+
42. *H. tchangi**			+			
43. *H. bleekeri*			+	+	+	+
44. *Hemiculterella sauvagei**			+	+		
45. *H. Wui*				+		
46. *Pseudohemiculter kweichowensis**			+			
47. *Cultrichthys erythropterus*			+	+	+	+
48. *Culter alburnus*			+	+	+	+
49. *C. mongolicus mongolicus*			+	+	+	+
50. *C. m. elongatus**			+			
51. *C. m. qionghaiensis**			+			
52. *C. Dabryi*			+	+	+	+
53. *C. Oxycephalus*			+	+	+	+
54. *C. oxycephaloides**			+	+	+	
55. *Parabramis pekinensis*			+	+	+	+
56. *Megalobrama skolkovii*			+	+	+	+
57. *M. pellegrini**			+	+		
58. *M. amblycephala**				+	+	+
59. *M. elongata**	CR		+			
60. *Xenocypris argentea*			+	+	+	+
61. *X. davidi*			+	+	+	+
62. *X. yunnanensis**	VU		+			
63. *X. fangi**	EN		+			
64. *X. microlepis*			+	+	+	
65. *Distoechodon tumirostris*			+	+	+	
66. *D. hupeinensis**	EN			+		
67. *Pseudobrama simony*			+	+	+	+
68. *Hypophthalmichthys molitrix*			+	+	+	+
69. *Aristichys nobilis*			+	+	+	+
70. *Hemibarbus labeo*			+	+	+	
71. *H. maculates*			+	+	+	+
72. *H. longirostris*				+	+	
73. *Paracanthobrama guichenoti**				+	+	+
74. *Belligobio nummifer*			+			
75. *B. pengxianensis**	EN		+			
76. *Pseudorasbora parva*			+	+	+	+
77. *P. elongata*	VU			+	+	
78. *Sarcocheilichthys sinensis*			+	+	+	+
79. *S. parvus*				+	+	
80. *S. kiangsiensis*				+	+	
81. *S. nigripinnis*			+	+	+	+
82. *S. davidi**			+			
83. *Gnathopogon herzensteini**			+	+		
84. *G. imberbis**			+	+		
85. *G. nicholsi**				+	+	

Order **Family** *Species*	Threat status	River-head	Upper reach	Middle reach	Lower reach	Estuary
86. G. taeniellus					+	
87. Squalidus argentatus			+	+	+	+
*88. S. nitens**			+	+	+	+
89. S. wolterstorffi			+	+	+	
90. Coreius heterodon			+	+	+	+
*91. C. guichenoti**			+	+		
92. Rhinogobio typus			+	+	+	
*93. R. hunanensis**				+		
*94. R. cylindricus**			+	+		
*95. R. ventralis**			+	+		
96. Platysmacheilus exiguus				+		
*97. P. longibarbatus**	VU			+		
*98. P. nudiventris**			+	+		
99. Abbottina rivularis			+	+	+	+
*100. A. obtusirostris**			+			
101. Microphysogobio fukiensis			+	+	+	
*102. M. microstomus**					+	
103. M. kiatingensis			+	+	+	
*104. M. tungtingensis**				+	+	
105. Huigobio chenhsinensis				+	+	
106. Pseudogobio vaillanti			+	+	+	
107. Saurogobio dumerili			+	+	+	
108. S. dabryi			+	+	+	+
*109. S. gracilicaudatus**				+		
110. S. xiangjiangensis				+		
*111. S. gymnocheilus**			+	+	+	+
*112. Gobiobotia (Progobiobotia) abbreviatea**			+			
113. G. (Gobiobotia) tungi				+	+	
114. G. (G.) meridionalis				+		
*115. G. (G.) filifer**			+	+		+
116. G. (G.) brevirostris				+		
*117. G. (G.) jiangxiensis**						
*118. Xenophysogobio boulengeri**			+	+		
*119. X. nudicorpa**			+			
120. Rhodeus ocellatus			+	+	+	+
121. R. lighti			+	+	+	+
122. R. fangi				+	+	+
*123. Acheilognathus gracilis**			+	+	+	
124. A. macropterus			+	+	+	+
*125. A. omeiensis**			+			
126. A. barbatus			+	+	+	+
127. A. barbatulus			+	+	+	+
*128. A. elongatus**	EN		+			
129. A. polylepis				+	+	
130. A. tonkinensis			+	+	+	+
131. A. tabira				+	+	
*132. A. hypselonotus**			+	+	+	
133. A. chankaensis			+	+	+	+

Order **Family** *Species*	Threat status	River-head	Upper reach	Middle reach	Lower reach	Estuary
134. *A. nanchongensis**						
135. *Paracheilognathus himantegus*				+	+	+
136. *P. imberbis*			+	+	+	+
137. *Spinibarbus hollandi*			+	+	+	
138. *S. sinensis**			+	+		
139. *Barbodes laticeps*			+			
140. *B. polylepis**			+			
141. *Percocyprinus pingi pingi**	VU		+			
142. *Sinocyclocheilus multipunctatus*			+			
143. *S. grahami grahami**	EN		+			
144. *Acrossocheilus labiatus*				+		
145. *A. fascitus*				+	+	
146. *A. jishouensis**				+		
147. *A. monticola**			+	+		
148. *A. elongatus*				+		
149. *A. yunnanensis*			+	+		
150. *Onychostoma macrolepis*			+	+		
151. *O. sima*			+	+		
152. *O. barbata*			+	+		
153. *O. lini*	VU			+		
154. *O. rara**	EN		+	+		
155. *O. angustistomata**			+			
156. *O. barbata*			+	+		
157. *O. daduensis**			+			
158. *O. brevis**			+			
159. *Tor brevifilis*			+	+		
160. *Sinilabeo rendahli**			+	+		
161. *S. tungting**				+	+	
162. *S. longibarbatus**			+			
163. *Rectoris luxiensis**			+	+	+	
164. *R. mutabilis**			+			
165. *Semilabeo notabilis*	VU		+			
166. *Pseudogyrincheilus procheilus**			+	+		
167. *Garra pingi pingi*			+	+		
168. *Sinocrossocheilus guizhouensis**	VU		+			
169. *Discogobio yunnanensis*			+	+		
170. *D. brachyphysallidos*			+			
171. *Schizothorax (Schizothorax) chongi**			+			
172. *S. (S.) wangchiachii**		+	+			
173. *S. (S.) dolichonema**	EN	+	+			
174. *S. (S.) sinensis**			+	+		
175. *S. (S.) grahami**	VU		+			
176. *S. (S.) cryptolepis**			+			
177. *S. (S.) heterochilus**			+			
178. *S. (S.) prenanti**			+	+		
179. *Schizothorax (Racoma) kozlovi**			+			
180. *S. (R.) yunnanensis weiningensis**			+			
181. *S. (R.) griseus*	EN		+			

Order **Family** *Species*	Threat status	River-head	Upper reach	Middle reach	Lower reach	Estuary
182. *S. (R.) parvus**	EW		+			
183. *S. (R.) ninglangensis**	EN		+			
184. *S. (R.) microstomus**	EN		+			
185. *S. (R.) labrosa**	EN		+			
186. *S. (R.) longibarbus**	CR		+			
187. *S. (R.) davidi**			+	+		
188. *Ptychobarbus kaznakovi*	VU	+	+			
189. *P.chungtienensis gezaensis**			+			
190. *P. c. chungtienensis**	EN		+			
191. *Gymnodiptychus pachycheilus*	EN		+			
192. *G. potanini potanini**			+			
193. *G. p. firmispinatus**			+			
194. *Schizopygopsis malacanthus malacanthus**		+	+			
195. *S. m. chengi**			+			
196. *S. kialingensis**			+			
197. *S. malacanthus baooxingensis**			+			
198. *Herzensteinia microcephalus**		+	+			
199. *Procypris rabaudi**	VU		+	+		
200. *Cyprinus (Mesocyprinus) micristius**	EN		+			
201. *C. (Cyprinus) chilia*			+			
202. *C. (C.) qionghaiensis**	EN		+			
203. *C. (C.) carpio*			+	+	+	+
204. *Carassius auratus auratus*			+	+	+	+
Homalopteridae						
205. *Vanmanenia stenosoma*				+	+	
206. *V. pinchowensis*				+		
207. *V. tetraloba*			+			
208. *Pseudogastromyzon fangi*				+		
209. *Paraprotomyzon multifasciatus*			+			
210. *P. niulanjiangensis**						
211. *P. lungkowensis**			+	+		
212. *Beaufortia szechuanensis**			+			
213. *B. niulanensis**						
214. *B. liui**			+			
215. *Lepturichthys fimbriata**			+	+		
216. *Hemimyzon yaotanensis**	VU		+			
217. *Jinshaia abbreviata**			+			
218. *J. sinensis**			+			
219. *Sinogastromyzon hsiashiensis**				+		
220. *S. sichuangensis**			+	+		
221. *S. szechuanensis**			+	+		
222. *Metahomaloptera omeiensis omeiensis**			+	+		
223. *M. o. hangshuiensis**				+		
Cobitidae						
224. *Triplophysa markehenensis**		+	+			
225. *T. stewarti*		+				
226. *T. orientalis*		+				
227. *T. bleekeri**			+	+		

Order **Family** *Species*	Threat status	River-head	Upper reach	Middle reach	Lower reach	Estuary
228. *T. pseudoscleroptera*			+			
229. *T. robusta*			+			
230. *T. microps*		+				
231. *T. stoliczkae*			+			
232. *T. rotundiventris*		+				
233. *T. ninglangensis**			+			
234. *T. stenura*		+	+			
235. *T. polyfasciata**			+			
236. *T. angeli**			+			
237. *T. venusta**			+			
238. *T. grahami**			+			
239. *T. xichangensis**			+			
240. *T. anterodorsalis**			+			
241. *T. brevicauda*			+			
242. *T. obscura*			+			
243. *T. leptosoma*			+			
244. *T. tanggulaensis**		+				
245. *T. xiangxiensis**				+		
246. *T. xiqiensis**			+			
247. *T. daqiaoensis**			+			
248. *T. brevibarba**			+			
249. *T. crassilabris**			+			
250. *T. yaopeizhii*			+			
251. *Nemacheilus obscurus*			+			
252. *Paracobitis variegates*			+	+		
253. *P. potanini**			+	+		
254. *P. wujiangensis**			+			
255. *Oreias dabryi*			+	+		
256. *Sphaerophysa dianchiensis**			+			
257. *Yunnanilus nigromaculatus**			+			
258. *Y. pleurotaenia**			+			
259. *Y. caohaiensis**			+			
260. *Y. sichuanensis**			+			
261. *Y. longibulla**			+			
262. *Schistura dabryi*			+			
263. *S. niulanjiangensis**						
264. *S. fasciolata*			+			
265. *S. incerta*				+		
266. *S. xiangxiensis**				+		
267. *Botia superciliaris*			+	+		
268. *B. reevesae**			+			
269. *Parabotia fasciata*			+	+	+	
270. *P. bimaculata**			+	+	+	+
271. *P. maculosa*			+	+	+	
272. *P. kiangsiensis**				+		
273. *P. banarescui**				+		
274. *Leptobotia elongata**	VU		+	+	+	
275. *L. taeniops**			+	+		

Order Family *Species*	Threat status	River-head	Upper reach	Middle reach	Lower reach	Estuary
276. *L. pellegrini*				+		
277. *L. guilinensis*				+		
278. *L. microphthalma**			+			
279. *L. tchangi**				+		
280. *L. tientaiensis hansuiensis**				+		
281. *L. rubrilabris**				+	+	
282. *L. orientalis*				+		
283. *Cobitis sinensis*			+	+	+	+
284. *C. taenia*				+	+	
285. *C. rarus*				+		
286. *C. macrostigma**				+	+	
287. *Misgurnus anguillicaudatus*			+	+	+	+
288. *Paramisgurnus dabrynus*			+	+	+	+
V. Siluriformes						
Bagridae						
289. *Pelteobagrus fulvidraco*			+	+	+	+
290. *P. eupogon**	VU		+	+	+	+
291. *P. vachelli*			+	+	+	+
292. *P. nitidus*			+	+	+	
293. *Leiocassis longirostris*			+	+	+	+
294. *L. crassilabris*			+	+	+	
295. *L. longibarbus**	CR		+			
296. *L. tenuifurcatus*			+	+		
297. *Pseudobagrus ussuriensis*			+	+	+	+
298. *P. pratti**			+	+	+	
299. *P. truncatus*			+	+	+	+
300. *P. ondon*				+	+	
301. *P. taeniatus*				+	+	+
302. *P. analis**				+		
303. *P. medianalis**	EN		+			
304. *P. tenuis*				+	+	+
305. *P. emarginatus*			+	+		
306. *P. brevicaudatus**			+	+		
307. *P. omeihensis**			+			
308. *P. adiposalis*				+	+	
309. *Mystus macropterus*			+	+	+	
Siluridae						
310. *Silurus asotus*			+	+	+	+
311. *S. meridionalis*			+	+	+	
312. *S. mento**	EN		+			
Amblycipitidae						
313. *Liobagrus marginatus**			+	+		
314. *L. anguillicauda*				+		
315. *L. kingi**	EN		+			
316. *L. nigricauda**				+	+	
317. *L. styani**				+	+	
318. *L. marginatoides**			+	+		
Sisoridae						

Order **Family** *Species*	Threat status	River-head	Upper reach	Middle reach	Lower reach	Estuary
319. *Glyptothorax fukiensis fukiensis*			+	+	+	
320. *G. sinense sinense**			+	+		
321. *Euchiloglanis kishinouyei**	CR	+	+			
322. *E. davidi**	EN		+			
323. *Pareuchiloglanis sinensis**	EN	+	+			
324. *P. feae*			+			
325. *P. sichuanensis**			+			
326. *P. robusta**			+			
327. *P. anteanalis**			+			
Clariidae						
328. *Clarias fuscus*			+	+	+	
VI. Osmeriformes						
Salangidae						
329. *Protosalanx chinensis*				+	+	+
330. *Hemisalanx brachyrostralis**				+	+	+
331. *H. prognathus*					+	+
332. *Salanx ariakensis*						+
333. *Neosalanx oligodontis*				+	+	+
334. *N. taihuensis*			+	+	+	+
335. *N. andersoni*						+
336. *Plecoglossus altivelis*	VU					
VII. Salmoniformes						
Salmonidae						
337. *Hucho bleekeri**	EN		+	+		
338. *Brachymystax lenok*	VU			+	+	+
VIII. Mugiliformes						
Mugilidae						
339. *Liza carinatus*						+
340. *L. haematocheila*						+
341. *Mugil cephalus*					+	+
342. *Osteomugil ophuyseni*						+
IX. Atheriniformes						
Atherinidae						
343. *Allanetta bleekeri*				+		+
X. Beloniformes						
Hemiramphidae			+	+		
345. *Hyporhamphus intermedius*			+	+	+	+
345. *H. sajori*						+
XI. Cypriodontiformes						
Oryziatidae						
346. *Oryzias latipes*			+	+	+	+
347. *O. latipes sinensis*	VU		+			
Poeciliidae						
348. *Gambusia affinis*			+	+		
XII. Synbranchiformes						
Synbranchidae						
349. *Monopterus albus*			+	+	+	+
XIII. Scorpaeniformes						

Order Family *Species*	Threat status	River-head	Upper reach	Middle reach	Lower reach	Estuary
Cottidae						
350. Trachidermus fasciatus	EN				+	+
XIV. Perciformes						
Serranidae						
351. Lateolabrax maculatus						+
352. Siniperca whiteheadi				+		
353. S. chuatsi			+	+	+	+
354. S. knerii			+	+	+	
355. S. scherzeri			+	+	+	+
356. S. undulats	VU			+	+	
357. Coreosiniperca roulei	VU			+	+	
358. Coreoperca obscura				+		
Belontiidae						
359. Macropodus chinensis			+	+	+	+
360. M. opercularis			+	+	+	
Channidae						
361. Channa argus			+	+	+	+
362. C. asiatica			+	+	+	
363. C. maculata				+		
Mastacembelidae						
364. Sinobdella sinensis				+	+	
365. Mastacembelus aculeatus				+	+	+
366. M. armatus				+	+	+
Polynemidae						
367. Eleutheronema tetradactylum						+
Terapontidae						
368. Terapon jarbua						+
Callionymidae						
369. Repomucenus olidus						+
Eleotridae						
370. Hypseleotris swinhonis			+	+	+	+
371. Odontobutis obscura				+	+	+
372. Bostrichthys sinensis						+
373. Eleotris oxycephala						+
374. Prionobutis koilomatodon						+
Gobiidae						
375. Mugilogobius myxodermus			+	+	+	+
*376. M. polylepis**	EN				+	+
377. Ctenogobius giurinus			+	+	+	+
378. C. multimaculatus					+	
*379. Ctenogobius shennongensis**	EN			+		
380. Acanthogobius elongata						+
381. Chaeturichthys stigmatias						+
382. Ctenogobius brunneus			+	+	+	
*383. C. szechuanensis**	EN		+			
384. C. cliffordpopei			+	+	+	+
*385. C. chengtuensis**			+			
386. Glossogobius giuris						+

Order Family *Species*	Threat status	River-head	Upper reach	Middle reach	Lower reach	Estuary
387. Lophiogobius ocellicauda						+
388. Odontamblyopus rubicundus					+	+
389. Synechogobius ommaturus						+
390. Tridentiger trigonocephalus					+	+
391. T. barbatus						+
392. Cryptocentrus filifer						+
393. Luciogobius guttatus						+
394. Amblychaeturichthys hexanema						+
395. Taeniodides cirratus						+
396. Oxuderces dentatus						+
397. Trypauchen vagina						+
398. Periophthalmus cantonensis						+
399. Boleophthalmus pectinirostris						+
400. Scartelaos histiophorus						+
XV. Pleuronectiformes						
Cynoglossidae						
401. Cynoglossus (Areliscus) semilaevis						+
402. C. (A.) abbreviatus						+
403. C. (A.) gracilis					+	+
404. C. (Cynoglossoides) robustus						+
XVI. Tetrodontiformes						
Tetrodontidae						
405. Takifugu obscurus				+	+	+
*406. T. coronoidus**	EN				+	+
407. T. flavidus	EN					+
408. T. ocellatus						+
409. T. vermicularis						+
410. T. xanthopterus						+
411. T. pseudommus						+
412. T. alboplumbeus						+
413. T. bimaculatus						+
414. T. porphyreus						+
415. T. reticularis						+
416. T. niphobles						+

*: Endemic fish species; EX: Extinct; EW: Extinct in the wild; CR: Critically endangered; EN: Endangered; VU: Vulnerable; +: Present

Appendix 1. List and distribution of fishes in the Yangtze River basin

7. References

Anonymous, Department of Ichthyology, Institute of Hydrobiology, CAS (1976). Fishes of the Yangtze River. Science Press, Beijing, China.

Anonymous (1980). The fishes of Hunan Province. Hunan Science & Technology Press, Wuhan, China.

Cao, W. (1992). Freshwater fish resources of China. In: Cultivation of the Chinese freshwater fishes. Liu, J. & He, B. (Eds). 30-64, Science Press, Beijing, China.

Cao, W.; Chen, Y.; Wu, Y. & Zhu, S. (1981). Origin and evolution of the Schizothoracinae and its relationship with uplift of the Tibetan Plateau. In: The era, range and shape of the Tibetan Plateau's uplift. Anonymous (Ed). 118-130, Science Press, Beijing, China.

Chen, Y. (1998). Fauna Sinica: Osteichthyes Cypriniformes II. Science Press, Beijing, China.

Chen Y.; Yang, J. & Li, Z. (1998). The diversity and present status of fishes in Yunnan Province. Chinese Biodiversity, 6: 272-277.

Chen, Z.; Li, J.; Shen, H. & Wang, Z. (2000). Yangtze River of China: historical analysis of discharge variability and sediment flux. Geomorphology 41: 77–91.

Chu, X. (1955). On fishes of Ichang, with notes on their distribution in the Yangtze River. Acta Hydrobiologica Sinica 2: 81-96.

Chu, X.; Zheng, B. & Dai, D. (1999). Fauna Sinica: Osteichthyes. Siluriformes. Science Press, Beijing, China.

Ding, R. (1994). The fishes of Sichuan Province. Sichuan Science & Technology Press, Chengdu, China.

Du, K. (1962). On the biology of the Chinese sneakhead, *Ophiocephalus argus*, in Liang-tse Lake. Acta Hydrobiologica Sinica, 2: 54-65.

Dudgeon, D. (2000). The ecology of tropical Asian rivers and streams in relation to bodiversity conservation. Annual Review of Ecology and Systematics, 31: 239-263.

Fu, C.; Wu, J.; Chen, J.; Wu, Q. & Lei, G. (2003). Freshwater fish biodiversity in the Yangtze River basin of China: patterns, threats, and conservation. Biodiversity and Conservation, 12: 1649-1685.

Groombridge, B. (1992). Global biodiversity: status of the earth's living resources. Chapman and Hall, London.

He, Y.; Wang, J.; Lek, S.; Cao, W. & Lek-Ang, S. (2011). Structure of endemic fish assemblages in the upper Yangtze River Basin. River Research and Applications, 27: 59-75.

Huang, G. & Xie, P. (1996). Changes in the structure of fish community with the analysis of the possible reasons in Lake Donghu, Wuhan. Acta Hydrobiologica Sinica, 20: 38-46.

Humphrey, C.; Faith, D. & Dostine, P. (1995). Baseline requirements for assessment of mining impact using biological monitoring. Australian Journal of Ecology, 20: 150-166.

Li, D.; Yang, C.; Xu, G. & Zhang, L. (1998). Biology of mandarin fish (*Siniperca chuatsi* Basilesky) in Poyang Lake. Acta Agriculturae Jiangxi, 10: 14-22.

Liang, Z.; Zhou, C. & Huang, H. (1981). Composition and seasonal changes of fishes in Lake Wuhu connected with Changjiang River. Oceanologia et Limnologia Sinica, 12: 468-478.

Liu, J. & Cao, W. (1992). Fish resources in the Yangtze basin and the strategy for their conservation. Resources and environment in the Yangtze Valley, 1: 17-23.

Liu, X. & Wang, H. (2010). Estimation of minimum area requirement of river connected lakes for fish diversity conservation in the Yangtze River floodplain. Diversity and Distributions, 16: 932–940.

Martens, K. (1997). Speciation in ancient lakes. Trends in Ecology and Evolution, 12: 177-182.

Minckley, W.; Hendrickson, D. & Bond, C. (1986). Geography of western North American freshwater fishes: description and relationship to intracontinental tectonism. In:

The zoogeography of North American freshwater fishes. Hocutt, C. & Wiley, E. (Eds). 519-614, John Wiley & Sons, New York.

Nelson, J.S., 2006. Fishes of the world (4th edition). Wiley, New Jersey.

Ni, Y. & Zhu, C. (2005). Fishes of the Taihu Lake. Shanghai Science & Technology Press, Shanghai, China.

Park, Y.; Chang, J.; Lek, S.; Cao, W. & Brosse, S. (2003). Conservation strategies for endemic fish species threatened by the Three Gorges Dam. Conservation Biology, 17: 1748-1758.

Song, T.; Zhang, G.; Chang, J.; Miao, Z. & Deng, Z. (1999). Fish diversity in Honghu Lake. Chinese Journal of Applied Ecology,10: 86-90.

Wang, S. & Xie, Y. (2009). China species red list. Vol. II Vertebrates - Part 1. High Education Press, Beijing, China.

Wei, Q.; Ke, F.; Zhang, J.; Zhuang, P.; Luo, J.; Zhou, R. & Yang, W. (1997). Biology, fisheries and conservation of sturgeons and paddlefish in China. Environmental Biology of Fishes, 48: 241-255.

Wu, H. & Zhong, J. (2008). Fauna Sinica: Osteichthyes. Perciformes V. Gobioidei. Science Press, Beijing, China.

Wu, Y. & Wu, C. (1992). The fishes of the Qinghai-Xizang Plateau. Sichuan Publishing Science & Technology Press, Chengdu, China.

Xie, C. & Yang, R. (2005). Changes of major fish population structure and environment in the Liangzi Lake. In: Fisheries resources and conservation of environment in lakes of the Changjiang River basin. Cui, Y. & Li, Z. (Eds). 193-229, Science Press, Beijing, China.

Xie, P. & Chen, Y. (1999). Threats to biodiversity in Chinese inland waters. Ambio, 28: 674-681.

Xie, Y., Li, Z., Gregg, W. & Li, D. (2001). Invasive species in China - an overview. Biodiversity and Conversation 10: 1317-1341.

Xu, H.; Wang, S. & Xue, D (1999). Biodiversity conservation in China: legislation, plans, and measures. Biodiversity and Conservation, 8: 819-837.

Ye, S. (2007). Studies on fish communities and trophic network model of shallow lakes along the middle reach of the Yangtze River. PhD Dissertation. Institute of Hydrobiology, Chinese Academy of Sciences, Wuhan, China.

Ye, S.; Li, Z. & Cao, W. (2007). Species composition, diversity and density of small fishes in two different habitats in Niushan Lake. Chinese Journal of Applied Ecology, 18: 1589-1595.

Ye, S.; Li, Z.; Lek-Ang, S.; Feng, G.; Lek, S. & Cao, W. (2006). Community structure of small fishes in a shallow macrophytic lake (Niushan Lake) along the middle reach of the Yangtze River, China. Aquatic Living Resources, 19: 349-359.

Yang, G. (1987). The fishes of Hubei Province. Hubei Science & Technology Press, Wuhan, China.

Yue, P. (2000). Fauna Sinica: Osteichthyes. Cypriniformes III. Science Press, Beijing, China.

Yue, P. & Chen, Y. (1998). China red data book of endangered animals: Pisces. Science Press, Beijing, China.

Zeng, X. (1990). Fishery Resources of the Yangtze River Basin. Marine Press, Beijing, China.

Zhang, G. (1991). Study on the population of the crucian carp (*Carassius auratus auratus*) in Lake Honghu. In: Studies on comprehensive exploitation of aquatic biological productivity and improvement of ecological environment in Lake Honghu. Honghu Research Group, Institute of Hydrobiology, CAS (Ed). 162-171, China Ocean Press, Beijing, China.

Zhang, T. & Li, Z. (2007). Fish resources and fishery utilization of Lake Poyang. Journal of Lake Sciences., 19: 434-444.

Zhang, X.; Cao, L.; Liu, J.; Bai, L. & Yu, C. (2008). Growth pattern of redfin culter, *Cultrichthys erythropterus* Basilewsky (1855) in Poyang Lake. Journal of Nanchang University (Natural Science), 32: 70-74.

Zhu, S. (1995). Synopsis of freshwater fishes of China. Jiangsu Science and Technology Press, Nanjing. 549 pp.

Zhuang, P.; Wang, Y.; Li, S.; Deng, S.; Li, C. & Ni, Y. (2006). Fishes of the Yangtze estuary. Shanghai Science & Technology Press, Shanghai, China.

4

Biodiversity and Conservation of the Estuarine and Marine Ecosystems of the Venezuelan Orinoco Delta

Patricia Miloslavich[1], Alberto Martín[1], Eduardo Klein[1], Yusbelly Díaz[1], Carlos A. Lasso[2], Juan José Cárdenas[3] and Oscar M. Lasso-Alcalá[4]
[1]*Universidad Simón Bolívar, Departamento de Estudios Ambientales, Caracas*
[2]*Instituto Alexander von Humboldt, Bogotá*
[3]*Fundatún, Caracas*
[4]*Fundación La Salle de Ciencias Naturales, Museo de Historia Natural La Salle, Caracas*
[1,3,4]*Venezuela*
[2]*Colombia*

1. Introduction

1.1 Oceanographic features and eco-regions of the Orinoco Delta

The Orinoco is one of the largest rivers in South America (2,150 Km). Globally, it is the third in water flow (yearly average of 39,000m^3/s), and the fifth in sediment transportation (150 million tons/year). The Orinoco basin extends for nearly a million square kilometers within Colombian and Venezuelan territories, and is characterized by at least ten large terrestrial and aquatic eco-regions including its opening into the Atlantic through the vast Orinoco Delta (located between 07° 46´ to 10° 00´ N, and 59° 50´ to 62° 35´ W) (Michelangeli, 2000; Bone et al., 2004; Lasso et al., 2010). These deltaic plains were formed by the combined action of sediment and freshwater discharges from the Orinoco, with the tides on a flat alluvial plain (Miloslavich et al., 2011) creating a dynamic ecosystem of fluvial and estuarine habitats. The boundaries between these two habitats are defined according to their altitude above sea level and tidal influence (Lasso & Sánchez-Duarte, 2011), with the fluvial delta corresponding to the mid and high delta according to Cervigón (1985). The lower limit of the fluvial delta varies due to seasonal annual fluctuations in freshwater discharge from the Orinoco, related to regional and, even local, precipitation regimes, and a salt wedge from the sea that moves inward with the incoming tide. In the estuarine delta, the ecological boundary of the fluvial delta is the limit at which the sea water influences the main stream and its channels, while the ecological boundary with the sea is a fringe, which is variable in size and forms after sand or mud bars at the opening of the seaward channels. Along this longitudinal zone from the upper to the lower delta, a geomorphological, physico-chemical, and therefore, biological gradient can be observed. In addition to this longitudinal gradient, a very important transversal gradient also exists, that creates particular lentic ecosystems with specialized biota. Within the Atlantic, the major areas of influence of the Orinoco are (1) the Gulf of Paria, a semi-enclosed area between the Araya Peninsula (Venezuela) and the west coast of Trinidad, connected to the Caribbean by Boca del Dragón, a ~8km strait, (2) the Orinoco Deltana Platform, the area within the Atlantic Ocean directly adjacent to the Orinoco deltaic plain, and (3) Boca de Serpiente, which connects the Deltana Platform with the Gulf of Paria. These combined areas cover 27,630 km^2 and shelter one of the major and best-preserved wetlands in South America (Figure 1).

In general, the Orinoco Delta and the marine area under its influence has a mean annual temperature between 25 and 28 °C (range: 15.7 to 37.2 °C), with two warm peaks, one occurring in May and the other in September. There are two seasons, the dry season occurs between January and April, and the rainy season between June and November. May and December are transitions months between the seasons. The rainy season occurs when the Intertropical Convergence Zone (ITCZ) moves from the Equator towards the Caribbean producing a mean of 1,500 to 2,000 mm of rainfall per year. At this time, winds usually come from the east at an average speed of 3.2 m/s. When the ITCZ moves back to the Equator, the area is influenced by the trade winds that blow from the north-northeast at an average speed of 6.2 m/s (Bone et al., 2004; Martín & Bone, 2007). The area is also affected by the discharge of the Orinoco, which introduces significant changes in salinity, temperature, nutrients, and water quality at both spatial and temporal levels. The productivity of this area is significant and one of the highest in comparison to the eastern Caribbean region. Such high productivity is due to the seasonal upwelling off the northeast coast of Venezuela, and the seasonal dispersal of the Orinoco plume, which covers an area of more than 300,000 km^2 in the adjacent Caribbean and provides for a significant increase in concentration of nutrients and photosynthetic pigments (Martín & Bone, 2007; Muller-Karger et al., 1989). The region is also influenced by the Guiana current, which flows parallel to the coastline towards the northwest (~106m^3/s). The Guiana current originates from the Brazilian current and forms high-energy whirlpool structures that may surpass in intensity the normal circulation and consequently cause transitory episodes of current reversal. The Amazon river also affects this area during the first six months of the year, when its waters extend to the northwest, increasing turbidity in coastal waters, and creating a large plume and sections of brackish water offshore. Tides are very important at the local level particularly because of associated currents. Tides in this region are mostly semidiurnal, but in the Gulf of Paria, tides are mixed. Within the Gulf, the height of the tide is magnified within an increasing gradient from Boca del Dragón to the south, reaching up to 2 m (Martín & Bone, 2007).

Based on ecological and biological features, and aiming toward a better conservation strategy, the area was classified into seven subregions (Figure 1), defined according to the internal variability of their ecosystems, communities, and populations, including those of the terrestrial areas when these are included in the landscape (Klein & Cárdenas, 2009). The subregions, along with the main marine or estuarine features that characterize them, are:

Subregion 1: Paria Peninsula. This zone is characterized by a high landscape heterogeneity, with beaches, coves, a few rocky shores, and wetlands (Mata Redonda and Bajo Alcatraz).

Subregion 2: Sac of the Gulf of Paria. This is a shallow water zone with a depth of 0-15 m characterized by extensive and tall mangrove formations that comprise a heterogeneous wetland with interconnected small channels into which the salt water penetrates. In this region, two estuarine systems (Caño Ajíes and Caño San Juan) are important fish reservoirs, with high biomass of fish larvae of the families Sciaenidae and Lutjanidae.

Subregion 3: Pedernales. This is a shallow water zone with a depth of 0-15 m with important extensions covered by mangroves, and some beaches characterized by boulders. The region has a high abundance of commercial species of fish and crustaceans and it is an important fish reservoir of juvenile stages.

Subregion 4: Gulf of Paria. This zone between 15 m of depth and the limit of international waters serves as an important migratory route for sea mammals, turtles and birds, and as a feeding area for turtles. The area was under intense pressure from trawling fisheries for many years, which were stopped in 2009 with the passage of a new environmental law in Venezuela that mandated that intensive trawling activities cease. While trawling fisheries still take place, they are conducted at the same intensity as artisanal fisheries.

Subregion 5: Boca de Serpiente. This zone with a depth between 0 and 60 m has a complex hydrodynamic system due to the acceleration of marine currents. It is an important migratory route for sea mammals, turtles and birds. Its marine and estuarine biodiversity is relatively high, with high abundance of phytoplankton and zooplankton, including fish larvae. The area contains wetlands with a high biodiversity of birds and a continuous, extensive, tall mangrove formation. Its beaches are formed by muddy sediments and its benthic community shows marked seasonal variations.

Subregion 6: Boca Grande. At a depth between 0 and 60 m, this zone is under the direct influence of the discharge of many of the delta channels, including the largest or Rio Grande. The zone contains numerous islands of different sizes with fragmented mangrove formations. Erosion and sedimentation processes are very active, and the area is significantly influenced by freshwater input. Biodiversity associated with this area is mainly euryhaline.

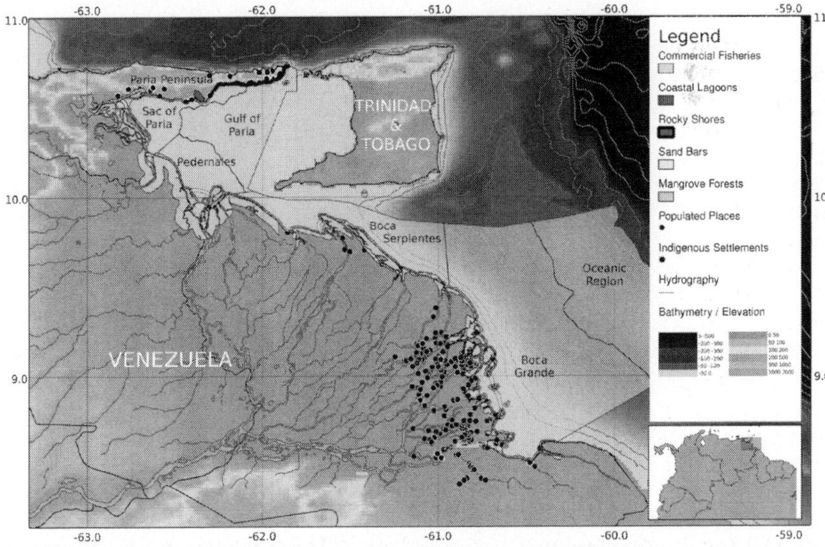

Fig. 1. Map of the Orinoco Delta and Gulf of Paria region in the Venezuelan Atlantic front. Map shows the main ecosystems and human settlements. Bathymetry and elevation scales in meters.

Subregion 7: Oceanic region. Extending from the 60 m isobar to the limit of the Venezuelan border with international waters, this zone is characterized by mostly oceanic species, with low benthic abundance. The area is within important migratory routes for marine mammals, turtles and birds, and is considered an important nursery ground for fishes. In contrast to the other regions, the waters here are oligotrophic, with low primary productivity and low content of particulate organic matter.

1.2 Local indigenous populations

Within the *Orinoquia* (area related to the Orinoco river covering Venezuelan and Colombian territories), a total of 17 ethnic groups and cultures have been reported, but within the Orinoco Delta, the dominant group are the *Warao* (the "canoe people", or "low-land people"). For the *Warao*, the exploitation of natural resources is a delicate balance between man, nature, and supernatural beings. Any intervention by man must be compensated so the harmony between these three dimensions is maintained. In this way, they alternate the use of natural resources through permanent and through temporal settlements, their habitat, or entourage being the most important spatial concept. The *Waraos* are natural sailors with their *curiaras* (long canoes made of wood). Their main source of protein comes from fishing, but they also grow wild plants (*moriche* palm) and other cultivated plants such as banana, sugar cane and yucca. Besides obtaining starch from the *moriche* palm, they work to obtain the fibers with which they build baskets, roofs, *chinchorros* (hammocks), and fishing gear. Their hunting of terrestrial animals is limited to rodents and birds. The *Warao* population in the Orinoco Delta was nearly 24,000 individuals in the 1990s (Michelangeli, 2000) and grew to 36,027, according to the Venezuelan National Census in 2001, of which more than 75% still speak the *Warao* language. This is the second largest ethnic group in Venezuela, and constitutes 25% of the population of the Delta Amacuro state. The *Warao* villages are built along the edges of the many delta channels, their wooden houses (or *palafitos*) elevated by wood pillars to avoid flooding. Each village is composed of an "extended" family comprising four to five generations (~120 people). These small communities driven by marriage and movement are considered sub-tribes (Michelangeli, 2000). About 202 *Warao* villages have been reported in four of the seven subregions mentioned above: Sac of the Gulf of Paria, Pedernales, Boca de Serpiente, and Boca Grande (Klein & Cárdenas, 2009). Throughout the 20th century and still at present, the Warao have been subject to several socio-cultural, religious, and environmental impacts that have brought important changes to their lifestyle, mostly with negative consequences (García Castro & Heinen, 1999; García Castro, 2000). In 2009, an area of 12 km^2 around each of the 202 villages was suggested as a conservation area to protect the *Warao* and their settlements, as well as the area under their direct use and influence, aiming to preserve their environment and the use they give to biodiversity, either for subsistence or for cultural reasons (Klein & Cárdenas, 2009).

2. Brief history of research and species discovery in the region

A bibliographic database containing 2,430 references within the time period of 1838 to 2008 was built and analyzed. The first recorded study was by Alexander von Humboldt (Humboldt, 1838) on the morphology, anatomy, distribution, and use of Orinoco manatees. During the late 1800s, a few works described the physical conditions of the Gulf of Paria (e.g. Guppy, 1895a) and provided records on mollusks (Guppy, 1895b, 1895c). Between 1900 and 1950, scientific literature related to the biodiversity of the area is very scarce with occasional reports mainly on fishes (Fowler, 1911, 1912, 1931), mollusks (Baker, 1923, 1925, 1927), and birds (Phelps & Phelps, 1950a, 1950b; Zimmer & Phelps, 1950). Between 1950 and 2000, a significant increase of studies in the region occurred with a commensurate amount of scientific literature was produced (Figure 2). The themes of these publications are variable

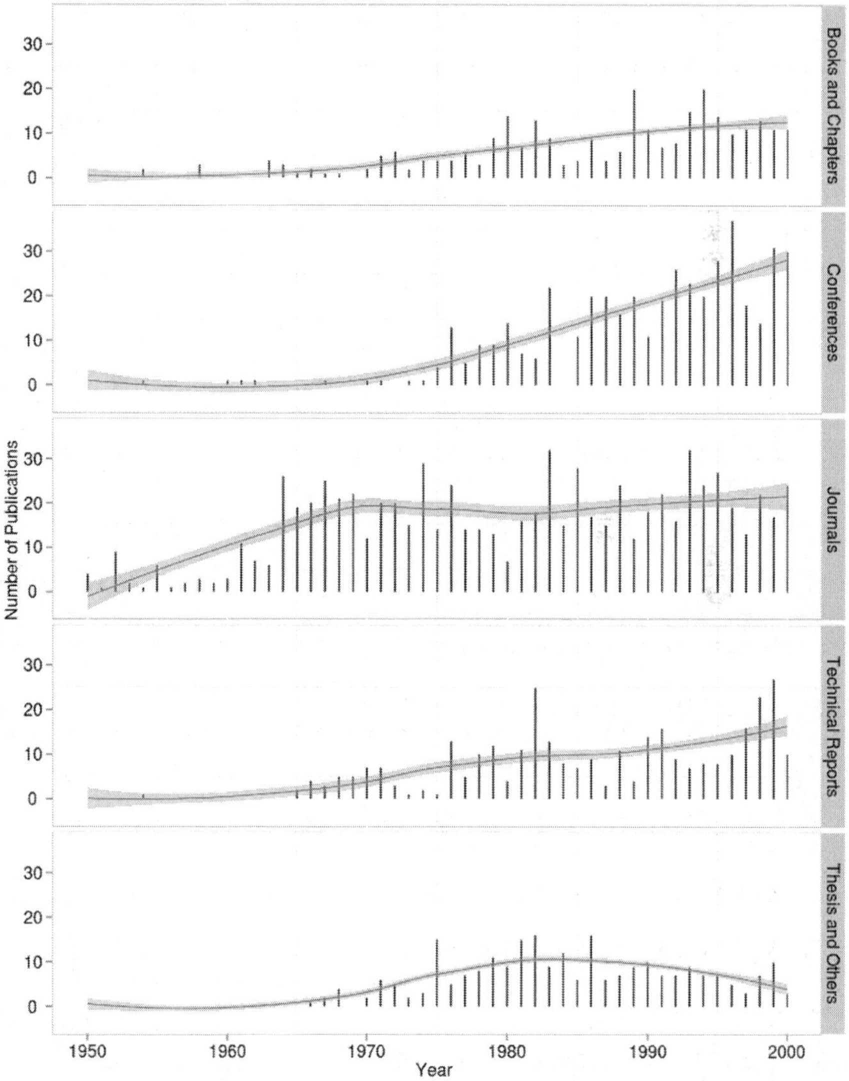

Fig. 2. Number of publications in books, conferences, journals technical reports, thesis and others between 1950 and 2008 related to the Orinoco Delta and the Gulf of Paria.

and range from the abiotic to the biotic. In general, the number of publications has increased for most of the themes, with studies related to fish and fisheries the most prolific (Figure 3).

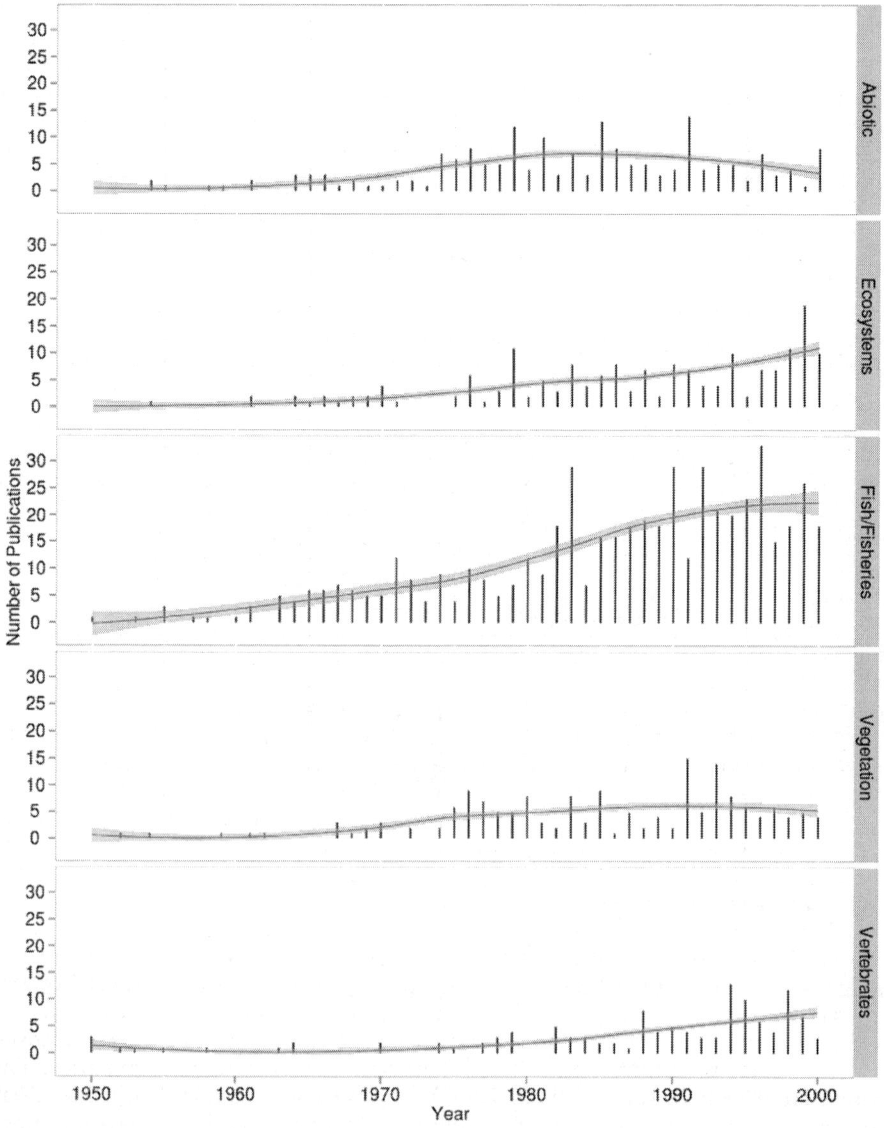

Fig. 3. Number of publications by themes between 1950 and 2008 related to the Orinoco Delta and the Gulf of Paria.

La Salle Foundation of Natural Sciences (FLSCN) has the longest tradition of research in this area (Cervigón, 1965; Ginés & Cervigón, 1968; Ginés et al., 1971, 1972; Lasso, 1993; Monente & Colonnello, 1997, 2004; Ponte & Mochcco, 1997; Jorgensen et al., 2000; Lasso et al., 2002, 2004a, 2004b, 2008, 2009; Lasso & Meri, 2003; Martínez-Escarbassiere et al., 2003; Colonnello, 2004; Lasso-Alcalá et al., 2005a, 2005b, 2008, 2009; Achury et al., 2006; Lasso-Alcalá & Lasso, 2007; Schneider et al., 2007; Lasso & Sánchez-Duarte, 2011),, and in more recent years, due to the pressure from gas and oil companies to carry out offshore exploitation in the area, more institutions have become involved. Despite the great research effort completed, projects carried out in this region have been very focused, geographically localized and punctual in time (no time series data) as well as very isolated. The area is very vast, so most of it remains largely unexplored. As for biodiversity, the best known groups are the fish, mollusks, and crustaceans, groups that are of economic importance, and other vertebrates such as birds and mammals, as well as plants (mangroves).

3. Marine and estuarine biodiversity of the region

The total number of species reported in the region is 2,118, of which 27% are planktonic species (not represented in the benthic community), 38% benthic, 21% fish, and nearly 15% other vertebrates. As in other 26 coastal regions of the world (Costello et al., 2010), half of all species richness is contributed by three groups: fish, crustaceans, and mollusks. Unlike these other 25 regions across the globe, where the contribution of other vertebrates, mainly mammals, reptiles and birds averaged only 2% and at most 8%, in the Orinoco Delta, the contribution of these groups is much higher, due to the extremely high bird species richness in this region (Martín & Bone, 2007). The following sections develop further the present knowledge of biodiversity of the different communities of the Orinoco Delta.

3.1 Invertebrates

3.1.1 Benthos

A total of at least 809 species-level taxa have been reported on benthic communities, from 5 animal phyla, 2 plant phyla (algae and angiospermae: mangroves and seagrasses) and only one group of protists (Foraminifera). Within these benthic communities, the group with the highest diversity is the crustaceans, with 310 species and 91 families reported, with amphipods and decapods comprising the greatest number (or highest percentage). The amphipod species are mainly marine, while decapods are represented by both marine and estuarine species. Mollusks represent the second most diverse group, with 189 species and 83 families, comprised of gastropods and bivalves (Table 1).
Phytobenthos are dominated by benthic algae, with 135 species and 33 families. Although eight seagrass species have been reported for the coast of Venezuela, none has been reported for the delta, only for the Gulf of Paria (Vera, 1992). The absence of seagrasses and benthic macroalgae might be a consequence of high loads of suspended matter and a lack of suitable substrates (Conde, 2000). The dominant terrestrial vegetation of the delta is the mangroves, with a mangrove formation between the Gulf of Paria and the Orinoco Delta representing more than 73% (183,500ha) of Venezuela's mangrove forest (Conde & Alarcón, 1993). Dominant mangrove species are the red mangrove *Rhizophora mangle*, the black mangrove *Avicennia germinans* and the white mangrove *Laguncularia racemosa*, though *R. harrisonii* and *R. racemosa*, the other two species of red mangrove have also been reported for the mid-delta (Pannier, 1979; Conde & Alarcón, 1993).

Kingdom	Phylum or Class	Class or Order	Described families	Described species	State of knowledge
Protista	Foraminifera		13	29	2
Plantae	Algae	Chlorophyta	7	23	3
	Algae	Phaeophyta	4	13	3
	Algae	Rhodophyta	22	99	3
	Angiospermae	(mangroves)	3	5	4
	Angiospermae	(seagrasses)	1	1	4
Animalia	Cnidaria	Scyphozoa	1	1	1
	Sipuncula		4	10	1
	Annelida	Polychaeta	39	94	3
	Mollusca	Bivalvia	30	83	3
	Mollusca	Cephalopoda	4	9	2
	Mollusca	Gastropoda	48	96	3
	Mollusca	Scaphopoda	1	1	1
	Crustacea	Amphipoda	32	143	3
	Crustacea	Cumacea	2	3	1
	Crustacea	Decapoda	43	142	3
	Crustacea	Isopoda	10	18	1
	Crustacea	Stomatopoda	1	1	1
	Crustacea	Tanaidacea	3	3	1
	Chelicerata	Pycnogonida	1	1	1
	Echinodermata	Asteroidea	4	13	3
	Echinodermata	Crinoidea	2	3	1
	Echinodermata	Echinoidea	4	9	2
	Echinodermata	Holothuroidea	3	2	1
	Echinodermata	Ophiuroidea	6	7	1
TOTAL			**288**	**809**	

Table 1. Number of benthic foraminifera, plant, and invertebrate species and families reported in the Orinoco Delta region. State of knowledge classified as: 5 = very well known (>80% described, and current taxonomic expertise); 4 = well known (>70% described, some taxonomic expertise); 3 = poorly known (<50% species described, no present expertise within region); 2= very poorly known (only few species recorded, no expertise); 1 = unknown (no species recorded, no expertise).

3.1.2 Plankton

The phytoplankton community in the area is comprised of Bacillariophyta with 150 species, Pyrrophycophyta (70 species), Cyanocophyta (39 species), Chlorophyta (24 species), Euglenophyta (6 species), and Crisophyta (3 species). Phytoplankton are distributed within two zones: the first dominated by an estuarine community, represented mainly by dinoflagellates, diatoms and chlorophytes, while the second consists of a marine habitat in which marine diatoms make up almost 80% of the community assemblage (Martín & Bone, 2007).

The zooplankton community in the area consists of 283 species of which 211 are crustaceans. Copepods dominate the holoplanktonic component with 116 species. Other groups are the gelatinous organisms, such as the hydrozoans (29 species) and the jellyfish (2 species).

Within the meroplankton, a total of 136 species have been reported, of which the decapod larval stages are the most important and diverse component. These larvae include species of shrimps with commercial value, such as species of the genus *Farfantepenaeus*, bivalves, barnacles and polychaetes (Table 2).

Kingdom	Phylum or Class	Class or Order	Described families	Described species	State of knowledge
Monera	Cyanophycota		8	39	1
Plantae	Bacillariophyta		33	150	4
	Chlorophyta		7	24	1
	Crisophyta		2	3	1
	Euglenophycota		1	6	1
	Pyrrophycophyta		14	70	3
Animalia	Cnidaria	Hydrozoa	15	29	1
	Cnidaria	Scyphozoa	2	2	1
	Ctenophora	Beroida	1	1	1
	Annelida	Polychaeta	5	9	1
	Mollusca	Gastropoda	6	10	1
	Crustacea	Amphipoda	21	54	3
	Crustacea	Branchiopoda	3	3	1
	Crustacea	Copepoda	30	116	3
	Crustacea	Decapoda	1	2	1
	Crustacea	Diplostraca	5	7	1
	Crustacea	Euphasiacea	1	12	2
	Crustacea	Isopoda	2	2	1
	Crustacea	Mysida	1	1	1
	Crustacea	Sessilia	1	6	1
	Crustacea	Stomatopoda	4	6	1
	Crustacea	Tanaidacea	2	2	1
	Echinodermata	Holothuroidea	1	1	1
	Chaetognatha		3	10	1
	Chordata	Appendicularia	2	8	1
	Chordata	Urochordata (Thaliacea)	2	2	1
TOTAL			**173**	**575**	

Table 2. Number of planktonic invertebrate species and families reported in the Orinoco Delta region. State of knowledge classified as in Table 1.

3.2 Vertebrates

3.2.1 Fishes

A total of 438 species have been reported in the Orinoco delta, grouped in 20 orders and 82 families. The best represented groups are the orders Characiformes, Perciformes, and Siluriformes with 132, 99 and 87 species respectively (Table 3). The five families with the highest species richness are the Characidae (73 species), Cichlidae (24 species), Pimelodidae (24 species), Sciaenidae (23 species), and Loricariidae (17 species) (Lasso et al., 2009; Lasso & Sánchez-Duarte, 2011). Of these identified species, 39 % are marine and estuarine, while 61% are strictly from freshwater habitats.

Orders	Families		Genera		Species	
	Number	%	Number	%	Number	%
Carchariniformes	2	2	2	0.7	2	0.5
Pristiformes	1	1	1	0.4	2	0.5
Myliobatiformes	7	9	10	3.6	15	3.4
Elopiformes	2	2	2	0.7	2	0.5
Anguilliformes	2	2	3	1.1	3	0.7
Clupeiformes	3	4	11	3.9	24	5.5
Characiformes	13	16	73	26.0	132	30.1
Siluriformes	11	13	62	22.1	87	19.9
Gymnotiformes	5	6	20	7.1	34	7.8
Batrachoidiformes	1	1	1	0.4	1	0.2
Lophiiformes	1	1	1	0.4	1	0.2
Atheriniformes	1	1	1	0.4	1	0.2
Cyprinodontiformes	3	4	7	2.5	12	2.7
Beloniformes	2	2	5	1.8	6	1.4
Syngnathiformes	1	1	1	0.4	1	0.2
Synbranchiformes	1	1	1	0.4	1	0.2
Scorpaeniformes	2	2	2	0.7	2	0.5
Perciformes	19	23	68	24.2	99	22.6
Pleuronectiformes	3	4	6	2.1	8	1.8
Tetraodontiformes	2	2	4	1.4	5	1.1
Total	**82**	**100**	**281**	**100**	**438**	**100**

Table 3. Number of fish families, genera, and species reported for the Orinoco Delta.

The ichthyological richness of the delta is third largest within the Orinoquia region, with the internal delta of the Ventuari-Orinoco (Venezuela) and the Estrella Fluvial de Inírida (Colombia) having the highest fish diversity (Lasso et al., 2010a). For the entire Orinoquia region, a total of 993 fish species (including freshwater and estuarine species) have been reported. These are grouped in 19 orders, 74 families, and 422 genera (Lasso et al., 2004c). For ichthyofauna, the Orinoco Delta is one of the best known regions of Venezuela compared to other remote and more continental areas of the Orinoco Basin that support a more complex fish fauna from a taxonomical point of view. Studies on fish span several

Theme	Author
Hydrology, vegetation and environmental Impact	Monente & Colonnello (1997, 2004), Colonnello (2004)
Exploratory fisheries Delta and Guyana	Ginés *et al.* (1971, 1972)
Exploratory fisheries Platform	Cervigón (1965), Ginés & Cervigón (1968)
Exploratory fisheries Delta	Ramos *et al.* (1982)
Artisanal fisheries Warao	Ponte & Mochcco (1997)
Warao ethno-ichthyology	Ponte (1995)
Effects of shrimp trawling fisheries	Novoa (2000a), Lasso *et al.* (2002, 2004b, 2008), Lasso & Sánchez-Duarte (2011)
History and socioeconomy of fisheries	Novoa (1982a, 1982b)
Shrimp and *curito* fisheries	Novoa (1982c, 1982d)
Fisheries catalog for the Delta	Novoa *et al.* (1982), Lasso & Sánchez-Duarte (2011)
Fisheries Delta and Gulf of Paria	Novoa (2000b), Bone *et al.* (2004), Lasso &Sánchez-Duarte (2011)
Fisheries resources Orinoquia	Novoa (2002)
Evaluation of fishing potential	Achury *et al.* (2006), Schneider *et al.* (2007), Lasso-Alcalá *et al.* (2008)
Biology of commercial species	Novoa & Ramos (1982), Lasso & Sánchez-Duarte (2011)
Exploratory fisheries	Ramos *et al.* (1982), Lasso & Sánchez-Duarte (2011)
Taxonomy and bio-ecology	Cervigón (1982), Lasso & Sánchez-Duarte (2011)
Taxonomy, biogeography and ecology	Cervigón (1985), Lasso *et al.* (2009), Lasso-Alcalá &Lasso (2007), Lasso-Alcalá *et al.* (2008), Lasso & Sánchez-Duarte (2011)
Trophic ecology	Ponte (1990)
Biogeography	Cervigón (1985), Lasso (1993), Lasso & Sánchez-Duarte (2011)
Biodiversity Orinoco Delta	Ponte *et al.* (1999), Lasso-Alcalá & Lasso (2007), Lasso *et al.* (2009), Lasso & Sánchez-Duarte (2011)
Biodiversity Morichal Largo	Antonio & Lasso (2003)
Biodiversity and community ecology	Ponte & Lasso (1994), Jorgensen *et al.* (2000), Lasso *et al.* (2002), Lasso-Alcalá *et al.* (2008), Lasso & Sánchez-Duarte (2011)
Community ecology Gulf of Paria	Lasso & Meri (2003), Lasso-Alcalá *et al.* (2008), Lasso & Sánchez-Duarte (2011)
Community ecology Delta	Lasso *et al.* (2004 a, 2004b), Lasso & Sánchez-Duarte (2011)
Biodiversity Biosphere Reserve Orinoco Delta	Bone *et al.* (2004), Ortaz *et al.* (2007)
Biodiversity Gran Morichal Reserve	Campo (2004)
Biodiversity and ecology	Lasso *et al.* (2008), Lasso & Sánchez-Duarte (2011)
Alien and introduced Species	Martínez-Escarbassiere *et al.* (2003), Lasso-Alcalá *et al.* (2005a, 2005b, 2009), Lasso & Sánchez-Duarte (2011)

Table 4. Summary of the main publications by theme related to the Orinoco Delta, the Gulf of Paria, and the Venezuelan Atlantic Front.

areas: biology, ecology, fisheries, biogeography, taxonomy, and biodiversity, among others (Table 4). The freshwater fish fauna of the Orinoco Delta results from the combination of the ancestral biota of the Guayanas with the more recent savannah (*"llanera"*) biota. On the other hand, the estuarine and marine ichthyofauna found from the Gulf of Paria and the South of Trinidad down to the Amazon river is very uniform (Cervigón, 1985). Toward the northern Venezuelan coast, this situation changes drastically. All the species that are characteristic of the estuarine zone of the delta almost disappear, with the exception of a few euryhaline species that inhabit either positive or negative estuaries, and other species that are mainly marine as adults, but their larvae and juveniles frequent brackish waters. In the Maracaibo lake, another large estuarine system located in the western Caribbean coast of Venezuela, these same estuarine species that had not been reported along more than 2,000 km of coast are present again, showing an important affinity in the fish fauna of the Maracaibo Lake and the Orinoco Delta. According to Cervigón (1985), this ichthyofaunal affinity in Venezuelan territory seems to extend to the Atlantic coast of Colombia, particularly to the south of Cartagena including the Gulf of Urabá.

3.2.2 Other vertebrates

Several aquatic mammals are widely distributed in the Orinoco Delta area. Amphibians are represented by terrestrial and freshwater species associated with mangrove forests. Birds are highly diverse with 255 species, including 21 orders and 54 families. There is only one endemic species of birds in this area, the black-dotted piculet (*Picumnus nigropunctatus*) (Martin & Bone, 2007). The aquatic birds are grouped in eight orders with 36 species of Charadriiformes, including jacanas, lapwings, plovers, sandpipers, snipes, skimmer, and gull terns; 21 species of Ciconiiformes, like herons, egrets, bitterns, storks, and ibises; 8 species of Procellariformes or petrels; 7 species of Pelecaniformes, including pelicans, boobies, cormorants, darters, and frigatebirds; 7 species of Gruiformes, such as limpkins, wood-rails, crakes, and gallinules; 3 species of Coraciiformes or kingfishers; 2 species of Anseriformes such as screamers and ducks, and one species of Phoenicopteriformes or flamingos.

Delta reptiles are represented by three orders, Crocodilia, Squamata, and Testudines. The most representative species of Crocodilia are the spectacular cayman (*Caiman crocodilus*), and the crocodiles *Crocodylus acutus* and *Crocodylus intermedius*, which are on the IUCN Red List of threatened species under the category of vulnerable and critically endangered, respectively. Marine reptiles are represented by five species of sea turtles, all reported along the Venezuelan coast. The five species are considered to be endangered and are protected by the Venezuelan wildlife protection law (Ojasti, 2005; Guada, 2000; Rodríguez & Rojas-Suárez, 2008). Four of these species nest in the Gulf of Paria, from Irapa to Punta Narizona. The leatherback turtle (*Dermochelys coriacea),* spawns in the Barra Mariusa and Tobejuba area, within the boundaries of the Delta del Orinoco National Park and Reserva de Biosfera Delta del Orinoco. As for mammals, the most common is the freshwater dolphin, *Inia geoffrensis* (tonina), while other species, like the river dolphin, *Sotalia fluviatilis,* the manatee, *Trichechus manatus,* the giant otter, *Petronura brasilienis,* and the long-tailed otter, *Lontra longicaudis,* are rare. The giant otter and the manatee have been reported as critically endangered (Linares, 1998). Only two species of the 13 Cetacea recorded (*Megaptera novaeangliae* and *Balaenoptera physalus*) are considered as vulnerable in the Red List of Venezuelan fauna (Table 5).

Kingdom	Phylum or Class	Class or Order	Order or Family	Described families	Described species	State of knowledge
Animalia	Chordata	Amphibia	Anura	3	6	1
		Aves	Accipitriformes	1	2	1
		Aves	Anseriformes	2	2	1
		Aves	Apodiformes	2	11	1
		Aves	Caprimulgiformes	2	4	1
		Aves	Charadriiformes	5	36	2
		Aves	Ciconiiformes	4	21	1
		Aves	Columbiformes	1	7	1
		Aves	Coraciiformes	1	3	1
		Aves	Cuculiformes	1	7	1
		Aves	Falconiformes	3	18	1
		Aves	Galbuliformes	1	3	1
		Aves	Galliformes	1	1	1
		Aves	Gruiformes	2	7	1
		Aves	Passeriformes	15	94	4
		Aves	Pelecaniformes	4	7	1
		Aves	Phoenicopteriformes	1	1	1
		Aves	Piciformes	2	9	1
		Aves	Procellariformes	3	8	1
		Aves	Psittaciformes	1	8	1
		Aves	Strigiformes	1	4	1
		Aves	Trogoniformes	1	2	1
		Reptilia	Crocodilia	2	4	4
		Reptilia	Squamata	6	12	4
		Reptilia	Testudines	8	14	4
		Mammalia	Carnivora	1	2	2
		Mammalia	Cetacea	4	13	3
		Mammalia	Sirenia	1	1	4
TOTAL				**79**	**307**	

Table 5. Number of vertebrate (amphibians, birds, reptiles, and mammals) species and families reported in the Orinoco Delta region. State of knowledge classified as in Table 1.

4. Threats to marine biodiversity

Any plan to achieve effective conservation of biodiversity and the functionality of the natural systems requires precise knowledge of the sources of actual or potential hazards or pressures and where and how those pressures could affect ecosystems and their components. Such pressures (pollution, resource exploitation, etc.) cause environmental disturbances in a given area, impacting the biological attributes of the biota or physical components of the ecosystem. Threats to biodiversity can then be described as a pressure source, by its spatial coverage, and the intensity of its actual or potential impacts. The Orinoco Delta and the Venezuelan Atlantic Front are very diverse collections of particular marine and coastal ecosystems (mangrove and palm forests, sand barriers, estuary and micro watershed nets, muddy/sandy marine bottoms, demersal fish and crustacean populations), with relatively limited coastal and marine area (35,000 km²). In general, all these ecosystems are healthy, with minimal disturbance. In this generally pristine landscape, there are some small urban and rural communities, whose presence determines the occupation and use of the area. The delta environment provides several types of resources for the people who live there (Creole and indigenous communities), and, in turn, the residents create different threats to the ecosystems, which in some cases, measurably impacts them.

In addition, in light of recent Venezuelan governmental strategies to boost the economy, the national oil industry has initiated a large scale project for offshore and nearshore gas exploitation, covering more than 25% of Venezuela's territorial sea and EEZ. This development could pose a significant threat to marine biodiversity in the region and to traditional livelihoods of local human populations. Few models have been developed to measure the impact of the human activities on the marine environment.

On a global scale, the model developed by Halpern et al. (2008), propose a 'low cumulative impact score' for the Golfo de Paria and Atlantic Front. In this estimation, impacts for the region generated by fisheries are minor and nonexistent for the oil and gas activities. In a more detailed model with higher spatial resolution (Klein & Cárdenas, 2009), fisheries and oil and gas activities (exploration, production, and decommissioning) have a high incidence in shaping the threats map (Figure 4).

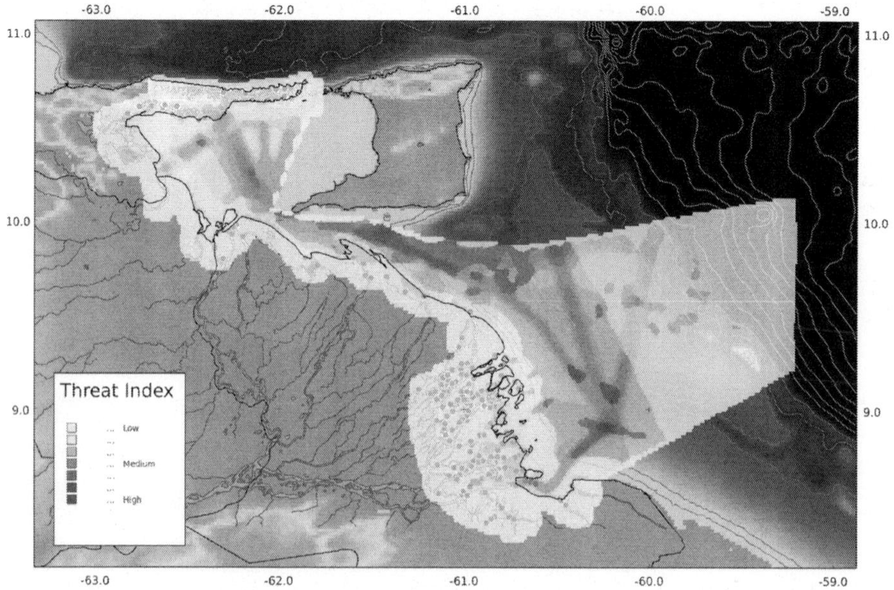

Fig. 4. Map of the Orinoco Delta and Gulf of Paria region in the Venezuelan Atlantic front showing the areas under threat by human activities.

The marine region of the Orinoco Delta supports important fisheries that provide 10,000 tons/year (around 5 % of the total marine national catch). The fishery targets mainly species

Threat	Description of impact
Industrial fisheries	Effects on fish and soft bottom communities produced by poorly controlled extraction of fish and some invertebrates organisms, and the disruption of the bottom sediments produced by trawling nets.
Artisanal fisheries	Effects on fish and invertebrates populations by poorly controlled extraction of organisms by different fishing techniques (traps, seines, small trawling nets, hooks).
Coastal urban development	Impacts generated by human activities such as habitat transformation and destruction, wildlife extraction and commercialization, introduction of non-native species, solid and liquid wastes, noise, etc. Effects depend on the size of human settlements and population density (i.e., large, dense towns, greater impacts).
Agricultural activities	Consider threats due to normal or poorly regulated agricultural activities like habitat transformation, wildlife extraction, introduction of non-native species, solid and liquid wastes, and chemical pollution of land and water bodies.
Channel dredging and flux control	Regular disturbance of soft-bottom benthic communities and increased suspended sediments due to major and maintenance dredging activities in the navigation channels. Regulation of water flux affects natural migration patterns of species.
Water pollution and upper river basin impacts	Water quality alteration due to high loads or regularly elevated loads of pollutants delivered by the river delta. Related to urban and agricultural activities in the upper basin. Effects could be propagated towards the open sea, depending on river discharge.
Mangrove deforestation	Though highly controlled as industrial activities, local extraction of construction material for housing and fishing gear could produce locally important habitat loss of mangrove and its associated communities.
Oil industry activities	Impacts produced by all stages of an oil industry project (exploration, development, exploitation, and decommissioning). Existing and planned activities equally considered.
Ports and marinas	Impact produced by regular activities of vessels at commercial ports including maneuvering, loading and unloading operations, and repairs. Fishing and sport marinas are also considered.
Maritime corridors	Major navigation routes. May serve as dispersion agents for alien species, and pollution sources from bilge water discharges.
Introduction of alien species	Competition, displacement and extinction of local species. Impact on fisheries due to predation of juveniles of commercial species, and on human health (transmission of diseases).

Table 6. Summary of threats to marine and coastal biodiversity and description of their impacts.

of Ariidae, Sciaenidae, Carangidae, Lutjanidae and Scombridae families in waters up to 60m depth. Industrial trawling, which fished mainly for shrimp, was banned in 2009, but artisanal trawling still persists near the coast. Poorly controlled, artisanal trawling represents a major threat to marine coastal biodiversity because of its intensity and the large swath of area it covers in highly sensitive coastal environments.

Although in early stages of development, offshore and nearshore gas reserves represent 2.4% of the total national gas reserves (6-10 TCF). Some structures are already in place (Pedernales subregion) and exploratory offshore drilling is ongoing (Oceanic subregion). Plans are that all gas production is directed to a new inland facility (CIGMA: Industrial Complex Gran Mariscal de Ayacucho) that will be built near Güiria (Península de Paria subregion). This transport and processing complex will directly affect 63 km^2 of coastal area and indirectly 110 km^2 of marine area. Its construction and operation will generate an elevated demand of services in what is now a poorly developed area. The environmental impacts of this complex have been evaluated, but little has been done to reduce or mitigate them.

At present, a number of major threats to Venezuelan marine biodiversity have been identified, quantified and mapped (Table 6). The threats and the potentially affected ecosystems can be viewed at http://paria.cbm.usb.ve/mapas/dinamicos , and Martínez-Escarbassiere et al. (2003), Lasso & Sánchez-Duarte (2011) and Lasso-Alcalá et al. (2005a, 2005b, 2009).

5. Conservation initiatives

The Orinoco Delta region has been recognized as one of the most pristine and biodiversity rich environments of Venezuela. Since 1979, three major National Parks have been established, covering a total of 4,385 km^2. All parks are classified as type II (IUCN, 1994), which mean they are managed mainly for ecosystem protection and recreation. None, however, includes a significant portion of their marine areas. To help rectify this, in 1991, the Orinoco Delta Biosphere Reserve was decreed by the Venezuelan Ministry of the Environment, and designated by UNESCO in 2009, to protect a land and water surface of 11,250 km^2. The reserve was created to protect the rich biodiversity of plants, animals and cultural features, and encompasses about 4,500 km^2 of marine and estuarine area. In 2002, a major effort also was made to evaluate the biodiversity of the reserve, and the investigations produced the first integrated and geo-referenced inventory of vegetation, mammals and aquatic and marine biota of the reserve (Bone et al., 2004).

Despite the establishment of the protected areas for the region, a new approach was initiated that considered present and future threats to the area's biodiversity (Klein and Cárdenas, 2010). Using an ecosystem-based model (Possingham et al., 2000), nine areas were proposed for implementation of special conservation strategies (priority conservation areas or PCA). This approach took into account the health status and vulnerability of the main ecosystems, and their populations, and the costs of implementing viable conservation actions, considering the cumulative potential threats for the region. A similar analysis was made for the Venezuela's Caribbean Sea (Klein, 2008). One of the most innovative characteristics of this ecosystem-based model is that it takes into account the future plans of offshore oil and gas development, while recognizing the coexistence of development and conservation. The proposed portfolio of priority conservation areas would add 17,681 km² of mostly marine areas to the formally protected areas (Figure 5). The proposed PCAs overlap to some extent with existing protected areas, but in many cases, areas not considered before are included. Although these additional PCAs have not been formally adopted by an official environmental agency, the associated conservation strategies are being used by the oil industry in planning their operations.

Considering the remoteness and limited accessibility of most of this region, adequate management of protected areas represents a challenge. In many cases, areas are accessible only by river or by air, although human activities are already taking place. Controlling and surveying the main threats is difficult at best, so their cumulative impacts may affect the biodiversity of the region. A major enforcement effort is necessary to provide basic services needed to guarantee effective implementation of conservation strategies for the region, which could be related to innovative management plans for the current and future protected areas. Such plans should not only regulate the landscape and its elements, but also consider the administration of trans-regional exploited populations, as are the widely distributed fisheries resources.

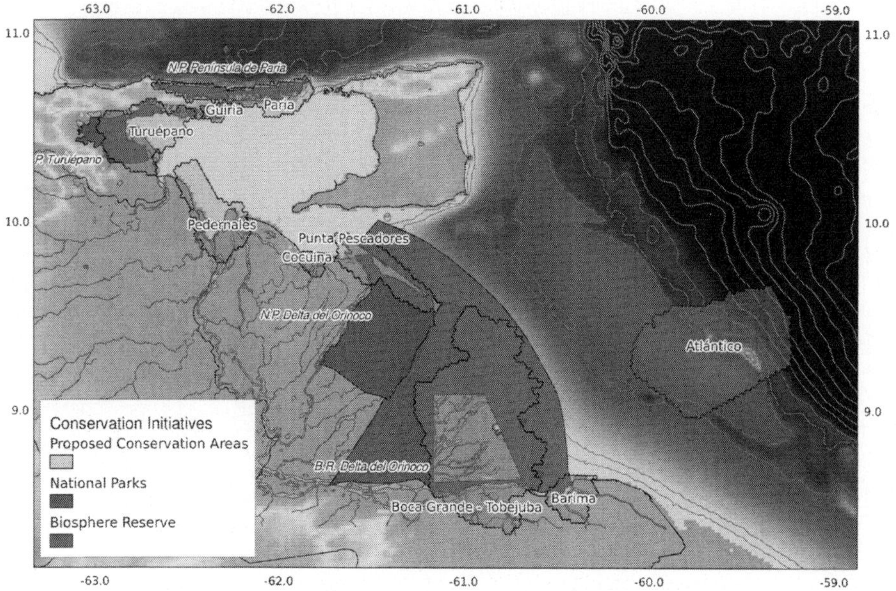

Fig. 5. Map of the Orinoco Delta and Gulf of Paria region in the Venezuelan Atlantic front showing the protected areas (NP= National Parks), the Orinoco Biosphere Reserve declared by UNESCO, and the proposed conservation areas considering present and future threats to biodiversity using an ecosystem based model.

6. Conclusion

The Orinoco Delta and the Gulf of Paria constitute one of the major wetlands in South America and are one of the best preserved ecosystems in the world. These wetlands were formed by the combined action of sediment and freshwater discharges from the Orinoco, one of the longest rivers in South America along with the tides on a flat alluvial plain. Based on physical and chemical characteristics and ecosystems that developed in this area, seven subregions were defined. In the Orinoco Delta, the dominant ethnic group are the *Warao* with a population of more than 36,000 people spread in at least 202 settlements which have been proposed as conservation targets.

The earliest studies of the region began in the 1800s, however, important production of publications did not begin until the mid 1950s and is still increasing today. In the late 1990s and early 2000s, baseline studies were carried out in the area in response to the interest of oil and gas companies in establishing offshore and coastal developments. Nevertheless, despite the great effort of research that has been done, projects carried out in this region have been very focused, geographically localized and punctual in time (no time series data) as well as very isolated, so most of the area remains largely unexplored. As for biodiversity, the best known groups are the fish followed by crustaceans and mollusks which contribute almost 50% of the total species richness, with other vertebrates contributing almost 15% of the total species richness, represented mainly by birds.

Major threats to biodiversity in this region are industrial and artisanal fishing, urban and agriculture development, dredging and flow navigation, water pollution (run off from the Orinoco basin), mangrove deforestation, activities related to oil and gas exploitation, port activities, and maritime transit.

The Orinoco Delta and Gulf of Paria region have two protected areas under special national conservation regulations. These are the Turuépano National Park in the Gulf of Paria, and the Orinoco Delta National Park. Both are under IUCN conservation category II, meaning they are managed mainly for ecosystem protection and recreation. An important area of the Orinoco Delta is also considered a Biosphere Reserve by UNESCO. Considering present and future threats to the area's biodiversity, nine additional priority conservation areas are proposed for implementation of special conservation strategies by using an ecosystem-based model. Although these additional PCAs have not been formally adopted by an official environmental agency, the associated conservation strategies are being used by the oil industry in planning their operations and further development.

7. Acknowledgement

We wish to acknowledge the Census of Marine Life Secretariat who supported this publication and Darlene Trew Crist for her editorial comments.

8. References

Achury, A.; Cárdenas, J. J. & Guillard, J. (2006). Evaluación por acústica submarina de los recursos ictiológicos en zonas estuarinas del delta del Orinoco. *Memoria de la Fundación La Salle de Ciencias Naturales,* Vol.165, pp. 103-113, ISSN 0037-8518

Antonio, M. E. & Lasso, C. (2003). Los peces del río Morichal Largo, estados Monagas y Anzoátegui, cuenca del río Orinoco, Venezuela. *Memoria de la Fundación La Salle de Ciencias Naturales,* Vol.156, pp. 5-118, ISSN 0037-8518

Baker, H. B. (1923). The Mollusca collected by the University of Michigan, Williamson Expedition in Venezuela, *Occasional Papers of the Museum of Zoology, University of Michigan,* Vol.137, pp. 1-158, ISSN 0076-8413

Baker, H. B. (1925). The Mollusca collected by the University of Michigan, Williamson Expedition in Venezuela, *Occasional Papers of the Museum of Zoology, University of Michigan,* Vol.156, pp. 1-56, ISSN 0076-8413

Baker, H. B. (1927). The Mollusca collected by the University of Michigan, Williamson Expedition in Venezuela, *Occasional Papers of the Museum of Zoology, University of Michigan,* Vol.182, pp. 1-50, ISSN 0076-8413

Bone, D.; Machado, A.; Spinello, P.; Ortaz, M.; Posada, J.; Molinet, R.; Yerena, E.; Rodríguez, C.; Klein, E. & Martín, A. (Eds.). (2004). *Conservación y uso sustentable de la diversidad biológica en la Reserva de Biosfera y los Humedales del Delta del Orinoco. Evaluación Ecológica Rápida de la Fauna Acuática,* Final Report, INTECMAR, Universidad Simón Bolívar, Caracas, Venezuela, 644 p.

Campo, M. (2004). Inventario preliminar de la ictiofauna de la Reserva de Fauna Silvestre Gran Morichal, Estado Monagas, Venezuela. *Memoria de la Fundación La Salle de Ciencias Naturales,* Vol.161-162, pp. 41-60, ISSN 0037-8518

Cervigón, F. (1965). Exploratory Fishing off the Orinoco Delta. *Proceedings of the Gulf and Caribbean Fisheries Institute,* 17th Session, pp. 20 - 23, ISSN 0072-9019

Cervigón, F. (1982). La ictiofauna estuarina del caño Mánamo y áreas adyacentes, In: *Los recursos pesqueros del río Orinoco y su explotación,* D. Novoa (Ed.), 205-260, Corporación Venezolana de Guayana, Editorial Arte, Legal deposit lf 82-3.784, Caracas, Venezuela

Cervigón, F. (1985). La ictiofauna de las aguas estuarinas del delta del río Orinoco en la costa atlántica occidental, Caribe, In: *Ecología de comunidades de peces en estuarios y lagunas costeras: hacia una integración de ecosistemas,* A. Yáñez-Arancibia (Ed.), 57-78, Universidad Nacional Autónoma de México, Programa Universitario de Alimentos, Instituto de Ciencias del Mar y Limnología, ISBN 968-837-618-3, México

Conde, J.E. (2000). The Orinoco River Delta, Venezuela, In: *Coastal Marine Ecosystems of Latin America,* U. Seeliger & B. Kjerve (Eds.), Springer-Verlag, ISBN 3-540-67228, Berlin, Germany

Colonnello, G. (2004). Las planicies deltaicas del río Orinoco y Golfo de Paria: aspectos físicos y vegetación. In: *Evaluación rápida de la biodiversidad y aspectos sociales de los ecosistemas acuáticos del delta del río Orinoco y Golfo de Paria, Venezuela,* C. A. Lasso, L. E. Alonso, A. L. Flores & G. Love (Eds.), 37-54, Boletín RAP de Evaluación Biológica 37, Conservation International, Washington DC, USA, ISBN 1-881173-76-3

Conde, J.E. & Alarcon, C. (1993). Mangroves of Venezuela, In: *Conservation and Sustainable Utilization of Mangrove Forest in the Latin America and Africa Regions. Part 1,* L.D. Lacerda (Ed.), 211-244, Serie Mangrove Ecosystems, Technical Report. Vol. 2, The International Society for Mangrove Ecosystems and The International Tropical Timber Organization, Okinawa, Japan

Costello, M. J.; Coll, M.; Danovaro, R.; Halpin, P.; Ojaveer, H. & Miloslavich, P. (2010). A Census of Marine Biodiversity Knowledge, Resources, and Future Challenges. *PLoS ONE,* Vol.5, No.8, e12110. doi:10.1371/journal.pone.0012110, ISSN 1932-6203

Fowler, H. W. (1911). Some fishes from Venezuela, *Proceedings of the Academy of Natural Sciences, Philadelphia,* Vol.63, pp. 419-437, ISSN 0097-3157

Fowler, H. W. (1912). Some fishes from Venezuela, *Proceedings of the Academy of Natural Sciences, Philadelphia,* Vol.67, pp. 203-243, ISSN 0097-3157

Fowler, H. W. (1931). Fishes obtained by the Barber Asphalt Company in Trinidad and Venezuela in 1930, *Proceedings of the Academy of Natural Sciences, Philadelphia,* Vol.83, pp. 391-410, ISSN 0097-3157

García Castro, A. (2000). Mendicidad indígena: Los Waraos urbanos, *Boletín Antropológico,* Vol.48, pp. 79-89, ISSN 1325-2610

García Castro, A. & Heinen, H. D. (1999). Planificando el desastre ecológico: Impacto del cierre del caño Manamo para las comunidades indígenas y criollas del Delta Occidental (Delta del Orinoco, Venezuela), *Antropológica,* Vol.91, pp. 31-56, ISSN 0003-6110

Ginés, H. & Cervigón, F. (1968). Exploración pesquera en las costas de Guayana y Surinam. *Memoria de la Fundación La Salle Ciencias Naturales,* Vol.28, pp.1-96, ISSN 0037-8518

Ginés, H., Cervigón, F. & Gómez, R. (1971). Pesca exploratoria en la costa N y NE de Sur-América, In: *Symposium on investigations and resources of the Caribbean Sea and adjacent regions,* FAO Fishery Resources Division, 57-93, FAO Fisheries Report 71.2, Paris, France, ISSN 0429-9337

Ginés, H., Angell, C., Méndez, M., Rodríguez, G., Febres, G., Gómez, R., Rubio, J., Pastor, G. & Otaola, J. (1972). *Carta pesquera de Venezuela. 1- Áreas del Nororiente y Guayana,* Fundación La Salle de Ciencias Naturales, Monografía 16, ISBN 84-399-1838-0, Caracas, Venezuela, 328 p.

Guada, H.J. (2000). Áreas de anidación e impactos hacia las tortugas marinas en la Península de Paria y lineamientos de protección. Master in Biological Sciences Thesis, Universidad Simón Bolívar, Caracas, Venezuela, 228p

Guppy, R. J. L. (1895a). Observations upon the physical conditions and fauna of the Gulf of Paria, *Proceedings Victoria Institute Trinidad,* Vol.2, pp. 105-115

Guppy, R. J. L. (1895b). The Mollusca of the Gulf of Paria, *Proceedings Victoria Institute Trinidad,* Vol.2, pp. 116-152

Guppy, R. J. L. (1895c). First sketch of a marine invertebrate fauna of the Gulf of Paria and its neighbourhood. Part I. Mollusca, *Proceedings Scientist Assessments Trinidad,* Vol.2, No.11, pp. 154-157

Halpern, B. S., Selkoe, K. A. , Micheli, F. & Kappel, C. V. (2007). Evaluating and Ranking the Vulnerability of Global Marine Ecosystems to Anthropogenic Threats, *Conservation Biology,* Vol.21, pp. 1301-1315, doi: 10.1111/j.1523-1739.2007.00752.x, ISSN 0036-8075

Humboldt, A. v. (1838). Über den Manati des Orinoko. Wiegmann's Arch. *Naturgesch,* Vol.4, No.1, pp. 1-10

IUCN. (1994). *Guidelines for Protected Area Management Categories,* CNPPA with assistance of WCMC. IUCN, Cambridge, UK and Gland, Switzerland, ISBN 978-2-8317-1086-0

Jorgensen, J.; Lasso, C. & Blanco-Belmonte, L. (2000). A method for the determination of fish biomass and density in macrophyte habitats in Orinoco floodplain lakes. *Memoria de la Fundación La Salle Ciencias Naturales,* Vol.153, pp. 35-50, ISSN 0037-8518

Klein, E. 2008. *Prioridades de PDVSA en la conservación de la biodiversidad en el Caribe Venezolano,* PDVSA, INTECMAR-USB, ISBN 978-9-8070-6215-2. Available at http://cbm.usb.ve/sv/prioridades-de-pdvsa-en-la-conservacion-de-la-biodiversidad-en-el-caribe-venezolano/.

Klein, E. & Cárdenas, J. J. (2009). Identificación de Prioridades de conservación asociadas a los ecosistemas de la Fachada Atlántica y su biodiversidad. Project LOCTI-INTECMAR, Technical Report, Universidad Simón Bolívar, The Nature Conservancy, Caracas, Venezuela, 313 p. Available at http://paria.cbm.usb.ve

Lasso, C. (1993). Primer registro de *Cleithracara maronii* (Steindachner 1882), (Teleostei, Cichlidae) para Venezuela: consideraciones biogeográficas. *Memoria Sociedad de Ciencias Naturales La Salle,* Vol.140, pp. 149-157, ISSN 0037-8518

Lasso, C. & Meri, J. (2003). Estructura comunitaria de la ictiofauna en herbazales y bosques inundables del bajo río Guanipa, cuenca del golfo de Paria, Venezuela. *Memoria de la Fundación La Salle Ciencias Naturales,* Vol.155, pp. 73-90, ISSN 0037-8518

Lasso, C.; Meri, J. & Lasso-Alcalá, O. (2002). Composición, aspectos ecológicos y usos del recurso íctico en el Bloque Delta centro, delta del Orinoco, Venezuela. *Memoria de la Fundación La Salle de Ciencias Naturales,* Vol.158, pp. 87-116, ISSN 0037-8518

Lasso, C.; Lasso-Alcalá, O. M.; Pombo, C. & Smith, M. (2004a). Ictiofauna de las aguas estuarinas del delta del río Orinoco (caños Pedernales, Mánamo, Manamito) y golfo de Paria (río Guanipa): diversidad, distribución, amenazas y criterios para su conservación, In: *Evaluación rápida de la biodiversidad y aspectos sociales de los ecosistemas acuáticos del delta del río Orinoco y Golfo de Paria, Venezuela,* C. A. Lasso, L. E. Alonso, A. L. Flores & G. Love (Eds.), 70-84, Boletín RAP de Evaluación Biológica 37, Conservation International, Washington DC, USA, ISBN 1-881173-76-3

Lasso, C.; Lasso-Alcalá, O.; Pombo, C. & Smith, M. (2004b). Composición, abundancia y biomasa de la ictiofauna béntica del golfo de Paria y delta del Orinoco, In: *Evaluación rápida de la biodiversidad y aspectos sociales de los ecosistemas acuáticos del delta del río Orinoco y golfo de Paria, Venezuela,* C. A. Lasso, L. E. Alonso, A. L. Flores & G. Love (Eds.), 85-102, Boletín RAP de Evaluación Biológica 37, Conservation International, Washington DC, USA, ISBN 1-881173-76-3

Lasso, C.; Mojica, J. I.; Usma, J. S.; Maldonado-Ocampo, J. A.; DoNascimiento, C.; Taphorn, D. C.; Provenzano, F.; Lasso-Alcalá, O. M.; Galvis, G.; Vasquez, L.; Lugo, M.; Machado-Allison, A., Royero, R.; Suárez, C. & Ortega-Lara, A. (2004c). Peces de la cuenca del río Orinoco. Parte I: lista de especies y distribución por subcuencas. *Biota Colombiana,* Vol.5, No.2, pp. 95-158, ISSN 0124-5376

Lasso, C., Lasso-Alcalá, O., Giraldo, A., Sánchez, P., González, K., Hernández-Acevedo, J. & Rodríguez, J. (2008). Evaluación rápida de la biodiversidad ictiológica de los caños Macareo y Mariusa, delta de río Orinoco Delta del Orinoco, Venezuela, In: *Biodiversidad Animal del Caño Macareo, Punta Pescador y Áreas Adyacentes, Delta del Orinoco,* C. A. Lasso & J. Señaris (Eds.), 73-126, StatoilHydro Venezuela AS, Fundación La Salle de Ciencias Naturales, Caracas, Venezuela, ISBN 978-980-7090-04-9

Lasso, C., Sánchez-Duarte, P., Lasso-Alcalá, O., Hernández-Acevedo, J., Martín, V., Samudio, H., González-Oropeza, K. & Mesa, L. (2009). Lista actualizada de los peces del delta del río Orinoco, Venezuela. *Biota Colombiana,* Vol.10, No.1-2, pp. 123-148, ISSN 0124-5376

Lasso, C. A., Usma, J. S., Trujillo, F. & Rial, A. (Eds). (2010). *Biodiversidad de la cuenca del Orinoco: bases científicas para la identificación de áreas prioritarias para la conservación y uso sostenible de la biodiversidad,* Instituto de Investigación de Recursos Biológicos

Alexander von Humboldt, WWF Colombia, Fundación Omacha, Fundación La Salle de Instituto de Estudios de la Orinoquia (Universidad Nacional de Colombia), ISBN 978-958-8554-13-6 , Bogotá, Colombia, 609 pp.

Lasso, C. A. & Sánchez-Duarte, P. (2011). *Los peces del delta del Orinoco. Diversidad, biología, uso y conservación,* Fundación La Salle de Ciencias Naturales - Chevron C.A, ISBN 978-980-7090-11-7, Caracas, Venezuela, 500 p.

Lasso-Alcalá, O., Lasso, C., Pezold, F. & Smith, M. (2005a). The mud sleeper *Butis koilomatodon* (Bleeker, 1879) (Pisces: Eleotridae): First record from the Western Central Atlantic. *Revista de Biología Tropical,* Vol.53(1-2), pp. 211, ISSN 0034-7744

Lasso-Alcalá, O., Lasso, C. & Smith, M. (2005b). The first record of the naked gobi *Gobiosoma bosc* (Lacepède, 1800) (Pisces: Gobiidae) from the north coast of South America. *Revista de Biología Tropical,* Vol.53(1-2), pp. 211, ISSN 0034-7744

Lasso-Alcalá, O. & Lasso, C. (2007). Taxonomía, distribución y aspectos ecológicos del género *Awaous* Valennciennes 1837 (Pises, Perciformes, Gobiidae), en Venezuela. *Memoria de la Fundación La Salle de Ciencias Naturales,* Vol.168, pp. 117- 140, ISSN 0037-8518

Lasso-Alcalá, O., Lasso., C. & Rodríguez, J. (2008). Comunidad de peces demersales del sector suroriental del golfo de Paria, Venezuela. *Memoria de la Fundación La Salle de Ciencias Naturales,* Vol.170, pp. 119-144, ISSN 0037-8518

Lasso-Alcalá, O., Lasso, C. & Posada, J. (2009). Especies Exóticas en el Mar Caribe: Introducción de *Omobranchus punctatus* (Valenciennes, 1836) (Perciformes, Blennidae) en las Costas de Venezuela. *Proceedings of the 61st Gulf and Caribbean Fisheries Institute,* pp. 391-395, ISSN 0072-9019

Linares, O.J. (1998). *Mamíferos de Venezuela,* Sociedad Conservacionista Audubon de Venezuela, ISBN 980-259-107-6, Caracas, Venezuela.

Martín, A. & Bone, D. (Eds.) (2007). *Línea Base Ambiental Plataforma Deltana,* Petróleos de Venezuela, S.A. Universidad Simón Bolívar, ISBN 978-980-7062-13-8, Caracas, Venezuela, 176 p.

Martínez-Escarbassiere, R., Capelo, J., García, J., Lasso, C., Pereira, G. & Lasso-Alcalá, O. (2003). Primer registro del mejillón dátil asiático *Musculista senhousia* (Benson 1842) (Bivalvia-Mytilidae): Especie introducida en Venezuela. *Memoria de la Fundación La Salle de Ciencias Naturales,* Vol.159 -160, pp. 309–312, ISSN 0037-8518

Michelangeli, F. (Ed.) (2000). *La Orinoquia,* Gerencia de Asuntos Públicos Operadora Cerro Negro, S.A. , ISBN 980-07-6372-4, Caracas, Venezuela, 357 p.

Miloslavich, P., Klein, E., Díaz, J. M., Hernández, C. E., Bigatti, G., Campos, L., Artigas, F., Castillo, J., Penchaszadeh, P. E., Neill, P. E., Carranza, A., Retana, M. V., Díaz de Astarloa, J. M., Lewis, M., Yorio, P., Piriz, M. V., Rodríguez, D., Yoneshigue-Valentin, Y., Gamboa, L. & Martín, A. (2011). Marine Biodiversity in the Atlantic and Pacific Coasts of South America: Knowledge and Gaps. *PLoS ONE* Vol.6, No.1, e14631. doi:10.1371/journal.pone.0014631, ISSN 1932-6203

Monente, J. A. & Colonnello, G. (1997). Hidroquímica del delta del Orinoco. *Memoria de la Fundación La Salle de Ciencias Naturales,* Vol.148, pp. 109-131, ISSN 0037-8518

Monente, J. A. & Colonnello, G. (2004). Consecuencia ambientales de la intervención del delta del Orinoco, In: *Evaluación rápida de la biodiversidad y aspectos sociales de los ecosistemas acuáticos del delta del río Orinoco y Golfo de Paria, Venezuela,* C. A. Lasso, L. E. Alonso, A. L. Flores & G. Love (Eds.), 114-124, Boletín RAP de Evaluación Biológica 37, Conservation International, Washington DC, USA, ISBN 1-881173-76-3

Muller-Karger, F. E., McClain, C. R., Fisher, T. R., Esaias, W. E., & Varela, R. (1989). Pigment distribution in the Caribbean Sea: observations from space, *Progress in Oceanography,* Vol.23, pp. 23–64, ISSN 0079-6611

Novoa, D. (1982a). Análisis histórico de las pesquerías del río Orinoco, In: *Los recursos pesqueros del río Orinoco y su explotación,* D. Novoa (Ed.), 21-50, Corporación Venezolana de Guayana, Editorial Arte, Legal deposit lf 82-3.784, Caracas, Venezuela

Novoa, D. (1982b). Diagnóstico socioeconómico de las pesquerías artesanales del río Orinoco, In: *Los recursos pesqueros del río Orinoco y su explotación,* D. Novoa (Ed.), 51-76, Corporación Venezolana de Guayana, Editoral Arte, Legal deposit lf 82-3.784, Caracas, Venezuela

Novoa, D. (1982c). El recurso camaronero en el delta del Orinoco, In: *Los recursos pesqueros del río Orinoco y su explotación,* D. Novoa (Ed.), 193-204, Corporación Venezolana de Guayana, Editorial Arte, Legal deposit lf 82-3.784, Caracas, Venezuela

Novoa, D. (1982d). La pesquería del busco o curito *Hoplosternum littorale* en el delta del río Orinoco, In: *Los recursos pesqueros del río Orinoco y su explotación,* D. Novoa (Ed.), 151-168, Corporación Venezolana de Guayana, Editoral Arte, Legal deposit lf 82-3.784, Caracas, Venezuela

Novoa, D. (2000a). Evaluación del efecto causado por la pesca de arrastre costera sobre la fauna íctica en la desembocadura del caño Mánamo, *Acta Ecológica del Museo Marino de Margarita* Vol.2, pp. 43-62,

Novoa, D. (2000b). *La pesca en el Golfo de Paria y Delta del Orinoco costero,* CONOCO Venezuela, Editorial Arte, ISBN 980-07-6491-7, Caracas, Venezuela, 140 pp.

Novoa, D. (2002). Los recursos pesqueros del eje fluvial Orinoco-Apure: presente y futuro, INAPESCA, Ministerio de Agricultura y Tierras, República Bolivariana de Venezuela, Caracas. 141 pp.

Novoa, D. & Ramos, F. (1982). Aspectos generales sobre la biología de las principales especies de peces de importancia comercial en el río Orinoco y su explotación, In: *Los recursos pesqueros del río Orinoco y su explotación,* D. Novoa (Ed.), 107-128, Corporación Venezolana de Guayana, Editorial Arte, Legal deposit lf 82-3.784, Caracas, Venezuela

Novoa, D., Cervigón, F. & Ramos, F. (1982). Catálogo de los recursos pesqueros del Delta del Orinoco, In: *Los recursos pesqueros del río Orinoco y su explotación,* D. Novoa (Ed.), 263-324, Corporación Venezolana de Guayana, Editorial Arte, Legal deposit lf 82-3.784, Caracas, Venezuela

Ojasti, J. (2005). Fauna Terrestre, In: *Conservación y usos sustentable de la diversidad biológica de la reserva de Biosfera y los humedales del Delta del Orinoco (RBDO),* Project VEN/99/G31, Ministerio del Ambiente y de los Recursos Naturales (MARN), Global Environmental Facility (GEF), United Nations Developmental Program (UNDP), Final Technical Report, Caracas, Venezuela

Ortaz, M., Machado-Allison, A. & Carrillo, V. (2007). Evaluación ecológica rápida de la íctiofauna en cinco localidades del delta del río Orinoco, Venezuela, *Interciencia,* Vol.32, No.9., pp. 601-609, ISSN 0378-1844

Pannier, F. (1979). Mangroves impacted by human-induced disturbances: a case study of the Orinoco Delta mangrove ecosystem, *Environmental Management,* Vol.3, pp. 205-216, ISSN 0364-152X

Phelps, W. H. & Phelps, W. H. J. (1950a). Lista de las aves de Venezuela con su distribución, Parte 1. Passeriformes, *Boletín de la Sociedad Venezolana de Ciencias Naturales*, Vol.12, pp. 1-427, ISSN 1335-5023

Phelps, W. H. & Phelps, W. H. J. (1950b). Seven new subspecies of Venezuelan birds, *Proceedings of the Biological Society of Washington*, Vol.63, pp. 115-126, ISSN 0006-324X

Ponte, V. (1990). *Recurso trófico utilizado por peces juveniles en dos áreas del delta interior del río Orinoco,* Undergraduate thesis, Escuela de Biología, Facultad de Ciencias, Universidad Central de Venezuela, Caracas, Venezuela, 101 pp

Ponte, V. (1995). Contributions of the Warao Indians to the ichtyology of the Orinoco Delta, In: *Nature and human ecology in the Neotropics,* H. Dieter Heinen, J. San José & H. Caballero Arias (Eds.), 371-392, Scientia Guaianae, ISSN 0798-1120, Caracas, Venezuela

Ponte, V. & Lasso, C. (1994). Ictiofauna del caño Winikina, Delta del Orinoco. Aspectos de la ecología de las especies y comunidades asociadas a diferentes hábitats, *Proceedings of the Segundo Congreso Venezolano de Ecología*, Guanare, Venezuela, February 20-26, 1994

Ponte, V. & Mochcco, O. (1997). Evaluación de las actividades pesqueras de la etnia Warao en el delta del río Orinoco, Venezuela, *Acta Biológica Venezuelica,* Vol.17, No.1, pp. 41-56, ISSN 0084-5906

Ponte, V., Machado-Allison, A. & Lasso, C. (1999). La ictiofauna del delta del río Orinoco, Venezuela: una aproximación a su diversidad, *Acta Biológica Venezuelica,* Vol.19, No.3, pp. 25-46, ISSN 0084-5906

Possingham, H. P., Ball, I. R. & Andelman, S. (2000). Mathematical methods for identifying representative reserve networks, In: *Quantitative methods for conservation biology*, S. Ferson & M. Burgman (Eds.), 291-305, Springer-Verlag, ISBN 978-0-387-94322-0, New York, USA

Ramos, F., Novoa, D. & Itriago, I. (1982). Resultados de los programas de pesca exploratoria realizados en el Delta del Orinoco, In: *Los recursos pesqueros del río Orinoco y su explotación,* D. Novoa (Ed.), 162-192, Corporación Venezolana de Guayana, Editorial Arte, Legal deposit lf 82-3.784, Caracas, Venezuela

Rodríguez, J. P. & Rojas-Suárez, F. (Eds.) (2008). *Libro Rojo de la Fauna Venezolana,* 3rd edition, Provita and Shell, Caracas, Venezuela, ISBN 980-210-011-0

Schneider, P., Achury, A., Guaiquirián, J. & Cárdenas, J. (2007). Análisis de la estructura y comportamiento de las comunidades ícticas en una zona del golfo de Paria con métodos acústicos. *Memoria de la Fundación La Salle de Ciencias Naturales*, Vol.168, pp. 25-44, ISSN 0037-8518

Vera, B. (1992). Sea grasses of the Venezuelan coast: distribution and community components, In: *Coastal plant communities of Latin America,* U. Seeliger (Ed.), 135-140, Academic Press, ISBN 978-0126345506, San Diego, USA

Zimmer, J. T. & Phelps, W. H. (1950). Three new Venezuelan birds, *American Museum Novitates*, No.1455, pp. 1-7, ISSN 0003-0082 abiotic and biotic factors (Hughes & Stachowicz, 2009). Similar results were obtained when genotypes of the green alga *Chlamydomonas reinhardtii* were set up in culture tubes in monocultures and all possible pair wise combinations: the mixtures were consistently more productive and less variable in their productivity under simulation of different environmental conditions (Bell, 1991).

To our knowledge, there are very few experimental studies on benthic animals investigating the role of genetic diversity on ecosystem function, two of which assessed the effects on intertidal invertebrates. Within-species diversity was manipulated in the barnacle *Balanus improvisus* by altering the number of parental broods (Gamfeldt et al., 2005). In a similar way, the number of maternal families was controlled in the amphipod *Gammarus duebeni* from intertidal rock pools (Gamfeldt & Kallstrom, 2007). Both experiments found positive effects of increased genetic diversity: larval settlement of *B. improvisus* increased and the more diverse amphipod assemblages exhibited more predictable (but not higher on average) survival when exposed to multiple stressors (Gamfeldt & Kallstrom, 2007). Apart from that, studies relating genetic diversity to some (though particular) measure of ecosystem functioning have mainly focused on tolerance to toxic substances, indicating that genetically diverse populations resist to or recover better from contamination (Pease et al., 2010; Phillips & Hickey, 2010).

The few studies relating genetic (genotypic) diversity to ecosystem functioning show that genetic diversity can be important for the stability and productivity of coastal ecosystems. Generally, increased genetic diversity enhances the continuity and reliability of ecosystems. The advantage of clonally reproducing organisms such as plants or algae is that they provide the relatively easy and thus straightforward possibility of setting up experimental groups of contrasting genotypic diversity. However, this method is not applicable to sexually reproducing organisms. For those, it is necessary to produce different levels of genetic diversity by controlled mating prior to the experiment. For example Gamfeldt & Kallstrom (2007) sampled natural populations of *G. duebeni* and then separated pairs of mating individuals to form different "families". With these families, they then formed inbred strains (lowest genetic diversity level) and strains of two to four different families (increasing diversity). This procedure provides reliable levels of contrasting genetic diversities, but it is time consuming and requires considerable effort and material. Nevertheless, it is probably the only reliable method of obtaining contrasting diversity levels for sexually reproducing organisms if natural inbred (representing low diversity) and outbred populations (high diversity) are unavailable. This is probably the main reason, why studies investigating the role of genetic diversity on the system functioning are still scarce.

3.2 Effects of species diversity on benthic ecosystem functioning

3.2.1 Macrophyte vegetations

Seagrass *Zostera marina*, pondweed or cordgrass *Spartina* ecosystems have been used widely to investigate the effect of species richness, species composition as well as food chain length and trophic cascades on ecosystem processes. For example all possible species combinations of four functionally and morphologically different submerged aquatic macrophytes (pondweed) were planted in artificial wading pools (Engelhardt & Ritchie, 2001). Although local wetlands are usually dominated by single vascular species, there was a clear effect of increased plant diversity leading to higher plant and algal biomass as well as lower loss of phosphorus (Engelhardt & Ritchie, 2001).

In several outdoor experiments, species diversity of different trophic levels was manipulated and the role of grazer diversity on plant biomass as well as grazer secondary production was investigated (Duffy et al., 2001). Grazers play an important role in the health of seagrass ecosystems because they feed preferentially on epiphytic algae and thus prevent eelgrass from becoming overgrown by epiphytes. In the experiment, the three grazer crustacean species (two isopods and one amphipod), which dominate the seagrass epifauna in the study area, were assembled in all possible species combinations (1 to 3 species and a grazer-free control), assessing the accumulation of epiphytic algae and eelgrass biomass as well as grazer secondary production. Results showed that all three grazer species differed substantially in their impact on ecosystem processes. For all processes measured, species composition was more important for eelgrass and grazer biomass accumulation than species richness. The study emphasized that species of one functional group need not necessarily be functionally redundant. In a follow-up experiment, six common grazer species (one gastropod, three amphipods and two isopods) were combined to further investigate the role of species diversity and species composition in relation to eelgrass ecosystem functioning (Duffy et al., 2003). Again, species were assembled in random combinations (one, three and six-species treatments) in outdoor mesocosms. The water supply was filtered, but allowed microscopic propagules of algae and sessile invertebrates to recruit. In addition to eelgrass, algal and grazer biomass, also measured was organic carbon in the sediment. Contrary to the previous study, increased biodiversity enhanced secondary production. Yet, increasing grazer diversity reduced total community diversity and allowed a sessile, grazing-resistant invertebrate to become dominant, presumably due to intraguild competition among grazer species.

In the previous experiment, interaction effects of species diversity and species composition were measured as deviations from expected additive effects based on single-species effects. In another attempt to separate the two effects, a mesocosm experiment was conducted where species richness and species composition were manipulated independently across multiple trophic levels including macrophytes, benthic grazers and carnivorous predators (Downing & Leibold, 2002). One, three or five species per functional group were assembled and within each level of richness, seven unique species compositions were nested and replicated. This being an extensive and statistically robust method, the results showed that species diversity as well as species composition had important effects on ecosystem functioning. Ecosystem productivity was highest at the highest diversity level, which implies that species within a functional group are not redundant, a result similar to that of Duffy et al. (2001). Not all measures of ecosystem functioning reacted similarly to species diversity or composition: phytoplankton biomass increased while zooplankton biomass decreased with total species richness. Moreover, there was a strong effect of species altering ecosystem processes indirectly by altering abundances of other species, implying that trophic interactions are highly important for the functioning of communities and ecosystems.

Experiments on macrophyte vegetations have successfully shown the effects of species diversity and identity on ecosystem processes. In general (but not always), a positive effect of biodiversity was found on ecosystem processes. However, these experiments were all conducted over a relatively short period of time. In contrast, the effects of seaweed species richness on biomass accumulation were stronger in long-term (3 years) than in short-term (2 months) field experiments (Stachovicz et al., 2008). Although short-term experiments are valuable in estimating the effects of species richness and identity on ecosystem processes, they are likely to be unable to capture seasonal environmental heterogeneity and population responses. Therefore, they are likely to underestimate the influence of diversity on ecosystem processes in natural ecosystems (Stachowicz et al., 2008).

3.2.2 Benthic invertebrate macrofauna from intertidal soft bottom areas

In a novel experimental approach designed to separate the effects of species richness and species identity, the dominant infaunal macrofaunal species of a mudflat community were collected in two distinct biogeographical areas: three species (one polychaete, an amphipod and a gastropod) in Scotland and five species (three gastropods, one ghost shrimp and one cockle) in South Australia. Null-models based on the response of monocultures (i.e., single species treatments) were constructed and then compared to the experimental results. It was found that individual species contributed idiosyncratically to the measured ecosystem process (ammonia release), and that the effects were biomass (density)-dependent (Emmerson & Raffaelli, 2000).

In a subsequent extended experiment, replicate species pools were sampled from three regions (Scotland, southwest Sweden and south Australia) and placed in mesocosms (Emmerson et al., 2001). The selected species (ranging from one to four species) were dominating the biomass at each of the three sites. Additionally, complete natural communities were sampled at three different sites in Scotland. These contained more species and included the small, delicate species that could not be manipulated directly. Thus two experiments were carried out: one with artificially assembled communities on one hand, and one with natural communities on the other. In the "artificial" assemblages, NH_4-N production became more predictable (i.e., less variable) with increased diversity. In the experiment where natural communities were used, there was a significant positive effect of species richness on nutrient flux; however, the process was driven disproportionately by individual, different species at each site (sampling effect).

The first field experiment on a benthic system was conducted on a mudflat in Scotland. Defaunated sediment was enclosed using cages with different mesh sizes. The cages were then allowed to be colonized by organisms from the surrounding sediments. Through this, groups of low, medium and high species richness as well as low, medium and high biomass of macrofaunal invertebrates were established (Bolam et al., 2002). In each experimental cage the following variables were measured as proxies of ecosystem functioning: phosphate and ammonium fluxes, community respiration, sediment shear strength, water content, water/silt content, organic content, redox potential, nitrate and nitrite. All but one variable remained unaffected by macrofaunal diversity or biomass; only oxygen consumption was positively related to both, but in fact it was driven by the largest species in the study.

Waldbusser et al. (2004) examined the relationship between functional diversity and ecosystem processes by manipulating three functionally diverse polychaete worms from nearshore sediments. They established microcosms of one and three species and kept them for 4 months. They found that species as well as functional diversity had an effect on the measured proxies of ecosystem processes (i.e., oxygen and phosphate fluxes, profiles of oxygen and pH). Although it is difficult to separate species from functional diversity when each species represents a different function, it was clear that single species had disproportionate effects on selected variables and that the three-species assemblage did not result in the additive effect one would have expected from single-species treatments.

In all the above-mentioned studies, a disproportionate effect of single species could be detected. Unlike the community members of macrophyte systems, benthic invertebrates from soft bottom intertidal areas seem to contribute individually to ecosystem functioning with their impact strongly depending on their functional role. Therefore, it can be essential to include such critical species in experiments assessing the effect of species richness on the functioning of the system. Since many experimental approaches include only the dominant species, conclusions drawn may not necessarily hold for the natural system, especially since some key species are not necessarily very abundant. Generally, natural benthic invertebrate assemblages from marine soft bottoms can be sampled easily, and as such provide a good system for BEF studies. Moreover, they can be established in micro- or mesocosms in the laboratory or in the field, and respond rapidly to experimental treatments. However, these organisms can reach very high densities, and especially the members of the meiofaunal component are often difficult to count and identify. Therefore, the choice of working with "artificial" (i.e., fewer species) or "natural" assemblages needs careful planning, which often means facing a trade-off between "natural" and "doable" with the available resources. The decision is most dependent on the scientific question, but also on the possibilities in terms of expertise and time.

3.2.3 Stream invertebrate larvae

Many insects develop as larvae on the substrate of streams or ponds, where they exhibit different feeding strategies and play different functional roles in the benthic system. As such, they provide an interesting system to test hypotheses regarding intraguild diversity. Caddisfly larvae, for example, construct silk nets in the pore spaces of the streambed, and passively feed on suspended particulate matter. Their structures generate topographical features, which influence patterns of water flow and therefore food availability (Cardinale et al., 2002). Often, several species with anatomically different feeding structures co-occur. This has led to the hypothesis that increased diversity leads to facilitation of food uptake. The hypothesis has been tested in stream mesocosms, where caddisfly larvae assemblages were established, either with a single species (18 larvae) or with 3 species (6 larvae per species; Cardinale et al., 2002). Results showed that higher diversity led to facilitative interactions and an increase in the uptake of organic matter. However, the species building the largest tubes had the strongest physical impact on streambed water flow.

A follow-up experiment tested whether the diversity-functioning relationship would be different under conditions of regular disturbance (Cardinale & Palmer, 2002). The hypothesis was that disturbance would induce mechanisms that would interfere with ecological processes. Disturbance was simulated by mechanically removing a given number of larvae in randomly selected pore spaces. Three ecological processes were measured at the end of the experiment: respiration of the benthic biofilm, primary productivity of benthic algae and the flux of particulate organic matter (POM) from the water column to the streambed. The disturbance treatment resulted in the suppression of a dominant taxon, which had a particularly low rate of nutrient excretion. This led to a negative correlation between primary production and species richness in the undisturbed streams, as this taxon was included in multi-species treatments. Concordantly, POM flux increased with species richness under disturbance conditions. Disturbance thus favoured the co-existence of competitively superior and inferior species, enhancing ecosystem processes (Cardinale & Palmer, 2002).

Other stream invertebrate larvae feed on leaf litter as "shredders" (Jonsson & Malmqvist, 2000). In a laboratory experiment, three species of stoneflies (*Plecoptera*), belonging to this leaf-eating feeding guild, were placed in one-, two- and three-species treatments to investigate the effect of species richness on leaf mass loss. Ecosystem process rates (leaf mass loss) increased significantly with species richness, but were not dependent on species identity, suggesting that the different species were not functionally redundant (Jonsson & Malmqvist, 2000).

An extension of the previous experiment investigated the effect of intraguild species diversity for other feeding guilds (Jonsson & Malmqvist, 2003). In several controlled laboratory experiments, multiple species of the following functional groups were collected: filter feeders (six blackfly larvae *[Diptera]*), grazers (two species of mayfly larvae *[Ephemeroptera]* and one snail), and predators (two stonefly species [*Plecoptera*] and one caddisfly species [*Trichoptera*]). Each functional group was set up in single-, two- and three species combinations and provided with their respective food source: dry yeast for the filter feeders, algae for the grazers and blackfly larvae for the predators. There was a strong effect of species identity, and species combinations, whereas species richness *per se* did not affect process rates in any of the three functional groups (Jonsson & Malmqvist, 2003).

These four examples show that stream invertebrate larvae form a complex system consisting of different feeding guilds exhibiting many different functions. Apparently the extent to which species richness affects ecosystem functioning depends greatly on the identity and the function of the species present. In general, however, like marine soft-bottom invertebrates, lotic larval communities seem to contain species exhibiting disproportionate effects on the system. The effects of species loss may be predictable when their functional role is accurately ascertained, but can become idiosyncratic when interactions such as competitive behaviour among species occur.

4. Two "special topics" in benthic BEF studies

As mentioned in the former sections, most BEF studies assess ecosystem functioning in terms of biomass production, effects on community composition or element cycling. However, there are other possible effects with important implications for the whole ecosystem. We choose two examples to illustrate this: trophic cascades and invasion success. We chose these two special topics because the former shows that slightly different levels of intraguild diversity can lead to completely opposed effects, whereas the latter illustrates that although experimental approaches meet hypotheses based on theory, they sometimes do not reflect what is actually happening in nature.

4.1 Trophic cascades

Trophic species-level cascades occur when a change in predator abundance induces changes in the biomass of primary producers, due to a control of the abundance of grazers, thereby releasing lower trophic levels from grazing pressure (Polis et al., 2000). Predator diversity may reduce trophic cascades and may therefore be important for the population development of lower trophic levels and primary production. The importance of such intraguild diversity was studied by manipulating predator species richness (1, 2 or 3 species). The predator species pool consisted of several invertebrates such as hunting and web-building spiders preying on an arthropod (planthopper) assemblage inhabiting *Spartina* cordgrass (Finke & Denno, 2005). The occurrence of trophic cascades was dependent on predator species richness, however the magnitude of the effect depended on whether the predators were a mixture of strict predators only, or of strict and intraguild predators combined. Higher intraguild predator richness led to antagonistic interactions among predators and dampened trophic cascades. As a consequence, it diminished herbivore suppression and therefore reduced primary productivity, indicating that predator diversity indirectly affects primary biomass production via trophic top-down control (Finke & Denno, 2005).

A similar outcome resulted from an experiment containing five species of predators, namely four carnivores (crabs, shrimp, blennies and killifish) and one omnivore (pinfish), preying on a herbivore assemblage dominated by amphipods and isopods, which in turn grazed on five species of macroalgae (Bruno & O'Connor, 2005). In an outdoor experiment, one, three or five predators were randomly assigned to mesocosms containing algae and herbivores. Predator diversity and identity had strong effects on the strength of the trophic cascade: when a generalist carnivore was present, grazers were significantly limited and algal biomass increased, whereas algal biomass decreased with increasing predator diversity, especially when the omnivorous fish was included.

Analogous to the previous examples, field observations in kelp forest ecosystems have revealed that predator diversity is negatively correlated with herbivore abundance and positively correlated with kelp density (Byrnes et al., 2006). To confirm the causality of this observation, predator richness was manipulated experimentally in kelp mesocosms. Accordingly, decreasing predator richness stimulated herbivore grazing leading to a decrease in giant kelp biomass. The underlying mechanism was not antagonistic behaviour among multiple predator species as observed in Finke & Denno (2005); instead, the herbivores changed their behaviour towards the different predators and therefore spent less time grazing.

Duffy et al. (2005) tested the effect of food chain length and trophic cascades on biomass of primary producers and consumers. In outdoor mesocosm tanks, experimental eelgrass communities were assembled in treatments of two (plants, grazers) and three (plants, grazers and predator) trophic levels. Grazer diversity was altered in the different treatments, and one omnivore predator (blue crab) was in- or excluded in order to modify the length of the trophic chain (Duffy et al., 2005). Results showed that the crabs had a strong effect on the grazers, resulting in higher algal biomass.

The results of these studies show that slight changes in predator diversity (e.g., one *versus* two species) can cascade to lower trophic levels. However, the exact outcome depends on the intraguild interactions at the predator's trophic level on one hand, and on the predator-prey response on the other. This implies that the loss of certain species at the top of the food chain can have unprecedented effects inducing fundamental changes to the whole system. In the case of seagrass and macroalgal systems, the loss of predators may induce the loss of key habitat-providing species with harmful consequences for the whole community (e.g. Estes et al., 1998).

4.2 Invasion success

Marine biological invasions are a major issue at regional and global scales. One of the major causes is the large number of transport vectors, like cargo vessels, but it has also been hypothesized that a depauperate native flora and fauna may facilitate the invasion success of exotic species (Elton, 1958). Non-native species can be a threat to regional biota due to competitive advantage, with uncontrollable consequences for the functioning of ecological

processes. To investigate the relationship between native species diversity and invasion success of exotic species, experiments were conducted on marine sessile epibenthic invertebrates of the southern rocky coasts of New England, exposed to the invasion of the colonial ascidian *Botrylloides diegensis*, native of the Pacific Ocean (Stachowicz et al., 1999, 2002). Experimental communities were composed of zero to four native species. Species richness was manipulated by transplanting juveniles or by allowing larvae to settle on 2x2 cm tiles that were (after the successful colonization) combined in a 5x5 grid (i.e., a complete plate consisted of 25 interchangeable subunits). During the course of the experiment all but the five target species were removed regularly. As a result, survival of the invasive species decreased with higher species richness, and it was hypothesized that reduced resource availability may be the underlying mechanism.
Contrary to the previous example, two observational studies on natural assemblages evidenced the opposite to what had been demonstrated in controlled experiments before. A long-term (11 months) survey of sessile invertebrates on a rocky jetty on the Tasmanian coast revealed a positive relationship between colonization rates of both native and introduced species and natural species richness (Dunstan & Johnson, 2004). Similarly, in a natural riparian assemblage, the most diverse assemblages were the most invaded by exotic plants; the same patterns were evidenced by an *in situ* experiment manipulating local diversity (Levine, 2000). A possible explanation therefore could be the different niche opportunities across varying environmental gradients (Shea & Chesson, 2002). Therefore at a larger observational scale, invasion success will be higher, given that more suitable niches are available, be it for native or exotic species, whereas invasion success would negatively relate to species richness at smaller, i.e. local scales (Shea & Chesson, 2002). On a more individual basis, the specific characteristics and functions of the resident species as well as the interactions among them and with the invasive species, rather than species richness *per se* may be the determinant for the invasion success (Dunstan & Johnson, 2004). These examples show that isolated experiments can sometimes be misleading in their outcomes. Therefore a combination of field observations, and field as well as laboratory experiments would certainly be the best approach to avoid such inherent biases. As this extensive approach is often not feasible, isolated laboratory experiments should be interpreted with caution.

5. Integrating species and genetic diversity in one experiment: an alternative approach

The preceding sections provided an overview of benthic experimental studies that alter either species or genotypic diversity separately. It has been argued that genetic diversity is only important in ecosystems that rely on or are dominated by one or a few key or habitat providing species, such as *Zostera marina* ecosystems (Hughes et al., 2008). However, all living organisms are hierarchically organized: genes make up genotypes, genotypes define populations and populations collectively constitute a species (Reusch & Hughes, 2006). If there is variation and heritability, as well as selection in ecologically important traits such as, e.g., growth rate or resistance to parasites, diversity at any level can have important ecological effects (Hughes et al., 2008). Moreover, it has been shown in terrestrial plant communities that genetic diversity can have direct consequences on species diversity: in a long-term experiment, genetically diverse communities reduced the rate at which species diversity declined (Booth & Grime, 2003), and genetic and species diversity maintained each other through mutual feedbacks (Lankau & Strauss, 2007). It is therefore fundamental to start to include changes in genetic (population) and species diversity in studies assessing the consequences of biodiversity decline.

Today, *Zostera marina* vegetations probably provide the best systems to investigate species and genetic diversity in concert (Reusch & Hughes, 2006). To our knowledge, it is also the only system where genotypic (seagrass) and species (grazer) diversity have been manipulated simultaneously (Hughes et al., 2010). One big advantage of this system is that seagrasses can reproduce clonally, which means that multiple shoots of one genotype can be handled as one unit when changing genetic (genotypic) richness just like a species is handled as one unit when changing species richness. This makes it possible to alter species and genotypic richness in one experiment. On the other hand, it is a method restricted to clonally reproducing organisms and cannot be applied to sexually reproducing animals. Another major advantage may be that seagrasses consist of relatively few but functionally diverse species. This is very valuable for the manipulation of the experimental units. However, the results cannot be extrapolated to systems with higher species richness, and questions concerning intraguild diversity may be addressed only to a limited extent. Besides, one major disadvantage of the seagrass system is that a lot of space is needed to set up outdoor (or indoor) mesocosms. Moreover, *Zostera marina* has to be available in order to establish the artificial plantings. Yet, its distribution is limited to the northern hemisphere, mainly along European and North American coasts.

As an alternative approach to test effects of species and genetic diversity on ecosystem functioning, we present a setup allowing experimental manipulation of multitrophic assemblages in very small and easy-to-replicate laboratory microcosms (Figure 1a). These comprise small amounts of sediment containing a benthic microbial food web of diatoms (i.e. primary producers), prokaryotes (i.e. decomposers) and nematodes and/or harpacticoid copepods (i.e. grazers of prokaryotes and of diatoms, as well as predators of other nematodes and copepods) (Figure 1b). These organisms dominate all soft sedimentary habitats such as beaches and mudflats around the world. They typically attain high local diversity, species numbers in 10 ml of sediment typically ranging in the order of hundreds to thousands, and several tens for prokaryotes, diatoms and nematodes, respectively. They are highly relevant to basic benthic ecosystem processes such as the decomposition of organic matter and the mineralization of nutrients. They respond more rapidly to environmental changes than most macrofauna (Bolam et al., 2002).

Former experiments with similar setups have proven to be very useful to study benthic processes. De Mesel et al. (2003, 2004, 2006) studied the impact of bacterivore nematodes on the bacterial community and the decomposition of cordgrass leaves. They found that four different bacterivore nematode species have a significant top-down effect on the structure of the bacterial community (De Mesel et al., 2004). However, species richness within the guild of bacterivore nematodes did not enhance decomposition rates, rather the process depended on species identity and on unexpected inhibitory and facilitative interactions among species within the same guild (De Mesel et al., 2003, 2006). Hubas et al. (2010) investigated the influence of microbial organisms, i.e., bacteria, diatoms and bacterivore nematodes, on the production of extracellular polymeric substances (EPS), a major sediment-stabilizing product (Stal, 2010). The results evidenced that the presence of bacterivore nematodes had a positive impact on microbial abundance and EPS production, which was highest at the highest community complexity, i.e., involving nematodes, diatoms and bacteria (Hubas et al., 2010). Dos Santos (2009) studied bottom-up and top-down controls of nematodes on the bacterial community. They established closed microcosms with bacterivore nematode species that colonize decaying organic matter, revealing that intraguild species interactions depend on food availability (Dos Santos et al., 2009). They also worked with semi-closed microcosms of sandy beach sediments with more natural diversity levels (up to 24 species). Results showed that bacterivore nematodes did not have a more pronounced effect on the structure of the bacterial community than nematodes from other trophic guilds; however, nematode species richness (across trophic levels) did (Dos Santos, 2009). Moreover, processes such as carbon production increased with species richness, and the most species-rich nematode assemblage had a positive effect on bacterial diversity (Dos Santos, 2009).

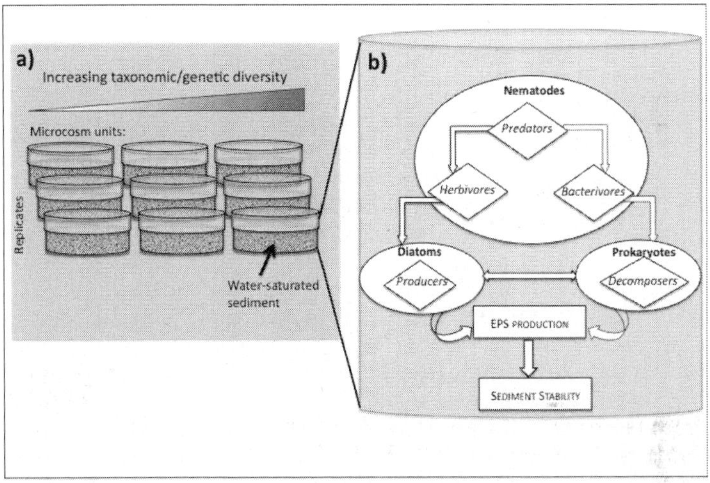

Fig. 1. The experimental system. a) The experimental unit. Microcosms can be setup in a variety of containers; for the illustration we chose a relatively small glass or plastic jar with a thin, water-saturated layer of sediment, covered by a plastic film or lid to avoid desiccation. Experimental microcosms contain increasing levels of nematode taxonomic and/or genetic diversity b) A schematic view of the benthic food web contained in the sediment of the microcosm. Predatory nematodes prey on herbivorous and bacterivore nematodes, which, in turn, graze on diatoms and prokaryotes respectively. In different experimental units, nematode genetic and species diversity are modified. The effects of the different diversity levels on the functioning of the food web are measured in terms of the production of extracellular polymeric substances (EPS) and sediment stability.

To address the BEF subject at different organizational levels simultaneously (i.e., species *and* genetic diversity) across different trophic guilds (i.e., grazers, bacterivores and predators), we follow the microcosm approach as applied in the study by Hubas et al. (2010). We suggest assessing components affecting directly the physical integrity of the habitat as a proxy of the functionality of our microcosms, i.e. the production of EPS and its effect on sediment stability (Figure 1b). However, other functions, such as community respiration, prokaryote enzymatic activities or the decomposition of a specific substrate, may be assessed in the same setup. Microbial EPS are increasingly recognized as a major stabilising factor. They are an ubiquitous component of aquatic ecosystems composed primarily of carbohydrates and proteins. They have multiple roles and functions: attachment to substrata, flotation and locomotion, feeding, protection against environmental factors such as desiccation, UV radiation and pollution, as well as the development of biofilms (Decho, 1990). Perhaps the most striking feature of these compounds is their ability to bind sediment grains together, enhancing their cohesive strength and thereby increasing resistance to erosion. Particularly diatoms are the major ecosystem engineers of this sediment bio-stabilisation (Stal, 2010), and prokaryotes are increasingly recognised as playing a key role in enhancing sediment surface adhesion and cohesive strength (Gerbersdorf et al., 2008, 2009; Lubarsky et al., 2010). The advantage of this alternative approach is that it offers the possibility of testing both effects of species and genetic diversity on ecosystem functioning.

5.1 Methodology

5.1.1 Microcosm setup

To setup the microcosm, a variety of small containers can be suitable, depending on the exact aim of the study. First, a specified amount of non-cohesive sedimentary substrate needs to be placed on the bottom. For example, acid-washed marine sand (Hubas et al., 2010) or non-cohesive glass beads (Lubarsky et al., 2010) can be used. The sediment needs to be fully hydrated by sterile seawater of a specific salinity (which depends on the organisms). Whether or not the seawater needs to cover the sediment depends on the specific aims of the study. Prokaryote and diatom cultures can be established by sampling sediment surface (0 - 5 mm) from natural tidal flats, and re-suspending the sediment slurry in culture media. Alternatively, or additionally, prokaryote and diatom strains can be obtained from established culture collections. Microbial assemblages should then be incubated for several weeks (e.g., Ribalet et al., 2008). The resulting assemblages can then be used for the experiment. An appropriate culture medium, which favours algal rather than bacterial growth, should be applied. F/2 medium is, for instance, widely used for growing marine algae, especially diatoms (Guillard, 1975). If the use of antibiotics cannot be avoided, a proper test of the effect of each antibiotic on the natural assemblage culture should be conducted.

Finally, nematodes can be placed into the microcosms, according to the experimental groups. There are two different approaches in order to obtain different taxonomic (i.e. species richness) diversity levels across trophic groups of nematodes: either from the field or from cultures. For the former, nematode assemblages can be sampled along diversity gradients from, e.g., an intertidal sandy beach (Gingold et al., 2010) or from different beaches (Dos Santos, 2009). They are extracted live through decantation over a 32µm sieve to obtain stock suspensions of different diversity levels. For the latter, different nematode species can be raised under artificial conditions (Moens & Vincx, 1998). Unfortunately, only a limited number of marine nematode species have hitherto been maintained in permanent monospecific cultures, and most of these belong to the same feeding guild, i.e. bacterivores. However, a majority of species from intertidal sediments can be easily extracted from freshly collected sediment and remain active in a variety of laboratory incubation conditions for several weeks (Moens & Vincx, 1998).

In order to establish microcosms containing nematodes of different genetic diversity, lab cultures and field-collected specimens of the same bacteria-feeding species from which permanent cultures exist can be used. Bacteria feeding nematode species such as *Rhabditis marina, Diplolaimelloides meyli, D. oschei,* and *Diplolaimella dievengatensis* have been cultured for over 15 years now (T. Moens, unpublished data). Therefore, the premise is that cultured nematodes have low genetic diversity due to inbreeding, unlike natural populations, which are expected to exhibit higher genetic diversity levels. This approach implies that the level of genetic difference has to be assessed before the experiment, which in view of the substantial cryptic diversity (e.g., Derycke et al., 2005) is not necessarily evident. Alternatively, contrasting genetic diversity levels can be obtained experimentally with cultures of selected nematode species. Cultures of species of the *Rhabditis marina* species complex, for instance, can be raised from single gravid females, yielding highly inbred lineages, which can be maintained easily for many generations. It is therefore possible to raise a culture collection of many inbred lineages and to establish experimental populations of very low to high genetic diversity. The advantage of this approach is that similar to the study by Gamfeldt & Kallstrom (2007), "families" can be obtained, and the level of genetic diversity (i.e. the "number of families") for each experimental group can be determined *a priori*.

5.1.2 Measuring EPS production and sediment stability

The main ecosystem function we concentrate on in this approach is measured in terms of EPS production (concentration and composition), and the stability of the sediments. To assess EPS concentration and composition, sediment samples are mixed with distilled water. Exopolymers are very diverse and complex molecules that are found in a wide variety of forms and structures in aquatic ecosystems. The selection of the protocol for their extraction, purification and subsequently for the measurement of their concentration in any sample will depend ultimately on the type of exopolymer studied (e.g., Panagiatopoulos & Sempéré, 2005; Danovaro, 2010). Exopolymers can be found dissolved in the water column or as gels in biofilms and aggregates so that their extraction and quantification is a primary step in assessing their potential availability (Decho, 1990). A description and evaluation of different methods and protocols is well beyond the scope of this paper, hence we will only briefly mention methods that are commonly used in the literature and which offer an acceptable ratio of information obtained to amount of time invested.

The water-extractable fraction (i.e. dissolved or colloidal EPS which are already present in pore-water and unbound to sediment grains) can be analyzed for carbohydrates and proteins following the phenolsulphuric acid protocol (DuBois et al., 1956) and the modified Lowry procedure (Raunkjaer et al., 1994). Carbohydrate and protein concentrations can be measured with a UV/VIS spectrophotometer and their concentrations are deduced from calibration curves assessed from glucose and bovine serum albumin, respectively.

Sediment stability can be estimated through the measurement of the erosion shear stress of the biofilm. Typically, it is possible to measure this using flumes (Jonsson et al., 2006) or *in situ* devices such as the Cohesive Strength Meter (CSM). This device has been used extensively in marine ecosystems since the pioneering work of Paterson (1989) and is an efficient way to estimate sediment erodibility. Mechanical properties of the biofilms in terms of surface adhesion have been developed relatively recently. The method is based on the magnetic attraction of magnetic particles (Magnetic Particle Induction - MagPI; Larson et al., 2009) and is both relatively easy to set up and inexpensive. Specifically, a known volume of ferromagnetic fluorescent particles is spread onto a defined area of the sediment surface. The particles are then recaptured by an overlying electromagnet and the force (magnetic flux) needed to retrieve the particles is determined as a measure of the retentive capacity of the substratum, a proxy for adhesion. This method is suitable for sensitive recording of changes of the surface adhesion of sediments or biofilms.

5.1.3 Assessing community composition

We suggest assessing nematode, prokaryote, and diatom community diversity and composition at a genetic and taxonomic level. Most studies identify nematode species morphologically, nematodes first being mounted on permanent slides (Platt & Warwick, 1988). Species are identified with a microscope, using pictorial (Platt & Warwick 1983, 1988; Warwick et al., 1998) taxonomic keys and primary literature on species descriptions, which is becoming increasingly available in online public databases such as NeMys (Deprez et al., 2005). Morphology based identification of diatoms is achieved by fixing the sampling cores in 4% glutaraldehyde and then embedding in Naphrax after chemically removing organic matter using potassium permanganate. The species composition of the microalgal community is assessed by light microscopy using identification keys (Krammer & Lange-Bertalot, 1986-1991; Lange-Bertalot, 1997; Pankow, 1990; Simonsen, 1962; Underwood et al., 1998; Witkowski et al., 2000). However, species diversity can be assessed using a metagenetic approach as well. The three sets of taxa (nematodes, prokaryotes and diatoms) may be pyrosequenced after distinct sampling procedures. Nematodes are extracted from sediment by sieving (45 μm mesh) and floatation techniques and genotyped using both small and large rDNA subunits (Creer et al., 2010; Porazinska et al., 2009, 2010) and/or mitochondrial genes such as part of the cytochrome oxidase subunit I gene (Derycke et al., 2010). Diatoms are collected on a filter before DNA extraction and amplification using the small rDNA subunit (Cuvelier et al., 2010; Medinger et al., 2010; Quaiser et al., 2010). Prokaryotes are assessed from direct DNA extractions of sediment samples and genotyped using 16S rDNA (Coolon et al., 2010; Xu, 2006). When studying well-documented communities, where sequence data is available for a majority of species, such an approach can yield information on species composition, relative abundances and diversity. In the absence of such a reference database (currently unavailable for any marine assemblage, but this will undoubtedly change in the near future), it can still provide an estimate of diversity in terms of both species richness and evenness. Metagenetic analyses yield information on the different genotypes present, and thus allow simultaneous assessment of species as well as intraspecific diversity, pending proper knowledge on barcoding gaps in the taxa of interest.

5.1.4 Advantages/disadvantages of our system

As a model system for BEF research, our setup combines many advantages. One major advantage from the practical point of view may be that it can be manipulated in small containers and yet contain diversity levels representative of natural communities. The microcosms can easily be incubated under controlled conditions and sufficiently replicated. Nematodes, diatoms and prokaryotes are key players in almost any coastal or stream soft sediment, performing similar essential ecosystem functions across a range of habitats. Therefore they are easily collectable, either at a sandy beach or tidal flat at seashore or from a stream or lake when working with freshwater organisms. A whole experimental setup fits in any average-sized laboratory. If enough space is available, it can even be expanded to bigger mesocosm units including macrofauna as an additional trophic and functional level, which is likely to impact on nematodes, diatoms and prokaryotes through trophic as well as physical interactions (Austen & Widdicombe, 1998; Braeckman et al., 2011). Other extensions are also possible, for example a temperature treatment can be applied to simulate temperature rise as a consequence of climate change. Although the duration of such a short experiment would not reflect absolute temporal scales of climate change, it may well reflect and therefore be representative for the rates of temperature change long-lived organisms might experience (Petchey et al., 1999). Knowing that genetic diversity enhanced resistance of seagrass *Zostera marina* to climate extremes (Reusch et al., 2005), it would be particularly interesting to plan hypotheses involving genetic diversity of nematodes and increased temperature of the environment.

A further advantage of our system is that it includes different trophic levels, from decomposers (prokaryotes) and producers (diatoms) to grazers (bacterivorous and herbivorous nematodes) and their predators (predatory nematodes), some of the latter being omnivorous rather than strictly predacious (Moens et al., 2004). Moreover, for some setups, we can integrate species of a natural community and not only the dominant species of a given habitat. This is important because even weak interactions can have important stabilizing effects on communities (Berlow, 1999). Contrary to the seagrass system, which alters diversity at the primary-producer level, we can assess the effect of species and genetic diversity also at the consumer level. This is important, as it has been suggested that the stronger top-down control in the sea relative to terrestrial habitats (Shurin et al., 2002) implies that in marine systems ecosystem functions such as primary production may be influenced more by herbivores and predators than by plant diversity (Duffy, 2003; Paine, 2002). In order to track and quantify trophic pathways, pulse-chase experiments which label a particular component and trace transfer to consumers can be designed (Middelburg et al., 2000; Moens et al., 2002; Van Oevelen et al., 2006). Such a characterization and quantification of the relationships among the different components of the community would be an additional improvement, since experimental research on multi-level food webs should not only assess the consequences of species richness and identity on ecosystem processes, but also evaluate trophic cascades and the distribution of interaction strengths within natural communities and how they change with community composition (Duffy, 2002).

However, there are probably three major disadvantages that have to be taken in account when planning the experiments: first, only a limited number of species of nematodes can be cultured with the present methods. Second, the identification of nematodes and diatoms needs considerable expertise, especially when working with field samples containing many different species. And third, the design essentially relies on closed microcosms, hence excluding the role of immigration and emigration. Moreover, by bottling aqueous sediments, a system is created which accumulates end products of decomposition and depletes oxygen. As a consequence, the microcosms can only be used for a limited duration, after which treatment effects are likely to become confounded by bottling effects.

6. Conclusions

Our experimental system provides a good alternative to investigate effects of species and genetic diversity in concert. It is easy to setup and contains organisms that are easily accessible. Moreover, it provides the possibility to be extended and address current challenges of BEF experiments. One of these challenges consists in that future BEF studies must go beyond experiments relating some selected species, singly and in combination, to some ecosystem process. Rather, they must address environmental heterogeneity in space or time, which can be captured in long-term studies on one hand, and working with natural communities on the other. Our experimental units as described here are closed systems, however, they can possibly be converted to semi-closed systems and be setup outdoors. Like this they would not suffer from the bottling effect, therefore allowing longer-term studies, and capturing the natural variation of environmental conditions. The exact methodology for this has still to be established, but such a system may yield promising results. As the experiment of, e.g., Cardinale & Palmer (2002), very elegantly shows, disturbance may dampen strong single-species effects. It is more probable that changing environmental conditions allow resource partitioning among species, and that under natural conditions, such disproportionate key species effects are greatly reduced allowing stronger diversity effects *per se* (Stachowicz et al., 2008). Moreover, there are many possibilities to include more organisms, species and/or trophic levels and functional groups in an extended version of our setup. Especially when including macrofauna, the system would more accurately reflect the complex interactions among species and trophic levels as they occur in nature.

Our experimental approach will contribute to the understanding of indirect mechanisms leading to biodiversity effects on ecosystem processes. The integrative insight in the interactions among the different levels and the consequences that changes in one level may have on the others and on the system as a whole, may offer the possibility for adequate decisions about conservational priorities. As shown by the experiment by Lankau & Strauss (2007), genetic and species diversity are inherently linked to one another. Therefore, conservation of species diversity may also depend on the maintenance of the processes that sustain genetic diversity. Although BEF studies are designed to understand general patterns and processes, rather than provide solutions to applied conservation problems (Duffy, 2009), they have provided fundamental inputs. The many studies on plant systems for example provided the background at Tijuana River National Estuarine Research Reserve, where a functional ecosystem needed to be restored (Callaway et al., 2003). Just as in many plant studies experimental plantings of differing species diversity were set up, and as a result, biomass and accumulation of nitrogen increased with increasing species diversity (Callaway et al., 2003). As such, our experimental system may offer valuable input for new conservation strategies.

We think that the integrative experimental system as we propose it can be a promising direction for future BEF studies focusing on benthic systems and potential applications for conservational efforts. Today's conservational effort concentrates mainly on identifying and protecting hotspots of species diversity. However, results of such integrative studies will show if the consideration of all organizational levels, as well as the functioning of the whole system, may provide an alternative approach in order to take the best conservation decision under given circumstances. Although BEF experiments conducted to date are likely to underestimate the importance of biodiversity to ecosystem functioning and the provision of ecosystem services in the real world (Duffy, 2009), they could offer constructive inputs for the sustainable management of entire communities and ecosystems.

7. Acknowledgements

RG thanks Tiago José Pereira for his generous help with the compilation of the literature, as well as Ian W. King and Tom Crystal for proof-reading earlier drafts.

8. References

Bell, G. (1991). *The Ecology and genetics of fitness in Chlamydomonas. 4. The properties of mixtures of genotypes of the same species*. Evolution 45, pp. 1036-1046

Bell, T.; Newman, J. A.; Silverman, B. W.; Turner, S. L. & Lilley, A. K. (2005). *The contribution of species richness and composition to bacterial services.* Nature 436, pp. 1157-1160

Benedetti-Cecchi, L. (2004). *Increasing accuracy of causal inference in experimental analyses of biodiversity.* Functional Ecology 18, pp. 761-768

Berlow, E. L. (1999). *Strong effects of weak interactions in ecological communities.* Nature 398, pp. 330-334

Bolam, S. G.; Fernandes, T. F. & Huxham, M. (2002). *Diversity, biomass, and ecosystem processes in the marine benthos.* Ecological Monographs 72, pp. 599-615

Booth, R. E. & Grime, J. P. (2003). *Effects of genetic impoverishment on plant community diversity.* Journal of Ecology 91, pp. 721-730

Braeckman, U.; Van Colen, C.; Soetaert, K.; Vincx, M. & Vanaverbeke, J. (2011) *Contrasting macrobenthic activities differentially affect nematode density and diversity in a shallow subtidal marine sediment.* Marine Ecology Progress Series 422, pp. 179-191

Brown, A. C. & McLachlan, A. (2002). *Sandy shore ecosystems and the threats facing them: some predictions for the year 2025.* Environmental Conservation 29, pp. 62-77

Bruno, J. F. & O'Connor, M. I. (2005). *Cascading effects of predator diversity and omnivory in a marine food web.* Ecology Letters 8, pp. 1048-1056

Byrnes, J.; Stachowicz, J. J.; Hultgren, K. M.; Hughes, A. R.; Olyarnik, S. V. & Thornber, C. S. (2006). *Predator diversity strengthens trophic cascades in kelp forests by modifying herbivore behaviour.* Ecology Letters 9, pp. 61-71

Callaway, J. C.; Sullivan, G. & Zedler, J. B. (2003). *Species-rich plantings increase biomass and nitrogen accumulation in a wetland restoration experiment.* Ecological Applications 13, pp. 1626-1639

Cardinale, B. J. & Palmer, M. A. (2002). *Disturbance moderates biodiversity-ecosystem function relationships: experimental evidence from Caddisflies in stream mesocosms.* Ecology 83, pp.1915-1927

Cardinale, B. J.; Palmer, M. A. & Collins, S. L. (2002). *Species diversity enhances ecosystem functioning through interspecific facilitation.* Nature 415, pp. 426-429

Constanza, R.; d'Arge, R.; de Groot, R.; Farber, S.; Grasso, M.; Hannon, B.; Limburg, K.; Naeem, S.; O'Neill, R. V.; Paruelo, J.; Raskin, R. G.; Sutton, P. & van den Belt, M. (1997). *The value of the world's ecosystem services and natural capital.* Nature 387, pp. 253-260

Coolon, J. D.; Jones, K. L.; Narayanan, S. & Wisely, S. M. (2010). *Microbial ecological response of the intestinal flora of Peromyscus maniculatus and P. leucopus to heavy metal contamination.* Molecular Ecology 19, pp. 67-80

Covich, A. P.; Austen, M. C.; Barlocher, F.; Chauvet, E.; Cardinale, B. J.; Biles, C. L.; Inchausti, P.; Dangles, O.; Solan, M.; Gessner, M. O.; Statzner, B. & Moss, B. (2004). *The role of biodiversity in the functioning of freshwater and marine benthic ecosystems.* Bioscience 54, pp. 767-775

Creer, S.; Fonseca, V. G.; Porazinska, D. L.; Giblin-Davis, R. M.; Sung, W.; Power, D. M.; Packer, M.; Carvalho, G. R.; Blaxter, M. L.; Lambshead, P. J. D. & Thomas, W. K. (2010). *Ultrasequencing of the meiofaunal biosphere: practice, pitfalls and promises.* Molecular Ecology 19, pp. 4-20

Cuvelier, M. L.; Allen, A. E.; Monier, A.; McCrow, J. P.; Messie, M.; Tringe, S. G.; Woyke, T.; Welsh, R. M.; Ishoey, T.; Lee, J.-H.; Binder, B. J.; DuPont, C. L.; Latasa, M.; Guigand, C.; Buck, K. R.; Hilton, J.; Thiagarajan, M.; Caler, E.; Read, B.; Lasken, R. S.; Chavez, F. P. & Worden, A. Z. (2010). *Targeted metagenomics and ecology of globally important uncultured eukaryotic phytoplankton*. Proceedings of the National Academy of Sciences USA 107, pp. 14679-14684

Danovaro, R.; Gambi, C.; Dell'Anno, A.; Corinaidesi, C.; Fraschetti, S.; Vanreusel, A.; Vincx, M. & Gooday, A. J. (2008). *Exponential decline of deep-sea ecosystem functioning linked to benthic biodiversity loss*. Current Biology 18, pp. 1-8

De Mesel, I., Derycke, S., Swings, J., Vincx, M. & Moens, T. (2003). *Influence of bacterivorous nematodes on the decomposition of cordgrass*. Journal of Experimental Marine Biology and Ecology 296, pp. 227-242

De Mesel, I.; Derycke, S.; Moens, T.; Van der Gucht, K.; Vincx, M. & Swings, J. (2004). *Top-down impact of bacterivorous nematodes on the bacterial community structure: a microcosm study*. Environmental Microbiology 6, pp. 733-744

De Mesel, I.; Derycke, S.; Swings, J.; Vincx, M. & Moens, T. (2006). *Role of nematodes in decomposition processes: Does within-trophic group diversity matter?* Marine Ecology Progress Series 321, pp. 157-166

Decho, A. W. (1990). *Microbial exopolymer secretions in ocean environments: their role(s) in food webs and marine processes*. Oceanography and Marine Biology: an Annual Review 28, pp. 73-153

Derycke, S.; Remerie, T.; Vierstraete, A.; Backeljau, T.; Vanfleteren, J.; Vincx, M. & Moens, T. (2005). Mitochondrial DNA variation and cryptic speciation within the free-living marine nematode *Pellioditis marina*. Marine Ecology Progress Series 300, pp. 91-103

Derycke, S.; Vanaverbeke, J.; Rigaux, A.; Backeljau, T. & Moens, T. (2010). *Exploring the Use of Cytochrome Oxidase c Subunit 1 (COI) for DNA Barcoding of Free-Living Marine Nematodes*. Plos One 5, pp. e13716

Deprez, T. & all (2005). NeMys. World Wide Web electronic publication. www.nemys.ugent.be, version (5/2011)

dos Santos, G. A. P. (2009). *Top-down and bottom-up controls on populations and assemblages of marine nematodes, and their effects on benthic ecosystem functioning: and experimental approach*. PhD Thesis, Ghent University

dos Santos, G. A. P.; Derycke, S.; Genevois, V. G. F.; Coelho, L.; Correia, M. T. S. & Moens, T. (2009). *Interactions among bacterial-feeding nematode species at different levels of food availability*. Marine Biology 156, pp. 629-640

Downing, A. L. & Leibold, M. A. (2002). *Ecosystem consequences of species richness and composition in pond food webs*. Nature 416, pp. 837-841

DuBois, M.; Gilles, K. A.; Hamilton, J. K.; Rebers, P. A. & Smith, F. (1956). *Colorimetric method for determination of sugars and related substances*. Analytical Chemistry 28, pp. 350-356

Duffy, J. E.; Macdonald, K. S.; Rhode, J. M. & Parker, J. D. (2001). *Grazer diversity, functional redundancy, and productivity in seagrass beds: an experimental test*. Ecology 82, pp. 2417–2434

Duffy, J. E. (2002). *Biodiversity and ecosystem function: the consumer connection*. Oikos 99, pp. 201-219

Duffy, J. E. (2003). *Biodiversity loss, trophic skew and ecosystem functioning*. Ecology Letters 6, pp. 680-687

Duffy, J. E.; Richardson, J. P. & Canuel, E. A. (2003). *Grazer diversity effects on ecosystem functioning in seagrass beds*. Ecology Letters 6, pp. 637-645

Duffy, J. E.; Richardson, J. P. & France, K. E. (2005). *Ecosystem consequences of diversity depend on food chain length in estuarine vegetation*. Ecology Letters 8, pp. 301-309

Duffy, J. E.; Cardinale, B. J.; France, K. E.; McIntyre, P. B.; Thébault, E. & Loreau, M. (2007). *The functional role of biodiversity in ecosystems: incorporating trophic complexity*. Ecology Letters 10, pp. 522-538

Duffy, J. E. (2009). *Why biodiversity is important to the functioning of real-world ecosystems*. Frontiers in Ecology and Environment 7, pp. 437-444

Dunstan, P. K. & Johnson, C. R. (2004). *Invasion rates increase with species richness in a marine epibenthic community by two mechanisms*. Oecologia 138, pp. 285-292

Elton, C. S. (1958). *The ecology of invasions by animals and plants*. Methuen, London, UK

Emmerson, M. C. & Raffaelli, D. G. (2000). *Detecting the effects of diversity on measures of ecosystem function: experimental design, null models and empirical observations*. Oikos 91, pp. 195-203

Emmerson, M. C.; Solan, M.; Emes, C.; Paterson, D. M. & Raffaelli, D. (2001). *Consistent patterns and the idiosyncratic effects of biodiversity in marine ecosystems*. Nature 411, pp. 73-77

Engelhardt, K. A. M. & Ritchie, M.E . (2001). *Effects of macrophyte species richness on wetland ecosystem functioning and services*. Nature 411, pp. 687-689

Estes, J. A.; Tinker, M. T.; Williams, T. M. & Doak, D. F. (1998). *Killer whale predation on sea otters linking oceanic and nearshore ecosystems*. Science 282, pp. 473-476

Finke, D. L. & Denno, R. F. (2005). *Predator diversity and the functioning of ecosystems: the role of intraguild predation in dampening trophic cascades*. Ecology Letters 8, pp. 1299-1306

Gamfeldt, L.; Wallén, J.; Jonsson, P. R.; Berntsson, K. M. & Havenhand, J. N. (2005). *Increasing intraspecific diversity enhances settling success in a marine invertebrate*. Ecology 86, pp. 3219-3224

Gamfeldt, L. & Kallstrom, B. (2007). *Increasing intraspecific diversity increases predictability in population survival in the face of perturbations*. Oikos 116, pp. 700-705

Gerbersdorf, S. U.; Manz, W. & Paterson, D. M. (2008). *The engineering potential of natural benthic bacterial assemblages in terms of the erosion resistance of sediments*. FEMS Microbial Ecology 66, pp. 282-294

Gerbersdorf, S. U.; Bittner, R.; Lubarsky, H.; Manz, W. & Paterson, D. (2009). *Microbial assemblages as ecosystem engineers of sediment stability*. Journal of Soils and Sediments 9, pp. 640-652

Gingold, R.; Mundo-Ocampo, M.; Holovachov, O. & Rocha-Olivares, A. (2010). *The role of habitat heterogeneity in structuring the community of intertidal free-living marine nematodes* Marine Biology 157, pp. 1741-1753

Glibert, P. M.; Harrison, J.; Heil, C. & Seitzinger, S. (2006). *Escalating worldwide use of urea – a global change contributing to coastal eutrophication*. Biogeochemistry 77, pp. 441-463

Grosholz, E. (2002). *Ecological and evolutionary consequences of coastal invasions*. Trends in Ecology and Evolution 17, pp. 22-27

Guillard, R.R.L. (1975). *Culture of phytoplankton for feeding marine invertebraes.* In: Smith, W. L. & Chanley, M. H. (eds.) *Culture of Marine Invertebrate Animals.* Plenum Press, New York, USA

Hector, A.; Schmid, B.; Beierkuhnlein, C.; Caldeira, M. C.; Diemer, M.; Dimitrakopoulos, P. G.; Finn, J. A.; Freitas, H.; Giller, P. S.; Good, J.; Harris, R. M.; Hogberg, P.; Huss-Danell, K.; Joshi, J.; Jumpponen, A.; Korner, C.; Leadley, P. W.; Loreau, M.; Minns, A.; Mulder, C. P. H.; O'Donovan, G.; Otway, S. J.; Pereira, J. S.; Prinz, A.; Read, D. J.; Scherer-Lorenzen, M.; Schulze, E. D.; Siamantziouras, A. S. D.; Spehn, E. M.; Terry, A. C.; Troumbis, A. Y.; Woodward, F. I.; Yachi, S. & Lawton, J. H. (1999). *Plant diversity and productivity experiments in European grasslands.* Science 286, pp. 1123-1127

Heip, C.; Vincx, M. & Vranken, G. (1985). *The Ecology of Marine Nematodes.* Oceanography and Marine Biology: An Annual Review 23, pp. 399-489

Hooper, D. U.; Chapin, F. S.; Ewel, J. J.; Hector, A.; Inchausti, P.; Lavorel, S.; Lawton, J. H.; Lodge, D. M.; Loreau, M.; Naeem, S.; Schmid, B.; Setälä, H.; Symstad, A. J.; Vandermeer, J. & Wardle, D. A. (2005). *Effects of biodiversity on ecosystem functioning: a consensus of current knowledge.* Ecological Monographs 75, pp. 3-35

Hubas, C.; Sachidhanandam, C.; Rybarczyk, H.; Lubarsky, H.; Rigaux, A.; Moens, T. & Paterson, D. M. (2010). *Bacterivorous nematodes stimulate microbial growth and exopolymeric production in marine sediments microcosms.* Marine Ecology Progress Series 419, pp. 85-94

Hughes, A. R. & Stachowicz, J. J. (2004). *Genetic diversity enhances the resistance of a seagrass ecosystem to disturbance.* Proceedings of the National Acadamy of Sciences USA 101, pp. 8998-9002

Hughes, A. R.; Inouye, B. D.; Johnson, M. T. J.; Underwood, N. & Vellend, M. (2008). *Ecological consequences of genetic diversity.* Ecology Letters 11, pp. 609-623

Hughes, A. R. & Stachowicz, J. J. (2009). *Ecological impacts of genotypic diversity in the clonal seagrass Zostera marina.* Ecology 90, pp. 1412-1419

Hughes, A. R. & Best, R. J., Stachowicz, J. J. (2010). *Genotypic diversity and grazer identity interactively influence seagrass and grazer biomass.* Marine Ecology Progress Series 403, pp. 43-51

Hughes, A. R. & Stachowicz, J. J. (2011). *Seagrass genotypic diversity increases disturbance response via complementarity and dominance.* Journal of Ecology 99, pp. 445-453

Humphrey, S.; Loh, J. & Goldfinger, S. (2008). *Living Planet Report.* WWF, Gland, Switzerland

Jonsson, M. & Malmqvist, B. (2000). *Ecosystem process rate increases with animal species richness: evidence from leaf-eating, aquatic insects.* Oikos 89, pp. 519-523

Jonsson, M. & Malmqvist, B. (2003). *Importance of species identity and number for process rates within different stream invertebrate functional feeding groups.* Journal of Animal Ecology 72, pp. 453-459

Jonsson, P. R.; van Duren, L. A.; Amielh, M.; Asmus, R.; Aspden, R. J.; Daunys, D.; Friedrichs, M.; Friend, P. L.; Olivier, F.; Pope, N.; Precht, E.; Sauriau, P. G. & Schaaff, E. (2006). *Making water flow: a comparison of the hydrodynamic characteristics of 12 different benthic biological flumes.* Aquatic Ecology 40, pp. 409-438

Krammer, K. & Lange-Bertalot, H. (1986-1991). *Süsswasserflora von Mitteleuropa*. Ettl, H.; Gerloff, H.; Heynig, H. & Mollenhauer, D. (eds.) Jena, Stuttgart

Lange-Bertalot, H. (1997). *A first ecological evaluation of the diatom flora in Central Europe*. Lauterbornia, pp. 117-123

Lankau, R. A. & Strauss, S. Y. (2007). *Mutual feedbacks maintain both genetic and species diversity in a plant community*. Science, pp. 1561-1563

Larson, F.; Lubarsky, H.; Gerbersdorf, S. U. & Paterson, D. (2009). *Surface adhesion measurements in aquatic biofilms using magnetic particle induction: MagPI*. Limnology and Oceanography: Methods 7, pp. 490-497

Levine, J. M. (2000). *Species diversity and biological invasions: Relating local process to community pattern*. Science 288, pp. 852-854

Loreau, M.; Naeem, S.; Inchausti, P.; Bengtsson, J.; Grime, J. P.; Hector, A.; Hooper, D. U.; Huston, M. A.; Raffaelli, D.; Schmid, B.; Tilman, D. & Wardle, D. A. (2001). *Biodiversity and ecosystem functioning: Current knowledge and future challenges*. Science 294, pp. 804-808

Lubarsky, H. V.; Hubas, C.; Chocholek, M.; Larson, F.; Manz, W.; Paterson, D. M. & Gerbersdorf, S. U. (2010). *The stabilisation potential of individual and mixed assemblages of natural bacteria and microalgae*. PLOS One 5, pp. e13794

McGrady-Steed, J.; Harris, P. M. & Morin, P. J. (1997). *Biodiversity regulates ecosystem predictability*. Nature 390, pp. 162-165

Medinger, R.; Nolte, V.; Pandey, R. V.; Jost, S.; Ottenwalder, B.; Schlotterer, C. & Boenigk, J. (2010). *Diversity in a hidden world: potential and limitation of next-generation sequencing for surveys of molecular diversity of eukaryotic microorganisms*. Molecular Ecology 19, pp. 32-40

Middelburg, J. J.; Barranguet, C.; Boschker, H. T. S.; Herman, P. M. J.; Moens, T. & Heip, C. H. R. (2000). *The fate of intertidal microphytobenthos carbon: An in situ C-13-labeling study*. Limnology and Oceanography 45, pp. 1224-1234

Moens, T. & Vincx, M. (1998). *On the cultivation of free-living marine and estuarine nematodes*. Helgoländer Meeresuntersuchungen 52, pp. 115-139

Moens, T.; Luyten, C.; Middelburg, J. J.; Herman, P. M. J. & Vincx, M. (2002). *Tracing organic matter sources of estuarine tidal flat nematodes with stable carbon isotopes*. Marine Ecology Progress Series 234, pp. 127-137

Moens, T.; Yeates, G. W. & De Ley, P. (2004). *Use of carbon and energy sources by nematodes*. International Congress of Nematology, Tenerife, Spain. In: Cook, R. & Hunt, D. J, (eds.). Brill Leiden-Boston, pp. 529-545

Naeem, S. & Li, S. B. (1997). *Biodiversity enhances ecosystem reliability*. Nature 390, pp. 507-509

Naeem, S.; Hahn, D. R. & Schuurman, G. (2000). *Producer-decomposer co-dependency influences biodiversity effects*. Nature 403, pp. 762-764

Páez-Osuna, F.; Guerrero-Galván, S. R. & Ruiz-Fernandéz, A. C. (1999). *Discharge of nutrients from shrimp farming to coastal waters of the Gulf of California*. Marine Pollution Bulletin 38, pp. 585-592

Paine, R. T. (2002). *Trophic control of production in a rocky intertidal community*. Science 296, pp. 736-739

Panagiotopoulos, C. & Sempéré, R. (2005). *Analytical methods for the determination of sugars in marine samples: A historical perspective and future directions*. Limnology and Oceanography: Methods 3, pp. 419-454

Pankow, H. (1990). *Ostsee-Algenflora*. Gustav Fischer Verlag, Jena

Paterson, D. M. (1989). *Short-term changes in the erodibility of intertidal cohesive sediments related to the migratory behavior of epipelic diatoms*. Limnology and Oceanography 34, pp. 223-234

Pauly, D.; Christensen, V.; Dalsgaard, J.; Froese, R. & Torres, F. J. (1998). *Fishing down marine food webs*. Science 279, pp. 860-867

Pease, C. J.; Johnston, E. L. & Poore, A. G. B. (2010). *Genetic variability in tolerance to copper contamination in a herbivorous marine invertebrate*. Aquatic Toxicology 99, pp. 10-16

Petchey, O. L.; McPhearson, P. T.; Casey, T. M. & Morin, P. J. (1999). *Environmental warming alters food-web structure and ecosystem function*. Nature 402, pp. 69-72

Phillips, N. R. & Hickey, C. W. (2010). *Genotype-dependent recovery from acute exposure to heavy metal contamination in the freshwater clam Sphaerium novaezelandiae*. Aquatic Toxicology 99, pp. 507-513

Platt, H. M. & Warwick, R. M. (1983). *Free-living Marine Nematodes Part I. British Enoplids*. Synopses of the British fauna, Vol. 28. Cambridge University Press, Cambridge

Platt, H. M. & Warwick, R. M. (1988). *Free-living marine nematodes. Part II. British Chromadorids*. Synopses of the British fauna (New Series), Vol. 38. E.J.Brill/W. Backhuys, Leiden

Polis, G. A.; Sears, A. L. W.; Huxel, G. R.; Strong, D. R. & Maron, J. (2000). *When is a trophic cascade a trophic cascade?* Trends in Ecology and Evolution 15, pp. 473-475

Porazinska, D. L.; Giblin-Davis, R. M.; Faller, L.; Farmerie, W.; Kanzaki, N.; Morris, K.; Powers, T. O.; Tucker, A. E.; Sung, W. & Thomas, W. K. (2009). *Evaluating high-throughput sequencing as a method for metagenomic analysis of nematode diversity*. Molecular Ecology Resources 9, pp. 1439-1450

Porazinska, D. L.; Sung, W.; Giblin-Davis, R. M. & Thomas, W. K. (2010). *Reproducibility of read numbers in high-throughput sequencing analysis of nematode community composition and structure*. Molecular Ecology Resources 10, pp. 666-676

Quaiser, A.; Zivanovic, Y.; Moreira, D. & Lopez-Garcia, P. (2010). *Comparative metagenomics of bathypelagic plankton and bottom sediment from the Sea of Marmara*. ISME Journal 5, pp. 285-304

Raunkjaer, K.; Hvitved-Jacobsen, T. & Nielsen, P. H. (1994). *Measurement of pools of protein, carbohydrate and lipid in domestic wastewater*. Water Research 28, pp. 251-262

Reusch, T. B. H.; Ehlers, A.; Hämmerli, A. & Worm, B. (2005). *Ecosystem recovery after climatic extremes enhanced by genotypic diversity* Proceedings of the National Academy of Science USA 102, pp. 2826-2831

Reusch, T. B. H. & Hughes, A. R. (2006). *The emerging role of genetic diversity for ecosystem functioning: Estuarine macrophytes as models*. Estuaries and Coasts 29, pp. 159-164

Ribalet, F.; Intertaglia, L.; Lebaron, P. & Casotti, R. (2008). *Differential effect of three polyunsaturated aldehydes on marine bacterial isolates*. Aquatic Toxicology 86, pp. 249-255

Schlacher, T. A.; Dugan, J.; Schoeman, D. S.; Lastra, M.; Jones, A.; Scapini, F.; McLachlan, A. & Defeo, O. (2007). *Sandy beaches at the brink*. Diversity and Distributions 13, pp. 556-560

Shea, K. & Chesson, P. (2002). *Community ecology theory as a framework for biological invasions*. Trends in Ecology and Evolution 17, pp. 170-176

Shurin, J. B.; Borer, E. T.; Seabloom, E. W.; Anderson, K.; Blanchette, C. A.; Broitman, B.; Cooper, S. D. & Halpern, B. S. (2002). *A cross-ecosystem comparison of the strength of trophic cascades*. Ecology Letters 5, pp. 785-791

Simonsen, R. (1962). *Untersuchungen zur Systematik und Ökologie der Bodendiatomeen der westlichen Ostsee*. Hydrobiologia 1, pp. 1-44

Snelgrove, P.; Blackburn, T. H.; Hutchings, P. A.; Alongi, D. M.; Grassle, J. F.; Hummel, H.; King, G.; Koike, I.; Lambshead, P. J. D.; Ramsing, N. B. & Solis-Weiss, V. (1997). *The importance of marine sediment biodiversity in ecosystem processes*. Ambio 26, pp. 578-583

Stachowicz, J. J.; Whitlatch, R. B. & Osman, R. W. (1999). *Species diversity and invasion resistance in a marine ecosystem*. Science 286, pp. 1577-1579

Stachowicz, J. J.; Fried, H.; Osman, R. W. & Whitlatch, R. B. (2002). *Biodiversity, invasion resistance, and marine ecosystem function: reconciling pattern and process*. Ecology 83, pp. 2575-2590

Stachowicz, J. J.; Bruno, J. F. & Duffy, J. E. (2007). *Understanding the effects of marine biodiversity on communities and ecosystems*. Annual Review of Ecology and Systematics 38, pp. 739-766

Stachowicz, J. J.; Best, R. J.; Bracken, M. E. S. & Graham, M. H. (2008). *Complementarity in marine biodiversity manipulations: Reconciling divergent evidence from field and mesocosm experiments*. Proceedings of the National Acadamy of Sciences USA 105, pp. 18842-18847

Stal, L. J. (2010). *Microphytobenthos as a biogeomorphological force in intertidal sediment stabilization*. Ecological Engineering 36, pp. 236-245

Tilman, D. & Downing, A. (1994). *Biodiversity and stability in grasslands*. Nature 367, pp. 363-365

Tilman, D.; Reich, P. B. & Knops, J. M. H. (2006). *Biodiversity and ecosystem stability in a decade-long grassland experiment*. Nature 441, pp. 629-632

Underwood, G. J. C.; Phillips, J. & Saunders, K. (1998). *Distribution of estuarine benthic diatom species along salinity and nutrient gradients*. European Journal of Phycology 33, pp. 173-183

van Oevelen, D.; Soetaert, K.; Middelburg, J. J.; Herman, P. M. J.; Moodley, L.; Hamels, I.; Moens, T. & Heip, C. H. R. (2006). *Carbon flows through a benthic food web: Integrating biomass, isotope and tracer data*. Journal of Marine Research 64, pp. 453-482

Waldbusser, G. G.; Marinelli, R. L.; Whitlatch, R. B. & Visscher, P. T. (2004). *The effects of infaunal biodiversity on biogeochemistry of coastal marine sediments*. Limnology and Oceanography 49, pp. 1482-1492

Warwick, R. M.; Platt, H. M. & Somerfield, P. J. (1998). *Free-living marine nematodes Part III: Monhysterids.* Synopses of the British fauna (new series), Vol 53. Field studies council, Shrewsbury

Witkowski, A.; Lange-Bertalot, H. & Metzeltin, D. (2000). *Diatom flora of marine coasts.* Gartner, A. R. G. (ed.), Koeltz Scientific, Vaduz, Königstein

Worm, B.; Barbier, E. B.; Beaumont, N.; Duffy, J. E.; Folke, C.; Halpern, B. S.; Jackson, J. B. C.; Lotze, H. K.; Micheli, F.; Palumbi, S. R.; Sala, E.; Selkoe, K. A.; Stachowicz, J. J. & Watson, R. (2006). *Impacts of biodiversity loss on ocean ecosystem services.* Science 314, pp. 787-790

Xu, J.P. (2006). *Microbial ecology in the age of genomics and metagenomics: concepts, tools, and recent advances.* Molecular Ecology 15, pp. 1713-1731

6

Microbiological Quality of River Sediments and Primary Prevention

Stefania Marcheggiani and Laura Mancini
Italian- National Institute of Health - Dep Environment and Primary Prevention Viale Regina Elena, Rome
Italy

1. Introduction

The preservation of aquatic ecosystems is fundamental because water quality, plant and animal biodiversity, industrial activities and human health rely on it. During recent centuries, the condition of aquatic ecosystems has become worse due to the increasing their use for irrigation and drinking, excessive land use and deforestation, hydro morphology alteration, riparian zone reduction and not least, climate change.
A strong interdependence between the health of the ecosystem and human health can be demonstrated. Microbiological risks for human health can occur through direct or indirect (fish, molluscs, recreational activities, algal bloom, vegetables, fruits) consumption of contaminated water (Tauxe, 1997; UNEP 1997, 1998; Noji, 1997; Ahem *et al.*, 2005). It is a priority to know the quality of the water and manage the data, in this sense it is necessary to identify the main variables that negatively impact human health.
Indicator organisms are commonly used to evaluate the *microbiological quality* of *aquatic ecosystems (*Berg 1978; Grabow 1996; EU 2006; Tyagi *et al.* 2006). Standard-based water quality assessment is an essential component of monitoring programs and also works to protect human health. As a rule, microbiological indicator detection (i.e. Enterococci and *Escherichia coli*) take place in the water column as necessitated by national and international laws.
In this context, it is appropriate underline the important role of sediment on aquatic ecosystems. This matrix is an extremely heterogeneous habitat characterised by a high microbial biodiversity due to the wide range of functional roles that they perform. Its origin is in the weathering and erosion of minerals, organic material, and soils in upstream areas usually after rainfall or the melting of snow. Transported downstream, it settles along the river bed and banks as sedimentation. Sediments consist of particulate matter that can be transported by fluid flow and which eventually is deposited as a layer of solid particles on the bed or bottom of water bodies. The suspended particle matter (SPM) which settle by sedimentation, include sediments whose diameter falls between 0.1 and 100 µm. Sediment heterogeneity (e.g. grain-size) in fresh and salt water, creates favorable conditions for biodiversity and is a source of life for a healthy river.
The particulate matter that plays a specific role in the aquatic ecosystems is the remaining material on a filter with a nominal porosity between 0.4 and 0.5 µm. The material with smaller dimensions is considered as colloidal and/or dissolved. When there have been

found particles with smaller dimensions than this limit, it is not certain if these are the products of aggregates derived from a process of fragmentation during sampling.

The structure and the functions of the microbial communities are strictly linked to the intrinsic peculiarities of the sediment. They are mainly responsible for the recovery of the sediment after the introduction of anthropic perturbations. Microbial processes are highly important for the regeneration of nutrients and nutrient cycles for the whole body of water. To date, estimates suggest that there are up to 10^4 bacterial species per g sediment, of which at least half (and perhaps as many as 95%) are yet unculturable. There is, therefore, the need to assess the contaminated sediment quality by means of standardised and reliable tools developed by advanced research investigation.

Their complex biochemical diversity enables them to exist in many different habitats everywhere on earth where they are essential for the geochemical cycle and the elimination of many pollutants.

By employing bio-indicators, consisting of organisms or communities of organisms which react to environmental effects by modification of their vital functions, it is possible to draw conclusions on the state of their environment. Due to the complexity of services and resources which are supplied by qualitative and quantitative differentiated ecosystems, it is very difficult to find general indicators that characterize the health of an ecosystem. A rich biodiversity, for example indicates a healthy system, but in some cases it can also be a symptom of disturbance when high amounts of nutrients in an aquatic ecosystem cause enhancement of growth. An indicator has to be relevant and useful.

Sulphate-reducing bacteria are a large group of anaerobic organisms that have an important role in many biogeochemical processes as the sulphur cycle and mineralization of organic matter in anoxic marine and freshwater environments and soil (Sitte *et al.* 2010). The sulphite reducing clostridium group, including *Clostridium perfringens*, has been shown to play an important role to assess faecal pollution in sediment ecosystems (Mancini *et al.* 2010, Marcheggiani *et al.* 2008, Marcheggiani *et al.*, 2004). Their presence can be influenced not only by organic matter but also by inorganic contaminants such as heavy metals (*Mancni et al.*, 2011; Mancini *et al.* 2008). *C. perfringens*, gram positive anaerobic spore-forming bacteria of the genus *Clostridum*, that does not carry out a dissimilatory reduction of sulphate, can be used as an alternative indicator for faecal contamination in aquatic ecosystems due to its adaptation to different habitats such as soils, sediments, and sewages . Furthermore, *C. perfringens* presence can be correlated to those of parasitic protozoan and enteric viruses in the water column as *Cryptosporidium sp., Aeromonas sp and Giardia*. In addition, the spores produced by *C. perfringens* are extremely resistant to disinfection and the WHO (1996) suggests that their presence in filtered supplies may be an indication of the need for treatment. Although the World Health Organization recommends *C. perfringens* as a useful indicator of fecal pollution in water quality surveys (WHO, 1978), this microorganism has been adopted in Europe exclusively as an additional source of water quality information (Cabelli , 1978; Olivieri, 1982; Rhodes *et al.*, 1999). The sequence of the 16S rRNA gene has been widely used as a phylogenetic marker to study genetic relationships between different strains of bacteria (phylogeny). The analysis of this gene can therefore be considered a standard method for the identification of bacteria at the family, genus and species levels (Woese , 1987; Weisburg *et al.*, 1991; Jeng *et al.*, 2001; Lehner *et al.*, 2004; Raju *et al.*, 2006; Johansson *et al.*, 2006), and has in fact been included in the latest edition of *Bergey's Manual of Systematic Bacteriology* (Bergey's Manual, 2005).

Therefore, the study of sedimentary microbial communities allows the sedimental damage/contamination state or the verification of the potential risk for human health and the need for primary prevention measures to be identified.

2. Interdependence of ecosystem health and human health

A strong correlation between ecosystem health and human health can be demonstrated and many new approaches to monitoring and environmental conduction are possible (Lackey RT, 2001).
There are four principle categories of ecosystem functions upon which human health is dependant (De Groot, 1992). The first category contains regulatory functions of ecological processes which deliver water, air, and clean soil through energetic and bio-chemical regulation processes, such as the recycling of organic material. The second category is the supply of space and appropriate substrates for human activities such as cultivation, recreation and living spaces. The third category is the production of numerous resources from which food and basic materials derive. The last category of functions has only an immaterial dimension: ecosystems have a large contribution in maintaining mental equilibrium by delivering opportunities for reflection, spiritual enrichment and aesthetic experience. By employing bioindicators, consisting of organisms or communities of organisms which react to environmental effects by modifying their vital functions, it is possible to draw conclusions on the state of their environment. Due to the complexity of services and resources which are supplied by qualitative and quantitative differentiated ecosystems, it is very difficult to find general indicators that characterize the health of an ecosystem. A rich biodiversity, for example indicates a healthy system, but in some cases it also can be a symptom of disturbance when high amounts of nutrients in an aquatic ecosystem cause the enhancement of growth. An indicator has to be relevant and useful.
These properties can be measured by its capacity to measure tendencies with cause of preoccupation, not only by the part of the scientists but also by the public opinion and policymakers. An ecological indicator has to have flexible and measurable characteristics.
Because of the dynamics of an ecosystem and the continuous increase of scientific information, an indicator has to be sufficiently extendable to incorporate new ecosystem components. The measurability of an indicator is determined by its disposability and cost, by the capacity to supply alert signals in respect to alteration, by the distribution in a wide geographic area, by the capacity to supply information about a wide set of stresses (Cecchi & Mancini, 2010). It is obvious that it is very difficult to find indicators or groups of indicators that satisfy all the characteristics mentioned above. The current methods of monitoring are studies about: populations, epidemiology, periodic sampling, toxicity tests and chemical analyses. All these techniques have to be used to supply exhaustive evaluations, but overall are not adequate to evaluate the integrity of the ecosystem in a conclusive way. More research is necessary to supply a truly conceptual picture for the definition of objectives for policy makers.

2.1 Indicators organisms

The acquirement of a reliable evaluation of the environmental quality of water bodies is generally the result of the synergic application of different analytical methods mainly including chemio-physical and biological parameters. The degree of contamination is commonly identified by the presence of a microbial community, which can easily proliferate in presence of particular compounds. For example, fecal bacterial indicators such as *Escherichia coli* and Enterococci are the most common indicators to detect water contamination by sewage pollution (Berg 1978; Grabow 1996; EU 2006; Tyagi *et al.* 2006). The figure showed the main microbiolgical indicators take into account by Italian regulation to assess water quality of main water typologies. (Tab 1).

Water typology	**Mcrobiolgical indicators**	**Regulation**
Surface water	*E. coli*	**D.Lgs 152/1999 (Italia, 1999)**
Surface water destined to drinking water productions	*Total Coliforms, Fecal Coliforms, Fecal Streptococci, Salmonellae*	**D.Lgs 152/2006 (Italia, 2006)** (2000/60/CE)
Wastewater (urban and indrustrial sewage in soil and surface water and fognature)	*E. coli*	**D.Lgs 152/2006 (Italia, 2006)** (2000/60/CE)
Waste used in shellfish culture	*Fecal Coliforms*	**152/2006 (Italia, 2006)** (2000/60/CE)
Bathing water (inland, costal and transitional waters)	*Intestinal Enterococci* and *Escherichia coli*	**D.Lgs 116/2008 (Italia, 2008)** **(2006/76/CE)**
Water for human consumption	*E. coli* and *Enterococci* , *C. perfringens* (spore includes) only for water originated from surface waters. *Pseudomonas aeruginosa, E. coli* and *Enterococci,* Colony count at 22° and 37° C only in packaged or bottle water	**D.Lgs 31/2001 (Italia, 2001)** **(98/83/CE)**

Table 1. Italian legislation ruling water management

Further studies have focused more attention on sediments due to their capacity to retain pollutants within the layers of the matrix thus giving rise to suitable habitats for the growth of particular species of microbes, mostly anaerobes (Davies *et al.* 1995; Robles *et al.* 2000; Desmarais,Solo-Gabriele and Palmer 2002).

The use of an appropriate set of microbiological indicators gives detailed information of pathogens present in freshwater and also represents the basis for emergency procedures. Micro-organisms connected to emerging diseases that are linked to contaminated water belong to different taxonomic groups of bacteria, viruses' protozoa and helminths (Fecham *et al.* 1983). The advantages of using a set of indicators during monitoring activities, is to obtain a more complete information about the state of the water quality; in this way it is possible to draw up emergency and prevention procedures (Tyagi *et.al*, 2006).

Set microbiological indicators should be included not only with *Escherichia coli*, an intestinal Enterococci whose presence indicates those of others pathogenic bacteria, but also with *Coliphages*, an indicator of enteric virus, and also with *Clostridium perfringens*, an obligate anaerobe which indicates the presence of parasitic protozoan and enteric viruses in the water column as *Cryptosporidium sp.,* Aeromonas sp e *Giardia* (Payment and Franco, 1993; Gleeson and Gray, 1996; U.S. EPA, 2007). In addition, the spores produced by *C. perfringens* are very resistant to disinfection and the WHO (1996) suggests that their presence in filtered water supplies may be an indication of the need for treatment.

2.2 Microbiological Indicators and sanitary significance for human health

Microbiological risks related to the consumption of contaminated water can occur directly or indirectly. Indirect consumption involves the consumption of fish, molluscs, vegetables and fruits contaminated through water used for irrigation, recreational activities and algal blooms (Tauxe, 1997; UNEP 1997, 1998). The risk of diarrhoeic disease outbreaks, due to enteric pathogens, is higher in developed than underdeveloped countries (Noji, 1997; Ahem *et al.*, 2005).

Several studies reported that the consumption of contaminated water and food are a probable source of Salmonellosis, Giardiasi, Hepatitis A and Criptosporidiosis outbreaks. (Mancini et al, 2010; Conio *et al.* 2000; Ballone *et al.*, 2001; Stroffolini *et al.* 1990; Leoni *et al.* 1998; Selvaggi *et al.* 1996). Indicator organisms are commonly used to evaluate the *microbiological quality* of *aquatic ecosystems (*Berg 1978; Grabow 1996; EU 2006; Tyagi *et al.* 2006). Standard-based water quality assessment is an essential component of monitoring programs and also works to protect human health. As a rule, microbiological indicator detection (i.e. Enterococci and *Escherichia coli*) takes place in the water column as necessitated by national and international laws.

2.3 Role of sediment in aquatic ecosystems

Sediments play an important role in the evaluation of the health of aquatic ecosystems (Carere *et al.*, 2008, Heise, 2008, Salomon, 2004, Heise, 2004, De Groot, 1992). Originating in the weathering and erosion of minerals, organic material and soil in upstream areas usually after rainfall on snow melt, sediments are then transported downstream, settling along the river bed and banks by sedimentation. Due to embankments and the loss of flooding areas naturally occurring sedimentation areas are very limited. Sediment heterogeneity, for example grain-size, provide favourable conditions for a high biodiversity and a good health of the river. Microbial processes are very important for the regeneration of nutrients and nutrient cycles for the whole water body (Sitte, 2010).

Sediments consist of particulate matter that can be transported by fluid flow and which eventually are deposited as a layer of solid particles on the bed or bottom of water bodies. Sedimentation is defined as the process of deposition by settling of a suspended material. The sediment will be transported as a suspended load if the flow is greater than the settling velocity. Sediments with a sufficient diameter to settle but still move are known as bed load and the particles are transported by mechanisms like saltation (jumping up into the flow, being transported a short distance then settling again). To characterize and identify pedological structures it is important to use a basic language to facilitate scientific communication. In the year 1926 the International Society of Soil Science developed, based on the research of Novak, the following conventional classification system (Table 2). The physical mechanical constitution of the sediment is a fundamental element. The particle dimensions can provide helpful information about to draining by superficial water, erosions and other forces. In additional, there may be a correlation between sediment grain-size composition and ecotoxicology (Benton, 1995).

Fraction number	**Name**	**Diameter (mm)**
1	Clay	<0.002
2	Silt	0.002-0.02
3	Sand (fine and very fine)	0.02-0.2
4	Sand (medium and big)	0.2-2
5	Gravel	2-20
6	Pebbel	>2

Table 2. Classification of the granulometric fractions after the system of Novak

The particulate matter that plays a specific role in the aquatic ecosystems is the remaining material on a filter with a nominal porosity between 0.4 and 0.5 µm. The material with smaller dimensions is considered as colloidal and/ or dissolved; even when there have been found particles with smaller dimensions than this limit (Gordon , 1970; Eisma *et al.*, 1970). It has not been proven that these are the products of aggregates derived from a process of fragmentation during sampling.
The upper dimensional limit of the particulate matter is not well defined, even when Visher (Visher, 1969) proposed the value of 100 µm as the limit between particles that move by saltation in laminar flow and these which are transported in suspension. Moss (Moss, 1963) indicates the limit between 70 and 100 µm. Baudo and Bertoni (Baundo & Bertoni, 1984) gave the dimensions of the following chemical and physical species present in the aquatic environment (Table 3).

Species	**Molecular Weight Dalton**	**Diameter (mm)**
Ions, inorganic complexes and small organic molecules	200	d≤1
Bigger, dissolved compound (Fatty acid, fulvic acids, polyhydoxid complexes, polysilicates, etc)	200 -10000	1≤d ≥10
Colloids, humic acids, proteins, metal hydroxid complexes and clayey minerals	106	10≤d ≥100
Suspended solid portions organic and inorganic particles		d≥100

Table 3. Dimensional indications of the chemical and physical species in the aquatic environment [Hunt, 1980]

The SPM (suspended particle matter) which settle by sedimentation, include sediments whose diameter falls between 0.1 up to 100 µm. Significant quantities of particulate matter, biogenic with low density, can be found as a subform of particles with d>100 µm, the so called floccules. Among these chemical compounds, present in the aquatic environment, a dynamic equilibrium with continuous changes and transformations by re-suspension and resettlement exists and dimensions of the particles can change in the long term. The composition of the particulate matter is extremely variable depending on the aquatic environment. The largest amount of sediment with continental origin is transported by the rivers into the oceans. The concentration of the particulate matter is not homogenous in the inner of the water body and is various due to the meteo-climatic characteristics of the catchment basin, the contribution of single tributaries and of hydraulic conditions. Concentrations of different particles can be measured over the course of one day or as variations on a monthly or seasonal scale at different water levels (high, medium, low).

To estimate the solid transport of a course of a river, the relation between the suspended solid fraction and the transported liquid fraction must be known. Even when this relation is not always linear the concentration of SPM is generally elevated during periods with a higher water level. Not all the SPM is transported to the ocean, in many cases it remains in areas of the river with a low energy level, deposition zones, lakes, or plains in artificial basins.

In their natural state, the particles are found in the form of aggregates called floccules with dimensions of 1-2 mm but occasionally in some cases even more than 10 mm. The flocculation of the SPM has a great influence on the transport of the particles and associated substances. The formation of floccules depends on an elevated number of factors such as the salinity of the water body, particle concentration (which augments the potentional contacts between the particles), turbulence, Brown movements (Hunt, 1980, McCave, 1988), polymerisation and adsorption to granules of dissolved organic substances (in particular carbon hydrates) (Sieburth J , 1965; Wassermann *et al.*, 1986), bacterial mucus and exudates or phytoplankton that reacts like adhesive, and aggregation of air or gas bubbles.

The transport of SPM in rivers is generally connected with the morphological characteristics of the basin and can be classified according to the length, the expansion and the erosionibility of the catchment basin. Furthermore, the concentration of the suspended solid fraction depends on hydraulic characteristics (e.g. the water content of the river and the roughness of the lower bed) as well as the meteorological conditions and the climate characteristics of drained air, and finally on the antrophic activity in the course of the river as well as the basin.

The SPM consists of an organic and inorganic fraction. **The organic fraction** consists of living organisms (phytoplankton, nanoplankton and bacteria), detrital material (products from the degradation of cells from plants or algae, fragments from diatoms) essences, secretions and faecal pellets (expression for the faecal products predominantly from invertebrates). The organic substances form a thin layer on the mineral particles consisting of clay (with a high specific surface), thus it can be concluded that the clay fraction has a high absorbance potential. As a consequence, the molecules or ions can be easily adsorbed into the SPM, and then transported and accumulated in the aquatic environment. The organic substances, which consist of a few chemical elements like oxygen, hydrogen, carbon, nitrogen, phosphor and sulphur are a source of food for a lot of heterotrophic aquatic organisms. Nevertheless, the biological oxidation of too much particulated organic substance by anaerobic microorganisms leaves insufficiently dissolved oxygen for the survival of the flora and fauna and causes a higher mortality due to eutrophication processes (Madej, 2005). The **suspended inorganic fraction** is derived by the modification and erosion of the continental crust, snow smelt and the atmosphere. Through volcanic activity the emission of inorganic substances can become locally and/or temporaneously important. The contribution of cosmic dust is nominal, but the content of cosmic particles in the oceanic sediments can provide important clues for the reconstruction of cosmic events (Grieve, 1987).

In conclusion, based on features of the sediment can be divided into two parts: an active part represented by the surface layer (approximately the first centimetre, silk , sand clay) an piece and quiet part represented by the deepest layer .

Several studies have highlighted the value of sediments as indicators of pollution (Mancini *et al.*, 2010 ; Marcheggiani *et. al.*, 2008, Mancini *et al.*, 2008). Indeed, sediments - as potential reservoirs for bacteria and viruses in aquatic environments - are able to provide information on past instances of pollution which are no longer detectable in water samples, as well as on the presence of pathogens, which may in some cases pose a future threat to human health. River sediment is ideal habitat of *C. perfringens* and of others species because in this matrix the main factors that enhance their capability for survival coexist.
In conclusion, indicator sets give useful information about microbiological risks for human health and can improve emergency and prevention plans

2.4 Sediment and microbiological indicators

Microorganisms are the main source of fertility and of degradation of organic matter and pollutants in sediments. Their complex biochemical diversity enables them to exist in various habitats throughout the planet where they are essential for the geochemical cycle of many elements and the elimination of many pollutants. Many of these reactions are performed by specialized organisms that cannot be easily substituted. Furthermore, microorganisms are indispensable for many symbiotic and pathogenic relationships with higher life-forms. For example the rhizosphere microflora gives plants additional competitive abilities and some microbial species serve as a food source for many soil animals. The disappearance of these communities could cause extinction of many species (van Beelen *et al.*, 1997). Due to their ubiquitous presence, microorganisms are very important as environmental indicators of contamination and provide an excellent subject for the establishment of quality guidelines (Mancini *et al.*, 2008). The sulphide content of soil is a leading influence on the bioavailability of metals, because trace metals are able to react with FeS (ferrous sulphide) (major component of acid-volatile sulphides) and form metal sulphides according to Me^{2+} FeS(s) $\leftrightarrow$ MeS (s) + Fe^{2+}.

2.5 Sulphate-reducing bacteria

Sulphate-reducing bacteria are a large group of anaerobic organisms that have an important role in many biogeochemical processes such as the sulphur cycle and mineralization of organic matter in anoxic marine and freshwater environments and soil (Sitte *et al.* 2010). The sulphite reducing clostridium group, including *Clostridium perfringens*, is important in the assessment of faecal pollution in soil ecosystems (Mancini *et al.* 2010, Marcheggiani *et al.* 2008, Marcheggiani *et al.*, 2004). Their presence can be influenced not only by organic matter but also by inorganic contaminants like heavy metals (Mancini *et al.* 2008). *C. perfringens*, gram positive anaerobic spore-forming bacteria of the genus Clostridum, that does not carry out a dissimilatory reduction of sulphate, can be used as an alternative indicator for faecal contamination in aquatic ecosystems due to its adaptation of different habitats such as soils, sediments and sewages. Furthermore, *C. perfringens* presence can be correlated to those of parasitic protozoan and enteric viruses in the water column as *Cryptosporidium sp.*, *Aeromonas sp* and *Giardia*.
The role of *C. perfringens* as an environmental fecal indicator has been recently acknowledged by several studies. Despite this, the ruling Italian National legislation (D.Lgs. 152/99) names *C. perfringens* exclusively as a 'supplemental' indicator in recreational waters. In contrast to running water, sediments are also able to provide evidence of former fecal contamination by the analysis of different layers at various depths. Especially in situations of discontinuous disposal events, water quality assessment can be misleading while

sediments are characterized by a higher stability favoring sludge accumulation. During the last decade molecular methods have been developed to study the diversity of indigenous microbial communities independent of the classical techniques such as cultivation and microscopic identification (Amann, Ludwig, and Schleifer 1995; Schäfer and Muyzer).

2.6 The microorganisms Clostridia

The genus Clostridium is considered a biological indicator because its presence in the sediment can be natural or caused by anthropogenic discharges.

The genus Clostridium was described for the first time in 1880 by Prazmowski who separated the genus from the genus Bacillus to include the Gram positive and obligatory anaerobic rods with central or subterminal heat resistant endospores. At that time there was less opportunity to separate the genus into smaller genera for a better. Bergey *et al.* made a slightly expanded definition of that of Prazmowski in 1923 (Cato *et al.*,1986; Cato *et al.*, 1989). The genus Clostridium is characterised by anaerobic or microaerophilic spore-forming rods that do not form spores in the presence of air, are usually Gram-positive and do not carry out a dissimilatory sulphate reduction. *C. perfringens* produces a variety of toxins that play an important role in the pathogenesis of infections. The strains are classified into five groups (types A-E) on the basis of their production of lethal toxins (Sterne & Warrack,1984). At the end of the 18th Century the first diseases of man and animal associated with *C. perfringens* were documented (Rood *et al.*, 1984). The five types of *C. perfringens* cannot be differentiated reliably on the basis of cellular or colonial morphology, biochemical reactions, or gas-liquid chromatographic analyses of fatty and organic acid metabolic end products (Cato *et al.*,1986).

Pathogenicity: Some species produce exotoxins and are pathogenic to humans, however, the infectious dose is 10^8-10^9 (Bitton, 1984; Pahren, 1987).

C. botulinum and *C. difficult*, responsible respectively for botulism and pseudomembranous colitis in humans and animals, other clostridia are associated mostly with wound infections, including the most important are: *C. tetani*, head of tetanus; *C. perfringens*, *C. novyi*, *C. septicum*, *C. histolyticum*, *C. brothels*, often associated with gas gangrene. *C. botulinum* and *C. perfringens* cause serious infectious phenomena associated to consumption of contaminated food. Other species of clostridia, *C. bifermentans* and *C. sporogenes*, are only rarely associated with infections and they are not consider pathogenic.

Bacilli are widespread in the environment can be divided into : Clostridia invasive: produce toxins less powerful (hystolitic enzymes), among these, *C. perfringens*, *C. novy*, *C. chauvoei*, *C. septicum* and *C. haemolyticum*. Clostridia non-invasive: the pathogenic action is due to the production of potent exotoxins and their dissemination in the body among these, *C. botulinum*.

Clostridium perfringens, a Gram-positive anaerobic spore-forming bacterium of the genus Clostridium, has been suggested (Bisson &C abelli,1980; Leeming *et al.*,1998) and successfully used as an alternative indicator of fecal contamination in aquatic environments due to its wide distribution in nature and to its adaptation to a variety of habitats such as soils, sediments and sewages. Moreover, this organism has been found in air, dust, water, and even food (Niilo, 1980, Van Metre *et al.*, 2000). Spores produced by this organism are very resistant to disinfection and their presence in filter units indicate treatment inefficiencies (WHO, 1996).

2.7 Sulphite reducing bacteria in river sediment of Italy

The health of river ecosystems can be assessed using indices and indicators, the study of which is oriented to detect health prevention actions. In this contest the aim of these works are to identify species of sulphite-reducing clostridia in river sediments. This can be obtained through a combined approach involving standard microbiological and molecular tools. In these paragraphs we have examined the available information deriving from studies specifically dedicated to assessing the genetic variability of the anaerobic microbial community in river sediments of Italy, and its relation to different fecal pollution sources (Study 1 and 2).

The *Clostridium perfringens* presence in sediments can be influenced not only by organic matter but also by inorganic contaminants such as heavy metals. Below is shown a study, performed in vitro to evaluate the potential effect of lead on the vitality of *C. pergringens* population (Study 3). At the end, a study was performed to evaluate the direct or indirect effect, of methyl mercury on sulphite reducing bacteria of lagoon sediments (Study 4).

2.7.1 Study 1

A study was performed to investigate the anaerobic community in river sediment samples of the lower Tiber catchment area, in central Italy, through a combined approach involving granulometric analysis of sediment samples, followed by microbiological and molecular (16S rRNA) analyses of bacterial strains isolated from these samples (Marcheggiani *et al.*, 2008). The study area includes the lower course of the Tiber river basin and the sampling sites have been selected to represent the upstream-downsteam gradient of pressure. The Analysis of 16S rRNA fragments was molecular tool used to investigated sulfate reducing bacteria community of sediment rivers. Eighty three PCR products were aligned and matched against NCBI database sequences. The resulting phylogenetic tree, based on the neighbor-joining method, is showed in Fig. 1. The genetic analysis assigned the bacteria to one of three clusters: *C. perfringens* (I), *C. bifermentans* (II) and *B. cereus* (III), grouped into two genuses (Clostridium and Bacillus). Cluster I included the taxa *C. perfringens, C. barati, C. thiosulfatireducens* and *C. butyricum*, and cluster II included *C. bifermentans, C. glycolicum* and *C. ghoni*. The cluster III was B. cereus. *C. perfringens* was found to be the most prevalent in all but one of the sampling sites. As expected, more biodiversity was observed downstream than upstream, both along the stem of the Tiber and along its tributaries.

Authors conclusion were the important role of *Clostridium perfringens* as a microbial indicator of fecal contamination in river sediments. The presence of this bacterium in all sediment sampling sites, as well as in both seasons (along the main stem of the Tiber, where samples were collected seasonally lend support to its suitability as an alternative indicator of fecal pollution in water quality surveys.

While further studies are still needed to explore possible relationships between the presence of specific microorganisms in sediments and the effects of such pollution on human health, one may safely assume that sulphite reducing bacteria, being useful sources of information, are destined to play a key role in the management of freshwater environments. Moreover, information derived from the analysis of sulphite-reducing bacteria in river sediments may prove valuable in water reclamation plans. In this context, the information may be used for purposes such as the evaluation of health risks in a given area, or quality control, to ensure that the health of the water ecosystem in question has in fact been restored or improved.

Further quantitative studies will be useful in order to state the suitability of this group of bacteria as indicators. Despite the usefulness of qualitative molecular tools, however, these cannot currently replace classical methods of routine water quality assessment, the latter being quantitatively more informative and easier to execute. An alternative, faster molecular methodology, which may be examined in future studies, is the direct extraction of microbial genome from the matrix and its amplification using selective primers specifically designed for each taxonomic unit.

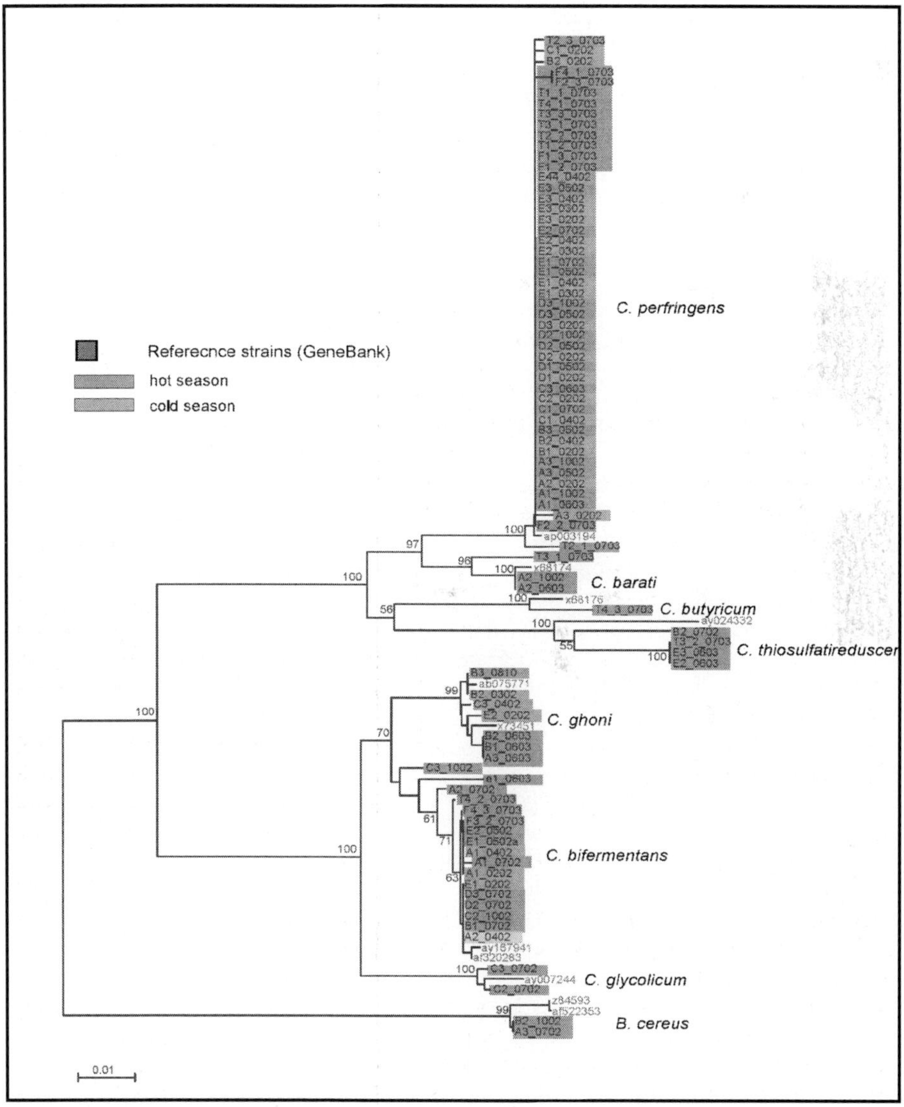

Fig. 1. Phylogenetic tree constructed with the Neighbor-Joining method-distance Kimura 2-, for a 702 bp fragment of the 16S rRNA coding region. Numbers above branches show bootstrap values expressed as percentages of 100 replications. and distribution of the genetic profiles observed in different seasons

2.7.2 Study 2

A study conducted on the river systems Foglia, Tevere, Astura, and Sitzerri located in three Italian regions, was performed to assess the genetic variability of the anaerobic microbial community in freshwater sediments as a preliminary step for future possible assessment of the relation of its members to different fecal pollution sources (Mancini *et al.*, 2010). Also, in this study was compared the concentration of Clostridia in sediments and *E. coli* in water column.

Results showed that the frequency of recorded haplotypes appeared to be almost heterogeneous for all the sampling sites. *Clostridium perfringens* was found to be the only ubiquitous species with frequencies of occurrence between 22% (Astura and Sitzerri Rivers) and 100% (Tiber). A relative higher biodiversity was recorded in summer samples. *C. perfringens* was the exclusive taxonomic unit identified in winter samples; up to four more species *(C. bifermentans, C. thiosulfatireducens, C. ghoni*, and *Bacillus cereus*) were recorded on late summer samples. Even the Sardinian samples showed a close correlation with seasonal changes but also with the typology of outfall. Samples collected in a brook running through a disused blende and galena minefield expressed a prevalence of *C. sporogenes,* while other water bodies were generally dominated *by C. perfingens*. Samples from the Foglia River were characterized only by *C. perfringens* in winter samples while those collected in summer expressed lower percentages of *C. bifermentans*. Genetic profiles detected on the only seasonal sample from the Astura River could be ascribed to the species *C. perfringens* and to a strain closely related *to C. thiosulfatireducens* identified on the NCBI database as 'swine manure gramþbacterium'. Two other strains, *Providencia stuartii* and *Proteus mirabilis,* optional anerobes, were identified but not included as results of a possible mismanagement in samples heat pretreatment.

The genetic relationships between the detected species are described by a dendrogram (Figure 2) that identifies six main clusters of which five are ascribed to the genus Clostridium and one to the genus Bacillus. The comparison with prototype sequences performed with BLAST, lead to the identification of the following taxonomic units: *C. perfringens, C. sporogenes, C. thiosulfatireducens, C. ghoni, C. bifermentans,* and *B. cereus*. The cluster referring to *C. perfringens* was found to be the most homogeneous in respect to the others. With the exception of *C. sporogenes,* none of the other systematic units showed complete homology with the respective prototypes.

The comparison between the concentration of Clostridia in sediments and *E. coli* in water for each site did not provide any significant correlations. According to the increment of pollution, relative concentrations of Clostridia and *E. coli* were characterized by an increasing gradient proceeding downstream on the water bodies.

In conclusion the lack of correlation between the occurrence of *E. coli* in water and *Clostridium sp* in sediments is to be attributed to the higher storage ability of sediments enhanced by accumulation and/or by the typology and seasonal regimes of the water bodies; the stability of sedimental layers is indeed inversely correlated with the irregularity of the water flow. Frequent changes are more common in small streams which make the settlement of suitable habitats for microbial growth more difficult. This may explain the higher number of profiles recorded in the Tiber River in respect to other streams. Second, the microbial composition of sediments can be temporarily influenced by particular climatic events such as floods or drought.

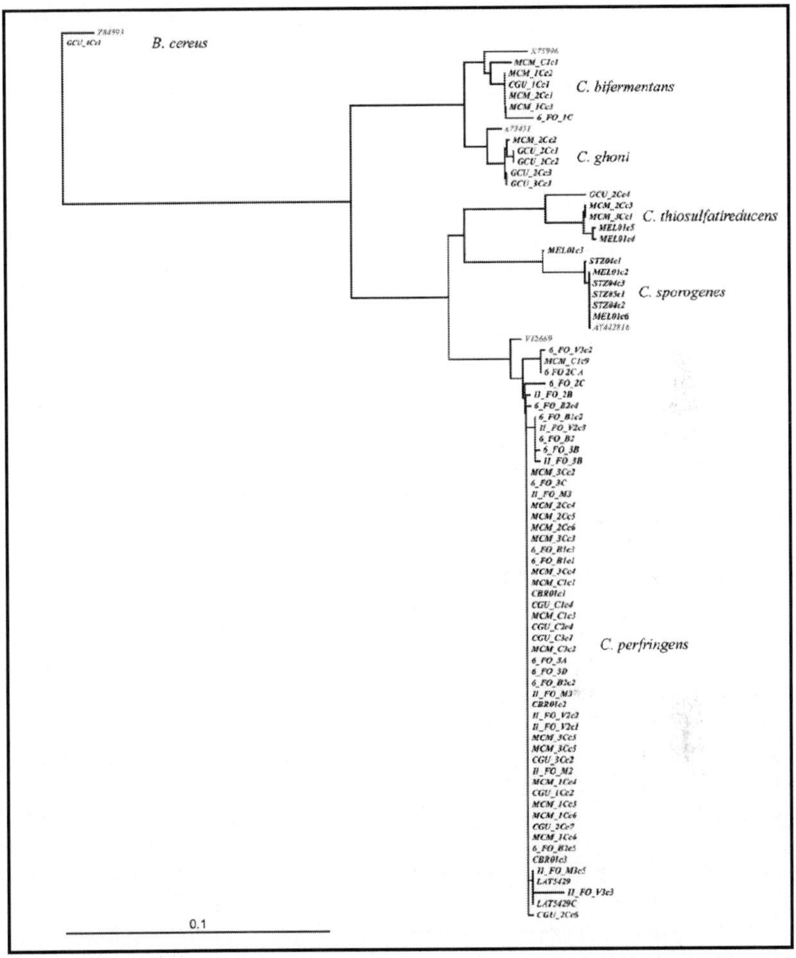

Fig. 2. Phylogenetic relationships observed between strains of sulphite-reducing isolated from sediment .genetic distance have been computed by CLUSTALX using Kimura evolutionary model (from Mancini *et al.*, 2011)

Finally, the disposal of industrial or urban waste waters could play an important role in the development of suitable habitats for the selective growth of particular species of microbes. The results of this study seem to confirm this hypothesis: *C. perfringens* and *C. bifermentans* resulted in being prevalent in urban and countryside areas where domestic and agricultural sewages are the most common impact source. The higher presence of *C. sporogenes* in the Sardinian brook running through a disused minefield suggests a role of mine wastes in providing suitable habitats for the selective growth of this species.

The evaluation of the relationships between the influence of particular impact sources and the selective growth of microbes should be investigated further in order to assess their function as indicators of pollution. Within this context the knowledge of the genetic profiles by molecular tools of the environmental microbial community is a precondition for a correct approach to assess possible linkages with pollution sources. However, further studies should be performed in this field in order to employ the anaerobes as indicators of environmental pollution.

2.7.3 Study 3

This study aimed at detecting the effect of lead on the vitality of *C. perfringens* population in a determined, experimentally controlled sediment matrix using a luminescent reaction (Mancini *et al.*, 2011).

Lead is a priority substance of the Water Framework Directive (European Parliament, 2000) and for this reason the emissions, releases and losses should be reduced within a certain deadline (European Parliament, 2008; 2000). Native lead occurs in nature very rarely and currently lead can be found bounded to zinc, silver and copper (Visher, 1969).Due to these activities, lead is so common in the environment that it can reach man via air, food, water, dust or soil. Deposition from atmosphere is major contributor to lead inputs to water and land. After deposition in water, depending on salt content of the water and the presence of organic complexing agents, lead splits between water and sediment. In order to contribute to the knowledge of these mechanisms, this study aimed at detecting the effect of lead on the vitality of *Clostridium perfringens* population in a determined, experimentally controlled sediment matrix using a luminescent reaction. Artificially polluted sediment was contaminated with a pure culture of *C. perfringens*. Bacterial vitality was measured for period of 40 hours by luminescence method using the BacTiterGlo™-, both in the blank sediment and in the sediments added with 50 ppm, 250 ppm and 760 ppm lead concentrations.

The results of the performed ecotoxicity test using the BacTiterGlo™-assay showed a significant decrease in bacterial vitality at 760 ppm concentration. This effect occurs within the first 16 hours and therefore the initial experimental time adopted in the protocol can be reduced to this range of time. The results showed that lead concentrations that influence *C. perfringens* vitality are similar to those found in hazardous waste sites.

Next steps, the effect of lead should be observed in this range of concentration and within this period of time with more replicates and shorter time distances between the measurements. In addition, the ecotoxicological effect observed in this way, should be tested also in different types of sediments and with different environmental strains to determine sensitivities through combination of sediment properties and properties of microorganisms.

2.7.4 Study 4

A study performed on evaluation of the quality of cotaminate sediment using sulphite reducing bacteria was performed in Orbetello lagoon –Italy (Marcheggiani *et al.*,2011)

Orbetello lagoon is a site of main ecological and biodiversity interest for its peculiar characteristics of brackish wetland and is located along the southern coast of Tuscany; the lagoon has also a relevant interest due to the activities of intensive and extensive aquaculture. Orbetello lagoon over last thirty years has showed an increase of the eutrophication phenomena (Giusti *et al.*, 2005). Additionally, an establishment of anaerobic conditions has formed in some parts of the lagoon (Giusti & Marsili Libelli, 2009). Ecotoxicological and chemical studies performed in this lagoon have highlighted the presence of methylmercury in the sediments and biota (Beccaloni *et al.*, 2011).

In order to improve the knowledge about the state of the sediment contamination of the Orbetello lagoon a study to investigate the composition of sulphite reducing bacteria from sediment samples will be performed. Sampling sites have been selected into critical point of the lagoon to detect the main pressures into the area; in the same sites ecotoxicological parameters, mercury and methylmercury will be also analysed (Beccaloni *et al.*, 2011). Each sediment sample will be submitted to granulometric (Shepard, 1954) and microbiological analysis.

Results showed an increase of bacteria concentration in the sites characterized by low concentration of methylmercury was observed particularly in the south-east sites and the north east sites. These sites are characterised by a silty clay sediment composition. Furthermore, the lagoon has a large amount of organic matter probably due to intensive fish farming and agriculture activities as well as the discharge of treated/untreated urban watstewater, which has significantly increased as a consequence of the tourist trade (Lenzi, 1992, Lenzi 1998). However, a low concentration of bacteria in sites with high metylmercury concentration was also observed (Beccaloni *et al.*, 2011). A probable reason for this phenomena could be the effect of metylmercury on the sulphite-reducing community. Other investigations performed in vitro for the evaluation the effect of heavy metals on these bacterial communities showed a decrease of their growth (Mancini *et al.*, 2011; Sitte *et al.*, 2010).

Preliminary results showed an increase in the rate of supply of organic matter probable due to the intensive fish farming and agriculture activities. The results of this study will be integrated with the ecotoxicological and chemical studies and will support also the knowledge related to the transformation of mercury in methylmercury in the sediments. The multicriteria approach, including a microbiological indicator, applied to a specific site of relevant ecological interest could represent a useful tool for the preservation of environmental resources, remediation actions and assessment of the risks for human beings. The authors concluded that The multi-criteria approach including a microbiological indicator which is applied to a specific site of relevant ecological interest could be a useful tool for the preservation of environmental resources, remediation actions and assessment of the risks to human beings. Further studies should be performed in order to consider the relationships between pollution and species occurrence in sediments. It is possible to assess the key role of sulphite reducing bacteria as useful information holders for the correct management of aquatic ecosystems. Moreover, this feature plays a special role in reclaiming plans such as the evaluation of the sanitary risks of a given area or to verify the improvement of the health of a water ecosystem.

3. Conclusion

The risk assessment of microbiological water sources is not an easy task Microbial risk assessment differs significantly from chemical risk assessment. Because microorganisms can proliferate in the host; they have different die-off rates in the environment; there are risks of secondary transmission; hosts may or may not acquire partial or complete immunity, and animal responses may be very different from human responses to the same microbe.

For these reason, often the mathematical models available are not an useful tool to predict microbiological risk. It is not easy to identify water-borne infectious diseases, it is possible only when at least 1% of the population of a community became ill within a few months

(epidemic). Furthermore, another important aspect is that microbiology risk depends on exposure to pathogens and not by accumulation.
The microbial risk analysis should be based on information on the infectious dose of the organism, the exposure response, the spread of the organism in water, the likelihood of infection and the secondary spread in the environment.
Indicator organisms are commonly used to evaluate the *microbiological quality* of *aquatic ecosystems* (Berg 1978; Grabow 1996; EU 2006; Tyagi *et al.* 2006). Standard-based water quality assessment is an essential component of monitoring programs and also works to protect human health. As a rule, microbiological indicator detection (i.e. Enterococci and *Escherichia coli*) take place in the water column as necessitated by national and international laws. Therefore, the study of sedimentary microbial communities permits both the sedimental damage/contamination state and the verification of the potential risk for human health to be evaluated and the need for primary prevention measures to be identified.

4. Acknowledgements

The authors wish to thank our unit colleagues for their own contributions to this activity during these years, AM D'Angelo, E. Pierdominici, R. Scenati, C. Puccinelli, M. Figliomeni and E. Volpi.

5. References

AAVV: Bergey's Manual of Systematic Bacteriology. 2nd edition (2005). Edited by: Garrity G. Springer, London, UK;

Amann, R.I., Ludwig, W. & Schleifer K.H. (1995). Phylogenetic identification and in situ detection of individual microbial cells without cultivation. *Microbiological Reviews* March 59: 143–69.

Ahem, M.; Kovats; R.S.; Wilkinson, P.; Few, R. & Matthies, F. (2005);. Global health impacts of floods: epidemiologic evidence. *Epidemiol Rev* 27:36-46.

Ballone, E. ; Fazii, P.; Riario Sforza, G.; Scassa, E. ; Di Nicola, M.; Ippolito, N. ; Di Mascio, C. & Schioppa, F. (2001). Indagine Sulla diffusione della giardiosi nell'area pescarese. *Ann Ig* ;13:111-120.

Baudo, R. & Bertoni R. (1984). Analisi del materiale particellato: significato e limiti. *Acqua-Aria*, 6: 722-727.

Beccaloni, M. R. ; Cicero, A.M.; Mancini, L.; Marcheggiani, S.; Miniero, R.; Scenati, R. ; Vendetti, C.; Ziemacki, G. & M. Carere.(2011) Approaches for the Derivation of Sediment Quality Criteria for the protection of Human Health in Italian water bodies. Proceedings *Sixth International Conference on Remediation of Contaminated Sediments* (New Orleans, Louisiana; February 7-10, 2011)-in press

Benton, M.J. ; Malott M.L.; Knight S.S.; Cooper C.M. & Benson, W.H. (1995). Influence of sediment composition on apparent toxicity in a solid-phase test using bioluminscent bacteria. *Environ. Toxic. and Chem.*, 14 (3): 411-414.

Berg, G. (!978). The indicator system. In Indicators of Viruses in Water and Food Edited by: Berg G. *Ann Arbor*, MI: *Ann Arbor Science*.1-13.

Bisson, J.W. & Cabelli, V.J. (1980). *Clostridium perfringens* as a water pollution 22. indicator. J *Water Pollut Control Fed* 1980, 52:241-248.

Bitton, G. ; Henis, Y. & Lahav, N. (1972). Effect of Several Clay Minerals and Humic Acid on the Survival of aerogenes Exposed to Ultraviolet Irradiation. *Appl Microbiol* 23:870-874

Cabelli, V.J. (1978). Obligate anaerobic bacterial indicators. In *Indicators of Viruses in Water and Food* Edited by: Berg G. Ann Arbor, MI: *Ann Arbor Science*; 1978:171-200.

Carere M. , Marcheggiani S., Miniero R., Pilozzi A. e Mancini L. Ed, (2008). Risk assessment elements for the management of contaminated sediments Ann Ist Super Sanità 44. pg 268

Cato, E.P. & George, W.L. (1986). Finegold. Genus clostridium. In: *Bergey´s Manual of Systematic Bacteriology*. Vol. 2. Streath PHA, Mair NS, Sharpe ME, Holt JG (Ed.). Baltimore: *Williams and Wilkins*; pp. 1141-200.

Cato, E.P. & Stackebrandt, E. (1989). Taxonomy and Phylogeny. In Clostridia- Biotechnology handbooks, III series Edited by: Minton NP, Clarke DJ. New York and London: *Plenum Press*; 1989:1-26.

Cecci & Mancini(2010). Salute degli ecosistemi come priorità della gestione ambientale. Roma: Istituto Superiore di Sanita (*Rapporti ISTISAN 06/10*).

Conio, O.; Palombo, F.; Borelli, E.; Pignata, C.; Gilli, G. & Carraio, E. (2000). Contaminazione da *Giardia* spp. e *Cryptosporidium* spp. nelle acque destinate al consumo umano in Italia. *Ig Mod*;114:77-100.

Davies, C.M.; Julian, A. ; Long, H. ; Donald, M. & Ashbolt, N.J. (1995). Survival of fecal microorganisms in marine and freshwater sediments. *Applied Environmental Microbiology* 61: 1888–96.

De Groot, R.S. (1992). Functions of Nature: evaluation of nature in environmental planning, management and decision-making. Groningen: *Wolters Noordhoff BV*.

Desmarais, T.J.; Solo-Gabriele, H.M. & Palmer. C.J. (2002). Influence of soil on fecal indicator organisms in a tidally influenced subtropical environment. *Applied Envirommnmental Microbiology* 68: 1165–72.

Eisma, D.; Kalf, J. & Veenhuis, M. (1980).The formation of small particles and aggregates in the Rhine Estuary. *Neth J Sea Res*;14:172-8.

EU, European Union: Directive 2006/7/EC 15 February 2006. Concerning the management of bathing water quality and repealing directive 76/160/EEC. Official Journal of the European Union 2006 L64/37: 1–15.

Eurometaux, 2000: Information and data on cadmium, lead and nickel in letter to the European Commission (DG ENV.E.1), 3 October 2000.

European Parliament and the Council. Official Journal of the European Communities, Directive 2000/60/EC of the 23 October 2000 establishing a framework for Community action in the field of water policy. L327. p. 1 ff. Official Journal of the European Communities 18 May 1976 L 129, 0023-0029.

European Parliament and the Council. Official Journal of the European Communities, Directive 2008/105/EC of 24 December 2008. Priority Substances and Certain Other Pollutants Annex II . OJ L348, p.84-97

European Parliament and the Council. Directive 2006/7/EC 15 February 2006. Concerning the management of bathing water quality and repealing directive 76/160/EEC. Official Journal of the European Union 2006 L64/37: 1–15.

European Parliament and the Council . Directive 98/83/EC of 3 November 1998 on the quality of water intended for human consumption Official Journal of the European Union 1998 L 330, 51- 32.

Feachem, R.G.; Bradley, D.J.; Garelick, H. & Mara D.D. (1983). In: *Sanitation and Disease: Health Aspects of Excreta and wastewater management*. New York, John Wiley and Sons, 1983.

Giusti, E. & Marsili-Libelli S. (2005). Modelling the interactions between nutrients and the submersed vegetation in the Orbetello Lagoon. *Ecological Modelling* 184: 141 - 161.

Giusti, E. & Marsili-Libelli, S. , (2009). Spatio-temporal dissolved oxygen dynamics in the Orbetello lagoon by fuzzy pattern recognition. *Ecological Modelling*, 220: 2415–2426

Gleeson, C. & Gray, N . (1996). In: *The coliform index and water borne disease*. London, E and F N Spon.

Gordon, DC. (1970). A microscope study of organic particles in North Atlantic Ocean. Deep-*Sea Res Ocean Abstract*;17:175-8.

Grabow. W.O.K. (1996). Waterborne diseases: Update on water quality assessment and control. Water SA, 22:193-202.

Heise, S. (2008). Risk management of sediments. In Risk assessment elements for the management of contaminated sediments. *Ann Ist Super Sanità* Vol. 44, No. 3: 224-232

Heise, S.; Apitz, S.E.; Babut, M.; Bergmann, H.; Besten, P.D.; Ellen, G.J.; Joziasse, J.; Katsiri A.; Vera Maaß, Oen A.; Slob, A. & White, S. (2004). Risk management of sediments and communication. Synthesis of the SedNet working group 5 outcomes. *J Soils & Sediments* 4:233-5.

Hunt, J.R. (1980). Prediction of ocean particle size distribution from coagulation and sedimentation mechanisms. *Am Chem Soc Washington* Adv Chem Ser;189:243-57.

Italia, (1999). Decreto legislativo 11 maggio 1999 n. 152. Disposizioni sulla tutela delle acque dall'inquinamento e recepimento della direttiva 91/271/CEE concernente il trattamento delle acque reflue urbane e della direttiva 91/676/CEE relativa alla protezione delle acque dall'inquinamento provocato fai nitrati provenienti da fonti agricole. *Gazzetta Ufficiale- Supplemento Ordinario* n. 124 del 29 maggio 1999.

Italia, (2006). Decreto Legislativo 3 aprile 2006, n. 152. Norme in materia ambientale. *Gazzetta Ufficiale Supplemento Ordinario* n. 88 del 14 aprile 2006.

Italia, (2008). Decreto Legislativo 30 maggio 2008, n. 116. Attuazione della direttiva 2006/7/CE relativa alla gestione della qualità delle acque di balneazione e abrogazione della direttiva 76/160/CEE. *Gazzetta Ufficiale - Supplemento Ordinario* n. 155 del 4 luglio 2008.

Italia, (2001). Decreto Legislativo 2 febbraio 2001, n. 31 .Attuazione della direttiva 98/83/CE relativa alla qualita' delle acque destinate al consumo umano. *Gazzetta Ufficiale Supplemento Ordinario* n. 41 n. 52 del 3 marzo 2001.

Jeng, R.S.; Svircev, A.M.; Myers, A.L.; Beliaeva, L.; Hunter, D.M. & Hubbes, M. (2001).The use of 16S and 16S-23S rDNA to easily detect and differentiate common Gram-negative orchard epiphytes. *J Microbiol Methods*, 44:69-77.

Johansson, A.; Aspan, A.; Bagge, E.; Båverud, V.; Engström Björn, E. & Johansson, K.E.(2006). Genetic diversity of *Clostridium perfringens* type A isolates from animals, food poisoning outbreaks and sludge. *BMC Microbiol*, 6:47.

Lackey, RT (2001). Value policy and ecosystem health. *BioScience*;51:437-44.

Leeming, R.; Nichols, P.D. & Ashbolt, N.J. (1998). Distinguishing sources of fecal pollution in Australian inland coastal waters using sterol biomarkers and microbial fecal indicators. In Research Report No. 204 Melbourne: *Water Services Association of Australia*.

Lehner, A.; Tasara, T. & Stephan, R, (2004). 16S rRNA gene based analysis of Enterobacter sakazakii strains from different sources and development of a PCR assay for identification. *BMC Microbiol*, 4(1):43.

Lenzi, M. & Mattei, N. (1998). Risultati di quattro anni di gestione dell'ecosistema lagunare di Orbetello attraverso la raccolta delle biomasse algali. *Biologi Italiani*, XXVIII, 2, 7-12.

Lenzi M.(1992). Experiences for the management of Orbetello Lagoon : eutrophication and fishing. Science of the Total Environment (suppl. 1992), 1189-1198. *Elsevier Science* Pubbl. B.V. Amsterdam.

Leoni, O.; Cosentino, M.; Michielotto, D.; Oria, C.; Massimo, E.; Lecchini, S. & Frigo, G.M (1998). La segnalazione spontanea delle reazioni avverse ai farmaci: indagine conoscitiva in ambito ospedaliero. *Le Basi Razionali della Terapia*, XXVIII (3-4): 233

Madej, M.A. (2005).The role of organic matter in sediment budgets in forsted terrain. Sediment budjets 2. In: *Proceedings of symposium 81 held during the Seventh IAHS Scienfific Assembly*. Foz do Iguacu, Brazil, April 2005). IAHS Publication 292.

Mancini, L.; Rosemann, S.; Puccinelli, C. ; Ciadamidaro, S. ; Marcheggiani, S. & Aulicino, F. A., (2008). Microbiological indicators and sediment management. *Annali dell'Istituto Superiore di Sanità*; 44(3):268-72.

Mancini L., Rosermann S., Aulicino A.F., Carere M., Miniero R., & Marcheggiani S., (2011). *Clostridium perfringens* Vitality as an Ecotoxicity Test for Measuring the Lead Concentration in Sediment. *Proceedings of Sixth International Conference on Remediation of Contaminated Sediments*- Battelle February 7-10 2011 New Orleans Louisiana.

Mancini, L.; Marcheggiani, S.; Puccinelli, C., Iaconelli, M.; D'Angelo, A.M.; Pierdominici, E.; Formichetti, ; Equestre, M.; Aulicino, F.A.; Floris, B.; Rosselli, P.; Ammazzalorso, P.; Le Foche, M.; Zaottini E. & C. Fabiani. (2010). "A molecular approach for the impact assessment of fecal pollution in river ecosystems." *Toxicological and Environmental Chemistry*, 92(3):581-591.

Marcheggiani S., Scenati R., Vendetti C., Mario C., Musmeci L.,. Cicero M. R, Beccaloni E., Mancini L.. (2011) Evaluation of the Quality of Contaminated Sediment using Sulphite-reducing bacteria - Orbetello lagoon (Italy)". *Sixth International Conference on Remediation of Contaminated Sediments* (New Orleans, Louisiana; February 7–10, 2011) (abstract and poster)

Marcheggiani, S.; Iaconelli, M.; D'Angelo, A. & Mancini, L. (2004). Health of river ecosystems: sulphite reducing clostridia as indicators of the state of sediments. Edited by ISS, 38 p. *Rapporti ISTISAN* 04/37.

Marcheggiani, S.; Iaconelli, M.; D'Angelo, A.M.; Pierdominici, E.; La Rosa, G.; Muscillo, M.; Equestre, M. & Mancini L. (2008). Microbiological and 16S rRNA analysis of sulphite reducing clostridia from river sediments in central Italy. *BMC Microbiology* 8:171.

McCave, IN. (1988). Particulate size spectra, behaviour and origin of nepheloid layers over the Nova Scotian Continaental Rise. *J Geophysical Res* (C12):7647-66.

Minton, N.P. & Clarke, D.J. (1989). Clostridia. Biotechnology handbooks. 2 Series. New York and London: *Plenum Press*;. pp. 2-28. 24.

Moss, A.J. (1963). The physical nature of common sandy and pebbly deposit. Part II. *Am J Sci*;260:97-343.

Niilo, L. (1980). *Clostridium perfringens* in animal disease: A review of current knowledge. *Can Vet J*, 21:141-148.

Noji, E.K. (1997). The Public Health Consequences of Disasters. New York, *Oxford University Press.*

Olivieri, V.P. (1982). Bacterial Indicators of Pollution. In *Bacterial indicators of pollution* Edited by: Minton NP, Clarke DJ. Boca Raton, FL: *CRC Press*;21-41.

Pahren, H. (1987), Microorganisms in municipal solid waste and public health implications. CRC *Critical Reviews in Environmental Control* 17, 187–228.

Payment, P. & Franco, E. (1993). *Clostridium perfringens* and somatic coliphages as indicators of the efficiency of drinking water treatment for viruses and protozoan cysts. *Appl Environ Microbiol*;59:2418-2424.

Raju, D.; Waters, M.; Setlow, P. & Sarker, M.R. (2006). Investigating the role of small, acid-soluble spore proteins (SASPs) in the resistance of *Clostridium perfringens* spores to heat. *BMC Microbiol*, 6:50.

Rhodes, M.W. & Kantor, H. (1999). Sorbitol-fermenting bifidobacteria as indicators of diffuse human faecal pollution in estuarine watersheds. *J Appl Microbiol*, 87:528-535.

Robles, S.;, Rodrìguez J.M.; Granados, I. & Guerrero, M.C. (2000). Sulfite-reducing clostridia in the sediment of high mountain lake (Laguna Grande, Gredos, Spain) as indicators of faecal pollution. Int Microbiol, 3:187-91

Rood, J.I.; Mc Clane, B.A.; Songer, J.G. & Tithall, R.W. (1997). The clostridia: molecular biology and pathogenesis. San Diego, California: *AP press*;. pp. 126-133

Salomons, W. & Brils, J. (2004). Contaminated sediments in European river basins. European Sediment Research Network SedNet. EC Contract No. EVKI-CT-2001-20002, Key Action 1.4.1 Abatement of Water Pollution from Contaminated Land, Landfills and Sediments. Den Helder, TNO. The Netherlands

Schäfer, H. & Muyzer, G. (2001). Denaturing gradient gel electrophoresis in marine microbial ecology. In Methods in microbiology, ed. J.H. Paul, Vol. 30, 425–68. London: *Academic Press.*

Selvaggi T.M.; Rezza, G.; Scagnelli, M.; Rigoli, R.; Massu, M.; De Lalla. F.; Pellizzer, G.; Tramarin, A.; Bettini, C.; Zampieri, L.; Belloni, M.; Dalla Pozza, E.; Marangon, S.; Marchioretto, N.; Togni, G.; Giacobbo, M.; Todessato, A. & Binkin, N. (1996).

Investigation of a Q fever outbreak in Northern Italy. *Eur J Epidemiol*;12:403–408.

Shepard, J. (1965) .Organic aggregation in seawaterby alkaline precipitazion of inorganic nuclei during the formation of ammonia by bacteria. *Jour Gen. Microbial* 41:20.

Sieburth, J. (1965). Organic aggregation in seawater by alcaline precipitation of inorganic nuclei during the formation of ammonia by bacteria. J Gen Microbiol;41:20.

Sitte, J.; Akob, D.M.; Kaufmann, C.; Finster, K.; Banerjee, D.; Burkhardt, E.M.;. Kostka, J.E; Scheinost, A.C.; Büchel, G. & Küsel, K. (2010). Microbial links between sulfate reduction and metal retention in uranium- and heavy metal-contaminated soil. *Appl. Environ. Microbiol.* 76(10):3143-52.

Solo-Gabriele, H.; Wolfert, M.; Desmarais, T. & Palmer, C. (2000). Sources of *E. coli* to a SubTropical Coastal Environment. *App Environ Microbiol*, 66(1):230-237.

Sterne, M. & Warrack G.M. (1984). The types of *Clostridium perfringens*. *J Pathol Bacteriol* 88:279-83.

Stroffolini, T.; Biagini, W.; Lorenzoni, L.; Palazzesi, G.P.; Divizia, M. & Frongillo, R. (1990). An outbreak of hepatitis a in young adults in central Italy. *Europ J Epidemiol*;6(2):156-159.

Tauxe, R.V. (1997). Emerging Foodborne Diseases: An Evolving Public Health Challenge. *Emerg Infect Dis*;3,4:425-434.

Tyagi VK; Chopra, A.K.; Kazmi, A.A. & Kumar, A. (2006) Alternative microbiol indicators of faecal pollution: Current Perspective. *(IAEH)Iran Ass Environ Health*, 3:205-216.

UNEP - United Nations Environment Programme. *Global Environment Outlook*, 1st ed., New York and Oxford, Oxford University Press, 1997.

UNEP - United Nations Environment Programme. *Human Development Report*. New York and Oxford, Oxford University Press.

US EPA - U.S. Environmental Protection Agency Ashbolt N., Fujioka R., Glymph T., McGee C., Schaub S., Sobsey M., Toranzos G. Chapter 2 "*Pathogens, Pathogens indicators and indicators of fecal contamination*" Pgs 35-56. In: Report of the expert scientific workshop on critical research needs for the development of new or revised recreational water quality criteria. EPA823-R-07-006. Washington, DC: US EPA , 2007.

van Beelen, P. & Doelman P. (1997). Significance and application of microbial toxicity tests in assessing ecotoxicological risks of contaminants in soil and sediment. *Chemosphere*;34:455-99.

Van Metre, P.; Mahler B. & Furlong, E. (2000). Urban sprawl leaves its PAH signature. *Environ Sci Technol*, 34:4064-4070.

Visher, G.S. (1969). Grain size distributions and depositional processes. *Jor. Of Sed. Petrol.*, 39 (3): 1074-1106.

Wassermann, P.; Naas, K.E. & Johannessen, PJ. (1986). Anual supply and loss of particulate organic carbon in Nordasvannet, a eutrophic, land-locked fjord in western Norway. Rapp P *Rèun Cs Int Explo Mer*;186: 423-31.

Weisburg, W.G.; Barns, S.M.; Pelletier, D.A. & Lane, D.J. (1991). 16S ribosomal DNA amplification for phylogenetic study. *J Bacteriol*, 17:697-703.

WHO - World Health Organization. *Guidelines for drinking water quality*. Second Edition, vol. 2, Health criteria and other supporting information. Geneva, World Health Organization, 1996

WHO - World Health Organization. *Guidelines for drinking water quality*. Second Edition, vol. 2, Health criteria and other supporting information. Geneva, World Health Organization, 1996.*Water Res* 1989, 23:191-197.34. International Hydrological Decade – World Health Organization (IHD-WHO): Water quality surveys; a guide for the collection and interpretation of water quality data. In *Studies and Reports in Hydrology (UNESCO), no. 23/International Hydrological Decade, 75 – Paris (France)* Geneva: World Health Organization; 1978:57-54.

Woese C.R. (1987). Bacterial evolution. *Microbiol Rev*, 51:221-271.

7

Biodiversity of Macroinvertebrates in Oxbow-Lakes of Early Glacial River Basins in Northern Poland

Krystian Obolewski
Pomeranian University in Słupsk, Department of Aquatic Ecology
Poland

1. Introduction

River basins are very complex systems which include both abiotic and biotic elements. Such a high number of components results in a situation that normal functioning of river with adjacent areas depends on a set of hydrological and geological processes. They influence biological diversity observed in river basins (Arscott et al., 2005; Marshall et al., 2006). Proper assessment of a lotic system should include not only the main watercourse but also wetlands flooded in spring and autumn. The diversity of habitats in natural river valleys increases biological diversity of aquatic ecosystems and thus the quality of environment (e.g., Boulton et al., 1992; Clausen & Biggs, 1997; Tockner et al., 1999; Gibbins et al., 2001; Sheldon et al., 2002; Arscott et al., 2003; Robinson et al., 2003, 2004; Arscott et al., 2005; Whiles & Goldowitz, 2005; Gallardo et al., 2008; Reese & Batzer, 2007; Obolewski, 2011a; Obolewski & Glińska-Lewczuk, 2011).

Each meandering, lowland River is forming its riverbed constantly. Often after rapid floods it turns out that a river flows in new riverbed and the cut-off fragments transform into oxbow-lakes (Amoros & Roux, 1988). They can be filled up during river rises and undergo succession (Junk et al., 1989; Tockner et al., 2000). Due to the diversity of river rise intensity, the connectivity between an oxbow-lake and a river can vary. Therefore, we distinguish lentic, semi-lotic and lotic oxbow-lakes. The first type is supplied by river waters under high water table level or by ground waters while the two remaining types are partly or totally connected to a river. Limited exchange of water in a river valley as well as its agricultural use causes that oxbow-lake drainage area often undergoes anthropopression which leads to quick eutrophication and massive phytoplankton blooms. As a result, water contains considerable amounts of biogenes, mineral salts but low oxygen content. Additional unfavourable factors are hydromorphological features of oxbows, i.e. small area (between a few hundreds squared meters and a few hectares) and depth which usually does not exceed 3 meters. As a result oxbows quickly react to changes in temperature and thanks to that they are perfect objects for the research on climate changes even in the global depiction (Klimaszyk, 2004).

The structure and functioning of wetland ecosystems, including oxbow-lakes, are directly and indirectly connected with the fluctuations in water table level of rivers during floods or flow pulsations (Junk et al., 1989; Tockner et al., 2000). According to Amoros & Roux (1988), hydrological connectivity determines the processes that take place both in a river and adjacent wetland systems. That concerns the transport of dissolved or suspended elements and organisms, environment reorganization, productivity and biodiversity of aquatic and land ecosystems (Amoros & Roux, 1988; Junk et al., 1989; Poff & Ward, 1989; Heiler et al., 1995; Ward & Stanford, 1995; Poff et al., 1997; Ward et al., 2002; Tockner et al., 2000). Limited hydrological connectivity as well as hydrotechnical appliances located on rivers of moderate river slope result in hydrological changes which can lead to the loss of many valuable ecosystems, including oxbow-lakes (Poff & Ward, 1989; Dynesius & Nilsson, 1994; Heiler et al., 1995; Poff et al., 1997; Tockner et al., 1999; Ward et al., 2002; Thorns & Sheldon, 2000). Consequently, the flora and fauna migration in a river valley occurs only during the periods of water exchange between wetlands and the main river channel, which deteriorates the river ecosystem biodiversity and functioning (Amoros & Roux, 1988; Heiler et al., 1995; Walker et al., 1995). Those changes can be observed by the monitoring of a selected hydrobiont group and its changes in terms of qualitative and quantitative structure that reflect the ecological state of a river valley.

Among various hydrobiont groups, benthic invertebrates are particularly attractive research objects, because their life cycles are short and they also easily adapt to environmental conditions due to, for instance, their shape and dispersion (Gasith & Resh, 1999). High abundance of benthic macroinvertebrates allows using them as indicator organisms in the monitoring of aquatic ecosystem quality (e.g. Boon, 1988; Obolewski et al., 2009). Moreover, some groups of macroinvertebrates periodically reach their maximum abundance and diversity in relation to spatial and environmental variables, however those phenomena have not been fully understood so far (e.g., Van der Brink et al., 1991; Tockner et al., 1999; Heino, 2000; Griffith et al., 2001; Arscott et al., 2005; Monaghan et al., 2005).

This study investigates the relationships between macrozoobenthos communities and the level of hydrological connectivity between wetlands and the main river channels in the river valleys of three lowland rivers located in northern Poland: Słupia, Łyna and Drwęca. The studied wetlands were selected according to the hypothesis, that the level of hydrological contact between oxbow-lakes and rivers influences the qualitative and quantitative structure of benthic macroinvertebrates inhabiting wetlands.

2. Study area

2.1 Characteristics of the studied river valleys in northern Poland

The study area covered the wetlands of middle Łyna, Pasłęka, Drwęca and Słupia rivers. They were chosen because of geomorphologic and hydrological similarities. They are also located in areas of similar climatic conditions and shows hydrological regime typical of lakelands.

Łyna River Valley

The studied oxbows are located in the middle section of the Łyna River Valley- the largest tributary of Pregoła which flows into the Zalew Wiślany reservoir. The drainage area of Łyna reaches 7126 km^2 and 5773 km^2 is within the territory of Poland while the lower part of drainage area with the estuary are located in Russia. Total length of the river is around 290 km. Łyna is characterized by varied, natural and valuable landscape of its valley thanks to polygenetic processes that occurred during ice-sheet recession in the Baltic glaciation (Michniewska - Szczepkowska & Szczepkowski, 1969). The river begins its course at the

elevation about 160 m.a.s.l., north-east from Łyna village, by the western edge of Lasy Napiwodzkie and 10 km to the north of Nidzica. The spring of Łyna is fairly abundant due to the presence of impermeable Pliocene loam. Spring streams formed a deep and branch ravine. Łyna flows through many lakes (Persing, Kiernoz Wielki and Kiernoz Mały, Łańskie, Ustrych), among which the Łańskie Lake is the largest. At the latitude of Kiernoz Wielki one of the largest tributaries- Marózka- flows into the Łyna River. Further course of Łyna is diverse in terms of its direction through morainic elevation and rich with numerous meanders. Just below the city of Olsztyn Łyna joints with the second largest tributary- Wadąg- which is 20 km long and with drainage area of 1218 km^2. The average slope of the upper course of the Łyna River is at the level of 1- 2 ‰.
In the middle course of the river, between Dobre Miasto and Lidzbark Warmiński, Łyna reveals the features of a meandering river, washes away river banks and transports considerable amounts of fine sediments. In the vicinity of Sępopol the river joints with such tributaries as Sunia and Kirsna, located in the area of the studied oxbows as well as Symsarna (drainage area of 276 km^2), Ekna (drainage area of 301 km^2) and Guber (length of 73 km, drainage area of 1682 km^2). Passing Sępopol the river crosses the border near Znamieńsk and flows into Pregoła at the altitude of 4 m.a.s.l. The slope in its lower course is around 0.4 ‰. The mean annual flow near the border is 38.5 m^3/s while the spread of water stages reaches about 4 m.

Parameter	Unit	River - gauging station		
		Łyna-Smolajny	Drwęca-Rodzone	Słupia-Słupsk
No. of kilometres (from estuary)	km	172.0	126.7	31.6
Drainage area	km^2	2290	2725	1450
Mean annual river flow SSQ	m^3 s^{-1}	14.7	11.2	15.0
$WQ_{10\%}$	m^3 s^{-1}	50.5	46.1	49.0
$NQ_{10\%}$	m^3 s^{-1}	4.49	4.50	6.02
Mean low river flow SNQ	m^3 s^{-1}	6.45	6.30	7.45
Flow coefficient *cv*	-	0.439	0.442	0.450
Oscillation of water stages	cm	207	191	171
Watercourse slope	m km^{-1}	0.32	0.30	0.18
Width of valley-bottom zone	m	372	618	315
River sinuosity	-	1.36±1.82	1.66±2.40	1.21±2.13
Mean number of oxbows per1 km of river valley length	no/1km	3	10	4

Table 1. Characteristics of the studied rivers at sections with meandering riverbed: Łyna (Smolajny), Drwęca (Rodzone) and Słupia (Słupsk)

The most interesting section of the Łyna River Valley is its middle part, rich with oxbow-lakes, which was included in this study. The mean annual flow over years 2007-2009 at the cross-section located in Smolajny (172 km) reached 14.7 m^3 s^{-1} while the mean low river flow was almost 6.5 m^3 s^{-1} (Table 1). Moreover, the highest water stage amplitudes were observed there comparing to the remaining rivers studied in this investigation.

Drwęca River Valley

The Drwęca River is a right-side, large tributary of Vistula, with the length of 207.2 km and drainage area of 5343.5 km^2. The spring of Drwęca is localized at the northern foot of the Dylewskie hills, to the south of Drwęck. The river flows into Vistula near Złotoria (vicinity of the city of Toruń). The upper section of the riverbed is a ravine, 20-30 m deep and 8 km long, called the "Czarci Jar". First the river flows to north-west as a small stream in a valley covered with forest and surrounded by agricultural areas, crosses the Ostrowin Lake and then joins with its tributary- the Grabiczek River. Further, the river changes its direction into west and flows into the Drwęckie Lake. Passing that lake the river is regulated and flows to south-west, receiving the waters of Poborska Struga, Gizela, Elszka (left-sided tributaries), Ruda and Iławka (right-sided tributaries). In the town of Bratian Drwęca joins with its largest tributary- the Wel River.

Starting from the town of Bratian the river is meandering and flows in a deep valley. After joining with the left-sided tributary- the Brybica River- Drwęca flows to north-west direction. Near Brodnica the river joins with its tributary- Skarlanka, then Rypienica and Struga Wąbrzeska. In the vicinity of Młyniec the river changes its characteristics. After building a dam in 1997 in Lubicz the river formed a reservoir with the area of 50 ha, where the water flow decreased to 0.2-0.3 $m^3\ s^{-1}$. In the lower section of the Drwęca River the most important tributaries are: Struga Kowalewska, Struga Rychnowska and Struga Lubicka. After passing the town of Nowa Wieś the river turns to north-west and flows into the Vistula River in Złotoryja (Mileska, 1992).

The drainage area of Drwęca is localised on clay, sand and gravel of glacial origin with peatlands in depressions. It was formed during the Baltic glaciation in the Toruń-Eberswald pra-valley. The studied part of valley is located 127 km from the estuary on ground moraine and 6 km wide. Typical of this section is a considerable number of oxbow-lakes (Table 1). The river flow of Drwęca is relatively stable thanks to the retention ensured by numerous lakes and forests. In general no catastrophic floods or low water stages are observed in the valley (Bralczyk, 1996). The average annual river flow in years 2007–2009 at Rodzone (127 km) was 11.2 $m^3\ s^{-1}$ while the low water flow slightly exceeded 6 $m^3\ s^{-1}$ (Table 1).

The Drwęca River Valley is under various forms of legal protection. In 1961 the whole river with lakes Drwęckie and Ostrowin, some of the tributaries and 5 m wide strip of river bank became a reserve in order to protect breeding ground for salmon fish. Moreover, the Area of Protected Landscape in the Lower Drwęca River Basin has been created with the area of 17 472.4 ha as well as the Area of Protected Landscape of the Upper Drwęca River Valley (area 8 039.5 ha). It is planned to include the river with some of its tributaries (area of 2 162.1 ha) to the European Ecological Network Nature 2000.

Słupia River Valley

The Słupia River flows into the Baltic Sea. The whole river basin is located in the Pomeranian Voivodeship. It borders the Wieprza River Basin (west), the Brda River Basin (south) as well as the Łeba and Łupawa river basins (east). The length of Słupia is 138.6 km (according to the „Hydrographic division of Poland", IMGW, 1983) while its basin covers the area of 1310 km^2. The spring of Słupia is in the Kaszuby Lake District near Sierakowska Huta at the height of 178 m.a.s.l. The river estuary is in Ustka and its average slope is around 1.3 per mill.

The characteristic features of Słupia formed around 15 000 - 10 000 lat BP. Melting ice-sheet created a net of sub-glacial gullies, marginal and riverine valleys. The Słupia's tributaries are strongly meandering water-courses. In the upper section the river flows through lakes Tuchlińskie, Pręgożyno, Skrzynka, Trzebocińskie, Gowidlińskie, Węgorzyno, Żukowskie.

The tributaries of the Słupia are asymmetrical, mostly located on the left side of the river (Skotawa- the largest tributary, Glaźna, Gnilna). On the right side there are Parchowska Struga, Stropna, Bytowa, Kamienica, Brodek, Kamienna, Żelkowa Woda, Kwacza.

Hydrological characteristic of the Słupia River is typical of the Pomeranian Region:

- domination of ground water supply (70-75% according to Paszczyk, 1976);
- even annual outflow determined by climatic conditions (stable precipitation, mild winters) and physiographic conditions (large forest area and flow-through lakes);
- low amplitude of water stages.

The average outflow in spring constitutes around 130-180% of average monthly outflow. The lowest outflow occurs in summer (mainly July). The floods are small, mostly in December, January or March and connected with melting snow. The ice phenomena in the Słupia River basin start in late November or early December. Due to the presence of fast currents the ice covers only part of the river (channels, still water bodies) and melts between 15 and 20 of March.

Numerous oxbow-lakes are located in the middle part of the Słupia River and they were mainly formed after river regulation at the beginning of XXth century. Nowadays they are within the borders of "Dolina Słupi" Landscape Park. Its area is the lowest but water outflow is the highest (Table 1).

2.2 Selected oxbow-lakes located in lowland river basins of northern Poland

The research covered 14 oxbow-lakes (four in the Drwęca river valley, five in Łyna and five in the Słupia river valleys)- 8 of them were cut-off, 4 were open and 2 were semi-open reservoirs. Those oxbows differed in habitat conditions and morphometric features (Table 2).

The oxbow-lakes in the Drwęca river valley were formed during regulation works and all of them are cut off from the main river channel (lentic). OLD 1, OLD 2 and OLD 4 are located on the right side of the river while OLD 3 is situated on the left side (Fig. 1). Their length varies from 140 to 1200 m (Table 2). The main shape of those reservoirs is determined by one river bend of sinuosity S around 5.5. In the valley there are also oxbows almost parallel to the riverbed (S=1.5) as well as better developed (S=9.8). Water table area of the studied reservoirs ranges from 0.1 to 1.2 ha and their volume from 5.3 to 16.7 thousand m^3 (in relation to the Drwęca water level H=100 cm on the watermark in Rodzone). Morphometric features change over time for each reservoir individually. The flat fragment of the river valley in the vicinity of Bratian does not limit the changes in water table level. The fluctuations in Drwęca water level cause that the volume and area of oxbows can be twofold different.

The oxbow-lakes in the Łyna river valley are located near the Smolajny village and were also formed as a result of regulation works. Three of them (OLŁ 2, OLŁ 3 and OLŁ 5) are connected with the river on both sides (lotic) and OLŁ 1 through one arm (semi-lotic) while OLŁ 4 is cut off from the river (lentic). The length of those reservoirs ranges from 420 to 700 (Table 2) but most of them reach 200-400 m (43% of all objects). Their main shape is determined by one bend of sinuosity S around 3.0. In the valley there are also oxbows almost parallel to the riverbed (S=1.2) as well as better developed (S=9.5). The area of oxbows varies between 0.3 and 1.5 ha while their volume between 2.2 and 25.3 thousand m^3 (in relation to the Łyna water level H=100 cm on the watermark in Smolajny). Morphometrical features change over time for each reservoir individually. Steep and high banks reduce the range of changes in water table area. The fluctuations in Łyna water level cause that the volume of oxbows can be 4-fold different while the area only 1.9-fold different.

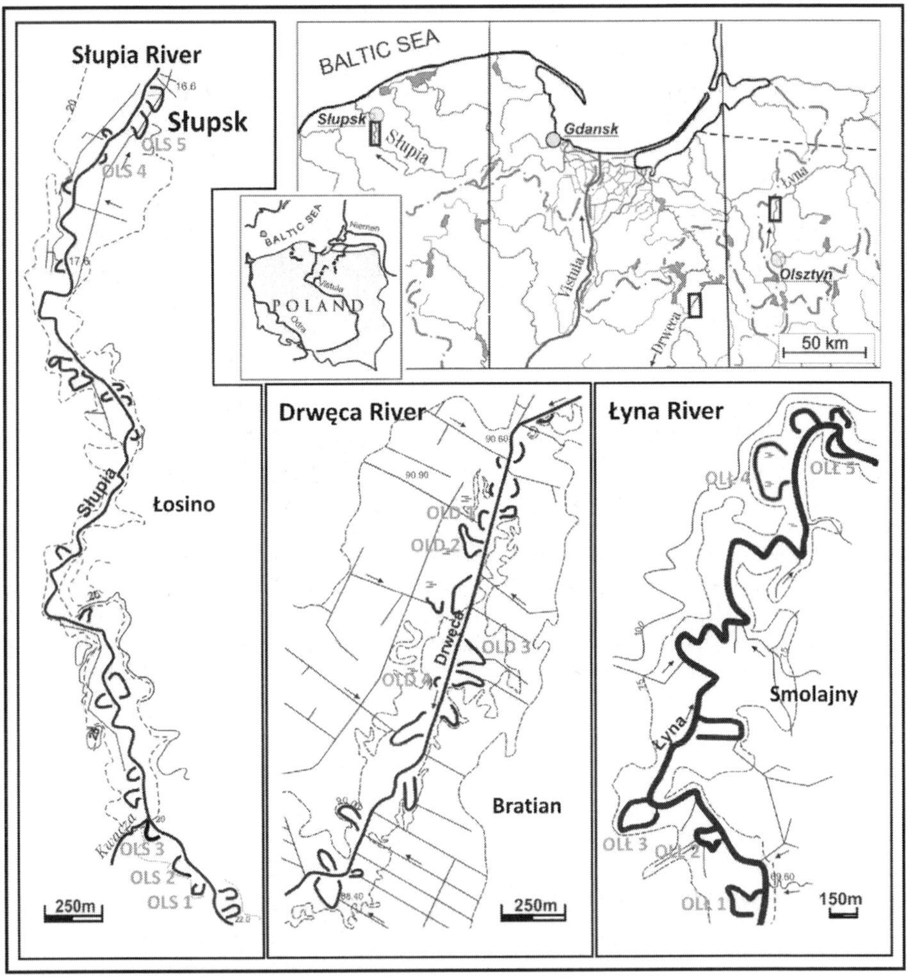

Fig. 1. Localisation of sampling sites

The studied oxbow-lakes in the Słupia river valley are located within the 15 km long river section. Three of them (OLS 1, OLS 2 and OLS 3) are near Kwakowo in the area of "Dolina

Oxbow-lake	River	Geographical coordinates	Type of connection with river	Length D [m]	Width S [m]	Area A [ha]	Sinuosity S=D/a-c [-]	Distance between arms a-c [m]	Distance from river:			Width		Depth		Volume
									Upper arm [m]	Lower arm [m]	Max. [m]	River S_{rz} [m]	River valley WWW S_{dol} [m]	$h_{śr}$ [m]	h_{max} [m]	$V_{śr}$ [thou m^3]
OLD-1	Drwęca	N 53°28′06.61″ E 19°35′56.96″	LE	481	11.2	0.49	9.8	49	37	26	237	11.1	296	1.07	3.11	5.29
OLD-2	Drwęca	N 53°27′51.98″ E 19°36′18.11″	LE	144	15.2	0.20	1.5	95	34	25	100	12.2	231	0.95	3.20	1.89
OLD-3	Drwęca	N 53°27′53.98″ E 19°36′34.16″	LE	1210	14.0	1.51	6.4	190	22	11	462	10.9	425	1.12	2.27	16.88
OLD-4	Drwęca	N 53°28′50.20″ E 19°36′04.74″	LE	347	19.2	0.60	4.3	80	44	14	193	14.0	434	1.16	1.71	6.92
OLŁ-1	Łyna	N 54°01′17.14″ E 20°24′00.39″	SLO	424	15.1	0.92	5.4	79	23	0	180	17.0	462	1.16	2.60	10.70
OLŁ-2	Łyna	N 54°01′38.33″ E 20°24′15.23″	LO	464	16.5	0.79	6.7	69	15	0	198	16.0	227	1.25	2.40	9.90
OLŁ-3	Łyna	N 53°57′49.68″ E 20°25′04.45″	LO	713	25.0	2.64	1.2	594	100	100	200	15.0	200	0.11	0.90	2.90
OLŁ-4	Łyna	N 54°05′40.38″ E 20°26′18.53″	LE	560	31.0	1.74	2.6	219	38	22	342	16.2	409	1.45	1.80	25.30
OLŁ-5	Łyna	N 54°05′26.41″ E 20°26′14.52″	LO	320	22.7	0.26	8.0	30	0	0	159	19.9	200	0.85	2.30	2.20
OLS-1	Słupia	N 54°23′07.31″ E 17°02′25.38″	LE	333	9.8	0.36	5.8	57	82	66	209	14.9	160	0.58	1.25	2.06
OLS-2	Słupia	N 54°23′14.57″ E 17°02′15.94″	LE	329	10.3	0.35	1.8	183	103	35	180	15.7	190	0.63	1.62	2.19
OLS-3	Słupia	N 54°23′25.62″ E 17°01′59.16″	SLO	143	10.5	0.14	1.1	131	124	0	118	12.4	217	0.6	1.30	0.82
OLS-4	Słupia	N 54°26′45.21″ E 17°01′38.70″	LO	125	9.1	0.13	1.6	77	16	23	59	15.2	374	0.88	1.82	1.14
OLS-5	Słupia	N 54°26′52.99″ E 17°01′58.41″	LO	512	10.3	0.71	2.8	186	37	22	180	14.5	308	0.79	1.65	5.51
Mean	-	-	-	**436.1**	**15.7**	**0.8**	**4.2**	**145.6**	**48.2**	**24.6**	**201.2**	**14.6**	**295.2**	**0.9**	**2.0**	**6.7**

Table 2. Morphological characteristics of the studied oxbow-lakes (LO –lotic, SLO- semi-lotic, LE-lentic)

Słupi" Landscape Park while OLS 4 and OLS 5 are situated near Słupsk (Fig. 1). OLS 1 and OLS 2 are cut off from the main river channel (lentic), OLS 3 is connected by one arm (semi-lotic) and the remaining two oxbows (OLS 4 and OLS 5) are connected to the river by pipes or melioration ditches (lotic). The Słupia oxbow-lakes are reservoirs of length ranging from 125 to 500 m (Table 1). Most of them are oxbows of not large width (around 10m). Their main shape is determined by one bend of sinuosity S around 3.0. In the valley there are also oxbows almost parallel to the riverbed (S=1.1) as well as moderately developed (S=5.8). The oxbow areas are similar and range from 0.1 to 0.7 ha while their volume from 0.8 to 5.5 thousand m^3 (in relation to the Słupia water level H=100 cm on the watermark in Słupsk). Morphometrical features change over time for each reservoir individually. The fluctuations in Słupia water level cause that the volume and area of oxbows can be 4-fold different.

3. Material and methods

In order to investigate the biodiversity of macroinvertebrates we chose 12 oxbow-lakes located in the valleys of three lowland rivers- Drwęca, Łyna and Słupia. The oxbows were classified into three types: lotic, semi-lotic and lentic. The sampling was performed in years 2007-2009 through four seasons (winter, spring, summer, autumn). Three sampling sites were locates in each oxbow-lake- in the middle of the reservoir and in both arms. Bottom sediments with macrozoobenthos were sampled using the Ekman's grab sampler (surface 225 cm^2) three times at each site and related to the bottom area of 1 m^2. The sediments were sieved through a 1 mm mesh size sieve and fixed in 4% formalin. Then, macrozoobenthos was separated from sediments in a laboratory and identified to taxonomic units. Further analysis consisted of the calculations of biocenotic indices.
The constancy of occurrence of the identified taxa was calculated according to the following formula (1):

$$C = N_a/n \cdot 100\% \tag{1}$$

where: N_a – number of samples with a given taxon
n – total number of samples
The index of constancy of occurrence is mostly used as the parameter which indicates the level of connection between a taxon and the environment. According to Tichler the index is classified as follows:
C_4 – euconstants (75.1 – 100%),
C_3 – constants (50.1 – 75.0%),
C_2 – accesoric taxa (25.1 – 50.0%),
C_1 – accidental taxa (<25.0%).
The domination index in terms of density is defined by the formula (2):

$$D_A = A/A_{av} \cdot 100\% \tag{2}$$

where: A – average density of a given taxon
A_{av} – total average density for the whole reservoir
It was classified after Biesiadko & Kowalik (1980) into:
eudominants (abundance above 10%),
dominants (5.01-10%),
subdominants (2.01-5%),

recedents (1-2%),
subrecedents (below 1%).
The ecological importance index Q was calculated according to the formula (3)

$$Q = \sqrt{C \cdot D} \tag{3}$$

where: D - domination index (in terms of density)
C - constancy of occurrence.
The following classification was applied in this study:
Q5 - very high >30.00%,
Q4 - high 15.01-30.00%,
Q3 - moderate 10.01-15.00%,
Q2 - low 5.01-10.00%,
Q1 - very low <5.00%.
The next parameter used in this investigation was the Shannon diversity index (4):

$$H' = -\sum_{i=1}^{S} p_i \log_2 p_1 \quad \text{where:} \quad p_i = \frac{n_1}{N} \tag{4}$$

where: n_i - species abundance
N - total density of taxonomic groups
S - number of taxonomic groups in the studied samples.

4. Results and discussion

4.1 Qualitative structure of benthofauna in the studied oxbow-lakes

4.1.1 Łyna River basin

Macrozoobenthos in the oxbows of the Łyna River was represented by 41 taxa belonging to 12 taxonomic groups. Benthofauna was predominated by Diptera of *Chaoborus* genus (Table 3). They were particularly abundant in lentic oxbow-lakes where they constituted over 50% of the total macrozoobenthos abundance and were classified as constants. In OLŁ1 they were accesoric species. That situation indicates a bad condition of the OLŁ1-OLŁ3 ecosystems (Glińska-Lewczuk, 2009). The share of Diptera of *Chaoborus* genus was much lower in the two remaining oxbow-lakes (OLŁ4 and OLŁ5). As for the other taxa, they were relatively abundantly observed in OLŁ1. Oligochaeta, Chironomidae larvae and Bivalvia reached there the status of accesoric species. Oligochaeta also predominated the bottom of OLŁ5, being a constant species.
The qualitative composition of benthofauna in the oxbows of the Łyna River indicated their considerable degradation through succession (Kajak, 1988, 2001). As a result the benthic invertebrates are eliminated (Lapmert & Summer, 2007) with stenobionts in particular because their adaptability to changing environmental conditions are very limited (Allan, 1995). As bioindicators they belong to accesoric species.
The pattern of benthofauna constancy of occurrence was similar in all the oxbow-lakes of the Łyna River. Extreme values were reached in OLŁ2 and OLŁ3, which was connected with the concentration of Diptera larvae represented by *Chironomus* sp. (Fig. 2). The median of benthofauna constancy of occurrence for 80% of the Łyna oxbow-lakes was equal to zero, probably because of the presence of many occasional species. Only the OLŁ1 oxbow lake differed in terms of median and no predominant species were observed there.

Taxa	Oxbow-lakes					average	classification
	OLŁ 1	**OLŁ 2**	**OLŁ 3**	**OLŁ 4**	**OLŁ 5**		
Chaoborus sp.	38.9*	20.8	58.6**	8.3	0.0	25.3	C_2
Oligochaeta	27.8*	9.0	2.2	8.3	52.9**	20.0	C_1
Chironomus sp.	27.8*	13.9	10.4	16.7	9.9	15.7	
Procladius sp.	44.4*	5.7	4.7	8.3	0.0	12.6	
Pisidium sp.	44.0*	0.0	0.0	8.3	5.1	11.5	
Erpobdella sp.	22.2	8.1	2.3	16.7	1.9	10.2	
Viviparus viviparus L.	22.2	6.7	3.4	8.3	10.5	10.2	
Sialis fuliginosa Pictet	22.2	9.3	1.7	8.3	0.0	8.3	
Asellus aquaticus L.	22.2	2.3	0.0	0.0	14.4	7.8	
Valvata pulchella Studer	0.0	0.0	0.0	25.0	0.0	5.0	
Glossiphonia sp.	22.2	0.0	1.7	0.0	0.9	5.0	
Bezzia sp.	5.6	5.4	7.5	0.0	0.2	3.7	
Cyrnus sp.	0.0	7.0	1.7	8.3	0.0	3.4	
Valvata piscinalis O.F. Müller	16.7	0.0	0.0	0.0	0.0	3.3	
Potamopyrgus antipodarum Gray	6.0	0.0	0.0	8.3	0.0	2.9	
Anodonta anatina L.	11.1	2.1	0.0	0.0	0.4	2.7	
Lymnaea stagnalis L.	11.1	0.0	0.0	0.0	1.5	2.5	
Limnephilus sp.	0.0	1.5	1.7	8.3	0.0	2.3	
Sphaerium sp.	11.0	0.0	0.0	0.0	0.0	2.2	
Dytiscus sp.	0.0	0.0	0.0	8.3	0.0	1.7	
Notonecta sp.	0.0	0.0	0.0	8.3	0.0	1.7	
Sergentia sp.	0.0	0.0	0.0	8.3	0.2	1.7	
Valvata cirstata O.F.Müller	5.6	0.0	0.0	0.0	0.0	1.1	
Galba truncatula O.F.Müller	5.6	0.0	0.0	0.0	0.0	1.1	
Anodonta cygnea L.	5.6	0.0	0.0	0.0	0.0	1.1	
Lepidostoma sp.	5.6	0.0	0.0	0.0	0.0	1.1	
Ecnomus tenellus Rambur	5.6	0.0	0.0	0.0	0.0	1.1	
Goera sp.	5.6	0.0	0.0	0.0	0.0	1.1	
Hydropsyche sp.	5.6	0.0	0.0	0.0	0.0	1.1	
Psychomidae	5.6	0.0	0.0	0.0	0.0	1.1	
Mochlonyx culiciformis Degeer	5.6	0.0	0.0	0.0	0.0	1.1	
Polypedilum sp.	5.6	0.0	0.0	0.0	0.0	1.1	
Limoniidae	5.6	0.0	0.0	0.0	0.0	1.1	
Helobdella stagnalis L.	0.0	1.5	1.7	0.0	0.0	0.6	
Chironomus sp. pupa	0.0	0.0	2.3	0.0	0.0	0.5	
Lestes sp.	0.0	2.1	0.0	0.0	0.0	0.4	
Epitheca bimaculata Charpentier	0.0	1.5	0.0	0.0	0.0	0.3	
Phryganaidae	0.0	0.0	0.0	0.0	1.5	0.3	
Dreissena polymorpha Pall.	0.0	1.5	0.0	0.0	0.0	0.3	
Caenis sp.	0.0	1.5	0.0	0.0	0.0	0.3	
Cloëon sp.	0.0	0.0	0.0	0.0	0.2	>0.1	

Table 3. Constancy of occurrence (C, %) of benthofauna representatives in the oxbow-lakes of the Łyna River with the classification of consecutive taxa (* - C_2, **- C_3)

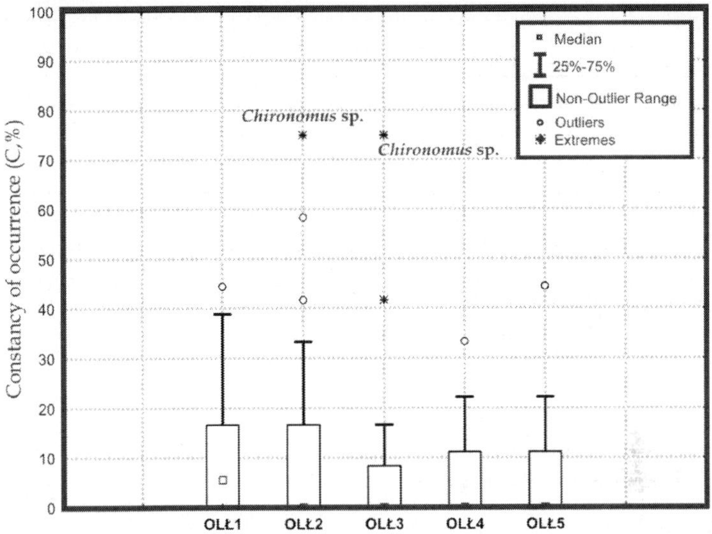

Fig. 2. Benthofauna constancy of occurrence (C, %) in the oxbow-lakes of the Łyna River

4.1.2 Drwęca River basin

Only 13 benthofauna taxa belonging to 7 taxonomic groups were observed in the oxbow-lakes of the Drwęca River. The most abundant were *Chaoborus* sp. representatives with very high constancy of occurrence (constant species, Table 4) reaching even 100% (euconstants).

Taxa	Oxbow-lakes				Mean	Classification
	OLD 1	OLD 2	OLD 3	OLD 4		
Chaoborus sp.	33.3*	100.0***	66.7**	33.3*	58.3	C_3
Chironomus sp.	8.3	75.0**	50.0*	66.7**	50.0	C_2
Erpobdella sp.	8.3	8.3	25.0	50.0*	22.9	C_1
Procladius sp.	8.3	33.3*	16.7	0.0	14.6	
Chironomus sp. pupa	0.0	25.0	33.3*	0.0	14.6	
Oligochaeta	0.0	25.0	16.7	0.0	10.4	
Bezzia sp.	0.0	8.3	8.3	0.0	4.2	
Cloëon sp.	8.3	0.0	8.3	0.0	4.2	
H. stagnalis	0.0	0.0	0.0	8.3	2.1	
Glossiphonia sp.	0.0	0.0	8.3	0.0	2.1	
Sergentia sp.	0.0	0.0	8.3	0.0	2.1	
A.aquaticus	8.3	0.0	0.0	0.0	2.1	
Cyrnus sp.	0.0	8.3	0.0	0.0	2.1	

Table 4. Benthofauna constancy of occurrence (C, %) in the oxbow-lakes of the Drwęca River with the classification of consecutive taxa (* - C_2, **- C_3, *** - C_4)

The second abundant were the larvae of *Chironomus* sp., being between accesoric and constant species. Considerable number of taxa were occasional species which occurred in some oxbow-lakes with higher abundance. Among accesoric species that situation particularly concerned leeches *Erpobdella* sp., which in OLD4 were present in half of the samples. That situation was caused by the abundance of Diptera larvae (Turoboyski, 1979). In OLD2 and OLD3 apart from the larvae of *Choborus* sp. and *Chironomus* sp., equally frequent were predatory *Procladius* sp. and chrysalis of *Chironomus* sp.

The oxbow-lakes of the Drwęca River were the most degraded ecosystems, which functioning depended on periodical floods (Tockner et al., 1999). The domination of migrating species indicated oxygen depletion in the bottom zone of those reservoirs. That suggests the need of reconnecting those oxbows to the river (Bornette et al., 1998; Gallardo et al., 2007; Ward & Stanford, 1995; Ward et al., 2002).

The constancy of occurrence of benthofauna varied between 0 and 50% (Me<10%). The only oxbow with more favourable conditions to invertebrates was OLD3 (Fig. 3). There was only one taxon which was the most often identified in each reservoir (*Chaoborus* sp.) and extreme values of constancy of occurrence were noted in OLD4.

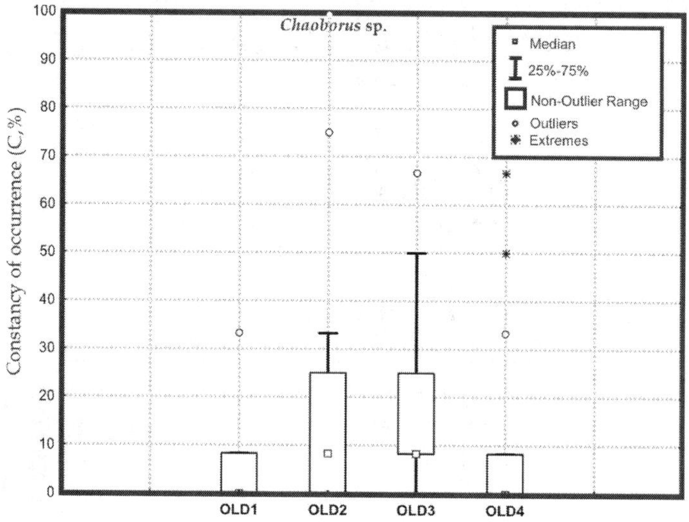

Fig. 3. Benthofauna constancy of occurrence (C, %) in the oxbow-lakes of the Drwęca River

4.1.3 Słupia River basin

31 benthofauna representatives belonging to 13 taxonomic groups were identified in the oxbow-lakes of the Słupia River (Table 4). The highest diversity of macrozoobenthos was noted in those reservoirs, probably due to regular spates connected with the functioning of hydro-power stations (Obolewski, 2011a).

The most frequently present were Crustaceans represented by *A. aquaticus* as well as Hirudinea represented by *Erpobdella* sp. (constant species). Both in the Łyna and Drwęca river basins those taxa were occasionally observed. The presence of *A. aquaticus* indicates β-mezosaprobic waters polluted with organic substances (Turoboyski, 1979). High abundance of *Erpobdella* sp. was caused by the presence of *A. aquaticus*, which they feed on. The representatives of Oligochaeta, Hirudinea, Insecta and Bivalvia were accessoric taxa. That taxonomic diversity was favoured by pulsating floods of the Słupia River (Junk et al., 1989; Tockner et al., 2000).

Taxa	Oxbow-lakes					Mean	Classification
	OLŁ 1	**OLŁ 2**	**OLŁ 3**	**OLŁ 4**	**OLŁ 5**		
A.aquaticus	58.3**	58.3**	58.3**	67.7**	83.3**	65.2	C_3
Erpobdella sp.	66.7**	66.7**	66.7**	42.0*	66.7**	61.8	
Oligochaeta	50.0*	58.3**	50.0*	16.7	67.0**	48.4	C_2
Procladius sp.	50.0*	58.3**	58.3**	41.7*	33.3*	48.3	
Chironomus sp.	33.3*	66.7**	33.3*	25.0	41.7*	40.0	
Sphaerium sp.	41.7*	33.3*	33.3*	41.7*	16.7	33.3	
Glossiphonia sp.	33.3*	8.3	58.3**	0.0	50.0*	30.0	
S.fuliginosa	8.3	33.3*	33.3*	58.0**	8.3	28.2	
Limnephilus sp.	33.3*	8.3	16.7	42.0*	33.3*	26.7	
Cloëon sp.	50.0*	25.0	8.3	8.3	25.0	23.3	C_1
H.stagnalis	25.0	25.0	58.3**	0.0	0.0	21.7	
Sergentia sp.	0.0	25.0	25.0	16.7	0.0	13.3	
Corixa sp.	16.7	8.3	16.7	8.3	16.7	13.3	
Gammarus sp.	0.0	0.0	16.7	33.3*	8.3	11.7	
Chaoborus sp.	8.3	0.0	8.3	25.0	8.3	10.0	
Lestes sp.	25.0	0.0	0.0	0.0	25.0	10.0	
A.grandis	8.3	0.0	0.0	8.3	25.0	8.3	
Dytiscus sp.	16.7	8.3	8.3	0.0	8.3	8.3	
V.viviparus	16.7	8.3	0.0	0.0	8.3	6.7	
Cyrnus sp.	0.0	0.0	0.0	17.0	8.3	5.1	
Piscola geometra L.	0.0	0.0	0.0	0.0	25.0	5.0	
Bezzia sp.	0.0	8.3	0.0	0.0	8.3	3.3	
Chironomus sp. pupa	0.0	0.0	8.3	8.3	0.0	3.3	
Platambus sp.	0.0	8.3	0.0	8.3	0.0	3.3	
E.bimaculata	0.0	0.0	0.0	8.3	0.0	1.7	
Ephemera sp.	0.0	0.0	8.3	0.0	0.0	1.7	
Notonecta sp.	0.0	0.0	0.0	0.0	8.3	1.7	
Libellula sp.	0.0	0.0	0.0	8.3	0.0	1.7	
Hydrovatus sp.	0.0	0.0	0.0	8.3	0.0	1.7	
Nepa cinera L.	0.0	0.0	0.0	8.3	0.0	1.7	
Tabanus sp.	8.3	0.0	0.0	0.0	0.0	1.7	

Table 5. Benthofauna constancy of occurrence (C, %) in the oxbow-lakes of the Słupia River with the classification of consecutive taxa (* - C_2, **- C_3, *** - C_4)

The pattern of constancy of occurrence was similar in all the oxbow-lakes of the Słupia River (Fig. 4). The medians were at the level of 10% which indicated that there were many occasional species present only during some seasons.

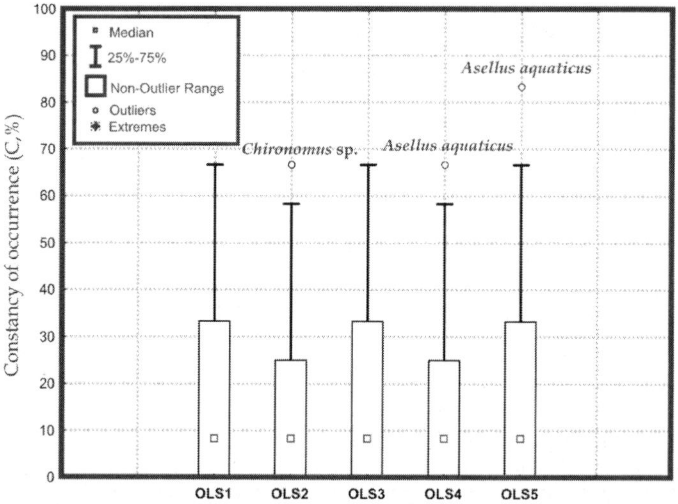

Fig. 4. Benthofauna constancy of occurrence (C, %) in the oxbow-lakes of the Słupia River

Collating the data from the studied river basins one can conclude that the most frequent were *Chironomus* sp. larvae, although they did not place the first position in any of the river basins (Table 6). *Erpodbella* sp. ($C_{av.}$=32%), *Chaoborus* sp. ($C_{av.}$=31%) and Oligochaeta ($C_{av.}$=26%) were also important elements of benthofauna structure. None of the identified invertebrate representatives reached the status higher than accesoric taxon.

Taxa	River basin		
	Łyna	**Drwęca**	**Słupia**
Chironomus sp.	15.7 (3)	50.0 (2)	40.0 (5)
Erpobdella sp.	10.2 (6)	22.9 (3)	61.8 (2)
Chaoborus sp.	25.3 (1)	58.3 (1)	10.0 (14)
Oligochaeta	20.0 (2)	10.4 (6)	48.4 (3)
Procladius sp.	12.6 (4)	14.6 (4)	48.3 (4)
A.aquaticus	7.8 (8)	2.1 (9)	65.2 (1)
Glossiphonia sp.	5.0 (9)	2.1 (9)	3.0 (7)
Cloëon sp.	0.1 (40)	4.2 (7)	23.3 (10)
Bezzia sp.	3.7 (11)	4.2 (7)	3.3 (19)

Table 6. Benthofauna constancy of occurrence (C, %) and their ranks in the studied river basins

4.2 Qualitative structure of benthofauna in oxbow-lakes with various type of connection to the main river

4.2.1 Lentic oxbow-lakes

The oxbow-lakes with limited hydrological connection to the main river were inhabited by 27 taxa. The most abundant were Chironomidae larvae (*Chironomus* sp. and *Chaoborus* sp.) as well as leeches of *Erpobdella* genus (Table 7). Those taxa are characterized by considerable adaptability to environmental conditions and therefore are common in various aquatic ecosystems, including oxbows (Galardo et al., 2008, Obolewski, 2011a). However, they were only accesoric taxa and their constancy of occurrence did not exceed half of the samples.

Taxa	Drwęca River Basin				Łyna River Basin	Słupia River Basin		Mean
	Oxbow-lakes							
	OLD 1	OLD 2	OLD 3	OLD 4	OLŁ 4	OLS 1	OLS 2	
Chironomus sp.	8.3	75.0	50.0	**66.7**	16.7	33.3	**66.7**	45.2
Chaoborus sp.	**33.3**	**100.0**	**66.7**	33.3	8.3	8.3	0.0	35.7
Erpobdella sp.	8.3	8.3	25.0	50.0	16.7	**66.7**	**66.7**	34.5

Table 7. Constancy of occurrence with classification of the most frequent benthofauna taxa ($C_{av.}$ > 25.0 %) in lentic oxbow-lakes

Both leeches and Diptera larvae are often observed in reservoirs polluted with organic substances, where the concentration of oxygen is low and limits the existence of other aquatic invertebrates. The remaining representatives of benthic fauna were classified as accidental taxa. The frequency of 5 taxa, including *Procladius* sp. larvae and Oligochaeta representatives, exceeded 10%,

The comparison of constancy of occurrence between the studied river basins gives interesting observations. The oxbow-lakes of the Drwęca River are closed and the refreshment of waters takes place during spring and autumn spates. Macrozoobenthos was represented there by 12 taxa and predominated by Diptera larvae (*Chironomus* sp., *Chaoborus* sp.), classified as constant taxa (OLD1) or euconstants (OLD2-OLD3). Particularly low macrozoobenthos diversity was observed in OLD1, where dense pleustonic fauna limited the mixing of waters and decreased the oxygenation (Glińska-Lewczuk, 2009). In the Łyna River Basin there was only one cut-off oxbow-lake, in which 14 taxa were observed and 85% of them occurred once. The most frequent was *V. pulchella* but it was still classified as accidental taxon (C<25%). The two closed oxbow-lakes in the Słupia river basin were rich with 20 macrozoobenthos taxa. 35% of them were constant taxa. Comparing to the oxbow-lakes of Drwęca and Łyna, *Erpobdella* and *A. aquaticus* occurred the most frequently in the Słupia river basin. No *Chaoborus* sp. larvae were observed which indicated good water oxygenation (Kajak, 2001).

4.2.2 Semi-lotic oxbow-lakes

Semi-open oxbow-lakes are hydrologically classified between cut-off and open reservoirs. They form environmental conditions favourable to fauna typical of both lotic and lentic ecosystems. The total number of taxa identified in the studied semi-lotic oxbow-lakes was equal to 40 which was reflected by the qualitative structure of benthofauna (Table 8)

The most frequent taxa in the oxbow-lake of the Łyna River were predatory larvae *Procladius* sp., bivalves *Pisidium* sp. and *Chaoborus* sp. larvae. However, the last two taxa were classified as accesoric, similarly to Oligochaeta and *Chironomus* sp. The remaining taxa were accidental. Different situation was observed in the oxbow-lake located in the Słupia River Basin, where the highest constancy of occurrence reached *Erpobdella* sp., *Procladius* sp., *A.*

aquaticus, H. stagnalis and *Glossiphonia* sp. (constant taxa). Oligochaeta, *Chironomus* sp., *S.fuliginosa* and *Sphaerium* sp. were accesoric taxa and the remaining were accidental.

Taxa	Łyna River Basin	Słupia River Basin	Mean
	Oxbow-lakes		
	OLŁ1	**OLS 3**	
Procladius sp.	**44.4**	58.3	51.4
Erpobdella sp.	22.2	**66.7**	44.4
A. aquaticus	22.2	58.3	40.3
Glossiphonia sp.	22.2	58.3	40.3
Oligochaeat	27.8	50.0	38.9
Chironomus sp.	27.8	33.3	30.6
H. stagnalis	0.0	58.3	29.2
S. fuliginosa	22.2	33.3	27.8

Table 8. Benthofauna constancy of occurrence with classification of the most frequent taxa ($C_{av.}$ >25.0 %) in semi-open oxbow lakes

4.2.3 Lotic oxbow-lakes

The most frequent taxon in lotic oxbow-lakes was *Chironomus* sp. which was classified as euconstants (C>50%). Quite often occurred *Erpobdella* sp., *A. aquaticus,* oligochaetes and *Chaoborus* sp. (Table 9).

Taxa	Łyna River Basin			Słupia River Basin		Mean
	Oxbow-lakes					
	OLŁ 2	**OLŁ 3**	**OLŁ 5**	**OLS 4**	**OLS 5**	
Chironomus sp.	**75.0**	**75.0**	**44.4**	25.0	41.7	52.2
Erpobdella sp.	41.7	16.7	33.3	42.0	66.7	40.1
A.aquaticus	16.7	0.0	22.2	**67.7**	**83.3**	38.0
Oligochaeta	41.7	16.7	44.4	16.7	67.0	37.3
Chaoborus sp.	58.3	**75.0**	0.0	25.0	8.3	33.3

Table 9. Benthofauna constancy of occurrence and classification of the most frequent taxa ($C_{av.}$ > 25.0 %) in lotic oxbow-lakes

The frequency of occurrence of other benthofauna representatives differed between the river basins. The open oxbow-lakes of the Łyna River were predominated by *Chironomus* sp. ($C_{av.}$=60%, euconstant taxon). The larvae of *Chaoborus* sp. also occurred quite often ($C_{av.}$=44%) but they were not present in OLŁ5. It was the youngest of the studied reservoirs while *Chaoborus* sp. prefers stagnant waters and therefore was absent in the discussed oxbow-lake (Kajak, 2001). The qualitative structure of benthofauna in the oxbows of the Łyna River Basin indicated decreased importance of *A. aquaticus* which was not observed in OLŁ3. This species is typical of oxbow-lakes in northern Poland (Obolewski, 2011a; Obolewski & Glińska-Lewczuk, 2011) rich with organic matter. They are absent in reservoirs which are uncovered or with high flow velocity washing out organic substances (Stańczykowska, 1986). Lotic oxbow-lakes of the Słupia River were predominated by *A. aquaticus* ($C_{av.}$=76%) and *Erpobdella* sp. ($C_{av.}$=54%) which were classified as euconstants.

4.3 Quantitative structure of benthofauna in oxbow-lakes of the studied river basins
4.3.1 Łyna River basin

The oxbow-lakes of the Łyna River were predominated by *Chaoborus* sp., Oligochaeta and *Chironomus* sp. which jointly constituted 66% of the total benthofauna abundance. They were classified as eudominant taxa (Table 10).

As for the other macrozoobenthos representatives, *V. viviparus* reached the status of dominant taxa and six taxa belonging to Crustacea, Hirudinea and Insecta were subdominants. Relatively large group of taxa was observed with low density which is typical of degraded aquatic ecosystems (Kajak, 2001).

The average density in the studied reservoirs was low and did not exceed 200 indiv. m^{-2}. However, each oxbow-lake was predominated by one taxa- OLŁ1-OLŁ3 by *Chaoborus* sp. (maximum density 1400 indiv.m^{-2}) and OLŁ4-OLŁ5 by Oligochaeta (maximum density 2600 indiv.m^{-2}), (Fig. 5).

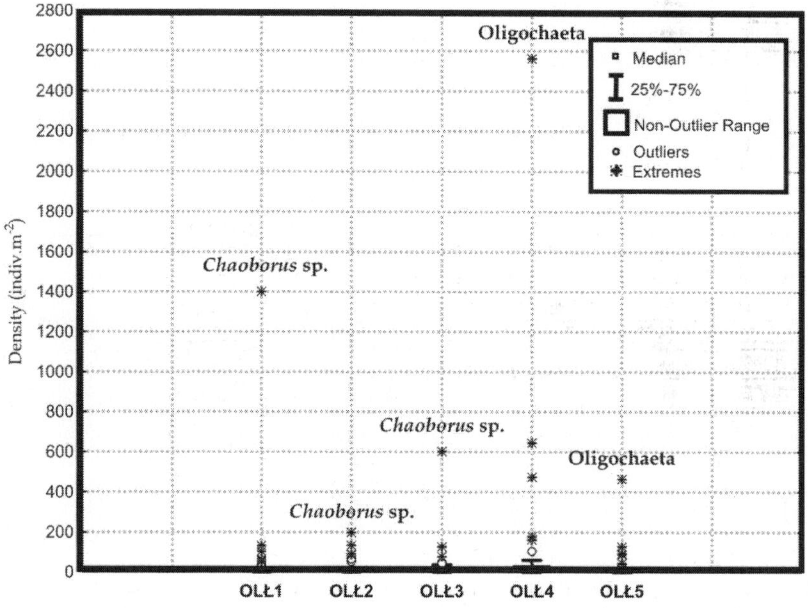

Fig. 5. Benthofauna density (indiv. m^{-2}) in the oxbow-lakes of the Łyna River

Such high abundance was reflected by the values of domination index (D, %). The share of *Chaoborus* sp. varied between 70% in OLŁ1 to 20% in OLŁ2, where they were eudominants. In OLŁ1 their domination was so high that no other taxa exceeded 10% of the total density. That situation indicates OLŁ1 as the most degraded reservoir which needs revitalization (Obolewski et al., 2009). The level of Oligochaeta domination in two oxbow-lakes was around 50-60% (Table 10).

Taxa	Oxbow-lakes					Mean	Classification
	OLŁ 1	**OLŁ 2**	**OLŁ 3**	**OLŁ 4**	**OLŁ 5**		
Chaoborus sp.	**69.1**	**20.8**	**58.6**	0.3	0.0	29.8	**ED**
Oligochaeta	6.5	9.0	2.2	**58.3**	**52.9**	25.8	
Chironomus sp.	2.3	13.9	10.4	14.7	9.9	10.2	
V. viviparus	3.7	6.7	3.4	1.3	10.5	5.1	**D**
A. aquaticus	2.2	2.3	0.0	0.0	14.4	3.8	**SD**
Limnephilus sp.	0.0	1.5	1.7	10.8	0.0	2.8	
Erpobdella sp.	1.0	8.1	2.3	0.8	1.9	2.8	
Bezzia sp.	0.0	5.4	7.5	0.0	0.2	2.6	
S. fuliginosa	0.6	9.3	1.7	0.3	0.0	2.4	
Procladius sp.	0.1	5.7	4.7	0.7	0.0	2.2	
Cyrnus sp.	0.0	7.0	1.7	1.0	0.0	1.9	**R**
Pisidium sp.	3.1	0.0	0.0	0.3	5.1	1.7	
M. culiciformis	5.9	0.0	0.0	0.0	0.0	1.2	
Dytiscus sp.	0.0	0.0	0.0	4.0	0.0	0.8	**SR**
Glossiphonia sp.	1.0	0.0	1.8	0.0	0.9	0.7	
Notonecta sp.	0.0	0.0	0.0	3.7	0.0	0.7	
H. stagnalis	0.0	1.6	1.7	0.0	0.0	0.7	
A. anatina	0.3	2.1	0.0	0.0	0.4	0.6	
Sergentia sp.	0.0	0.0	0.0	2.5	0.2	0.5	
Chironomus sp. pupa	0.0	0.0	2.3	0.0	0.0	0.5	
Lestes sp.	0.0	2.1	0.0	0.0	0.0	0.4	
L. stagnalis	0.5	0.0	0.0	0.0	1.5	0.4	
E. bimaculata	0.0	1.5	0.0	0.0	0.0	0.3	
Phryganaidae	0.0	0.0	0.0	0.0	1.5	0.3	
D. polymorpha	0.0	1.5	0.0	0.0	0.0	0.3	
Caenis sp.	0.0	1.5	0.0	0.0	0.0	0.3	
V. pulchella	0.0	0.0	0.0	0.9	0.0	0.2	
Polypedilum sp.	0.9	0.0	0.0	0.0	0.0	0.2	
V.piscinalis	0.4	0.0	0.0	0.0	0.0	0.1	
P. antipodarum	+	0.0	0.0	0.4	0.0	0.1	
V. cirstata	0.3	0.0	0.0	0.0	0.0	0.1	
A. cygnea	0.3	0.0	0.0	0.0	0.0	0.1	
E. tenellus	0.4	0.0	0.0	0.0	0.0	0.1	
Goera sp.	0.5	0.0	0.0	0.0	0.0	0.1	
Psychomidae	0.4	0.0	0.0	0.0	0.0	0.1	
G. truncatula	0.2	0.0	0.0	0.0	0.0	+	
Sphaerium sp.	0.4	0.0	0.0	0.0	0.0	+	
Lepidostoma sp.	0.2	0.0	0.0	0.0	0.0	+	
Hydropsyche sp.	+	0.0	0.0	0.0	0.0	+	
Limoniidae	+	0.0	0.0	0.0	0.0	+	
Cloëon sp.	0.0	0.0	0.0	0.0	0.2	+	

Table 10. Domination index (D, %) of benthofauna in the oxbow-lakes of the Łyna River with classification (ED - eudominants, D - dominants, SD - subdominants, R - recedents, SR - subrecedents) +<0.1%

4.3.2 Drwęca River basin

Macrozoobenthos in the oxbow-lakes of the Drwęca River was predominated by Diptera larvae: *Chaoborus* sp., *Chironomus* sp. and *Procladius* sp. which jointly constituted 76% of the total density. *Chaoborus* sp. was the most abundant in OLD1 and OLD4 while *Chironomus* sp. in OLD2 and OLD3 (Table 11). The increase in *Chironomus* sp. density was accompanied by higher abundance of predatory *Procladius* sp. but they were not observed in the reservoirs predominated by *Chaoborus* sp. In terms of domination, larvae of *Chironomus* sp.

Taxa	Oxbow-lakes				Mean	Classification
	OLD 1	OLD 2	OLD 3	OLD 4		
Chaoborus sp.	**37.3**	26.5	22.6	**68.3**	38.7	ED
Chironomus sp.	14.5	**46.3**	**30.2**	19.4	27.6	
Procladius sp.	4.8	17.1	18.7	0.0	10.2	
Erpobdella sp.	24.1	1.4	2.0	9.8	9.3	D
A. aquaticus	14.5	0.0	0.0	0.0	3.6	SD
Glossiphonia sp.	0.0	0.0	8.7	0.0	2.2	
Cloëon sp.	4.8	0.0	2.6	0.0	1.9	R
Oligochaeta	0.0	2.7	3.9	0.0	1.7	
Sergentia sp.	0.0	0.0	6.1	0.0	1.5	
Chironomus sp. pupa	0.0	1.8	3.5	0.0	1.3	
Bezzia sp.	0.0	1.5	1.7	0.0	0.8	SR
Cyrnus sp.	0.0	2.7	0.0	0.0	0.7	
H. stagnalis	0.0	0.0	0.0	2.5	0.6	

Table 11. Benthofauna domination index (D, %) in the oxbow-lakes of the Drwęca River with classification (ED - eudominants, D - dominants, SD - subdominants, R - recedents, SR - subrecedents)

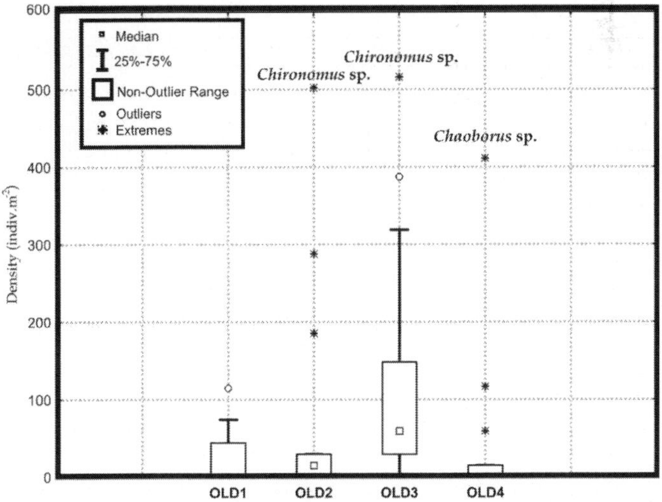

Fig. 6. Benthofauna density (indiv. m^{-2}) in the oxbow-lakes of the Drwęca River

predominated and reached even 70% of benthofauna (OLD4) while *Chironomus* sp.- 30-50% (OLD2-OLD3). The only exception was OLD1 where leeches of *Erpobdella* sp. as well as *Chironomus* sp. appeared (Table 11).
The density of benthofauna did not exceed the average value of 50 indiv.m^{-2} and more abundant macrozoobenthos was observed only in OLD3 (Fig. 6). The larvae of *Chironomus* sp. appeared massively (density above 500 indiv.m^{-2}) in that reservoir as well as in OLD2. OLD4 seems to be the most degraded oxbow-lake in the Drwęca River Basin due to low benthofauna density and the predomination of *Chaoborus* sp. (density above 400 indiv.m^{-2}). The only reservoir with no predominant taxa was OLD1 and represented different zoocenotic type.

4.3.3 Słupia River Basin

The oxbow-lakes in the Słupia River Basin were abundantly inhabited by taxa belonging to Oligochaeta, Crustacea and Insecta (Table 12). They constituted 70% of benthofauna and the share of *A. aquaticus* reached 40%. That species predominated in three oxbows (D=30-36%). *A. aquaticus*, particularly large individuals are typical of β-mezosaprobic waters (Turoboyski, 1979). Almost half of benthofauna density in OLS1 constituted the larvae of *Chironomus* sp. but their share was lower in the remaining reservoirs. In turn, the density of Oligochaeta was similar in all the oxbows (Table 12).
Quite atypical domination structure was observed in OLS5, where benthofauna was predominated by *Cloëon* sp. (25% of total density). That taxon is regarded as bioindicator and its presence is a sign of good environmental conditions, particularly high water oxygenation (Turoboyski, 1979).
Average macrozoobenthos density in the oxbow-lakes of the Słupia River was at the level of 100 indiv.m^{-2} with the domination of *A. aquaticus* (density between 400 and 700 indiv.m^{-2}), *Chironomus* sp. (maximum density over 1500 indiv.m^{-2}) or *Cloëon* sp. (1600 indiv.m^{-2}), (Fig. 7).

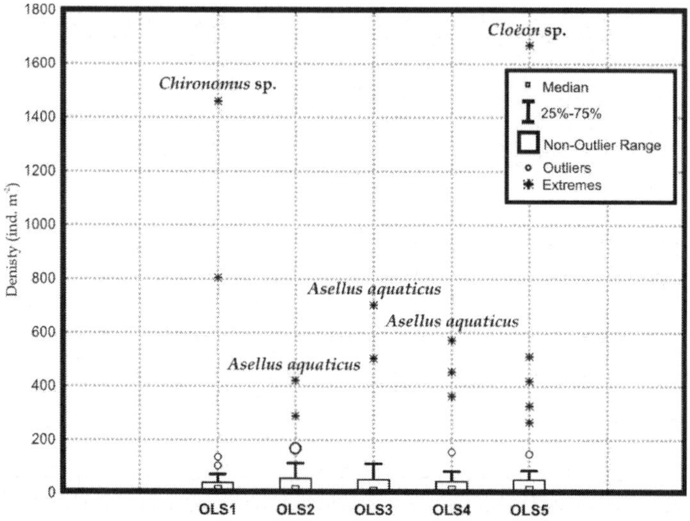

Fig. 7. Benthofauna density (indiv. m^{-2}) in the oxbow-lakes of the Słupia River

The comparison of averaged domination indices and ranks of the consecutive taxa in the studied river basins shows the frequent occurrence of *Chaoborus* sp. larvae which predominated in the oxbow-lakes of Drwęca and Łyna (Table 13). Its importance was lower in the Słupia River Basin. Opposite situation was observed in case of *A. aquaticus* and *Cloëon* sp. which were not abundant in the Drwęca and Łyna river basins but constituted the most importand quantitative component of macrozoobenthos in the Słupia River Basin. High ranks were also reached by *Chironomus* sp. and Oligochaeta.

Taxa	Oxbow-lakes					Mean	Classification
	OLS 1	OLS 2	OLS 3	OLS 4	OLS 5		
A. aquaticus	27.1	**29.0**	**36.3**	**29.3**	11.4	26.6	ED
Chironomus sp.	**49.2**	11.6	4.3	23.3	0.7	17.8	
Oligochaeta	2.4	19.9	26.1	7.9	13.5	14.0	
Cloëon sp.	3.5	7.8	0.4	0.8	**44.1**	11.3	
Procladius sp.	4.6	7.6	2.1	4.4	1.4	4.0	SD
Gammarus sp.	0.0	0.0	0.8	18.1	0.4	3.9	
Erpobdella sp.	1.3	1.8	2.4	1.5	8.6	3.1	
S. fuliginosa	2.3	4.6	3.8	2.4	0.4	2.7	
Sphaerium sp.	2.3	3.8	1.5	3.8	0.8	2.4	
Limnephilus sp.	1.2	2.0	4.8	1.3	1.1	2.1	
Corixa sp.	0.7	1.0	0.5	0.8	7.3	2.1	
Sergentia sp.	0.0	1.7	5.7	2.4	0.0	2.0	
Glossiphonia sp.	0.7	1.0	3.2	0.0	2.3	1.4	R
V. viviparus	0.7	4.2	0.0	0.0	0.4	1.1	
Bezzia sp.	0.0	1.0	0.0	0.0	3.9	1.0	
H. stagnalis	0.7	1.0	2.7	0.0	0.0	0.9	SR
Chaoborus sp.	0.5	0.0	0.4	1.6	0.6	0.6	
Dytiscus sp.	0.5	1.0	0.4	0.0	0.9	0.6	
Hydrovatus sp.	0.0	0.0	1.9	0.0	0.0	0.4	
Lestes sp.	0.8	0.0	0.0	0.0	0.8	0.3	
Tabanus sp.	1.0	0.0	0.0	0.0	0.0	0.2	
Cyrnus sp.	0.0	0.0	0.0	0.8	0.4	0.2	
Chironomus sp. pupa	0.0	0.0	0.4	0.8	0.0	0.2	
Platambus sp.	0.0	1.0	0.0	0.0	0.0	0.2	
E. bimaculata	0.0	0.0	0.0	0.8	0.0	0.2	
Ephemera sp.	0.0	0.0	0.8	0.0	0.0	0.2	
Notonecta sp.	0.0	0.0	0.0	0.0	0.5	0.1	
Libellula sp.	0.0	0.0	0.0	0.4	0.0	0.1	
P. geometra	0.0	0.0	0.0	0.0	0.5	0.1	
A. grandis	0.5	0.0	0.0	0.0	0.0	0.1	
N. cinera	0.0	0.0	0.4	0.0	0.0	0.1	

Table 12. Benthofauna domination index (D, %) in the oxbow-lakes of the Słupia River with classification (ED – eudominants, SD – subdominants, R – recedents, SR – subrecedents)

Taxa	River basin		
	Łyna	**Drwęca**	**Słupia**
Chaoborus sp.	29.8 (1)	38.7 (1)	(0.6) 17
Chironomus sp.	10.2 (3)	27.6 (2)	17.8 (2)
Oligochaeta	25.8 (2)	1.7 (8)	14.0 (3)
A.aquaticus	3.8 (5)	3.6 (5)	26.6 (1)
Procladius sp.	2.2 (10)	10.2 (3)	4.0 (5)
Erpobdella sp.	2.8 (6)	9.3 (4)	3.1 (7)
Cloëon sp.	0.1 (36)	1.9 (7)	11.3 (4)
Glossiphonia sp.	0.7 (15)	2.2 (6)	1.4 (13)
Bezzia sp.	2.6 (8)	0.8 (11)	1.0 (15)
Sergentia sp.	0.5 (19)	1.5 (9)	2.0 (12)

Table 13. Domination index (D, %) and ranks of benthofauna representatives in the oxbow-lakes of consecutive river basins

4.4 Qualitative structure of benthofauna in oxbow-lakes with various type of connection to the main river

4.4.1 Lentic oxbow-lakes

Limited hydrological connectivity between cut-off oxbow-lakes and the river causes that those reservoirs are predominated by taxa belonging to eurybionts (Kajak, 2001). The main components of benthofauna in the studied lentic oxbows were Diptera larvae (*Chironomus* sp. and *Chaoborus* sp.), Oligochaeta and Isopoda (*A. aquaticus*) – 8 taxa in total. They formed the group of eudominants and constituted 71% of benthofauna abundance (Table 14). They were accompanied by predatory larvae of *Procladius* sp. as well as leeches *Erpobdella* sp. (dominants). As for the other benthofauna representatives, the presence of *Cloëon* sp. *Limnephilus* sp. should be noticed.

Benthofauna structure in lentic oxbow-lakes depended on the specific features of consecutive river basins. In the Drwęca River Basin the inflow of fresh water occurs during floods and therefore benthofauna was predominated by taxa resistant to low oxygen concentration in water- some of the Diptera larvae (Mikulski, 1974; Stańczykowska, 1986). The cut-off oxbows of Łyna and Słupia are additionally refreshed by regular, twenty-four

Taxa	Oxbow-lakes							Mean	Classification
	OLD 1	**OLD 2**	**OLD 3**	**OLD 4**	**OLŁ 4**	**OLS 1**	**OLS 2**		
Chironomus sp.	14.5	**46.3**	**30.2**	19.4	14.7	**49.2**	11.6	26.6	**ED**
Chaoborus sp.	**37.3**	26.5	22.7	**68.3**	0.3	0.5	0.0	22.2	
Oligochaeta	0.0	2.7	3.9	0.0	**58.3**	2.4	19.9	12.5	
A.aquaticus	14.5	0.0	0.0	0.0	0.0	27.1	**29.0**	10.1	
Procladius sp.	4.8	17.1	18.7	0.0	0.7	4.6	7.6	7.6	**D**
Erpobdella sp.	24.1	1.4	2.0	9.8	0.8	1.3	1.8	5.9	
Cloëon sp.	4.8	0.0	2.6	0.0	0.0	3.5	7.8	2.7	**SD**
Limnephilus sp.	0.0	0.0	0.0	0.0	10.8	1.2	2.0	2.0	

Table 14. Domination index of the most import ant benthofauna representatives (mean D>2 %) in lentic oxbow-lakes with classification (ED – eudominants, D – dominants, SD – subdominants)

hour oscillation of water level caused by the work of hydroelectric power stations (so called effect of Intensive Flood Pulse IFP) (Obolewski, 2011b). As a result, benthofauna abundance is higher with the domination of Oligochaeta and Crustacea.

4.4.2 Semi-lotic oxbow-lakes

Semi-open oxbow lakes were predominated by the larvae of *Chaoborus* sp. which were accompanied by *A. aquaticus* and Oligochaeta. They formed the group of eudominants and constituted 70% of total benthofauna density (Table 15). Other important macrozoobenthos components were 6 taxa belonging to Insecta (Diptera, Trichoptera, Megaloptera) and Hirudinea.

The structure of benthofauna domination in Łyna and Słupia river basins was specific. The diversity of lotic and lentic habitats causes that those biocenosis are unique biological systems (Tockner et al., 1999; Obolewski, 2011a). *Chaoborus* sp. and *M. culiciformis* predominated in the oxbow-lake of the Łyna River, opposite to the Słupia River Basin. That indicates possible differences in the level of water refreshment. Moreover, water currents may wash away *Chaoborus* sp. (Mikulski, 1974) but favour the presence of *A. aquaticus* thanks to the inflow of organic matter carrying by water. As for other taxa, *Limnephilus* sp. and leeches *Glossiphonia* sp. were classified as subdominants. Their occurrence indicates β-mezosaprobic waters (Turoboyski, 1979).

Taxa	Oxbow-lakes		Mean	Classification
	OLŁ1	**OLS 3**		
Chaoborus sp.	**69.1**	0.4	34.7	ED
A. aquaticus	2.2	**36.3**	19.2	
Oligochaeta	6.5	26.1	16.3	
Chironomus sp.	2.3	4.3	3.3	SD
Sergentia sp.	0.0	5.7	2.9	
M. culiciformis	5.5	0.0	2.8	
Limnephilus sp.	0.0	4.8	2.4	
S.fuliginosa	0.6	3.8	2.2	
Glossiphonia sp.	1.0	3.2	2.1	

Table 15. Domination index of the most import ant benthofauna representatives (mean D>2 %) in semi-lotic oxbow-lakes with classification (ED - eudominants, SD - subdominants)

4.4.3 Lotic oxbow-lakes

Lotic oxbow-lakes are reservoirs with constant inflow of fresh water which intensity depends on water flow in the river (Ward et al., 2002). The studied open oxbows were predominated by Oligochaeta, Diptera larvae and Crustacea. Their share in benthofauna qualitative structure reached 56% (eudominants). The higher was hydrological connectivity the lower was the percentage share of eudominants but higher number of important benthofauna components. Regardless the level of contact between oxbows and rivers, the same taxa prevailed. However, the importance of Oligochaeta increased and they were the most abundant in OLŁ5 (Table 16).

The domination structure also depended on hydrological connectivity. The oxbow connected with the main river by melioration ditches was predominated by *Cloëon* sp.- bioindicator of oligotrophic waters.

Taxa	Oxbow-lakes					Mean	Classification
	OLŁ 2	**OLŁ 3**	**OLŁ 5**	**OLS 4**	**OLS 5**		
Oligochaeta	9.0	2.2	**52.9**	7.9	13.5	17.1	**ED**
Chaoborus sp.	**20.8**	**58.6**	0.0	1.0	0.4	16.2	
Chironomus sp.	13.9	10.4	9.9	23.3	0.7	11.6	
A.aquaticus	2.3	0.0	14.4	**29.3**	11.1	11.4	
Cloëon sp.	0.0	0.0	0.2	0.8	**44.1**	9.0	**D**
Erpobdella sp.	8.1	2.3	1.9	1.2	8.6	4.4	**SD**
V.viviparus	6.7	3.4	10.5	0.0	0.4	4.2	
Gammarus sp.	0.0	0.0	0	18.1	0.4	3.7	
Bezzia sp.	5.4	7.5	0.2	0.0	3.9	3.4	
Procladius sp.	5.7	4.7	0	4.2	1.4	3.2	
S.fuliginosa	9.3	1.7	0	2.4	0.4	2.8	
Cyrnus sp.	7.0	1.7	0	0.8	0.4	2.0	

Table 16. Domination index of the most import ant benthofauna representatives (mean D>2 %) in lotic oxbow-lakes with classification (ED - eudominants, D - dominants, SD - subdominants)

4.5 Assessment of benthofauna diversity

4.5.1 Diversity of benthofauna in oxbow-lakes of the studied river basins

Łyna River Basin

The Shannon diversity index for the Łyna River Basins varied between 0.148 and 0.438 with average value equal to 0.293 (Table 17). Considerable standard deviations indicate the instability of habitat conditions (Ward, 1998). The highest average value of H' was observed in semi-lotic OLŁ2 and the lowest in cut-off OLŁ5.

Oxbow-lakes				
OLŁ 1	**OLŁ 2**	**OLŁ 3**	**OLŁ 4**	**OLŁ 5**
0.388±0.220	**0.438±0.246**	0.316±0.209	0.174±0.234	0.148±0.147

Table 17. Average values of Shannon diversity index (H'± SD) for benthofauna in the oxbow-lakes of the Łyna River

The index of ecological importance reached the highest values for *Chaoborus* sp. (Table 18). Oligochaeta and *Chironomus* sp. were also important components of benthofauna in the Łyna River Basin. The ecological importance of many taxa was low.

Drwęca River Basin

The Shannon diversity index for the oxbow-lakes in the Drwęca River Basin varied between 0.06 and 0.29 (0.152 on average) (Table 19). High standard variations, often exceeding mean values indicate unstable habitat conditions (Ward, 1998). The highest H' value was observed in OLD2 and the lowest in OLD1.

The analysis of ecological importance indicated *Chaoborus* sp. and *Chironomus* sp. as the most important taxa (Table 20). Among other taxa, the role of *Procladius* sp. and *Erpobdella* sp. should be noticed.

Taxa	Oxbow-lakes					Mean	Classification
	OLŁ 1	**OLŁ 2**	**OLŁ 3**	**OLŁ 4**	**OLŁ 5**		
Chaoborus sp.	**51.8**	**34.8**	**66.3**	1.9	0.0	31.0	Q5
Oligochaeta	13.4	19.3	6.0	**25.4**	**48.5**	22.5	Q4
Chironomus sp.	8.0	32.3	30.4	18.0	21.0	22.0	
V. viviparus	10.9	13.0	7.7	3.9	10.8	9.3	Q1
Erpobdella sp.	5.8	18.3	6.0	4.3	4.6	7.8	
Bezzia sp.	0.5	13.4	17.7	0.0	1.4	6.6	
A. aquaticus	7.8	6.2	0.0	0.0	17.9	6.4	
Pisidium sp.	11.8	0.0	0.0	1.9	15.0	5.7	
Procladius sp.	0.7	11.9	6.0	2.7	0.0	4.3	
S. fuliginosa	4.2	8.8	3.5	1.9	0.0	3.7	
Limnephilus sp.	0.0	3.6	3.5	10.9	0.0	3.6	
Cyrnus sp.	0.0	10.8	3.5	3.3	0.0	3.5	
A. anatina	1.4	7.2	0.0	0.0	2.9	2.3	
Phryganaidae	6.8	0.0	0.0	0.0	4.1	2.2	
Glossiphonia sp.	0.0	0.0	4.9	0.0	3.2	1.6	
M. culiciformis	7.5	0.0	0.0	0.0	0.0	1.5	
Lestes sp.	0.5	7.2	0.0	0.0	0.0	1.5	
V.piscinalis	2.1	0.0	0.0	5.6	0.0	1.5	
H. stagnalis	0.0	3.6	3.5	0.0	0.0	1.4	
Notonecta sp.	0.0	0.0	0.0	6.4	0.0	1.3	
L. stagnalis	2.3	0.0	0.0	0.0	4.1	1.3	
Dytiscus sp.	0.0	0.0	0.0	6.7	0.0	1.3	
Sergentia sp.	0.0	0.0	0.0	5.1	1.4	1.3	
Chironomus sp. pupa	0.0	0.0	6.0	0.0	0.0	1.2	
E. bimaculata	0.0	3.6	0.0	0.0	0.0	0.7	
D. polymorpha	0.0	3.6	0.0	0.0	0.0	0.7	
Caenis sp.	0.0	3.6	0.0	0.0	0.0	0.7	
P. antipodarum	0.5	0.0	0.0	1.9	0.0	0.5	
Polypedilum sp.	2.2	0.0	0.0	0.0	0.0	0.4	
Goera sp.	2.0	0.0	0.0	0.0	0.0	0.4	
Sphaerium sp.	1.6	0.0	0.0	0.0	0.0	0.3	
Cloëon sp.	0.0	0.0	0.0	0.0	1.4	0.3	
V. pulchella	0.0	0.0	0.0	0.9	0.0	0.2	
Lepidostoma sp.	0.8	0.0	0.0	0.0	0.0	0.2	
Psychomidae	0.8	0.0	0.0	0.0	0.0	0.2	
G. truncatula	1.0	0.0	0.0	0.0	0.0	0.2	
V. cirstata	0.3	0.0	0.0	0.0	0.0	0.1	
A. cygnea	0.3	0.0	0.0	0.0	0.0	0.1	
E. tenellus	0.7	0.0	0.0	0.0	0.0	0.1	
Limoniidae	0.5	0.0	0.0	0.0	0.0	0.1	
Hydropsyche sp.	0.1	0.0	0.0	0.0	0.0	<0.1	

Table 18. Index of ecological importance (Q, %) for benthofauna representatives in the oxbow-lakes of the Łyna River and its classification (Q5 - very high, Q4 - high, Q1 - very low)

Oxbow-lakes			
OLD 1	**OLD 2**	**OLD 3**	**OLD 4**
0.056±0.157	**0.293**±0.164	0.166±0.216	0.092±0.131

Table 19. Average values of Shannon diversity index (H'± SD) for benthofauna in the oxbow-lakes of the Drwęca River

Taxa	Oxbow-lakes				Mean	Classification
	OLD 1	**OLD 2**	**OLD 3**	**OLD 4**		
Chaoborus sp.	**35.3**	49.8	**38.9**	**47.7**	42.9	Q5
Chironomus sp.	11.0	**57.0**	**38.9**	35.9	35.7	
Procladius sp.	6.3	23.1	17.6	0.0	11.8	Q3
Erpobdella sp.	14.2	3.3	7.0	22.2	11.7	
Chironomus sp. pupa	0.0	6.5	10.8	0.0	4.3	Q1
Oligochaeta	0	8.0	8.1	0	4.0	
Glossiphonia sp.	0.0	0.0	8.5	0.0	2.1	
Cloëon sp.	6.3	0.0	4.7	0.0	2.7	
Sergentia sp.	0.0	0.0	7.1	0.0	1.8	
Bezzia sp.	0.0	3.3	3.8	0.0	1.8	
A.aquaticus	6.3	0	0	0	1.6	
Cyrnus sp.	0.0	4.6	0.0	0.0	1.2	
H. stagnalis	0.0	0.0	0.0	4.5	1.1	

Table 20. Index of ecological importance (Q, %) for benthofauna representatives in the oxbow-lakes of the Drwęca River and its classification (Q5 – very high, Q3 – moderate, Q1 – very low)

Słupia River Basin

Benthofauna diversity in the Słupia River Basins was the highest among the studied basins (0.412 on average) (Table 21). Low values of standard deviations indicate more stable habitat conditions which favour taxonomic diversity (Ward, 1998; Gallardo et al., 2008; Obolewski, 2011a; Obolewski & Glińska-Lewczuk, 2011).

Oxbow-lakes				
OLS 1	**OLS 2**	**OLS 3**	**OLS 4**	**OLS 5**
0.354±0.163	0.377±0.291	**0.488**±0.207	0.377±0.232	0.464±0.180

Table 21. Average values of Shannon diversity index (H'± SD) for benthofauna in the oxbow-lakes of the Słupia River

The analysis of Q index indicated *A. aquaticus* as the most important component of benthofauna in the Słupia River Basin. High Q values were reached by Oligochaeta and *Chironomus* sp. while moderate by *Procladius* sp., *Cloëon* sp. and *Erpobdella* sp. (Table 22). Such ecological structure, with the presence of oligotrophic water bioindicators (*Cloëon* sp., *Gammarus* sp., *Ephemera* sp.) is a sign of good environmental conditions (Lapmert & Summer, 2007; Turoboyski, 1979).

Taxa	Oxbow-lakes					Mean	Classification
	OLS 1	**OLS 2**	**OLS 3**	**OLS 4**	**OLS 5**		
A.aquaticus	39.8	**41.1**	**46.0**	**44.2**	30.4	40.3	Q5
Oligochaeta	11.0	34.1	36.0	11.5	30.0	24.5	Q4
Chironomus sp.	**40.5**	27.8	11.9	24.1	5.4	22.0	
Procladius sp.	15.1	21.0	11.1	13.3	6.8	13.5	Q3
Cloëon sp.	13.2	14.0	1.8	2.5	**33.2**	12.9	
Erpobdella sp.	9.4	10.9	12.1	6.7	24.0	12.6	
Sphaerium sp.	9.8	11.3	7.1	12.3	3.6	8.8	Q2
S.fuliginosa	4.1	12.4	11.3	11.5	1.8	8.2	
Limnephilus sp.	6.5	4.1	9.0	6.7	6.0	6.4	
Glossiphonia sp.	5.0	2.9	13.6	0.0	10.7	6.4	
Gammarus sp.	0.0	0.0	3.6	24.9	2.6	6.2	
Corixa sp.	3.5	2.9	2.9	2.5	13.8	5.1	
Sergentia sp.	0.0	6.5	12.0	6.2	0.0	4.9	Q1
H.stagnalis	4.1	5.0	12.5	0.0	0.0	4.3	
V.viviparus	3.5	5.8	0.0	0.0	1.8	2.2	
Chaoborus sp.	2.0	0.0	1.8	5.0	1.8	2.1	
Dytiscus sp.	2.9	2.9	1.8	0.0	2.6	2.0	
A. grandis	2.0	0.0	0.0	2.5	4.4	1.8	
Lestes sp.	4.6	0.0	0.0	0.0	4.4	1.8	
Bezzia sp.	0.0	2.9	0.0	0.0	5.7	1.7	
Platambus sp.	0.0	2.9	3.1	0.0	1.8	1.6	
Cyrnus sp.	0.0	0.0	0.0	3.6	1.8	1.1	
P. geometra	0.0	0.0	1.8	0.0	3.6	1.1	
Chironomus sp. pupa	0.0	0.0	1.8	2.5	0.0	0.9	
Hydrovatus sp.	0.0	0.0	4.0	0.0	0.0	0.8	
Tabanus sp.	2.9	0.0	0.0	0.0	0.0	0.6	
E.bimaculata	0.0	0.0	0.0	2.5	0.0	0.5	
Ephemera sp.	0.0	0.0	2.5	0.0	0.0	0.5	
Libellula sp.	0.0	0.0	0.0	2.5	0.0	0.5	
Notonecta sp.	0.0	0.0	0.0	0.0	1.8	0.4	
N. cinera	0.0	0.0	1.8	0.0	0.0	0.4	

Table 22. Index of ecological importance (Q, %) for benthofauna representatives in the oxbow-lakes of the Słupia River and its classification (Q5 - very high, Q4 - high, Q3 - moderate, Q2 -low , Q1 - very low)

The comparison of average values of ecological importance index Q between river basins with the classification of consecutive taxa shows, that the most important role was played by *Chironomus* sp. larvae (Table 23). Its average ecological importance reached $Q_{av.}$=27%, similarly to *Chaoborus* sp. larvae ($Q_{av.}$=25%). Oligochaeta and *A. aquaticus* also revealed considerable ecological importance ($Q_{av.}$ equal to 17% and 16%, respectively). It is interesting, that the importance of *Chaoborus* sp. is very high in the Łyna and Drwęca river basins but very low for Słupia oxbow-lakes.

Taxa	River basin		
	Łyna	**Drwęca**	**Słupia**
Chironomus sp.	22.0 (3)	35.7 (2)	22.0 (3)
Chaoborus sp.	31.0 (1)	42.9 (1)	2.1 (16)
Oligochaeta	22.5 (2)	4.0 (6)	24.5 (2)
A.aquaticus	6.4 (7)	1.6 (11)	40.3 (1)
Procladius sp.	4.3 (9)	11.8 (3)	13.5 (4)
Erpobdella sp.	7.8 (5)	11.7 (4)	12.6 (6)
Cloëon sp.	0.3 (31)	2.7 (8)	12.9 (5)

Table 23. Index of ecological importance (Q, %) for benthofauna representatives in the studied river basins and their ranks

4.5.2 Diversity of benthofauna in oxbow-lakes with various type of connection to the main river

Benthofauna diversity changed depending on hydrological connectivity (Fig. 8). The Shannon index for closed oxbow-lakes did not exceed 0.7 ($H'_{av.}$=0.217±0.194). In case of semi-lotic reservoirs H' values ranged from 0.8 in the Łyna River to 1.1 in the Słupia River Basin with 0.438±0.214 on average. Higher diversity comparing to closed oxbow is caused by more often fresh water inflow (Amoros & Bornette, 2002; Obolewski et al., 2009). Full hydrological connectivity did not increase considerably benthofauna diversity. In open oxbow-lakes the Shannon index reached 0.349±0.203 on average, which is 1.2-fold less than in semi-lotic oxbows and 1.6-fold higher than in lentic reservoirs. One can conclude, that

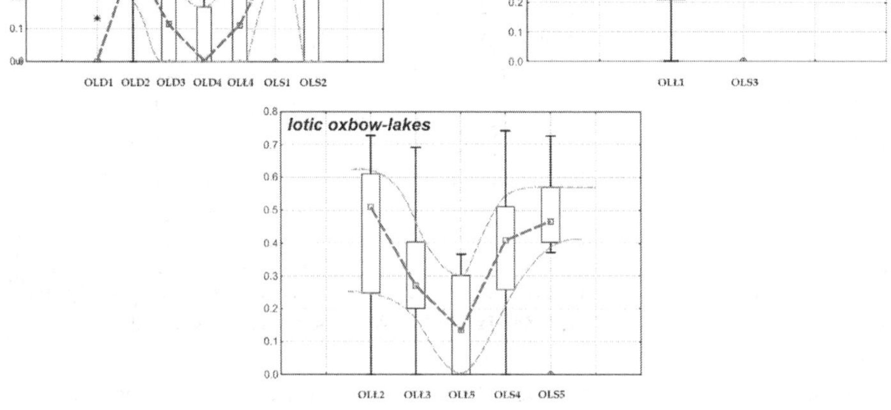

Fig. 8. Shannon diversity index (H') in oxbow-lakes with different types of connection to the main river course

semi-lotic oxbow-lakes are optimal habitat for macrozoobenthos in terms of diversity (Gallardo et al., 2008; Obolewski, 2011a and b).
The analysis of ecological importance (Q) for consecutive hydrodynamic types of oxbow-lakes indicates that it evens out for higher hydrological connectivity (Table 24). A unimodal pattern with the domination of *Chironomus* sp. was observed in cut-off reservoirs while in semi-lotic oxbows three taxa were important (*Chaoborus* sp., *A. aquaticus* and Oligochaeta). Lower frequency of occurrence for *Chironomus* sp. there is caused by fish foraging (Penczak et al., 2007). In lotic oxbow-lakes the ecological importance of main benthofauna components was at similar level. Four taxa reached high values (Oligochaeta, *Chironomus* sp., *Chaoborus* sp. and *A. aquaticus)* while only for *Erpobdella* the Q level was moderate. The role of Oligochaeta increases with hydrological connectivity, opposite to *Chaoborus* sp. due to washing out by water movement (Stańczykowska, 1986) and ichthyofauna pressure.

Taxa	Oxbow-lakes		
	lentic	**semi-lotic**	**lotic**
Chaoborus sp.	28.18	**28.65**	23.20
Chironomus sp.	**34.64**	10.05	24.65
Oligochaeta	16.78	25.14	**25.24**
A.aquaticus	13.40	28.20	20.83
Erpobdella sp.	14.26	8.43	13.28

Table 24. Index of ecological importance ($Q_{av.}$>10.01%) for benthofauna representatives in oxbow-lakes with different types of connection to the main river course

5. Conclusions

This study confirms that hydrological connectivity between the main river and its wetlands strongly influences benthofauna composition and abundance. Limited hydrological connection as well as hydrotechnical buildings on rivers of moderate slope cause permanent hydrological changes. As a result many valuable aquatic ecosystems may be degraded because water exchange is sporadic. Recreation of hydrological connectivity in river basins favours its protection and revitalization as well as protection against rapid floods. Reconstruction of biodiversity after such events is initiated in wetlands where diverse habitats favour benthofauna survival and further recolonization.
Regardless of the distinguished hydrodynamic types of oxbow-lakes those reservoirs are important ecological centres (so called "hot spots") within a river valley or even a region which form various habitats for many fauna and flora species. Relationships between the main water-course and the rest of valley include production, decomposition and consumption which depend on periodical floods and oscillations of water table level. According to the theory of Junk et al. the alternating flood and low water stage periods increase the decomposition and nutrient circulation which trigger biological diversity and productivity. Those processes may be reinforced by intensive flood pulses caused by hydroelectric power stations and therefore river valleys with hydrotechnical buildings show higher benthofauna diversity.

6. Acknowledgments

I would like to thank Katarzyna Glińska-Lewczuk, Szymon Kobus and Paweł Burand for their help in morphometric studies (Dept. of Land Reclamation and Environmental

Management, University of Warmia and Mazury) as well as Ernest Murza for his help in collecting and preparation of biological material (Dept. Aquatic of Ecology, Pomeranian University in Słupsk). This study was supported financially by the Polish Ministry of Education and Science. grant no. NN 305 3247 40.

7. References

Allan, J.D. (1995). *Stream ecology. Structure nad function of running waters.* Chapman & Hall, ISBN 92-826-2929-5, London, UK

Amoros, C. & Roux, A.L. (1988). Interaction between water bodies within the floodplain of large rivers: function and development of connectivity. *Miinstersche Geographische Arbeiten*, Vol. 29, pp. 125 -130

Amoros, C. & Bornette, G. (2002). Connectivity and biocomplexity in waterbodies of riverine floodplains. *Freshwater Biology,*Vol. 47, pp. 761-776, ISSN 0046-5070

Arscott, D.B.; Tockner, K. & Ward, J.V. (2003). Spatio-temporal patterns of benthic invertebrates along the continuum of a braided Alpine river. Archiv *für Hydrobiologie*, Vol. 158, 431 - 460, ISSN 1863-9135

Arscott, D.B.; Tockner, K. & Ward, J.V. (2005). Lateral organization of aquatic invertebrates along the corridor of a braided floodplain river. *Journal of the North American Benthological Society,* Vol. 24, pp. 934 - 954, ISSN 0887-3593

Biesiadka, E. & Kowalik, W. (1980). Water mites (Hydracarina) of the western Bieszczady Mountains. 1. Stagnant waters. *Acta Hydrobiologica*, Vol. 22, pp. 279-298, ISSN 0065-132X

Boulton, A.J., Peterson, C.G., Grimm, N.B. & Fisher, S.G. (1992). Stability of an aquatic macroinvertebrate community in a multiyear hydrologie disturbance regime. *Ecology,* Vol. 73, pp. 2192 -2207, ISSN 0012-9658

Bonn, P.J. (1988). The impact of river regulation on invertebrate communities in the UK. *Regulated Rivers: Research and Management*, Vol. 2, pp. 389–409, ISSN 0886-9375

Bornette, G.; Amoros, C.; Piégay, H.; Tachet, J. & Hein, T. (1998). Ecological complexity of wetlands within a river landscape. *Biological Conservation*, Vol. 85, pp. 35-45, ISSN 0006-3207

Bralczyk, M. (1996). Reserve protection of the Drwęca River- reality or fiction? *Komunikaty Rybackie IRS*, Vol. 5, pp. 23-27, ISBN 1230-641X

Clausen, B. & Biggs, B.J.F. (1997). Relationships between benthic biota and hydrological indices in New Zealand streams. *Freshwater Biology*, Vol. 38, pp. 327 - 342, ISSN 0046-5070

Dynesius, M. & Nilsson, C. (1994). Fragmentation and flow regulation of river systems in the northern 3rd of the world. *Science,* Vol. 266, pp. 753 - 762, ISSN 0036-8075

Gallardo, B.; Garcia, M.; Cabezas, A.; Gonzalez, E.; Gonzalez, M.; Ciancarelli, C. & Comin, F.A. (2008). Macroinvertebrate patterns along environmental gradients and hydrological connectivity within a regulated river-floodplain. *Aquatic Sciences*, Vol. 70, pp. 248 - 258, ISSN 1015-1621

Gibbins, C.N.; Dilks, C.F.; Malcolm, R.; Soulsby, C. & Juggins, S. (2001). Invertebrate communities and hydrological variation in Cairngorm mountain streams. *Hydrobiologia*, Vol. 462, pp. 205 - 219, ISSN 0018-8158

Glińska-Lewczuk, K. (2009). Water quality dynamics of oxbow lakes in young glacial landscape of NE Poland in relation to their hydrological connectivity. *Ecological Engineering*, Vol. 35, pp. 25 - 37, ISSN 0925-8574

Gasith, A. & Resh, V.H. (1999). Streams in mediterraneanclimate regions: abiotic influences and biotic responses to predictable seasonal events. *The* Annual *Review of* Ecology, *Evolution, and Systematics*, Vol. 30, pp. 51-81, ISSN 1545-2069

Griffith, M.B.; Kaufmann, P.R.; Herlihy, A.T. & Hill, B.H. (2001). Analysis of macroinvertebrate assemblages in relation to environmental gradients in Rocky Mountain streams. *Ecological Applications*, Vol. 11, pp. 489 - 505, ISSN 1051-0761

Heiler, G.; Hein, T.; Schiemer, F. & Bornette, G. (1995). Hydrological connectivity and flood pulses as the central aspects for the integrity of a river-floodplain system. *Regulated Rivers: Research and Management,* Vol. 11, pp. 351 - 361, ISSN 0886-9375

Heino, J. (2000). Lentic macroinvertebrate assemblage structure along gradients in spatial heterogeneity, habitat size and water chemistry. *Hydrobiologia,* Vol. 418, pp. 229 - 242, ISSN 0018-8158

Junk, W.J.; Bayley, P.B. & Sparks, R.E. (1989). The flood pulse concept in river–floodplain systems. In: *Canadian Special Publications of Fisheries and Aquatic Science,* D.P. Dodge. (Ed.), 110–127, ISBN 0660132591, Ontario, Canada

Kajak, Z. (1988). Considerations on benthos abundance in freshwaters, its factors and mechanisms. International Review *of Hydrobiology,* Vol. 73, pp. 5-19, ISSN 1434-2944

Kajak, Z. (2001). *Hydrobiology- limnology: Island aquatic ecosystems.* PWN, Warsaw, ISBN 8301-12537-3

Klimaszyk. P. (2004). Oxbow-lakes and natural eutrophic aquatic reservoirs *Nympheion, Potamion*. In: Herbich J. (ed.). Fresh waters and peatlands. Handbook pf species and habitat protection Nature 2000. Ministerstwo Środowiska, Warszawa. Vol. 2. pp. 59-71. ISBN 83-86564-43-1.

Lampert, W. & Sommer, U. (2007). *Limnoecology: The Ecology of Lakes and Streams.* 2nd edition. Oxford: Oxford University Press, ISBN 97-80199213-93-1.

Marshall, J.C.; Sheldon, F.; Thorns, M. & Choy, S. (2006). The macroinvertebrate fauna of an Australian dryland river: spatial and temporal patterns and environmental relationships. Australian *Journal of Marine and* Freshwater *Research,* Vol. 57, pp. 61 - 74, ISSN 0067-1940

Michniewska-Szczepkowska, B. & Szczepkowski, B. (1969). *Olsztyńskie Voivodeship: geographical conditions.* Olsztyn, Pojezierze

Mikulski, J., St. (1982). *Ecology of Island waters.* PWN, Warsaw.

Mileska. M. (1992). *Geographical and Wight-seeing dictionary.* PWN, Warsaw. ISBN 83-01- 09822-8

Monaghan, M.T.; Robinson, C.T.; Spaak, P. & Ward, J.V. (2005). Macroinvertebrate diversity in fragmented Alpine streams: implications for freshwater conservation. *Aquatic Sciences,* Vol. 67, pp. 454 - 464, ISSN 1015-1621

Obolewski, K. (2011a). Macrozoobenthos patterns along environmental gradients and hydrological connectivity of oxbow lakes. *Ecological Engineering,* 37, 796-805, ISSN 0925-8574

Obolewski, K. (2011b). *Influence of environmental parameters on selected zoocenosis of oxbow-lakes in northern Poland.* PhD msc, Słupsk-Olsztyn

Obolewski K., Glińska-Lewczuk K. (2011). Effects of oxbow reconnection based on the distribution and structure of benthic macroinvertebrates. *Clean – Soil, Air, Water,* Vol. 39 (9), pp. 853-862, ISSN 1863-0669

Obolewski, K.; Glińska - Lewczuk, K. & Kobus, Sz. (2009). The effect of flow on the macrozoobenthos structure in a re-opened oxbow lake – a case study of the Słupia river, northern Poland. In: *Ecohydrology of Surface and Groundwater Dependent Systems: Concepts, Methods and Recent Developments.* M. Thoms, K. Heal, E. Bøgh, A. Chambel & V. Smakhtin (Eds.), Vol. 328, 13 – 23, ISBN 978-1-907161-00-1, Oxfordshire, UK

IMGW, (1983). *Hydrographical division of Poland.* Warsaw

Paszczyk, J. (1976). *Role of ground waters in riverine outflow water balance of Poland.* PhD, UMCS, Lublin

Penczak, T.; Galicka, W.; Głowacki, Ł.; Koszaliński, H.; Kruk, A.; Zięba, G.; Kostrzewa, J. & Marszał, L. (2004). Fish assemblage changes relative to environmental factors and time in the Warta River, Poland, and its oxbow lakes. *Journal of Fish Biology*, Vol. 64, pp. 483–501, ISSN 0022-1112

Poff, N.L. & Ward, J.V. (1989). Implications of streamflow variability and predictability for lotic community structure -a regional-analysis of streamflow patterns. Canadian Journal *of Fisheries and* Aquatic *Sciences*, Vol. 46, pp. 1805 - 1818, ISSN 0706-652X

Poff, N.L.; Allan, D.; Bain, M.B.; Karr, J.R.; Prestegaard, K.L.; Richter, B.D.; Sparks, R.E. & Stromberg, J.C. (1997). The natural flow regime. *BioScience*, Vol. 47, pp. 769 - 784, ISSN 0006-3568

Reese, E.G. & Batzer, D.P. (2007). Do invertebrate communities in floodplains change predictably along a river's length? *Freshwater Biology*,Vol. 52, pp. 226 - 239, ISSN 0046-5070

Robinson, C.T.; Uehlinger, U. & Monaghan, M.T. (2003). Effects of a multi-year experimental flood regime on macroinvertebrates downstream of a reservoir. *Aquatic Sciences*, Vol. 65, pp. 210 - 222, ISSN 1015-1621

Robinson, C.T.; Uehlinger, U. & Monaghan, M.T. (2004). Stream ecosystem response to multiple experimental floods from a reservoir. *River Research and Applications*, Vol. 20, pp. 359 - 377, ISSN 1535-1459

Sheldon, F.; Boulton, A.J. & Puckridge, J.T. (2002). Conservation value of variable connectivity: aquatic invertebrate assemblages of channel and floodplain habitats of a central Australian arid-zone river. Cooper Creek. *Biological Conservation*, Vol. 103, pp. 13 - 31, ISSN 0006-3207

Stańczykowska, A. (1986). *Invertebrates inhabiting our waters*. WSiP, Warsaw, ISBN 83-02-00153-8

Tockner, K.; Schiemer, F.; Baumgartner, C.; Kum, G.; Weigand, E.; Zweimuller, I. & Ward, X.V. (1999). The Danube restoration project: Species diversity patterns across connectivity gradients in the floodplain system. *Regulated Rivers: Research and Management*, Vol. 15, pp. 245 - 258, ISSN 0886-9375

Tockner, K.; Malard, F. & Ward, J.V. (2000). An extension of the flood pulse concept. *Hydrological Processes*, Vol. 14, pp. 2861-2883, ISSN 1099-1085

Thorns, M.C. & Sheldon, F. (2000). Water resource development and hydrological change in a large dryland river: the Barwon-Darling River, Australia. *Journal of Hydrology*, Vol. 228, pp. 10 - 21, ISSN 0022-1694

Turoboyski, L. (1979). *Technical hydrobiology*. PWN, Warsaw, Poland, ISBN 83-01-00739-7

Van den Brink, F.W.B. & Van der Velde, G. (1991). Macrozoobenthos of floodplain waters of the rivers Rhine and Meuse in the Netherlands: a structural and functional analysis in relation to hydrology. *Regulated Rivers: Research and Management*, Vol. 6, pp. 265–277, ISSN 0886-9375

Walker, K.F.; Sheldon, F. & Puckridge, J.T. (1995). A perspective on dryland river ecosystems. *Regulated Rivers: Research and Management*, Vol. 11, pp. 85 - 104, ISSN 0886-9375

Ward, J.V. (1998). Riverine landscapes: Biodiversity patterns, disturbance regimes, and aquatic conservation. *Biological Conservation*, Vol. 83, pp. 269 - 278, ISSN 0006-3207

Ward, J.V. & Stanford, J.A. (1995). Ecological connectivity in alluvial river ecosystems and its disruption by flow regulation. *Regulated Rivers: Research and Management*,Vol. 11, pp. 105– 119, ISSN (printed) 0886-9375

Ward, J.V.; Tockner, K.; Arscott, D.B. & Claret, C. (2002). Riverine landscape diversity. *Freshwater Biology*. Vol. 47. pp. 517 - 539, ISSN 0046-5070

Whiles, M.R. & Goldowitz, B.S. (2005). Macroinvertebrate communities in Central Platte River wetlands: Patterns across a hydrologie gradient. *Wetlands*, Vol. 25, pp. 462 - 472, ISSN 0277-5212

8

Mnemiopsis leidyi Invasion and Biodiversity Changes in the Caspian Sea

Aboulghasem Roohi[1] and Ameneh Sajjadi[2]
[1]*Caspian Sea Research Institute of Ecology, Khazarabad Boolvar, Sari,*
[2]*Meteorology Dept, Islamic Azad University Rasht Branch, Rasht, Guilan*
Iran

1. Introduction

1.1 Caspian Sea

The complex history of the Caspian Sea formation has lead to a variety of different habitats. Like Australia, the Caspian Sea became isolated thousands of years ago (Plate 1). This isolation led to the speciation of many rare animals in particular the sturgeon.

The Caspian Sea is the biggest enclosed body of water on Earth, having an even larger area than that of the American Great Lakes or that of Lake Victoria in East Africa. It is situated where the South-Eastern Europe meets the Asian continent, between latitudes 47°.07́N and 36°.33́N and longitudes 45°.43́E and 54°.20́E. It is approximately 1,030 km long and its width ranges from 435 km to a minimum of 196 km. It has no connection to the world's oceans and its surface level at the moment is around _26.5 m below MSL. At this level, its total coastline is some 7,000 km in length and its surface area is 386,400 km^2. The water volume of the lake is about 78,700 km^3.

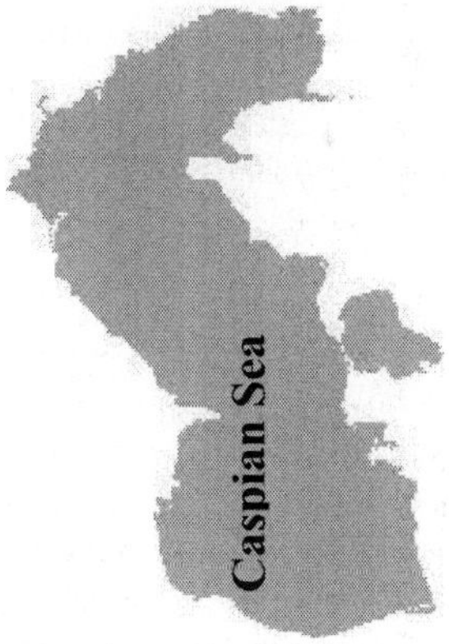

Plate 1. Caspian Sea area

The Caspian can be considered as divided into three parts, the northern, middle and southern parts. The border between the northern and middle parts runs along the edge of the North Caspian shelf (the Mangyshlak threshold), between Chechen Island (near the Terrace River mouth) and Cape Tiub-Karagan (at Fort Shevchenko). The border between the middle and southern parts runs from the Apsheron threshold connecting Zhiloi Island in the west to Cape Kuuli in the east (north of Turkmenbashi). The northern part covers about 25% of the total surface area, while the middle and southern parts cover around 37% each. However, the water volume in the northern part accounts for a mere 0.5%, while the volume in the middle part make up 33.9%, and in the southern part 65.6% of Caspian waters. These volumes are a reflection of the bathymetry of the Caspian. The northern part is very shallow, with average depths of less than 5m. In the middle part, the main feature is the Derbent Depression with depths of over 500m. The southern part includes the South Caspian Depression with its deepest point being 1025m below the surface (plate 2).

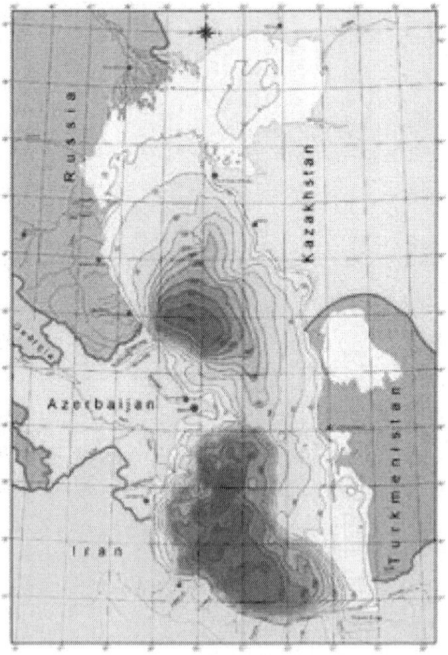

Plate 2. Caspian Sea riparian countries

Approximately 130 large and small rivers flow into the Caspian, nearly all of which flow into the north or west coasts. The largest of these is the Volga River that drains an area of 1,400,000 sq. km and runs into the northern part of the Caspian. Over 90% of the inflowing freshwater is supplied by the 5 largest rivers: Volga - 241 km3, Kura - 13 km3, Terek - 8.5 km3, Ural - 8.1 km3 and Sulak 4 km3. The Iranian rivers and the smaller streams on the western shores supply the rest, since there are no permanent inflows on the eastern side.

Apart from the extensive shallows of the northern part, the other two physical features that characterize the Caspian are the Volga and the Kara Bogaz Gol gulf.

The Volga Delta is situated in the Prikaspiisk lowlands covering around 10,000 km2 and the delta has a width of about 200 km. A feature of the delta region are the so-called Baer knolls which are hillocks, between 3m and 20m in height, formed by the action of onshore winds on the river sediments. These sediments are discharged into the delta at a rate of 8 million tones per year. Numerous small lakes can be found between the knolls and there is a complex system of channels with many islets. The Volga-Caspian shipping canal traverses the delta and is dredged to maintain a depth of no less than 2m (Aladin and Plotnikov, 2004).

2. Biodiversity in the Caspian

The biodiversity of the Caspian aquatic environment is a product of thousands of years of isolation from the world's oceans, allowing ample time for speciation. The biological diversity of the Caspian and its coastal zone makes the region one of the most valuable ecosystems in the world. The Caspian harbors some 147 species of fish, 450 species, varieties, or forms of phytoplankton, 87 species of algae, and 315 species of zooplankton. One of the most important features of the Caspian's biodiversity is the relatively high level of endemism among its fauna. Recent studies suggest the actual endemism may be even higher than what is already known. To date, there are 331 known endemic species in the Caspian. They are represented by the following: UNDP, www.caspianenvironment.org/newsite/Data-MajorDocuments.htm.

Four (4) species of *Spongia*	One (1) species of *Isopoda*
Two (2) species of *Coelenterata*	Sixty-eight (68) species of *Amphipoda*
Twenty-nine (29) species of *Turbellaria*	Nineteen (19) species of *Cumacea*
Three (3) species of *Nematoda*	One (1) species of *Decapoda*
Two (2) species of *Rotatoria*	Two (2) species of *Hydracarina*
Two (2) species of *Oligochaeta*	Fifty-three (53) species of *Mollusca*
Four (4) species of *Polychaeta*	Fifty-four (54) species of fish
Nineteen (19) species of *Cladocera*	One (1) species of marine mammal
Three (3) species of *Ostracoda*	Twenty (20) species of *Mysidacea*
Twenty-three (23) species of *Copepoda*	One (1) species of *Isopoda*
Twenty (20) species of *Mysidacea*	

Tran boundary Diagnostic Analysis for the Caspian Sea, Caspian Environment Programme. September 2002. Baku. www.caspianenvironment.org/newsite/Data-MajorDocuments.htm.

Table 1. Known endemic species in the Caspian Sea.

The decline in bioresources and biodiversity are closely linked through food chains and feeding patterns. A disturbance in the phytoplankton-zooplankton and benthic communities caused by invasive species for instance may impact species at higher trophic levels, such as sturgeon or seals. With the invasion of ML (*Mnemiopsis leidyi*) as well as introductions of other species the naturally occurring food web may have undergone or be undergoing potentially significant disruptions particular when under concurrent stresses.

The sturgeon species existed 200 million years ago at the same time as dinosaurs and can therefore be called living fossils. At that time sturgeon inhabited many ancient seas. Later on in the process of evolution, possibly due to competition with bony fish species, the sturgeons started to become extinct but managed to survive in the Caspian Sea. This gigantic lake contains more than 90% of the world resources of sturgeon. Furthermore, the Caspian Sea is also home to many other rare species of crustaceans and mollusks (Birstein et al., 1968).

Due to its unique and diverse habitats, the Caspian Sea has become home to many rare species of flora and fauna. In connection with an increase of the Caspian Sea level during the period of 1994 - 1996, habitats for rare species of aquatic vegetation have drastically decreased. This can be attributed to a general lack of seeding material in newly formed coastal lagoons and water bodies (Aladin and Plotnikov, 2004).

Many rare and endemic plant species of Russia are associated with the intra-zonal communities of the Volga delta and riparian forests of the Samur River delta as well as to the Sarykum barkhan which is a unique refuge for flora adapted to the loose sands of the ancient Central Asian Deserts. The principal limiting factors to successful establishment of plant species are hydrological imbalances within the surrounding deltas, water pollution, and various land reclamation activities. The water level change within the Caspian Sea is an indirect reason for which plants may not get established. This affects aquatic plants of the Volga delta, such as: *Aldrovanda veiculosa* and *Nelumbo caspica*. About 11 plant species are found in the Samur River delta, of which some form a unique liana forest that dates back to the Tertiary period.

Different factors are involved in decline of biodiversity in the Caspian Sea.

1. One of the factors contributing to depleted fisheries and ecosystem resilience is the separation of anadromous fish from their natal river systems in the Caspian. Reduced access to sturgeon spawning sites began in the 1930s with the construction of irrigation weirs, followed by the construction of large dams on the Kura River in the 1950s, the Volga River in the 1960s, and the Sefidrud River in the early 1970s. In the past 50 years, anadromous fish migrations have been blocked to up to 90% of natural spawning grounds on rivers like the Volga and the Kura. As summarized above, anadromous fish such as sturgeon, salmon or herring develop genetically distinct sub-populations in response to environmental variability. Dams without fish passages block migration up rivers for spawners and down rivers for fingerlings. This loss of connectivity and natural selection cannot be replaced by hatcheries and has had the effect of drastically reducing the biological diversity of the Caspian's fish species and populations. It has led to reduced numbers of fish overall and reduced numbers of genetically distinct populations of fish (Aladin and Plotnikov, 2004).
2. Invasive species are also factors thought to be contributing to ecosystem stress, loss of biodiversity and depleted fisheries. Invasive species have been shown the world over to have direct and indirect impacts on many ecosystem components, including productive fisheries and the economy. Ecosystems often contain cascading feeding interactions that respond in unpredictable ways to introductions. Invasive species affect individuals, populations, and assemblages of populations in the ecosystems where they occur. One assemblage-level impact is a substantial shift in relative abundances, resulting in declines and losses among native fishes, for example. This is widely believed to have happened in the Caspian with respect to the native species of fish called the Kilka among others. *Mnemiopsis lediyi*, an invasive species of jellyfish, is thought to have affected the cascading feeding interactions that the Kilka relied upon, possibly causing the Kilka populations to decline dramatically, which in turn is thought

to have impacted the Caspian seal, for whom Kilka are an important food source. Clearly, to restore depleted fisheries, ecosystems and the processes and interactions that occur within them must be protected.

3. The presence of POPs (in particular pesticides) and PTS from exploitation of oil in some parts of the Caspian Sea is a major source of concern, especially their accumulation in the long-lived species – mollusks, seals, and sturgeons (UNDP, www.caspianenvironment.org/newsite/Data-MajorDocuments.htm).

3. *Mnemiopsis leidyi* problem in the Caspian

In the early 1980s, the comb jelly *Mnemiopsis leidyi,* a ctenophore that normally resides off the eastern United States, was accidentally introduced into the Black Sea via ballast waters from cargo ships. This voracious zooplanktonic predator (with extremely high rates of reproduction and growth) reached enormous biomass levels (a few hundreds million tons for the entire basin!) devastating the pelagic (i.e. in water column) food chain in the entire Black Sea basin by the end of 1980s (Vinogradov et al., 1989). Inevitably, such high biomass of this comb jelly consumed a considerable fraction of the zooplankton that had been the food for pelagic fish and their larvae before its arrival. One of the dramatic consequences of the *M. leidyi* invasion was the sharp drop (from about 630,000 tons in 1988 to steadily 150,000 tons in 1991) in commercial catches of planktivorous fish (mainly the anchovy Engraulis encrasicolus L.) in the Black Sea (Kideys 1994; Prodanov et al., 1997). The yearly economical damage to the fisheries sector alone were estimated to be about 250-500 million USD during this period. Although merely one or two researchers pointed out the overfishing as the major cause, the concurrent sharp decrease in zooplankton quantity from different regions in the Black Sea (Kovalev et al., 1998; Gubanova et al., 2002; Gordina et al., 2004) was a conclusive evidence. Indeed one would expect much higher quantities of zooplankton (due to decreased predation) at low levels of planktivorous fish occurence. The decreased levels of the pelagic fish must have also affected the abundance of top predators (several species of predator fish as well as the three species of dolphins) in the Black Sea. Although there are no systematic data on dolphins, they were noted to be scarcer by fishermen and mariners at this period.

M. leidyi did not only affected the quantity of animals but also of plant organisms, known as phytoplankton. These (mainly) photosythetic organisms are the food for zooplankton. Due to decreased levels of zooplankton, phytoplankton had a chance to over-grow in the Black Sea (Yunev et al., 2002) during the peak period in *M. leidyi* quantity. Such increase was deleterious particularly for some shallow regions in the Black Sea ecosystem (e.g. off Danube River) already badly suffering from eutrophication.

The situation in the Black Sea has been one of the most striking examples in marine bioinvasion history. Due to scale of the problem, UNEP intervened and gathered international experts in Geneva in 1994, for investigating methods for solving this problem (GESAMP, 1997). The futility of physical and chemical methods for this problem were noted and therefore, biological control seemed the only workable remedy. And, based on the literature knowledge of feeding specificity, another ctenophore species (*Beroe ovata*) rose as the best candidate for dealing with *M. leidyi* problem. Indeed, B. ovata reported feeding only on other ctenophore species (Kremer and Nixon 1976), most notably on *M. leidyi*. However, scientists from the Geneva meeting could not stress on using a new predator species for dealing with the problem in the Black Sea, due to risk of unexpected problems: What if *B. ovata* start feeding on other species rather than *M. leidyi?*.

A warning that *M. leidyi* might also invade the Caspian Sea had been voiced during the Geneva meeting as well as by Dumont (1995). Unfortunately, at the end of the 1990s the invasion of *M. leidyi* in the Caspian Sea was already being reported (Esmaeili et al., 2000; Ivanov et al., 2000; Roohi, 2000). It must have also been transported in the ballast waters of ships traveling from the Black Sea (salinity 18 ppt) to the Caspian Sea (max. salinity 13-14 ppt) through the Volga Don Canal. Investigations in the Caspian Sea showed by September 2000, it was found everywhere including the northern Caspian where the salinity can be as low as 4 ppt (Shiganova et al., 2001a).

The impact of *M. leidyi* on the Caspian Sea ecosystem has been even worse than in the Black Sea due to the greater sensitivity of this enclosed basin. Adverse impacts from *M. leidyi* could be listed as the following:

1. Again the fish collapse was the most apparent problem in the ecosystem. Striking decreases were observed in the pelagic (mainly sprat *Clupeonella* spp.) fishery of all countries bordering the Caspian Sea: almost a 50% decrease in the kilka catches of both Iranian, Azerbaijan and Russian fisheries had occurred during 1999 and 2001. During spring and summer of 2001, mass (estimated as 250,000 tons, or 40% of the population) mortalities of sprat were reported at the sea surface (Davis et al., 2003). The fish catch value was halved again in 2002, resulting in great economic losses (Kideys et al., 2004, 2005). Fishermen even stopped fishing during most part of 2003, due to lack of fish (Fazli and Roohi 2003).
2. Sharp decrease in fish catch became a big problem for thousands people earning livelihood from sprat fishery. The economical loss from sprat fishery alone is hundreds million Euros per year. Most of the fishermen in Iran, who once took loans from banks for starting to a business with promising outlook, cannot now pay their debts and may even end up in prison. Their problem was even at headlines on BBC World TV in 23rd July 2001.
3. Not only pelagic fishes, but also some large predators feeding on these fish such as white sturgeon *Huso huso* and the endemic Caspian seal *Phoca caspica* are also suffering from significant population decrease. As reported by the media, the mass deaths of Caspian seals (*Phoca caspica*) occurred in the northern Caspian Sea during the spring of 2000. There is strong evidence that the epizootic disease observed in seals during the spring of 2000 was caused by under nourishment (Davis et al., 2003). Significantly decreased pregnancy and fat content inseal population were also reported. The white sturgeon, that is famous for the quality of its caviar, mainly depend on sprat as food (Hashemian and Roohi 2004).
4. Biodiversity of the Caspian is important as most of species occur only in this sea all over the world (i.e. endemic). Not only the quantity of zooplankton is reported to decrease sharply, but also the number of species. For example, number of zooplankton (copepod and cladocerans) species during 2001-2002 was only 3 compared to 22 species in 1995 or 1996!. The consequences of such reduction could be very significant for the ecosystem (Roohi et al., 2010)
5. Due to decreased levels of zooplankton, eutrophication (to much plant production) started to be a significant problem for this ecosystem. Global chlorophyll distribution

obtained via remote sensing display the Caspian Sea as one of the most eutrophic regions in the world in recent years, in contrast to years before *M. leidyi* invasion (Roohi et al., 2008a, b)

4. General aspect of *Mnemiopsis*

Mnemiopsis leidyi - is the lobate ctenophore. Two oral lobes are derivatives of the ctenophore body (spherosome). Four smaller lobes -auricules are situated under the principal two oral lobes. During their movements the lobes in fold completely its buccal orifice. The oral lappets carry tentacular rings. Its central part is situated above the lips of the mouth crevice. Both "lips" are extremely contractible (Agassiz, 1860; Seravin, 1994, plate 3).
Mnemiopsis **characteristics** in a glance are as follows:

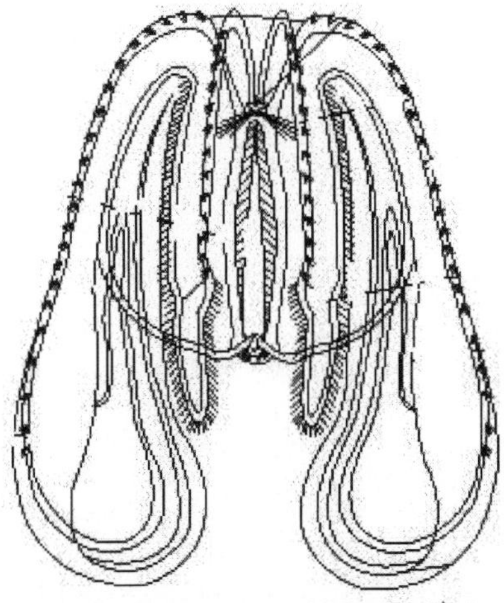

Mnemiopsis leidyi (A. Agassiz, 1965)
(Drawing by Tamara Shiganova)

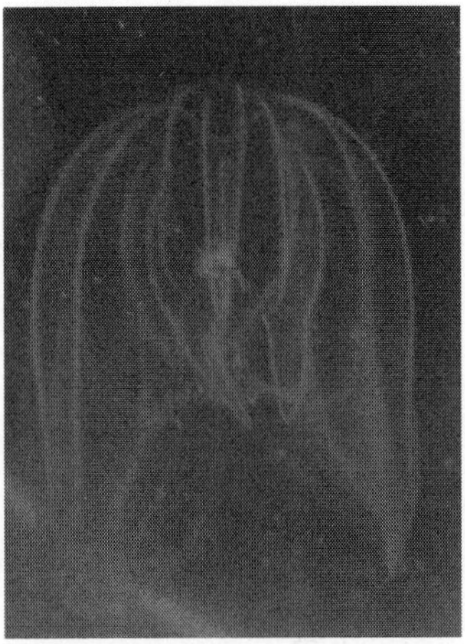

Mnemiopsis leidyi photo by ROOHI, A.

Plate 3. *Mnemiopsis leidyi* images of the Caspian Sea

Luminescence- *Mnemiopsis* is remarkably phosphorescent. The seat of the phosphorescence is confined to the rows of locomotive flappers.
Ecological group- Macrozooplankton
Origin: North American species might be brought into the Black Sea with ballast water by Russian tankers driving oil to the ports at eastern coast of USA. From the Black Sea Mnemiopsis might be transferred into the Caspian Sea also by tankers driving oil though the Volga-Don Canal.
World distribution: The native habitat of the ctenophore, *Mnemiopsis,* is in temperate to subtropical estuaries along the Atlantic coast of North and South America (Harbison et al., 1978). In the early 1980s, it was accidentally introduced to the Black Sea (Vinogradov et al., 1989), where it flourished and expanded into the Azov, Marmara, eastern Mediterranean, and Caspian Seas (Studenikina et al., 1991, Shiganova et al, 2001a, Shiganova et al, 2001b).

Habitat: *Mnemiopsis leidyi* inhabits coastal areas and surface layers (above thermocline) open sea. Some large ctenophores can spread deeper and even can be found near the bottom in the coastal areas of the Caspian Sea.
Migrations: Transferred with the currents. Dial vertical migrations were not recorded, although it is more abundant near the surface at night where they feed and reproduce
Relation to salinity: Euryhalinic species. Salinity range from 2 to 38 (Kremer, 1993). In the seas of Mediterranean basin *M .leidyi* occurs in waters with salinities ranging from 3 in the Sea of Azov to 39 in the eastern Mediterranean. In the Caspian Sea its distribution is limited isohalines of 4 ‰.
Relation to temperature: Eurythermic species. Temperatures range from 0°C in northern native locations in the winter, to 32°C in the southern estuaries during the summer.
Feeding type: Heterotrophic, carnivorous
Feeding behavior: The larvae of *Mnemiopsis* can retract entirely their two tentacles into the tentacular sheaths on either side of the body, between the oral and aboral poles.
Reproduction type: *Mnemiopsis leidyi*- is a self-fertilizing hermaphrodite
Relation to environmental factors: The main factors, which are important for reproduction, are temperature and food concentration

5. Highlights of *Mnemiopsis* monitoring data in the Caspian

Invasion of the Caspian Sea by the comb-jelly *Mnemiopsis leidyi* (ML) since late 1990s has become one of the main environmental issue of this unique ecosystem. The adverse effects of this ctenophore was first visible on the pelagic fishery but also evident on other major compartments of the ecosystem, including, phytoplankton, zooplankton, benthos, Caspian Seal and even on some sturgeon species. Some endemic zooplankton species appear to have completely disappeared from samples of ongoing monitoring programs. ML invasion has had major impact on fisheries industry causing considerable economic damage, mostly to the coastal communities which depend on pelagic fisheries for their livelihood. The case of ML in the Caspian Sea is one of the largest invasion impacts ever occurred in a marine ecosystem all over the world.

6. A review of Mnemiopsis investigations of the Caspian Sea over the last decade

After *Mnemiopsis* invasion into the Caspian Sea via the ballast water from the Black Sea and/or the Sea of Azov in 1999 (Roohi et al., 2008a), some objectives of this alien ctenophore was taken into account in several local or national projects such as follows:

- Distribution and abundance of *Mnemiopsis leidyi* in the Caspian Sea (Iran- Russia – Azerbaijan)- in 2001-2004 and 2009
- Feeding, respiration, reproduction of *Mnemiopsis leidyi* in the Caspian Sea- in 2001-2009
- Comparative feeding study of *Mnemiopsis leidyi* and Kilka in the Caspian Sea- in 2003 - 2004 and 2008/9
- Zooplankton and phytoplankton changes after ML invasion

Mnemiopsis monitoring with the spatial and temporal investigations were conducted along the inshore and offshore of the Caspian Sea in Iran- Russia- Azerbaijan and Turkmenistan coasts. Fortunately, the main two countries (Iran and Russia) had established the favorable framework of the jelly study and achieved the appropriate results in which most of the discussion were based on two countries data analysis.

Ctenophore samples were collected with an METU (Medalist Technology University) net having a mouth opening of 0.2 m^2 and a screen with a mesh size of 500 m, from the same depths as the Juday net (Vinogradov et al., 1989; Kideys et al., 2001). On completion of each tow, the cod end was immediately passed into a container and ctenophores counted by eye. The body length of each individual with lobes was measured lying flat (out of water) onboard, and the density of *Mnemiopsis leidyi* (per m^2 and m^3) was calculated from the net diameter and tow depth. The ctenophores were sorted in length groups of 5-mm intervals to determine the abundance of different size groups. Length measurements were converted to wet weight using an appropriate equation (Kideys et al., 2001). Samples of M*nemiopsis* were collected from 20001 along few semi- transects perpendicular to the Iranian coast of the Caspian Sea (Fig. 1).

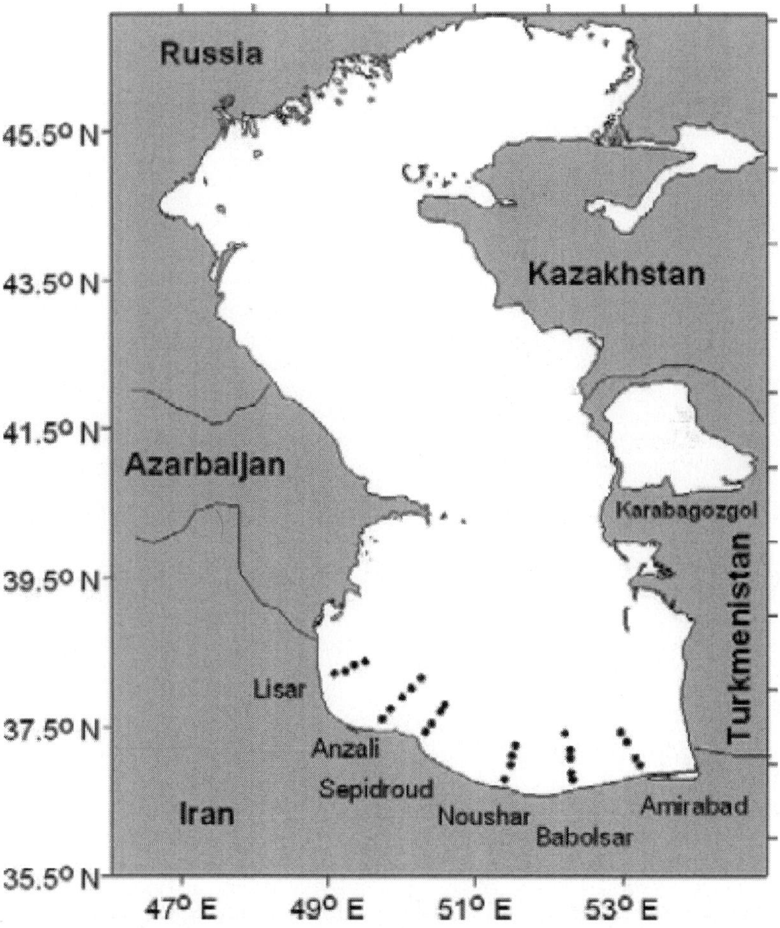

Fig. 1. Distribution of sampling stations in the southern Caspian Sea.

The ctenophore *Mnemiopsis leidyi* was found at all stations from 2001–2009. There was a seasonal succession of ctenophore densities every year, the maximum being observed in August and September, and the minimum density in the winter months. A significant correlation was found between the water temperature and the abundance of Mnemiopsis leidyi (P <0.005). The highest summer- autumn average of Mnemiopsis leidyi abundance was observed in 2002 (760 ± 1148 ind.m^3), although the biomass during this period (23.2 ± 23.3 g.m^3) was lower than in 2001 (41.5 ± 44.3 g.m^3). In terms of monthly averages, October 2001 was the month of the maximum abundance and biomass (1157 ± 1614 ind.m^3 and 58.9 ± 40.0 g.m^3).

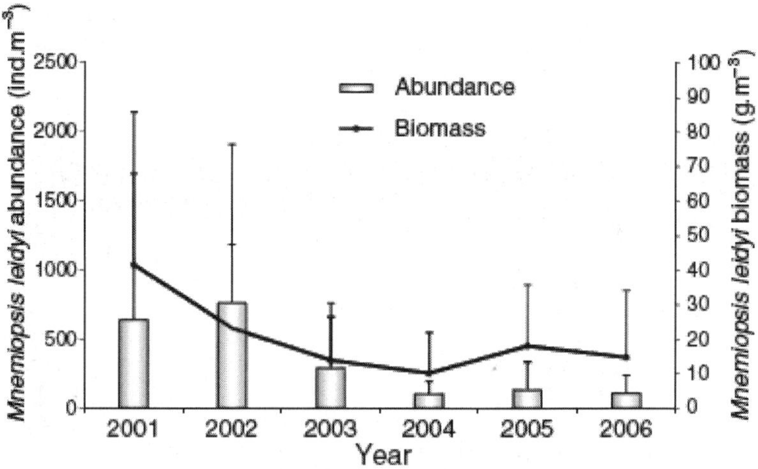

Fig. 2. Spatial and depth means of *Mnemiopsis leidyi* abundance and biomass in the Southern Caspian Sea during summer–autumn 2001–2006

In terms of spatial distribution, in spring the maximum abundance of *Mnemiopsis leidyi* (141 ind.m^3) was recorded in the coastal area of the southeastern Caspian Sea (Amirabad) and the minimum (3–14 ind.m^3) in waters 100 m deep. In summer, the highest abundance was again noted in the southeastern Caspian Sea, with values 763 ind.m^3. Likewise, in autumn, the maximum abundance was in the southeast at Babolsar (at the shallowest station of 5 m depth), with a value of 1235 ind.m^3. In addition, abundance was high at a station with a depth <20 m (500–700 ind.m^3. In winter the maximum abundance was recorded in the Anzali region, with a value of 653 ind.m^3 (Fig. 3).

In the Northern Caspian, *M. leidyi* was first found only in September 2000; its abundance increased in October, but values were not high: 108 ±65 ind. m-2 (21.6 ±9 ind. m^{-3}), biomass 140.4 ±42 g m^{-2} (28.1 ±8 g m^{-3}).

In May 2001, *M. leidyi* was recorded only in the Southern Caspian (Fig. 6A), where its abundance was 1972 ± 683 ind. m^{-2} (100 ± 34 ind. m^{-3}) and biomass 128 ± 57.5 g m^{-2} (6.4 ± 2 g m^{-3}) and in the southwestern part of the Middle Caspian, up to 43° N, abundance was 230 ± 144 ind. m^{-2} (12 ± 20 ind. m^{-3})and biomass was 20.0 ± 37 g m^{-2} (1.4 ± 2 g m^{-3}) (Fig. 4). *M. leidyi* was most abundant in the western and middle areas of the Southern Caspian, with maximum abundance at the Apsheron Swell and in the western slope waters. Mean size was

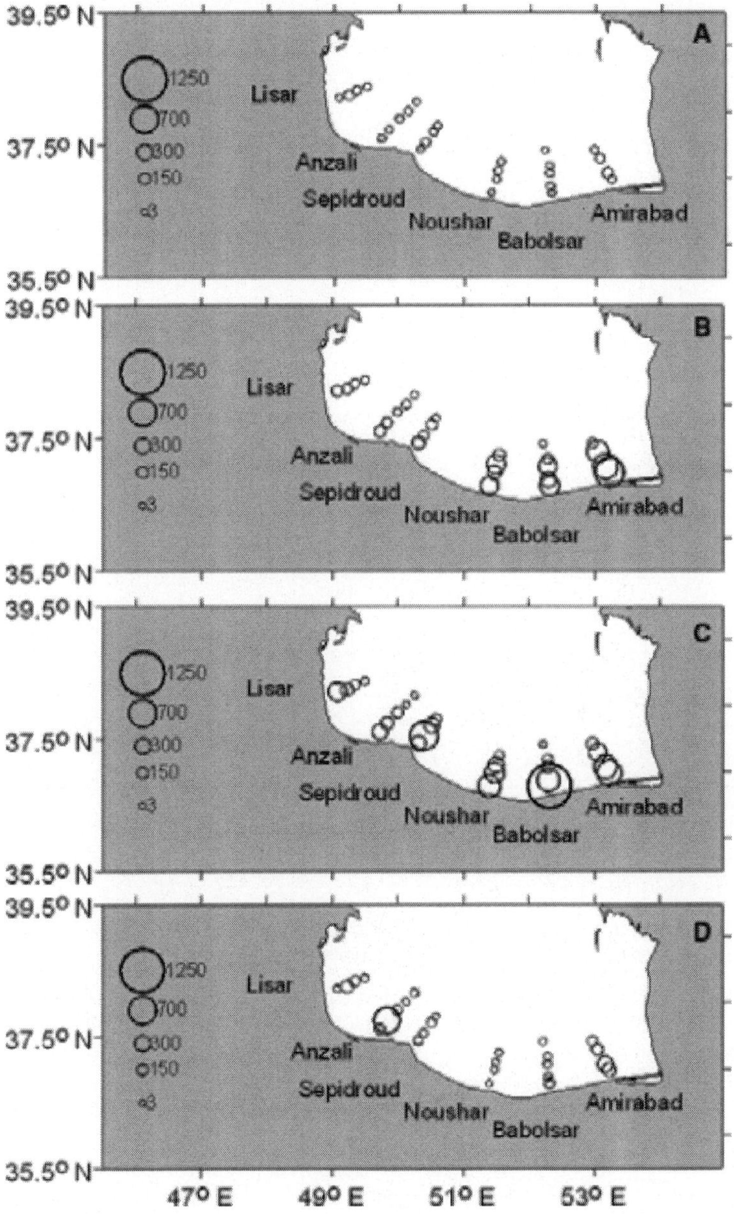

Fig. 3. Seasonal distribution of Mnemiopsis leidyi abundance (depth averages of available months) in the Southern Caspian Sea during 2001–2006. (A) Spring. (B) Summer. (C) Autumn. (D) Winter.

very small: up to 3.6 mm in the Southern Caspian and 4.2 mm in the Middle Caspian. It is well known that *Mnemiopsis* shrinks in unfavorable conditions; here, salinity, food, or a combination of both may have been strongly suboptimal.

In May, a few eggs and larvae were found in the Southern and Middle Caspian, but mass reproduction did not start yet because of scarcity of reproducing adults and probably, low spring temperatures (16°C in the Southern and 15°C in the Middle Caspian) (Fig.4).

In June 2001, *Mnemiopsis leidyi* began to reproduce and continued its expansion towards the north: in the Southern and south Middle Caspian (Fig. 4), its average abundance was 680 ± 16.8 ind. m^{-2} (34 ± 2 ind. m^{-3}), and biomass 88.3 ± 7.78 g m^{-2} (4.3 ± 1 g m^{-3}) (Fig. 4). The highest abundance and biomass, found in the Southern Caspian, represented values of 2005 ± 1248 ind. m-2 (100 ± 62 ind. m-3) and 230 ± 197.66 g m^{-2} (10.2 ± 9 g m^{-3}), respectively (Shiganova et al., 2004).

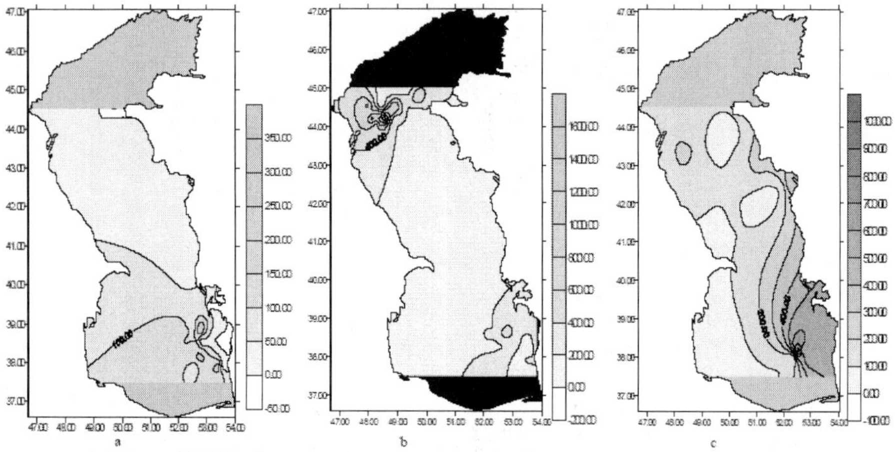

a) February- March b) August- September c) October- November

Fig. 4. Seasonal distribution of the Caspian population of *Mnemiopsis leidyi* (ind.m^3) in 2003 (shiganova et al., 2004)

7. A review of zooplankton investigations of the Caspian

Investigations performed in the last decade indicate that there have been important changes in the zooplankton composition and structure in the Caspian Sea. However, contrasting events taking place in different regions of the Caspian Sea indicate a non-uniform structure of its ecosystem. Several fodder zooplankton species have either disappeared from or substantially decreased in number at different sampling sites of the Caspian Sea over the last decade. Some other species adapted to thrive in eutrophic conditions have either appeared or increased in quantity especially meroplankton. Meanwhile the biomass of the fodder zooplankton has also fluctuated considerably through the years. However, there seems to be a reverse trend in the long-term variation of fodder zooplankton between the shallow western and deep eastern areas. Over the last decade the abundance of fish larvae has decreased significantly when compared either to past records or with larval abundances of other seas. This was shown to be due mainly to malnutrition of larvae. One of the most striking changes in the Ichtyoplankton has been the shift in the spawning areas of the main fish species in Caspian Sea. Even the invading ctenophore *Mnemiopsis* were found to be starving. The condition of other species (*Calanipeda aquae dulcis* and *Limnocalanus grimaldii*) disclosed the fact that cyclonic regions where chlorophyll and nutrient concentrations are high provide better nutrition than anticyclonic regions.

8. Species composition of zooplankton

A total of 18 zooplankton species (mero- and holozooplankton) were found. Among them there were 13 species of merozooplankton and only five species of holozooplankton. The latter belonged to Copepoda (four species) and Cladocera (one species) (Table 2). The only Cladocera species was *Podon polyphemoides*. Four copepod species were found, with the predominant calanoid *Acartia tonsa* present in all stations and every season. In 2006, a slightly higher diversity of Copepoda was seen; *Eurytemora grimmi*, absent in 2001–2005, was then observed for the first time at 50 m depth of the 100-m-deep station off Anzali (49°N and 37°E) in 2006.

Zooplankton group	Before *Mnemiopsis* Hossieni	After *Mnemiopsis leidyi* invasion					
Year	*et al.* (1996)	2001	2002	2003	2004	2005	2006
Copepoda	7	3	3	3	3	2	4
Cladocera	24	0	1	1	1	1	1
Merozooplankton	5	5	11	9	7	7	8
Total	36	8	15	13	11	10	13

Table 2. Species number of zooplankton before and after *Mnemiopsis leidyi* invasion in the Southern Caspian Sea.

9. Zooplankton frequency

In general, low zooplankton abundance and biomass (wet weight) were observed in summer months from 2001 to 2006. The highest abundance and biomass of zooplankton along the whole water column were not regularly found in the same season each year (Fig. 5). The maximum abundance recorded was 22,088 ± 24,840 ind.m^3 (average of stations and depths) in December 2001, whereas the highest biomass was 64.1 ± 56.8 mg.m^3 (average of stations and depths) in August 2004. Monthly variations of zooplankton biomass were similar to the fluctuations in abundance except in some summer–autumn periods when large-sized specimens dominated. The minimum zooplankton abundance and biomass were 397 ± 567 ind.m^3 and 1.8 ± 2.6 mg.m^3, respectively, in September 2002. The annual mean zooplankton abundance varied between 3361 and 8940 ind.m^3 during 2001–2006. The average zooplankton abundance and biomass for all months and years were calculated as 7015 ± 11.959 ind.m³ and 32.8 ± 57.6 mg. m³, respectively.

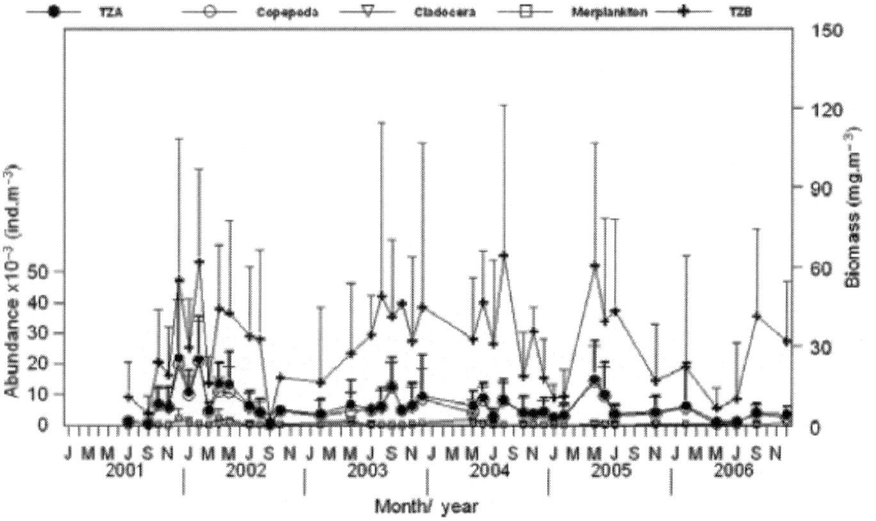

Fig. 5. Monthly variations in spatial and depth averages of Copepoda. Cladocera merozooplankton and total zooplankton abundance and biomass in the Southern Caspian Sea during 2001-2006. TZA, total zooplankton abundance; TZB, total zooplankton biomass

The maximum seasonal mean of zooplankton abundance was recorded in spring and the minimum in summer (Table 3, Fig. 6A–D). In spring, the greatest zooplankton abundance was observed at the Sephidroud River inlet (9×10^4 ind.m³) (Fig. 6A). In summer, the abundance of zooplankton decreased compared with the values reported in spring, and an almost even distribution was found along the coastal regions (max. 8– 11 · 103 ind.m³) decreasing towards the open sea (Fig. 6B). In autumn, zooplankton concentration was slightly greater than in summer; again the highest abundance (2×10^4 ind.m$\times 10^4$) was reported at the Sephidroud River inlet of 5 m depth (Fig. 6C). In winter, abundance was greater than in autumn and the maximum was observed at the Babolsar stations (3×10^4 ind.m³, Fig. 7D).

Comparison among different groups of zooplankton showed that Copepoda accounted for the maximum abundance and biomass every year from 2001 to 2006 (Fig. 6). Among Copepoda, different developmental stages of the calanoid species A. tonsa dominated during the study period. Copepoda, Cladocera and merozooplankton constituted 88%, 4% and 8% of total zooplankton abundance, respectively.

Hossieni et al. (1996) reported 36 zooplankton species (86% holoplankton and 14% meroplankton) in the southern Caspian Sea, consisting of 24 species of Cladocera, seven species of Copepoda and meroplankton such as larvae of Bivalvia and Balanidae (Table 3).

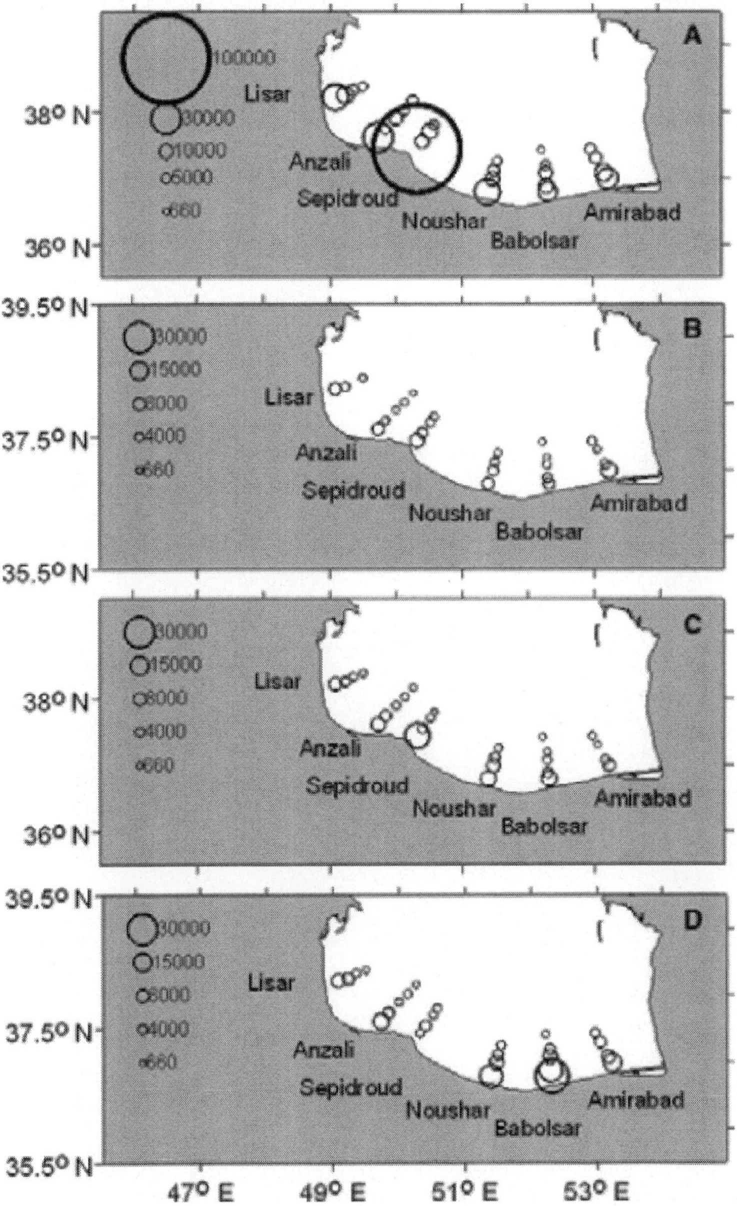

Fig. 6. Spatial distribution of zooplankton abundance in different seasons (depth averages of available months) in the Southern Caspian Sea during 2001–2006. (A) Spring. (B) Summer. (C) Autumn. (D) Winter.

	Hossieni	After *Mnemiopsis leidyi* invasion												
	et al. (1996)	2001		2002		2003		2004		2005		2006		
Zooplankton	A	SD	A	SD	A	SD	A	SD	A	SD	A	SD	A	SD
A)														
Acartia tonsa	5766	4935	6999	12,544	10,360	8831	6176	6758	5370	4676	8144	8793	3975	8944
Limnocalanus grimaldii	5	11	0	0	0	0	0	0	0	0	0	0	0	0
Calanipeda aquae dulcis	85	115	1	0	0	0	1	0	0	0	0	0	0	0
Eurytemora grimmi and *Eurytemora minor*	2172	1774	0	0	0	0	0	0	0	0	0	0	5	0
Nauplii Calanoida	2763	3296	0	0	0	0	0	0	0	0	0	0	0	0
Halicyclops sarsi (Cyclopoida)	170	168	5	0	2	0	120	2	5	1	22	3	1	0
Ectinosoma consimum (Harpacticoida)	0	0	0	0	3	3	0	0	1	0	11	19	0.4	0
Total Copepoda	10,961	10,300	7005	12,544	10,364	8834	6296	6760	5375	4678	8177	8814	3981	8944
Podon polyphemoides	237	442	0	0	450	672	1205	822	228	342	328	398	367	415
Polyphemus exicus	83	129	0	0	0	0	0	0	0	0	0	0	0	0
Total Cladocera	320	571	0	0	450	672	1205	822	228	342	328	398	367	415
Lamellibranchia larvae	5459	13,156	0	0	282	642	33	23	365	777	55	75	430	934
Cirripedia larvae (*Balanus* spp.)	439	648	86	378	661	1446	387	708	723	1359	328	617	277	430
Arachnida larvae	0	0	14	25	170	488	3	1	8	11	0	0	0.3	0
Mysidaceae larvae	NQD	0	0	0	0	0	0	0	0	0	0	0	2	0
Nematidae	NQD	0	0	0	214	361	163	137	0	0	2	2	0.5	0
Nereididae larvae	0	0	573	1838	435	877	12	18	13	23	38	52	9	0
Ostracoda larvae	NQD	0	5	8	32	106	2	2	58	117	0	0	0	0
Chironomidae larvae	0	0	0	0	0	0	5	8	0	0	0	0	4	0
Oligochaeta larvae	0	0	0	0	13	1	133	284	208	241	150	281	75	244
Total Meroplankton	5898	13,804	677	2248	1806	3920	738	1181	1374	2527	572	1027	799	1608
Total Zooplankton	17,180	24,675	7682	14,792	12,620	13,426	8239	8763	6977	7547	9077	10,239	5147	10967

	Hossieni	After *Mnemiopsis leidyi* invasion												
	et al. (1996)	2001		2002		2003		2004		2005		2006		
Zooplankton	B	SD	B	SD	B	SD	B	SD	B	SD	B	SD	B	SD
B)														
Acartia tonsa	37.2	36.5	23.0	32.0	35.4	26.6	30.1	33.6	36.9	34.2	42.0	38.6	22.9	33.4
Limnocalanus grimaldii	0.4	0.6	0.0	0.0	0.0	0.0	0.0	0.0	0.0	0.0	0.0	0.0	0.0	0.0
Calanipeda aquae dulcis	0.8	1.0	0.0	0.0	0.0	0.0	0.0	0.0	0.0	0.0	0.0	0.0	0.0	0.0
Eurytemora grimmi and *Eurytemora minor*	20.7	20.1	0.0	0.0	0.0	0.0	0.0	0.0	0.0	0.0	0.0	0.0	0.1	0.0
Nauplii Calanoida	7.6	7.9	0.0	0.0	0.0	0.0	0.0	0.0	0.0	0.0	0.0	0.0	0.0	0.0
Halicyclops sarsi (Cyclopoida)	1.0	1.0	0.0	0.0	0.0	0.0	0.0	0.0	0.0	0.0	0.1	0.0	0.004	0.0
Ectinosoma consimum (Harpacticoida)	0.0	0.0	0.0	0.0	0.0	0.0	0.0	0.0	0.0	0.0	0.1	0.1	0.003	0.0
Total Copepoda	68	67	23	32	35.4	26.6	30.1	33.6	37.0	34.2	42.2	38.8	23.0	33.4
Podon polyphemoides	1.6	2.7	0.0	0.0	1.8	2.7	4.8	3.3	0.8	1.3	0.9	1.3	1.5	1.7
Polyphemus exicus	0.5	0.8	0.0	0.0	0.0	0.0	0.0	0.0	0.0	0.0	0.0	0.0	0.0	0.0
Total Cladocera	2.1	3.4	0.0	0.0	1.8	2.7	4.8	3.3	0.8	1.3	0.9	1.3	1.5	1.7
Lamellibranchia larvae	26.9	65.0	0.0	0.0	1.4	3.2	0.2	0.1	1.8	3.9	0.3	0.4	2.152	4.6704
Cirripedia larvae (*Balanus* spp.)	1.0	1.5	0.2	0.8	1.4	2.9	1.4	3.7	4.2	8.5	1.4	1.8	1.114	2.0674

Table 3. Annual variations (A) in abundance (ind.m3) and (B) in biomass (wet weight, g.m3) of total zooplankton, Copepoda, Cladocera and merozooplankton species in the Southern Caspian Sea (before and after Mnemiopsis invasion). (C) Seasonal variations in abundance and biomass of zooplankton species in 1996 from Hossieni et al. (1996). (D) Full list of Cladocera species reported by Hossieni et al. (1996).

10. Phytoplankton population

In the present study, a total of 226 phytoplankton species were identified. While diatoms constituted 45% of the total species number, chlorophytes, cyanophytes, dinoflagellates and euglenophytes formed 20, 17, 11 and 8% of phytoplankton species, respectively (Fig. 7). Number of species in spring (91 species) and summer (101 species) were higher than in autumn (86 species) and winter (77 species).

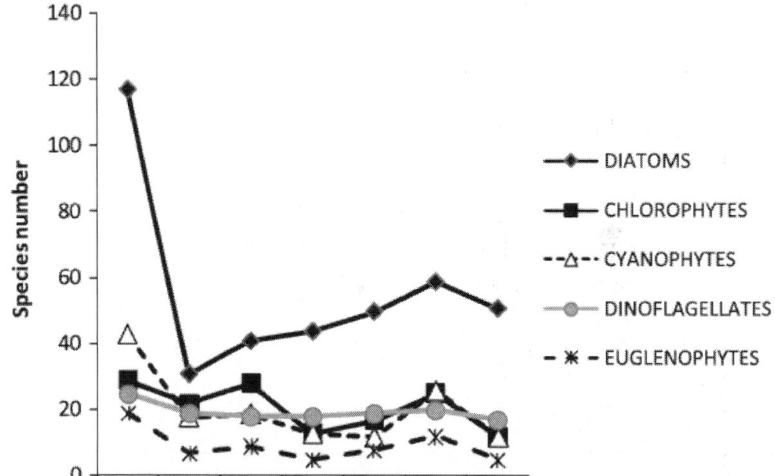

Years	1986-1994	2001	2002	2003	2004	2005	2006	2001-2006
DIATOMS	117	31	41	44	50	59	51	101
CHLOROPHYTES	29	22	28	13	17	25	12	45
CYANOPHYTES	43	18	19	13	12	26	12	38
DINOFLAGELLATES	25	19	18	18	19	20	17	25
EUGLENOPHYTES	19	7	9	5	8	12	5	17

Fig. 7. Variations in species number of different phytoplankton groups in the southern Caspian Sea during 2001–2006 and 1986–1994

The highest monthly mean phytoplankton abundance and biomass were 396 × 10^6 ± 299 × 10^6 cells m-3 in January 2002 and 1,789 ± 1,761mg m-3 in May 2002 (Fig. 8). Minimum abundance and biomass values were observed in August 2003 (1 × 10^6 ± 1 × 10^6 cells m-3 and 7 ± 5 mg m^{-3}) (Fig. 8).The overall average cell abundance and biomass of phytoplankton during 2001–2006 were 64 × 10^6 ± 76 × 10^6 cells m-3 and 250 ± 360 mg.m^{-3}, respectively. While diatoms were the most abundant phytoplankton group during 1996, after the introduction of *M. leidyi* the abundances of cyanophytes (in autumn) and dinoflagellates (in winter) exceeded diatom abundance in 2001 and 2002 (Fig. 8 and 9). Excluding 2005, diatom abundance was again high during 2003–2006. An unprecedented bloom of the toxic cyanophyte *Nodularia* sp. was observed between the second half of August and the end of

September in 2005. The bloom area covered ~20,000 km2 (CEP 2006, Fig. 9). According to the sampling on 20 September 2005, in addition to *Nodularia* sp., another cyanophyte Oscillatoria sp. was also high in abundance. Abundance of Nodularia sp. was 18 $\times 10^6$ cells m^{-3} at 7 m depth and 1,006 $\times 10^6$cells m^{-3} at 20 m depth. Average cyanophyte abundance and biomass at 7 and 20 m depths were 582 9 106 cells m-3 (of which 512 cells m^{-3} was Nodularia sp.) and 1,655 mg m^{-3}. The highest seasonal means of phytoplankton abundance and biomass were 179 × 106 cells m^{-3} and 880 mg m-3 in winter during 2001- 2006.

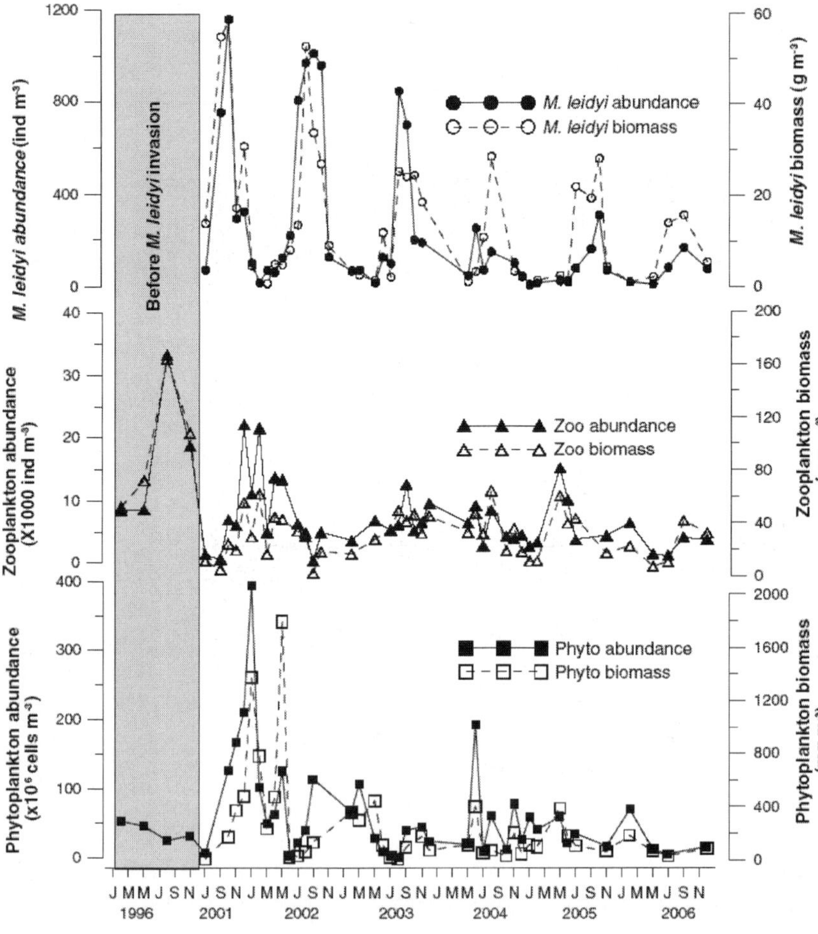

Fig. 8. Annual variations in the abundance and biomass of phytoplankton, zooplankton and Mnemiopsis leidyi in the southern Caspian Sea during 2001–2006 (values are depth and station averages). 1996 values are from Hossieni et al. (1996), spring 2001 values are from Kideys et al. (2001)

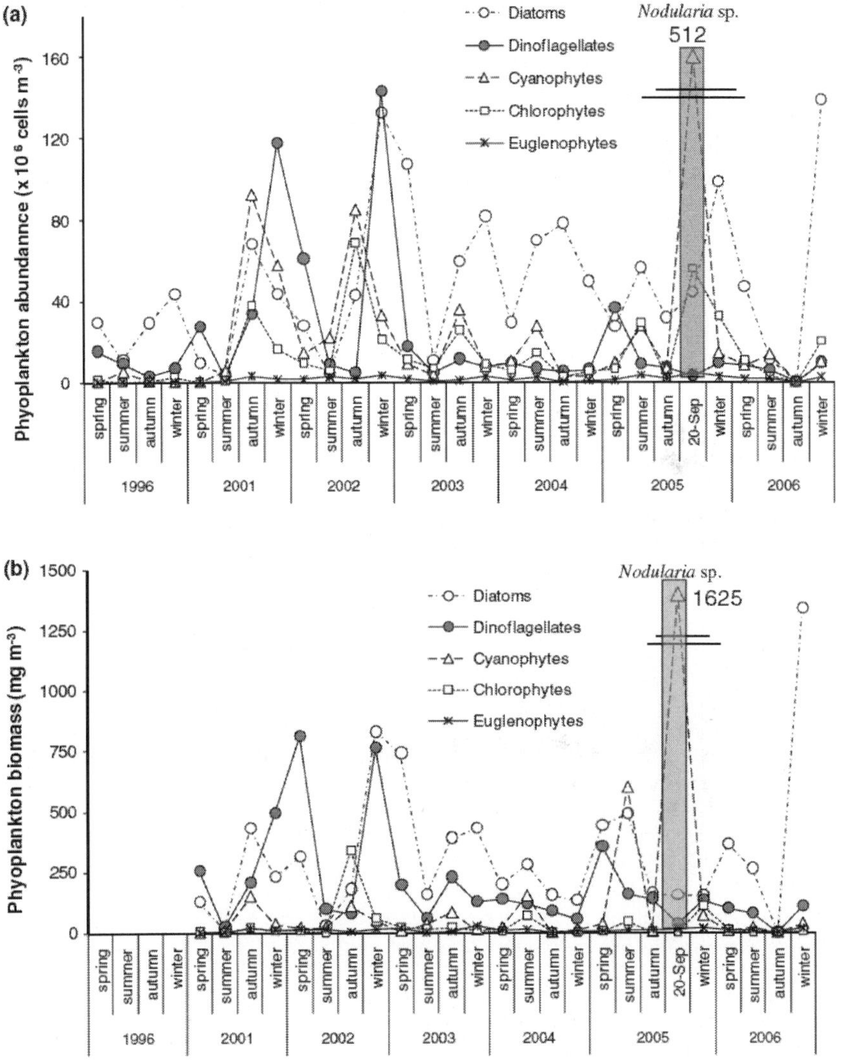

Fig. 9. Seasonal changes a in abundance and b in biomass of different phytoplankton groups before (Hossieni et al. 1996) and after Mnemiopsis leidyi invasion in the southern Caspian Sea (values are depth and station averages)

11. Other factors that have to be considered

11.1 Increased chlorophyll levels in the southern Caspian Sea after ML invasion

A significant correlation was observed between satellite derived chlorophyll *a* (Chl *a*) concentrations and the biomass of the invasive comb jellyfish *Mnemiopsis leidyi* in the southern Caspian Sea. By consuming the herbivorous zooplankton, the predatory ctenophore *M. leidyi* may have caused levels of Chl *a* to rise to very high values (~9mg m^{-3}) in the southern Caspian Sea. There might also be several other factors concurrent with predation effects of *M. leidyi* influencing Chl *a* levels in this region, such as eutrophication and climatic changes which play major roles in nutrient, phytoplankton, and zooplankton variations (kideys et al., 2008). The decrease in pelagic fishes due to overfishing, natural, and anthropogenic impacts might have provided a suitable environment for *M. leidyi* to spread throughout this enclosed basin (Fig. 10).

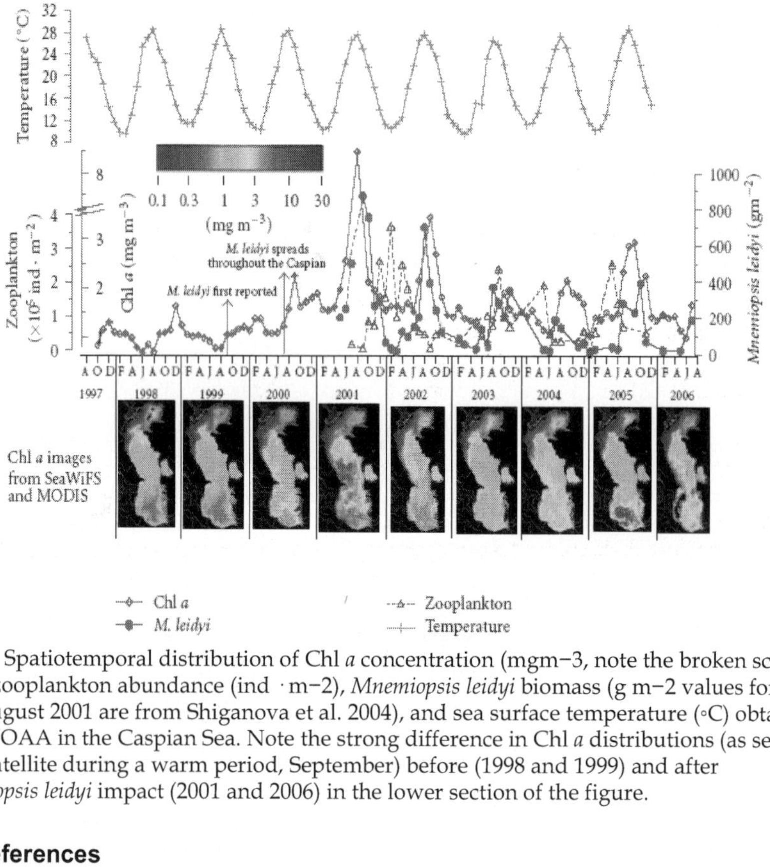

Fig. 10. Spatiotemporal distribution of Chl *a* concentration (mgm−3, note the broken scale here), zooplankton abundance (ind · m−2), *Mnemiopsis leidyi* biomass (g m−2 values for June and August 2001 are from Shiganova et al. 2004), and sea surface temperature (°C) obtained from NOAA in the Caspian Sea. Note the strong difference in Chl *a* distributions (as seen from satellite during a warm period, September) before (1998 and 1999) and after *Mnemiopsis leidyi* impact (2001 and 2006) in the lower section of the figure.

12. References

Agassiz L. 1860. Contribution of the natural history of the United States of America. Boston: Truber and Co. V.3.p.1-321.

Aladin, N. B. and Plotnikov, I. S. 2004. The Caspian Sea, Lake Basin Management Initiative, the Caspian Bulletin 4: 112–126.

Birstein, Y. A., Vinogradov, L.K., Kandakova, N., Kon, M. S., Stakhovaya, T. V. and Romanova, N. N. 1968. Atlas of the Caspian Sea invertebrates. Food industrial Co. Moscow, 415 pp.

CEP. 2006. A Study on the Harmful Algal Bloom in the Southwestern Basin of the Caspian Sea, Caspian Environment Programme, 15 pages

Davis, B.A.S., Brewerb, S., Stevensona, A.C., Guiotc, J. and Data Contributors. 2003. The temperature of Europe during the Holocene reconstructed from pollen data, Quaternary Science Reviews 22 (2003) 1701–1716.

Esmaeili, A., Abtahi, B., Khoda bandeh, S., Talaeizadeh, R., Darvishi, F. and Terershad, H. 2000. First report on occurrence of a combjelly in the Caspian Sea. *Journal of Environmental Sciences and Technology, Islamic Azad University*, 3, 63–69.

Fazli, H. and Roohi, A. 2003. The impacts of *Mnemiopsis leidyi* on species composition, catch and CPUE of Kilka in Iranian commercial catch. UNESCO, Caspian Floating University Research bulletin, Astrakhan (KaspNIRKh) No. 3: 99-104.

GESAMP. 1997. Opportunistic settlers and the problem of the ctenophore *Mnemiopsis leidyi* invasion in the Black Sea. Reports and Studies no. 58, GESAMP (Joint Group of Experts on the Scientific Aspects of Marine Environmental Protection), London.

Gordina, A. D.; Tsytsugina, V. G.; Ovsyanyj, E. I.; Romanov, A. S.; Kemp, R. B., 2004: Condition of pelagic fish spawn in coastal waters of the Black Sea near Sevastopol. *Hydrobiol. J.* 40, 43–55.

Gubanova AD, Polikarpov IG, Saburova MA, Prusova IYu. 2002. Long-term dynamics of mesozooplankton by the example of the Copepoda community in Sevastopol Bay (1976–1996) Oceanology 42 (4): 512-520.

Harbison, G. R., Madin, L. P. and Swanberg, N. R. 1978. On the natural history and distribution of oceanic ctenophores. Deep-Sea Research 25, 233-256.

Hashemian, A. and Roohi, A., 2004. A survey of *Mnemiopsis leidyi* impacts on Sturgeon fish feeding in the Southern Caspian Sea, 2sd International applied biological Congress, Mashhad Free University, Mashhad, Iran.

Ivanov P. I., A. M. Kamakim, V. B. Ushivtzev, T. Shiganova, O. Zhukova, N. Aladin, S. I. Wilson, G. R. Harbison and H. J. Dumont. 2000. Invasion of Caspian Sea by the comb jellyfish *Mnemiopsis leidyi* (Ctenophora). *Biological Invasions* 2:255–258.

Kideys A. E., Ghasemi S., Ghninejad D., Roohi A., Bagheri S. 2001. Strategy for Combatting *Mnemiopsis* in the Caspian Waters of Iran. A report prepared for the Caspian Environment Programme, Baku, Azerbaijan, Final Report, July 2001.

Kideys, A. E. 1994. Recent dramatic changes in the Black Sea ecosystem: The reason for the sharp decline in Turkish anchovy fi sheries. Journal of Marine Systems 5:171-181.

Kideys, A. E., Roohi, A., Elif, E. D., Melin, F.and Beare, D. 2008. Increased Chlorophyll Levels in the Southern Caspian Sea Following an Invasion of jellyfish, Hindawi Publishing Corporation, Research Letters in Ecology, Article ID 185642, 4 pages, and doi:10.1155/2008/185642.

Kideys, A., Finenko, F., Anninsky, B., Shiganova, T., Roohi, A., Tabari, M., Youseffyan, M., Rostamian, M., Rostami, H. and Negarestan, H. 2004. Physiological characteristics of the ctenophore *Beroe ovata* in Caspian Sea water, *Marine Ecology*, Vol. 266: 111–121.

Kideys A.E., A. Roohi, S. Bagheri, G. Finenko, L. Kamburska. 2005. Impacts of Invasive Ctenophores on the Fisheries of the Black Sea and Caspian Sea. *Oceanography-Black Sea Special Issue* 18(2):76-85.

Kovalev AV, Gubanova AD, Kideys AE, Melnikov VV, Niermann U, Ostrovskaya NA, Prusova IYu, Skryabin VA, Uysal Z, Zagorodnyaya JuA. 1998. Long-term changes in the biomass and composition of fodder zooplankton in coastal regions of the Black Sea during the period 1957–1996. In: Ivanov LI. and Oguz T (eds.) Ecosystem Modelling as a Management Tool for the Black Sea. Kluwer Academic Publishers, Dordrecht/Boston/London, Vol. 1, pp.209–219

Kremer, P. and Nixon, S. 1976. Distribution and abundance of the ctenophore *Mnemiopsisleidyi* in Narragansett Bay. Estuar. coast. mar. Sci. 4: 627-639.

Kremer, P. M. 1993. Ctenophore population dynamics: patterns of abundance for *Mnemiopsis leidyi* in U.S. coastal waters. ICES Statutory meeting L.36: 1-9.

Prodanov, K., Mikhailov, K. G., Daskalov, K., Chashchin, A., Arkhipov, A., Shlyakhov, V. and Ozdamar, E. 1997. Environmental impact on fish resources in the Black Sea. In: Ozsoy E, Mikaelyan A (eds) Sensitivity of the North Sea, Baltic Sea and Black Sea to anthropogenic and climatic changes. Kluwer, Dordrecht p 163–181.

Roohi, A. 2000. An introduction to the some ecological aspects of the new alien ctenophore *Mnemiopsis leidyi* in the Southern Caspian Sea, Caspian Sea Research Institute of Ecology, Tehran, Iran.

Roohi A., Zulfigar Y., Kideys A., Aileen T., Ganjian A., and Eker-Develi E. 2008a. Impact of a new invader ctenophore *Mnemiopsis leidyi* on the zooplankton of the southern Caspian Sea, *Marine Ecology,* Vol. 29, No. 4, pp. 421-434.

Roohi A., Zulfigar Y., Kideys A., Aileen T., Eker-Develi E., Ganjian A., Nasrollazadeh H. 2008b. Environmental changes due to a new allein comb jelly, *Mnemiopsis leidyi* in the Southern Caspian Sea, International conference on environmental research and technology (ICERT 08), 28-30 MAY 2008, Penang, Malaysia.

Roohi A., Kideys A., Sajjadi A., Hashemian A., Pourgholam R., Fazli H., Ganjian Khanari A. and Eker-Develi E. 2010. Changes in biodiversity of phytoplankton, zooplankton, fishes and macrobenthos in the Southern Caspian Sea after the invasion of the ctenophore Mnemiopsis Leidyi, Biological Invasions, 12: 2343- 2361

Seravin, L. N. 1994. The systematic revision of the genus Mnemiopsis (Ctenophora, Lobata). Zool. Zhurnal 73: 9-18 (in Russian).

Shiganova, T. A., Bulgakova, J. V, Volovik, S. P, Mirzoyan, Z. A. and Dudkin, S. I. 2001b. A new invader, *Beroe ovata* Mayer 1912 and its effect on the ecosystems of the Black and Azov Seas in August-September 1999. Kluwer Ac.pub . Hydrobiologia, 451: 187-197.

Shiganova, T., Dumont, H., Sokolsky, A., Kamakin, A., Tinenkova, D. and Kurasheva, A. 2004. Aquatic Invasions in the Black, Caspian, and Mediterranean Seas, 71–111. Kluwer Academic Publishers, Netherlands.

Shiganova, T., Mirzoyan, Z., Studenikina, E., Volovik, S., Siokoi-Frangou, I., Zervoudaki, S., Christou, E., Skirta, A. and Dumont, H. 2001a. A review of the invader ctenophore *Mnemiopsis leidyi* (A.Agassiz) population development in the Black Sea and in the other seas of the Mediterranean basin. Marine biology, 139:431-445.

Studenikina, E. I., Volovik, S. P., Miryozan, I. A. And Luts, G. I. 1991. The ctenophore *Mnemiopsis leidyi* in the Sea of Azov. Oceanology 3: 722-725.

UNDP Project Document, Governments of: Azerbaijan, Islamic Republic of Iran, Kazakhstan, Russian Federation & Turkmenistan, United Nations Development Programme, PIMS #4058, The Caspian Sea: Restoring Depleted Fisheries and Consolidation of a Permanent Regional Environmental Governance Framework, "CaspEco", www.caspianenvironment.org/newsite/Data-MajorDocuments.htm.

Vinogradov, M. E., E. A. Shushkina, E. I. Musaeva & P. Yu. Sorokin,.1989. Ctenophore *Mnemiopsis leidyi* (A. Agassiz) (Ctenophora: Lobata) - new settlers in the Black Sea. Oceanology 29: 293-298.

Yunev OA, Carstensen J, Moncheva S, Khaliulin A, Aerteb- jerg G, Nixon S. 2002. Nutrient and phytoplankton trends on the western Black Sea shelf in response to cultural eutrophication and climate changes. Estuar Coast Shelf Sci 74:63-76

9

Main Ecosystem Characteristics and Distribution of Wetlands in Boreal and Alpine Landscapes in Northern Sweden Under Climate Change

J. Jeglum, S. Sandring, P. Christensen, A. Glimskär,
A. Allard, L. Nilsson and J. Svensson
Swedish University of Agricultural Sciences
Sweden

1. Introduction

Wetlands and peatlands are integral parts of many of the world's biomes, forming important transition zones between upland and aquatic systems. These habitats have a high degree of complexity of hydrology, edaphic conditions, and vegetation composition, contributing to the biodiversity of landscapes and species richness. They act to influence and modify the movement of runoff and groundwater from uplands into streams and lakes, by laying down organic remains (peats), and absorbing and releasing elements, compounds, gases, and particulate and dissolved organic matter. They therefore act as hydrological water retainers and biological filters in the landscape.

Many kinds of wetlands and peatlands can be found, each with a particular hydrology and surface form, moisture and chemical regime, and range of vegetation types and associated biota. Owing to their hydrological characteristics, predominantly peat soils and hydrophytic plants, wetlands and peatlands are key habitats to indicate climate change, particularly changes towards drying (e.g., decreased precipitation, increased runoff from melting glaciers and snow pack). Changes in moisture regime will effect changes in the processes of peat accumulation and decomposition, release of nutrients and dissolved organic matter, and vegetation and species. Drainage for agriculture and forestry, peat harvesting, and development have already caused considerable areas of peatlands to decrease in depth and area. Owing to drying, some peatlands adjacent to uplands have decreased in depth to less than 30 cm, the defined depth for peatlands in Sweden, and thus the total area of peatland has decreased.

Drying also has caused changes in vegetation, for examples, advances of trees and shrubs from the margins into the centres of peatlands (e.g., Fig. 1; cf. Hebda et al., 2000; Linderholm & Leine, 2004), and the dying of Sphagnum by lowered water levels and being covered over by leaf litter. Hebda *et al.* estimated the zone of influence of water lowering in Burns Bog, a bog on the Fraser River Delta in southern British Columbia, Canada, to extend over 100 m from a peripheral ditch. The Swedish Wetland Inventory, VMI, (Gunnarsson & Löfroth, 2009) was conducted during 25 years and generated results that indicate that about 15% of mires in the northern part of Sweden are strongly influenced or even destroyed, and 55% are weakly influenced while the rest are considered uninfluenced by human impact. Impacts are mainly from machines used for forestry and the digging of ditches. Small mires and mires in the Scandinavian Mountain Range were not covered by the VMI, however, and it may be assumed that processes such as increased tree cover or successional shifts to other vegetation types are more evident in such habitats. The main threats to the mountainous mires are tracks from all-terrain vehicles and snowmobiles, and much concern has been expressed from County Boards and the Swedish Environmental Protection Agency about this issue (e.g. Renman, 1989; Allard et al., 2004).

Fig. 1. A significant proportion of the arable land in Sweden has been created through drainage of wetlands. The figure (Color Infra Red aerial photo) shows an area in southern Sweden. The farm buildings are placed on the uplands in the SE corner and the wetlands have been drained and form long rectangular arable fields (light tones). Trees and shrubs (pink tones) now invade along the ditches and into the center of some fields as a result of decreasing land use intensity. The area above the drained fields shows invasion of trees and shrubs on a previously open string flark fen.

It is generally anticipated that climate change will cause considerable impacts and changes on landscapes and ecosystems. Dale et al. (2001) provide a review of the impacts of climate change on forests, which included altering the frequency, intensity, duration, and timing of fire; drought; introduced species; insect and pathogen outbreaks; hurricanes, windstorms, ice storms; and landslides. Similar impacts may be expected in peatlands. Several reviews on predicted impacts of climate change on wetlands and peatlands have been published (Gorham, 1991; 1995; Gorham et al., 2001; Strack et al., 2004; Tarnocai, 2006; Warner &

Asada, 2006; Strack, 2008). European climate models predict increases in warm indices and decreases in cold indices (Persson et al., 2007; Lind & Kjellström, 2008). Higher summer temperatures will promote higher evapotranspiration in the summer, resulting in deeper water tables in the wetlands. Relatively higher warming is expected for the interior in northern Sweden. On the other hand, there are predicted to be increases in wet indices in the north and dry indices in the south, and this may lessen the impacts of drying on wetlands in the north, and accentuate impacts of drying on wetlands in the south. Trends and trajectories are largely unknown, however, as are the influences on natural conditions and biodiversity.

The National Inventory of Landscapes in Sweden (NILS) has undertaken to monitor landscape changes owing to natural and anthropogenic disturbances, and ecological processes at the landscape scale (Svensson et al., 2009; Ståhl et al., 2011). NILS encompasses all terrestrial habitats in Sweden, and includes mapping of peatland and wetland types and characterizing the tree, shrub, field and ground vegetation, and surfaces. One of the main incentives behind NILS is to capture relevant data and provide analyses to answer the national Environmental Quality Objectives, of which the "Thriving wetlands" objective concerns the conservation and restoration of peatlands (e.g. Government of Sweden, 2009).

The aims of this chapter are to elucidate multiple-scale biodiversity aspects - i.e. landscape, community-ecosystem, population-species (Noss, 1990) - through area and distribution of wetlands and peatlands in northern Sweden; area and distribution of the Hydrotopographic types; and associated vegetation strata and ground conditions associated with the Hydrotopographic types. We include a discussion of the value of the NILS monitoring of the wetlands for assessing climate changes and other changes owing to anthropogenic causes in the context of biodiversity on multiple scales.

2. Study area

The study area is situated in the northern part of Sweden, coinciding with the two northernmost counties of Västerbotten and Norrbotten (Fig. 3). The landscape ranges from the coastal boreal areas consisting of often flat forested areas, up through the interior with a mixture of boreal forest and mires, to the mountains and birch-forested valleys of the Scandes Mountains. The study area has been classified into Natural Geographic Regions, sometimes referred to as Biogeographical Regions, according to the Nordic Council of Ministers (1977), Helmfrid (1996), and Lennartsson & Stighäll (2005). These regions were chosen as they take into account both east-west and north-south variations. As we wanted to study the distribution in the context of occurrences at different elevations as well as the climatic aspect, the regions were regrouped into five Elevation Zones which then were used as the basis for summarizing wetland and peatland data (Fig. 2).

2.1 Characteristics of the elevation zones

Zone 1. The Arctic/Alpine zone has steep mountains with glaciers and vegetation zones at lower altitudes. The prevailing Atlantic wind and precipitation provide strong climatic differences between westward slopes and eastward facing slopes with more continental climates. Nutrients in the bedrock and soil are varied and the soil layers are usually very thin. The zone encompasses the gradual shift from forest to treeless alpine habitats, and, hence, the alpine tree line.

Zone 2. The Northern boreal zone has wide heaths and forests of mountain birch or coniferous trees, mainly Norway Spruce (*Picea abies*) but also Scots Pine (*Pinus sylvestris*), mixed with mires. The relief includes scattered mountains and plateaus.
Zone 3. The Upper middle boreal zone has mixed coniferous forests in flat or hilly terrain and is rich in mires of mixed types, together with willow and birch swamps. The zone is predominantly above the highest post-glacial coastline and the metamorphic bedrocks are covered by till. Few and scattered human settlements of late date can be found.
Zone 4. The Lower middle boreal zone is vegetated by coniferous forest and swamp forest, in a hilly terrain rich in lakes. The transition from above to below the highest post-glacial coastline occurs predominantly in this zone.
Zone 5. The Coastal boreal zone is characterised by sandy or morainal islands and coastal boreal plains, with hilly terrain and valleys dominated by marine sediments back from the coast. The vegetation ranges from the marshlands, meadows and lowland forests along the Gulf of Bothnia coast to the inhabited parts where cultivated land and towns are situated (sparsely), which brings open grassland and deciduous forest between the coniferous parts. The precipitation is somewhat higher than the inland forest zones.

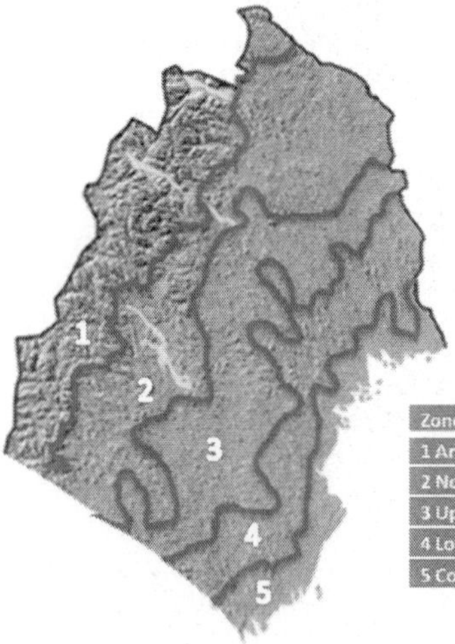

Zone	Mean height	Min height	Max height	Mean STD
1 Arctic/Alpine	743	352	1352	40
2 Northern boreal	495	252	712	14
3 Upper middle boreal	364	225	568	13
4 Lower middle boreal	168	65	275	12
5 Coastal boreal	49	1	173	7

Fig. 2. The five Elevation Zones from west to east: 1 Arctic/Alpine, 2 Northern boreal, 3 Upper middle boreal, 4 Lower middle boreal, and 5 Coastal boreal zone. The zones are based on the Natural Geographic Regions of Sweden (Helmfrid, 1996). The table in the figure shows the mean, minimum and maximum of elevation heights (m above sea level), and the standard deviation of the mean height.

3. Methods

The data are collected from the NILS programme which uses a strategically and systematically placed gridwork of 5×5 km squares (Ståhl et al., 2011). There are in total 631 squares and a full rotation of the inventory consists of 5 years. Two types of inventory are undertaken within a central, 1×1 km square, i) inventory by aerial photo interpretation in stereo models and ii) field-based inventory of sample plots and sample transects. This chapter will deal only with the aerial photo part of the inventory. The data set consists of the first set of air photo samples collected in 2003 to 2005, and is restricted to the northern part of the country. The data used here consists of 116, 1×1 km squares, which contain a total of 3229 registered polygons interpreted from the Colour Infra-Red imagery (Fig. 3).

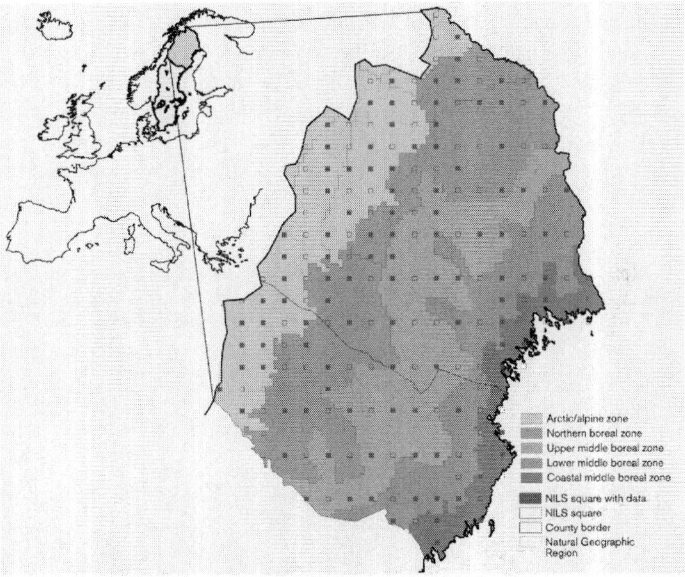

Fig. 3. The study area with five Elevation Zones, and with the distribution of the 1x1 km squares. The grid locations of the permanent NILS squares are indicated in red, the yet unsampled squares are not filled.

3.1 Aerial photo interpretation

The scale of aerial photos allows mapping based on vegetation structure and overall vegetation composition (e.g. Ihse, 2007; Morgan et al., 2010). A basic level of classification utilizes the fundamental division into bog (rain-nourished) or fen (mineral soil nourished). Each of the finer divisions is based on different hydrotopographic conditions – slope (e.g., flat, sloping, raised), location in landscape, and distinctive vegetation patterning on the surface. At the polygon level the types are described for general dominance of main vegetation in the understory. Within polygons smaller units, such as hummocks, hollows, and flarks, can be identified. This means that one can use the structure of the vegetation and peat surface, along with several other ecological indications, to distinguish between the different main types of mires and vegetation elements inside them. On the landscape scale the different hydrotopographic types of mires can be put together as mire complexes (Rydin et al., 1999). We hypothesize that the geographic distributions and areal extents of types will vary from east to west and north to south in northern Sweden, and also when comparing north with south Sweden.

The interpretation methods are described in detail in Allard et al. (2003, 2010). Strict rules are applied for spatial mapping accuracy and timing during the vegetation season. The technology is based on viewing the digital images in stereo in a computer-based photogrammetric system, with ancillary data such as maps and geology in a geographic information system (GIS program). The NILS program uses quantitative and qualitative variables in a context-dependent variable flow, instead of the usual predefined classes. This allows for aggregation of variables in several different ways, making several different classifications possible, after the data is registered into a geographical database. Similar basic variables are used both for the aerial photo interpretation and for the field survey, although the entire square is inventoried in the aerial photos, and sample plots and line inventories are registered in the field. A decision tree was used to define criteria for the polygon delineation in the aerial photo inventory, with a smallest mapping unit of 0.10 ha (31.6×31.6 m). When this study was made, complete data from the interpretation was available for the years 2003, 2004 and 2005 of the first rotation in NILS.

3.1.1 Variables in NILS relevant to mires

The NILS program operates on variables, in total 356 of which 156 are documented in the aerial photo interpretation. The main variables documented in wetlands and peatlands come from the recognition of the division between fen and bog; the recent emphasis in peatland classification on physiography, morphology, and hydrotopography; and vegetation elements (e.g., National Wetlands Working Group, 1988; Löfroth, 1991; Gunnarsson & Löfroth, 2009).

There are 17 Hydrotopographic mire types recognized for Swedish conditions (Gunnarsson & Löfroth, 2009). In this study, the mires were the primary focus, but we added a 'Non-mire wetland' type by using the category of semiaquatic land (occasionally flooded by fresh, brackish or salt water). Swamp forests are not included as one of the Hydrotopographic mire types, although they are registered in the aerial photo interpretation. The Hydrotopographic mire types are registered in the aerial photo interpretation of polygons as uniform hydrotropographic units, and can be classified together as large-scale mire complexes.

Thirteen Field and bottom layer vegetation types are recognized. Only one type of field and bottom layer type is registered as the dominant class for each polygon, since several types may occur in one polygon. Seven elements in a Microtopographic series are registered as percentages inside each polygon. The Microtopographic elements are defined on physiognomic as well as vegetation composition aspects of the field and bottom layer, ground level, and firmness of the substrate, following the Microtopographic series tradition in Swedish mire ecology (Rydin & Jeglum, 2006).

3.1.2 Influences and disturbances

Disturbance descriptors are important when explaining the kind of vegetation and ground cover. In the NILS program they are registered for example as peat harvesting, ongoing or historical, and influences including such disturbances as vehicle tracks, ditching, dredging, and mechanical site preparation. Other facts that can be derived from the data are forestry impacts, mowing, agriculture, erosion by wind or water, and burning. Of particular interest for climate change is the designation of productive forest land, fjell and tree line, and tree-limiting climatic impediments below the tree line.

3.1.3 Tree and shrub cover

The variables dealing with tree and shrub cover are important indicators of community-ecosystem and population-species biodiversity, land influences, disturbances, and climate change. These variables are important to identify those polygons that have had some influence by forestry, for example, tree planting being indicated by the patterns of tree distribution. In the aerial photo inventory, the distinction between trees and shrubs/small trees are as follows: all woody vegetation above 3 m height and in mountain birch forest above 2 m height are considered trees, regardless of species, and all woody vegetation below these heights falls into the shrub/small tree variable.

Variables in NILS relevant to mires

Hydrotopographic Mire Type

Fen
- Flat or slightly sloping fen
- Limnogenous fen
- Sloping fen
- Hill fen
- String flark fen
- Fen, strongly influenced

Bog
- Plateau bog
- Domed bog
- Sloping bog
- Flat or weakly raised bog
- Net bog
- Blanket bog
- Bog strongly disturbed

Mixed mire
- String mixed mire
- Mosaic mixed mire
- Palsa mire
- Mire, weakly raised bog or fen

Microtopographic Series
- Hummock (Dwarf shrub dominated)
- Lawn
- Carpet
- Mud-bottom
- Flark pools
- Bog pools
- Marsh fen, Shore fen

Field / Bottom Layer
- Field cover/bottom layer missing
- Graminoid- or herb-dominated
- Graminoid-dwarf shrub type
- Dwarf shrub dominated
- Lichen-dwarf shrub type
- Lichen type
- Belt formation of e.g. *Phragmites* spp.
- Tall sedge
- Low sedge
- Sphagnum dominated
- Other moss dominated
- Logging residues
- Field layer/Ground layer in shadow

Trees and Shrubs
- Tree cover in 4 types
- Tree height
- The pattern of tree distribution in 8 subclasses
- Tree species mixture in 6 subclasses
- Height dispersal in 4 subclasses
- Occurrence of broadly crowned trees, in percentage
- Occurrence of shrubs in 4 subclasses
- Shrub cover, deciduous and needle-leaved.
- The pattern of shrub distribution in 8 subclasses

4. Wetland area and distribution

The main factors of peatland formation are climate, relief (topography), parent material, biota, and time. The key factor is relief, with flatness or concavity being the most important types of relief. Within the climate regime, the two climatic factors temperature and moisture (humidity) are interrelated in terms of control on wetland and peatland biota and community development.

The total number of wetland polygons in 116 sampled squares was 3229, and the average number of polygons per square was 28.9 (SE 2.1, median 25, minimum 1, maximum 105). Across the Elevation Zones the highest numbers of squares were 34 in the Arctic/Alpine zone, and 32 in the Northern boreal zone, decreasing to 23 in the Upper middle boreal zone, 14 in Lower middle boreal, and 13 in the Coastal boreal zone (Table 1). The higher numbers in the Arctic/Alpine, Northern boreal, and Upper middle boreal are because of the NILS sampling strategy to establish a denser gridwork of squares in the Alpine and adjacent interior land where landscape data are redundant or missing (Ståhl et al., 2011).

Calculation procedure for determining wetland areas for zones:

1. for each stratum*, divide the total wetland area per inventoried 1×1 per km square in the zone by the number of squares sampled to get *'mean wetland cover / square'*;
2. multiply by 25 to get 'wetland cover / 5×5 per km square'
3. multiply by the total number of squares from which the sample was taken in the stratum to get *'estimate of the total wetland area per stratum'*
4. for each zone, calculate the sum all strata areas to get *'estimate of the total wetland area per zone'*
5. obtain the *'estimate of the total terrestrial area per zone'* by doing step 1 to 4 for terrestrial land cover instead of wetland cover.
6. obtain the *'wetland proportion'* by dividing the estimated total wetland area (from step 4) by the estimated total terrestrial area (from step 5);
7. multiply the wetland proportion by the corresponding terrestrial area of the zone to obtain *'ratio estimate of total wetland area'*.

* The distribution of sample squares follows a stratified design with least sampling effort in the central boreal parts of Sweden (Fig. 3; Ståhl et al. 2011). Stratum borders do not coincide with zone borders.

4.1 Wetland areas in squares and zones

The total wetland areas compared amongst Elevation Zones were significantly different (One-way ANOVA, $F = 4.76$ and $P = 0.001$; Fig. 4). The Northern and Upper middle boreal zones achieved the highest total wetland areas, and the Coastal boreal zone the lowest. The minimum cover achieved among squares was 1180 m^2 in the Coastal boreal zone; the maximum was 898163 m^2 in the Upper middle boreal zone, which is close to filling a full 1×1-km^2 square (1 million m^2). Expressed as percentages of the 1×1 km^2 square (10^6 m^2) these are 0.12 and 89.8%, respectively.

Total wetland area depends on the areal sizes of the elevation zones, which are not equal. To obtain a truer picture of the relative area covered by wetlands, we calculated the relative percentages of land covered by wetland in each zone (Table 1). The highest percentages of wetlands are in the Upper middle boreal zone followed closely by the Northern boreal zone. The Arctic/Alpine and the Coastal boreal zones have the lowest percentages, with a slightly higher value in the Lower middle boreal zone. The percentage cover of wetlands in the whole study area was 20.56% (SE 1.87). This matches well with other estimates of about 20% for Swedish wetlands (Hånell, 1989; Joosten & Clarke, 2002; Paavilainen & Päivänen, 1995; Christensen et al., 2007; Gunnarsson & Löfroth, 2009). However, all these references refer to the whole of Sweden, whereas the present paper deals only with Northern Sweden. A summary of wetlands comparable to the present one could be done for all southern Sweden below the study area.

Elevation zone	Sample size (Number of squares)	Total terrestrial land area (km²)*	Estimated percentage wetland of terrestrial area, with standard errors		Relative standard error
1 Arctic/Alpine	34	33905	12.39	(2.42)	0.20
2 Northern boreal	32	45084	25.86	(3.57)	0.14
3 Upper middle boreal	23	37769	26.54	(4.10)	0.15
4 Lower middle boreal	14	23860	16.35	(6.02)	0.37
5 Coastal boreal	13	12127	14.36	(3.51)	0.24
Total	116	152747	20.56	(1.87)	0.09

* Terrestrial land area (excluding water) obtained from the Swedish road map series (GSD-Vägkartan, vektor, 1:100 000, Swedish Land Survey)

Table 1. Wetland statistics in the 116 1×1-km² squares. Estimated wetland area as a proportion of the terrestrial land area.

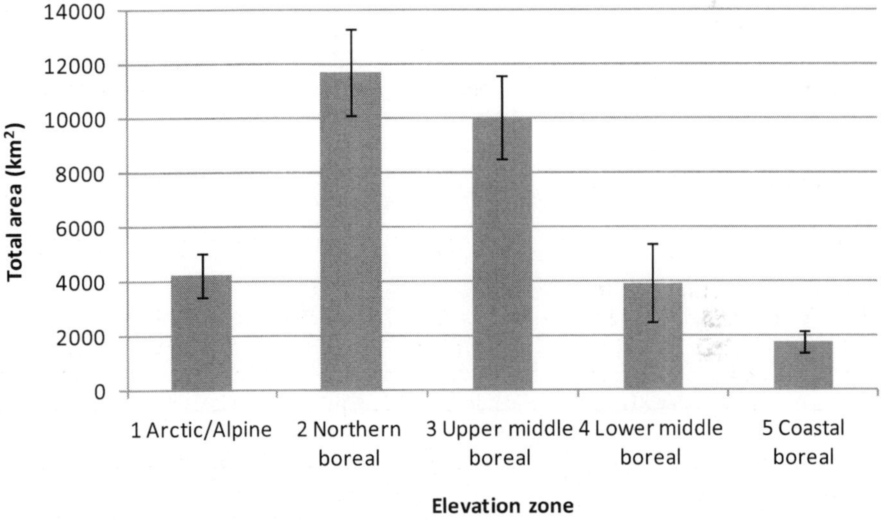

Fig. 4. Estimated total areas of wetlands in the five Elevation zones: Arctic/Alpine, Northern boreal, Upper middle boreal, Lower middle boreal, and Coastal boreal. Error bars indicate standard errors.

4.1.1 Variation and adequacy of sample

There was high variation among squares owing in part to the variation of wetlands within the zones, and also to the variation in numbers of squares amongst the zones. Relative standard error (Standard error divided by the estimate) indicates the quality of the estimate and the adequacy of sample (Table 1).

The lowest numbers for relative standard error for the wetland area estimate were obtained for the Northern and Upper middle boreal zones, indicating an adequate sample size of 32 and 23, respectively. The other zones had higher relative standard errors for estimated wetland areas indicating that the numbers of required squares for adequacy exceeded the actual numbers of squares sampled. However, for all 116 squares combined, the relative standard error was below 15%. In other words, for sampling the whole study area the sample was adequate, but for the five individual zones, only the two zones with the highest wetland covers were adequately sampled.

4.2 Sizes of wetland polygons

Mean number of wetland polygons varied amongst squares and zones (Table 2). The occurrence of wetland polygons was highest in the Upper middle and Northern boreal zones, with 39.6 and 31.5 polygons per km^2, respectively. The other three zones had similar number of polygons per km^2. The mean areas of individual polygons were largest in the Arctic/Alpine zone and decreased progressively towards the Coastal boreal zone. However, mean polygon size among zones was not significantly different, (ANOVA, $F = 0.76$ and $P = 0.552$; Table 2). The smallest wetland polygon was 331 m^2, sampled in the Coastal boreal zone. The largest wetland polygon was 428620 m^2 sampled in the Arctic/Alpine zone.

The number and area statistics for wetland polygons (Table 2) give an estimate of the configuration of the landscape with respect to wetland components, and hence are indicative for landscape biodiversity. The complexity of wetlands in the landscape, as indicated by the size of uniform areas for mapped polygons, is least in the Arctic/Alpine zone, and is greater progressing towards the Coastal boreal zone. These results indicate that the landscape configuration along the coast generally provides the highest potential for wetland biodiversity, and that the potential decreases westwards towards the alpine zone.

It is known that the landscape near the Coastal boreal zone is more influenced by man's activities, and is ecologically younger owing to the ongoing glacio-isostatic land uplift (e.g. Svensson, 2002), consisting of patches of wetlands in earlier successional stages. The polygon sizes also indicate smaller areas (Table 2). These small patches tend to combine and merge together over time, such that in the oldest exposed parts of Northern Sweden, and the

Elevation zone	Mean number of wetland polygons per km^2 with standard errors	Minimum number of polygons per sample	Maximum number of polygons per sample	Mean wetland polygon area (m^2) with standard errors
1 Arctic/Alpine	19.3 (3.1)	1	73	11644 (3856)
2 Northern boreal	31.5 (3.7)	1	105	9208 (1233)
3 Upper middle boreal	39.6 (4.9)	4	91	7160 (821)
4 Lower middle boreal	22.0 (5.3)	1	55	6169 (1502)
5 Coastal boreal	25.3 (4.7)	1	47	6119 (763)
Total	28.9 (2.1)	1	105	8422 (957)

Table 2. Polygon statistics in the data set of 116 squares. Given are the mean, minimum and maximum number of polygons in a square, estimates of the number of wetland polygons per km^2, and the mean area per polygon in m^2, in the five Elevation Zones and the total study area.

parts with the greatest amounts of weakly sloping, flat and concave topographies, the wetlands would have had the longest time to develop and form larger homogeneous units. The mean size of polygons in each square is positively correlated with the total wetland area ($r = 0.330$, $P = 0.000$, $n = 116$) which indicates that when there is higher wetland cover, it is formed by increasingly large sizes of polygons.

4.3 Relation of wetland area to height (elevation) above sea level

The Elevation Zones reflect a gradient in height above sea level, with the Arctic/Alpine zone having the highest mean height and the Coastal boreal zone having the lowest. Standard deviation shows how the variation in heights within squares varies. Standard deviation values are highest in the Arctic/Alpine zone, and lowest in the Coastal boreal zone (Fig. 2). The three interior zones have standard deviations that are quite similar, increasing only slightly in the Lower middle boreal zone. This indicates a similar degree of elevational variability of the landscapes there.

Simple Pearson correlations were done for the mean wetland areas with means of height (above sea level), standard deviation of height, and mean area of polygons per square. Mean wetland area is strongly negatively correlated with mean standard deviation of height ($r = -0.367$, $P = 0.000$, $n=116$), in other words, the lower the relief within a square, the greater the amount of wetland. Mean wetland area is not significantly correlated with height ($r = -0.096$, $P = 0.305$); this is explained by the bell-shaped relationship of wetland area to elevation, low in the Arctic/Alpine, higher in Northern boreal and Upper middle boreal, and dropping lower in Lower middle boreal and Coastal boreal (Fig. 3).

Even though non-significant the mean size of the polygons increased steadily from a low in the Coastal boreal to a high in the Arctic/Alpine zone (Table 2). However, polygon size was not correlated to elevation ($r =0.011$, $P = 0.515$). Polygon size was largest at intermediate elevations. In fact we would expect smaller polygons at higher elevations. The high value for polygon size in the Arctic/Alpine zone can be explained by the high variation of height within this zone. In addition to high elevation squares with smaller polygons it also contains a number of intermediate elevation squares with large mean polygon sizes.

5. Hydrotopographic types distribution and area

Each polygon was classified into Hydrotopographic wetland types, slightly modified after the wetland types used in the Swedish Wetland Inventory (Gunnarsson & Löfroth, 2009). Of the 18 wetland types (compared with 17 types in Swedish VMI), 14 were recorded in the 116 squares for northern Sweden. There were five types that were relatively common in terms of area and percentage of total wetlands: Flat (topogenous) fen with 44%, Sloping fen with 14%, Flat or weakly raised bog with 12%, String flark fen with 11%, and Non-mire wetland with 6% (Table 3, Fig. 5). Three other types achieved moderately low levels: Limnogenous fen 4%, String mixed mire 5%, and Mosaic mixed mire 3%. The remaining six types were quite uncommon. It is significant that Flat fens are by far the most abundant wetlands. It is also noteworthy that the Non-mire wetlands are among the more common of the wetlands.

The wetland types missing from the inventory were Plateau bog, Domed bog, Blanket bog, and Palsa mire. Plateau bogs and Domed bogs are more common in the south of Sweden where there is more heat and more precipitation, allowing these bogs to grow upwards into distinct raised bodies (similar to Finland, Seppä, 1996). Blanket bogs are missing in our sample because the precipitation is too low (ca. 600 mm). They are rare in Sweden, but are found in Jämtland County where precipitation is high enough to allow their development (Götbrink & Haglund, 2010). Palsa mires were not found in the sample, but they do occur in the subalpine zone in the study area. Palsas from near Hemavan in Västerbotten (Zuidhoff & Kolstrup, 2000), and in three locations northwards in Norrbotten (Zuidhof, 2003), have been studied from the points of view of climatic data, soil temperatures, and geomorphological properties. Palsas are frozen peat structures that occur as the raised mound features of a mixed mire. They are difficult to recognize from aerial photos owing to their similarity to the small bog mound features in mixed mires at lower elevations.

Sloping bog and Net bog were both recorded only once in the sample in northern Sweden. Most sloping peat bodies in northern Sweden are minerotrophic sloping fens, sloping mixed mires with strings and flarks, or flark fens, by virtue of receiving some amount of input from mineral soil. However, large weakly sloping peatlands can become quite acidic and poor in base cations in their centres, and may only have scattered minerotrophic indicators (e.g., Carex rostrata). Net (patterned) bogs may be the result of the conflux of two weak flow patterns coming together approximately at right angles, which cause ridges to form at right angles to each other, eventually merging into net-shaped ridges. The flarks in the net-shaped ridges become isolated from mineral soil water influx, and eventually the whole net and flark complex becomes ombrogenous.

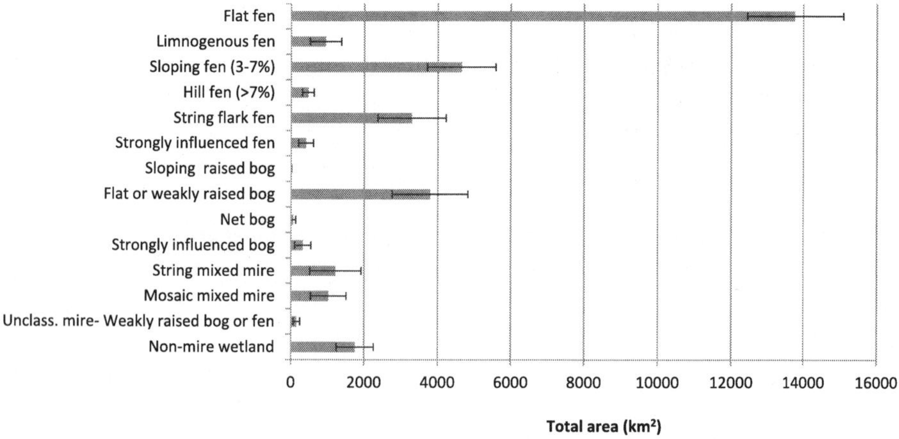

Fig. 5. Estimates of the total areas of the 14 Hydrotopographic mire types in the study area of Northern Sweden. Error bars indicate standard errors.

5.1 Hydrotopographic wetland types across five zones

In Table 3 we present the area percentage for the 14 types across the zones. We found that a majority of the types (eight) peaked in the Northern and Upper middle boreal zones. These two zones had also the highest proportion of wetland cover in total (Table 1). Most of the other six types had highest peaks in adjacent zones, and Hill fen which was most frequent in Arctic/Alpine had highest mean area in Upper middle boreal. Since there were only three Hill fen polygons sampled in the Upper middle boreal, at least one of these had a very large area.

The main wetlands characterizing the zones are:

1. **Arctic/Alpine** - Limnogenous fen, Hill fen, Unclassified mire, Weakly raised bog or fen - northern, Non-mire wetland.
2. **Northern boreal** - Flat fen, Sloping fen, String mixed mire.
3. **Upper middle boreal** - String flark fen, Flat or weakly raised bog, Mosaic-mixed mire.
4. **Lower middle boreal** - None of the Hydrotopographic types peak in this zone, but common types are Flat fens and Flat or weakly raised bogs.
5. **Coastal boreal** - Limnogenous fen and Strongly influenced (disturbed) fen.

Some of these distributions are as expected, e.g., the Sloping fens occur in terrains that have more relief and higher effective moisture regimes, and we expect the steeper Hill fens to be more common where there are steep hills and mountains with groundwater discharges and springs on lower and toe slopes. Strongly influenced fens peak in the Coastal boreal zone, but also occur in Lower and Upper middle boreal zones where the anthropogenic influences

Hydrotopographic wetland type	Arctic/ Alpine	Northern boreal	Upper middle boreal	Lower middle boreal	Coastal boreal	Total study area
Flat fen	4.49 (1.08)	**11.92** (1.89)	11.36 (1.86)	6.98 (1.96)	7.46 (2.01)	8.99 (0.87)
Limnogenous fen	**1.58** (1.10)	0.28 (0.13)	0.14 (0.09)	0.20 (0.16)	1.5 (1.47)	0.63 (0.28)
Sloping fen (3-7%)	2.85 (0.87)	**5.56** (1.75)	1.93 (0.58)	1.64 (1.26)	0.48 (0.36)	2.90 (0.58)
Hill fen (>7%)	**0.72** (0.34)	0.39 (0.23)	0.11 (0.09)	0	0	0.29 (0.10)
String flark fen	0.08 (0.08)	2.13 (0.83)	**4.53** (1.82)	2.19 (1.84)	0.65 (0.32)	2.19 (0.62)
Strongly influenced fen	0	0.06 (0.06)	0.36 (0.33)	0.19 (0.11)	**1.60** (1.15)	0.31 (0.15)
Sloping raised bog	0	0	**0.02** (0.02)	0	0	0.01 (0.01)
Flat or weakly raised bog	0.06 (0.05)	1.42 (0.93)	**4.90** (1.36)	4.72 (3.35)	1.29 (0.66)	2.54 (0.69)
Net bog	0	0.14 (0.14)	0	0	0	0.04 (0.04)
Strongly influenced bog	0	0	**0.56** (0.56)	0.30 (0.30)	0.35 (0.18)	0.23 (0.16)
String mixed mire	0	**2.28** (1.60)	0.26 (0.12)	0.07 (0.07)	0.60 (0.39)	0.76 (0.44)
Mosaic mixed mire	0.12 (0.13)	0.76 (0.52)	**1.69** (1.08)	0	0	0.68 (0.32)
Unclassified mire-Weakly raised bog or fen	**0.21** (0.21)	0.08 (0.07)	0.12 (0.11)	0	0.01 (0.01)	0.10 (0.06)
Non-mire wetland	**2.65** (1.32)	1.02 (0.52)	0.75 (0.29)	0.09 (0.04)	0.72 (0.41)	1.12 (0.33)

Table 3. Estimates of the percentage of terrestrial area for 14 Hydrotopographic types in five Elevation Zones and the total study area with standard errors in parentheses. The largest areas for each Hydrotopographic type across zones are indicated in bold.

have been greatest. The greatest frequencies of Limnogenous fen and Non-mire wetlands are in the Arctic/Alpine, which may relate to higher areas of floodplains in the upper reaches of rivers and along lakes there than in interior regions.

The Hydrotopographic types are defined on the basis of slope of the topography, flow patterns of the water, surface patterns of vegetation and Microtopographic elements, and the mineralogical characteristics of the water. A fundamental division is **minerogenous (fen)**, influenced to any degree by water derived from mineral soil, versus **ombrogenous (bog)** nourished only by rain water. Minerogenous is further subdivided into flat (**topogenous)** versus sloping **(soligenous)**. Surface movement of water and other complex interrelationships with the surface vegetation and near surface peat causes the formation of string or net patterns. One must take into account the influence of adjacent open water systems and occasional flooding, and this is done in the categories of **Limnogenous fens and Marsh.** Obviously, it is essential to develop diagnostic features for these conditions, and much of the NILS training consists of interpreting these features from features on the remote imagery. First and foremost is separating ombrogenous from minerogenous. This means using peatland form and vegetation to recognize bog as rounded or elongated peat bodies that are isolated from mineral soil water influence, and are flat or slightly raised in their centres.

We recognize several lines of variation or gradients of Hydrotopographic types within the classification:

1. Flat fens → Sloping fens→ Hill fens
2. Flat fens→Flat or weakly raised bogs
3. Sloping fens→Sloping bogs
4. Sloping fens →String flark fens→ String mixed mires→Net bog
5. String mixed mires→ Mosaic mixed mires→ Palsa mixed mires

It must be noted that the NILS inventory thus far does not distinguish the wetlands-peatlands-mires that are well-covered with trees, known as swamp forest, treed peatlands, and treed wetlands. These conditions have been assessed by Hånell (1989) and should be added to the NILS air photo interpretation scheme for a more complete representation of the total wetland-mire-peatland area. Many of the forested peatlands have been influenced or disturbed by drainage owing to long history of forest drainage to improve growth of trees. It has yet to be determined how the open wetlands-mires-peatlands included in NILS will merge with the treed wetlands. Undoubtedly the area of wetlands-peatlands will be considerably expanded when the treed wetland areas are added.

Some north-south gradients in distribution of types across the study area are discernible, indicating that there are higher frequencies of Sloping fen, Hill fen, Strongly influenced fen, Strongly influenced bog, and Mosaic mixed mire in the south. Higher percentages for the Strongly influenced fens and bog types in the south can most likely be explained by more frequent farming and forestry activities. Higher percentages of Limnogenous fens and Non-mire wetlands in the north may be related to a higher frequency of flooded lake and river margins and floodplains. There are also higher frequencies of String flark fens and String mixed mire types in the north, suggesting a more common occurrence of soligenous flow in sloping drainage ways and higher flooding stages during spring melt, which is associated with more frequent string and flark patterns.

6. Field and bottom cover

General classes of vegetation physiognomy are recognized under Field and bottom cover. In the photo interpretation, the most dominant class was assigned to each polygon. In Table 4 we summarize the occurrence of the Field and bottom cover types in the 3229 polygons. The most common types are Sphagnum (33.5%), Low sedge (21.1%), Dwarf shrub (17.6%), Graminoid/dwarf shrub (11.8%), Graminoid- and/or herb (6.8%), and Tall sedges/graminoids (3.3%). Five percent of the polygons were missing a field/bottom layer, probably owing to coverage by shadows, water, peat extraction, or dense cover of shrub or tree vegetation (the field/bottom layer is not registered when the trees have > 50 % crown cover).

In Table 4 we give a short, selected list of some dominant species that characterize the different Field and bottom layer types. Since the Field and bottom types were developed for air photo interpretation, they are necessarily quite broad and reflect main appearance and life form of the vegetation. Finer community types of course can be identified with the more detailed ground survey work that is done in each of the permanent squares. Many other species lists and descriptions of the wetland vegetation in northern Sweden can be found (e.g., Gunnarsson & Löfroth, 2009).

Field and bottom layer type	Occurrence (% of polygons)	Typical species
Field/bottom layer missing	5.0	Peat extraction or water, and polygons with > 50 % tree cover.
Graminoid and/or herb	6.8	Graminoids: *Carex rostrata, Carex lasiocarpa, Calamagrostis purpurea, Eriophorum angustifolium, Molinia coerula;* Herbs: *Dactylorrhiza* spp., *Menyanthes trifoliata, Potentilla palustris.*
Graminoid-dwarf shrub	11.8	See Graminoids above and Dwarf Shrubs below.
Dwarf shrub	17.6	*Andromeda polifolia, Betula nana, Calluna vulgaris, Empetrum nigrum, Ledum palustre, Vaccinium uliginosum.*
Lichen-dwarf shrub	0.1	See Lichen and Dwarf Shrub.
Lichen	0.2	*Cetraria islandica, Cladonia mitis, C. rangiferina, Racomitrium lanuginosum.*
Reeds	0.1	*Phragmites australis, Typha latifolia, Iris pseudacorus.*
Tall sedges/graminoids	3.3	*Equisetum fluviatile, Scirpus lacustris, Carex acuta, C. aquatilis, C. rostrata, C. lasiocarpa.*
Low sedge	21.1	*Carex chordorrhiza, C. livida, Eriophorum vaginatum, Rhynchospora alba, Trichophorum alpinum, T. cespitosum.*
Sphagnum mosses	33.5	*Sphagnum balticum, S. capillifolium, S. cuspidatum, S. fuscum, S. papillosum, S. squarrosum, S. tenellum.*
Other mosses	0.6	*Scorpidium scorpioides, Campylium stellatum, Drepanocladus revolvens, Tomenthypnum nitens.*
Logging residues	0.0	Branches, twigs and leaves.
Layer cannot be interpreted	0.1	Shadows from structures on the ground or from clouds obstruct the view.

Table 4. Field and bottom cover Types, and percent frequency of the types in the data set (3229 total polygons), and typical abundant species for each type.

The field and bottom cover types can be broadly related to the moisture and water table gradient. Dwarf shrubs, Lichen-dwarf shrubs, and Lichen types are well developed on hummock and mound levels of the microtopographic gradient. Graminoid and/or herb, and Graminoid-dwarf shrub are common types of lawns in fens or bogs. Low sedges are in the carpet phases of fens and bogs. Sphagnum occurs widely across the whole wetland gradient, except for frequently flooded sites and strongly shaded sites (trees, dense shrubs). Reeds and tall sedges/graminoids are on shores and frequently flooded locations, often with higher pH and base-richness. Often there is a Sphagnum bottom layer beneath the dwarf shrub, graminoid, and low sedge types.
There are also general relationships of these Field and bottom cover types with the pH-base-richness gradient. Dwarf shrub and dwarf shrub-lichen tend to have quite low pH-base status and are in bogs or in the bog phase of mixed mires. The sequence tall graminoids, graminoids, and low sedges tend to follow a gradient of high to low pH-base levels. *Sphagnum*-rich sites have a range of species that differentiate both on a moisture-water table gradient and a pH-base richness gradient from rich to poor fen and into ombrogenous bog. Tall reeds and emergents, floating plants, and submerged plants are in water at edges of rivers and lakes, and vary from circumneutral to weakly acid and moderate to low base levels.

6.1 Relationships of Field and bottom cover to Hydrotopographic types

To explore relationships of the field and ground layer vegetation with hydrotopographic types we reduced the data set to the six most common Field and bottom types, and the eight

Hydrotopographic wetland type	Fewer occurrences than expected of Field/bottom layer	Higher occurrences than expected of Field/bottom layer
Flat fen (all types of field/bottom occur)	Graminoid/herb, Graminoid/Dwarf shrub, and Dwarf shrub.	Low sedges and Sphagnum mosses.
Limnogenous fen	Graminoid/Dwarf shrub, Dwarf shrub, and Sphagnum mosses.	Graminoid/herb, Tall and Low sedges.
Sloping fen	Graminoid/herb, Dwarf shrubs and Tall sedges.	Graminoid/dwarf shrub and Low sedges.
String flark fen	Graminoid mixtures with Herb Graminoid/Dwarf shrub, Dwarf shrub.	Low sedges and Sphagnum mosses.
Flat or weakly raised bog	Graminoid/herb, Tall sedges, and Sphagnum mosses.	Graminoid/Dwarf shrub and Dwarf shrub.
String mixed mire	Graminoid/herb, Graminoid/Dwarf shrub, and Tall sedges.	Dwarf shrub and Sphagnum mosses.
Mosaic mixed mire	Graminoid/herb, Tall sedges, and Sphagnum mosses.	Graminoid/Dwarf shrub and Dwarf shrub.
Non-mire wetland	Dwarf shrub, Low sedges and Sphagnum mosses.	Graminoid/herb, Graminoid/Dwarf shrub, and Tall sedges.

Table 5. The relations between the Hydrological mire type and associated Field/bottom layers in the dataset, from analyses with Chi-square test.

most common Hydrotopographic types. By summarizing the results from a Chi-square analysis, the relation of Field/bottom types to different Hydrotopographic wetland types can be seen (Table 5). The relationship is such that if the observed is higher than the expected by random, there is a positive association, while if it is lower there is a negative association. For example, Graminoid herb was rather strongly and positively associated with Limnogenous fen and Non-mire wetland, while Sphagnum type was strongly positively associated with Flat Fen, and strongly negatively associated with Flat or weakly raised bog.

From this analysis it is possible to indicate, in a general way, the main kinds of Field and bottom layers associated with each Hydrotopographic mire type. However, it is concluded that one cannot very reliably use Field/bottom layer type to indicate Hydrotopographic type, or vice versa. Instead one must use such defining attributes as, flow patterns, peatland shape, and physiographic location to classify polygons. And for Field / bottom layer classification one must rely on the dominant life form.

7. Microtopographic series

The definition of these types was based upon the microtopographic series - hummock, lawn, carpet, mud-bottom, pool series – introduced by Sjörs (1948) and now used internationally (e.g., Rydin and Jeglum 2006). Seven elements are recognized in the NILS system: Dwarf shrub dominated hummock, lawn, carpet, mud-bottom, flark pool, bog pool, and marsh. (Marsh as described in Rydin and Jeglum, 2006, is very similar to Sumpkarr as described in Allard, 2005). In conducting the photo interpretation, these types together should add up to 100% cover. In the data explored here hummock or lawn were the most common types, and often together totalled 100%. There was a strong inverse correlation between the values for these two types, with a Pearson r value of -0.635, P-Value < 0.0001.

The area data for Microtopographic series in the polygons are given in Table 6, showing that the most common elements are, in decreasing order: Lawn, Hummock, Carpet, and Mud-bottom. Flark pool, Bog pool, and Marsh were very infrequent.

Microtopographic element	Total area (km^2) with standard errors		Proportion of total mire area with standard errors	
Hummock asterisk	8423	(1196)	30.51	(2.64)
Lawn	11272	(1149)	40.83	(2.47)
Carpet	5614	(816)	20.33	(1.72)
Mud-bottom	1886	(557)	6.83	(1.61)
Flark pool	329	(180)	1.19	(0.56)
Bog pool	82	(30)	0.30	(0.10)
Marsh	62	(32)	0.23	(0.10)

*Small raised mounds usually built-up by Sphagnum and characterized by dwarf shrubs.

Table 6. Estimates of total areas of mires of seven Microtopographic elements, and proportions of the total mire area studied, with standard errors (in parentheses).

We performed a chi-square between Microtopographic series and Hydrotopographic wetland types to determine relationships between the two classifications. Those comparisons of observed and expected with enough values showed some interesting relationships. In Flat or weakly raised bog, Strongly influenced bog, and String mixed mire, hummocks were more abundant than lawns, probably because of more of the raised bog phase in these types. Carpets were more frequent than the mud-bottoms in all the Hydrotopographic types. Flark pools were distributed unevenly, and observed counts were lower than expected in Flat fen, Sloping fen, and Flat or weakly raised bog, but higher than expected in String flark fen and String mixed mire. Clearly the sample method was able to pick up the flark element. Pools were more common than expected in Flat fen whereas Marsh was less common than expected in Flat fen. However, Marsh was more common in Limnogenous fen, just as one would expect along lake and stream margins.

Because it is deemed so important for interpreting climate change to be able to accurately determine the proportions of each Microtopographic element, we give detailed descriptions of the elements which follow and augment the descriptions of Rydin & Jeglum (2006):

Hummocks - Hillocks or small mounds with rounded, convex form, usually formed of hummock *Sphagnum* species (e.g. *S. fuscum, S. capillifolium*) rising to ca. 30-60 cm above adjacent hollows. Water tables range from 20-50 cm below the tops of the hummock. Sphagnum may be the main cover viewed from above, without any overtopping vegetation. However, usually the Sphagnum is covered with dwarf shrubs (e.g., *Calluna vulgaris, Ledum palustre, Vaccinium uliginosum*), and sometimes may be partially covered by patches of lichen (e.g., *Cladonia* spp.). With climate warming and lowering of water tables dwarf shrubs and lichens will probably increase, as will sapling and tree cover, on the hummock phase. Hummocks can occur as the dominant microtopographic element in raised bogs, with smaller amounts of hollow types (Rydin & Jeglum, 2006), or they may form raised linear strings or miniature bog islands surrounded by lawns and other wetter elements in patterned bogs, net bogs, and mixed mires.

Lawns - Level surfaces covered with fairly dense cover of graminoids (tall sedges, grasses, e.g., *C. lasiocarpa, C. rostrata, Eriophorum vaginatum*). Most of the time lawns surfaces are 5-20 cm above the water table, but during spring melt and heavy rains water can be at or slightly above the surface. Because of the strong rooting systems of the graminoids, lawns are so firm that footprints rapidly disappear. The moss cover beneath the graminoids is continuous, and lawns seem to have the greatest bryophyte diversity.

Carpets - Level surfaces with moderate to sparse cover of low sedges (e.g., *Carex livida, Rhynchospora alba, Scirpus cespitosus)* and Sphagnum and/or brown mosses in the bottom layer. Most of the time lawns surfaces are from 5 cm below to 5 cm above water table, but during spring melt and heavy rains water can be more than 5 cm above the surface. The lack of strong graminoid rooting systems makes carpets so soft that a footprint remains visible for a long time.

Mud-bottoms - Mud-bottoms are often inundated by water and then are dark to black and hard to separate from the adjacent flark pools or bog ponds. However, in summer drought, mud-bottoms may be exposed bare peat, and may have thin covering of algae, or a gelatinous layer of decomposing peat and microorganisms, or a layer of dark green to black liverwort (e.g., *Cladopodiella fluitans*) with some scattered horizontal shoots of mosses.

Flark pools, bog pools: These are water-covered, and appear very dark to black. They have elongated, crescentic or more or less rounded shapes. Flark pools are the elongated pools in minerogenous fens with unidirectional flow, whereas bog pools are either crescentic or rounded shapes in ombrogenous bogs (Rydin Pers. Comm.). Even though flark pools and bog pools are supposed to be in fens and bogs respectively, sometimes it is difficult to determine positively whether the type in which the pool occurs is a bog or a fen. Hence, there may be misidentifications of flark pools (and other fen pools) and bog pools.

Only one Field/bottom type was assigned to each polygon, whereas the Microtopographic series had a range of seven classes which were given percentages if present in a polygon. Therefore the detail of sampling is greater for the Microtopographic elements. The accuracy of the estimates of covers for different elements should be explored further by conducting assessments of well-mapped wetland polygons, and comparing several photo interpreters. Such exercises are conducted for the staff of NILS as thematic blocks, and are extremely valuable for training and to develop more reliable estimates. This is particularly important because the Microtopographic elements are key attributes in documenting effects of climate change and anthropogenic influences. In addition, it would be of value to define an additional element, i.e., hummock with trees (treed hummock) to allow for documentation of tree

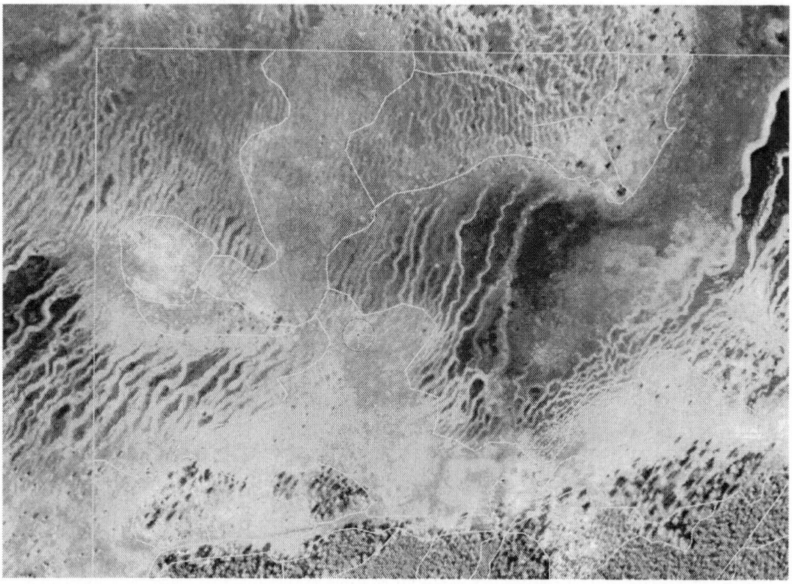

Fig. 6. A portion of a NILS square (yellow lines left and top) in the mid-south of the study area rich in String flark fens. Yellow lines encircle the mapped polygons. The diameter of the circle with the cross hair (centre left) is 20 m. Inside the flark pools can be only water, or water plus mud-bottoms, or only mud-bottoms; therefore flark pools and mud-bottoms could be combined in assessing climate change. In the upper left are some areas in which the unpatterned sedge fens are either very light or medium grey-pink; combinations of lawns and carpets could be more reliable indicators of climate change.

encroachment onto peatlands (e.g., Fig. 1). We also suggest that it may be useful in future analyses to combine i) lawn and carpet, and ii) mud-bottom with flark pools (e.g., Fig. 6). These elements individually have low accuracy of estimation, but by combining these pairs we may achieve more accurate and reliable estimations of cover changes.

8. Use of monitoring data for biodiversity and climate change

The NILS program covers all terrestrial habitats in Sweden, including wetlands and peatlands. Moreover, NILS also includes the alpine area of the Scandinavian Mountain Range and the various shallow-peated wetlands that harbour a rich diversity of habitats and flora. The long-term series of data that will be made available have a substantial potential for various purposes. With the 5-year inventory cycle of the NILS programme it should be possible to observe changes of vegetation that may relate to climate change, as well as anthropogenic influences such as various land use activities and atmospheric pollution. Below we present a list of possible wetland and peatland ecosystem changes that may be observed, having in mind future improvements and supplementary research within the NILS monitoring system.

1. **Shifts in temperature and moisture regimes will cause biological responses**: With increases in temperature, we expect increased evapotranspiration, decreases in water level, and increased decomposition in warmer acrotelms. However, there are predicted to be net positive or negative changes in precipitation, which may move different habitats toward wetter or drier conditions (Lind et al., 2008). If the net effect of the changes is towards warming, the vegetation will have net increases in biomass. Tall woody vegetation, dwarf shrubs, *Carex* and other graminoids, herbs, *Sphagnum*, other mosses, liverworts, and lichens will all respond to the effects of climate changes, with changes in metabolic processes (photosynthesis, productivity, respiration), and this will translate to changes in decomposition rates and net accumulation or loss of peat (Mooney, 1991).
2. **Changes in the Field and bottom layers:** The changes will not be uniform, but rather will change in different directions depending on the pH-base saturation status. Dwarf shrub types will develop denser cover with increased shading and litterfall over *Sphagnum*. In the case of Graminoid types, it is probable that productivity will increase and canopies become taller and more lush owing to response to increased release of nutrients from more rapid decomposition and cycling in deeper acrotelms. For these same reasons, tall graminoids may gain at the expense of low graminoids. Sphagnum will probably decrease in general owing to being covered by litterfall of more rapidly growing field layers. Lichen cover will probably increase. Tall sedge, Limnogenous fen, and Non-mire wetlands could increase if precipitation increases, causing higher/more frequent flooding.
3. **Changes in relative abundances of Sphagnum:** Hummock species of Sphagnum will increase over lawn and carpet Sphagnums. Mauquoy *et al.* (2002) showed from peat core analyses that in periods with lower temperature, due to decreased solar activity, there was a shift in representation from lawn and hummock species to hollow species in ombrotrophic mires in Denmark and the UK. Breeuwer et al. (2008) performed greenhouse warming studies with species from sites in southern and northern Sweden.

They predict that in northern Sweden, hollow species such as *S. balticum* will lose competitive strength relative to hummock species such as *S. fuscum* and southern species such as *S. magellanicum*.

4. **Changes in proportions of Microtopographic elements**: Most of these changes will be owing to the changes in Sphagnum dominance just mentioned. With higher temperatures, increasing evapotranspiration, and lowered water tables, bogs will change in the direction of more hummock and treed hummock vegetation elements. Hummocks with dwarf shrubs will invade into the lawns. Soft carpets may become drier and develop towards lawns, while mud-bottoms may develop towards soft carpets. The area of open water flarks and pools may decrease. The Hydrotopographic types that are transitional between bog and fen such as Flat or weakly raised bog and Weakly raised bog or fen-northern, may develop toward more hummock-preferring Sphagnum and develop a more ombrogenous bog ecosystem, owing to less influence of minerogenous waters.
5. **Changes owing to drainage:** Drainage effects can be used to predict what will happen with climate warming and drying. Many of the same changes listed under Items 2-4 may occur by lowering water tables. This has and still is occurring extensively owing to forest drainage in Sweden, Finland, and the Russian Federation. Drainage of Flanders Moss in Scotland has shown conversion of most of its previous lawn vegetation to hummock vegetation, and only a little of the carpet and pool vegetation is left in small pockets (Pers. Obs.). Hebda *et al.* (2000) estimated the zone of influence of water table lowering in Burns Bog, a bog on the Fraser River Delta in southern British Columbia, Canada, to extend over 100 m from a peripheral ditch.
6. **Expansion of the density and cover of trees into peatlands.** Such expansions are often owing to lowered water tables, best illustrated by the expansion of trees in drained peatlands (e.g., see Fig. 1). Several studies have documented advances of trees and shrubs from the margins into the centres of peatlands (e.g., Hebda et al., 2000; Linderholm and Leine, 2004), and the dying of Sphagnum at the margins by lowered water levels and being covered by forest litter. The NILS monitoring in 5-year intervals is ideally suited to recording changes in canopy cover of trees and saplings.
7. **Changes of vegetation and invasive species.** Striking examples are the invasion by bamboo, *Sasa* sp., into drained mires in Japan (Iqbal et al., 2005); and Reed Canary Grass (*Phalaris arundinacea*) into bog owing to marginal flooding by mineral-rich waters (Burns Bog, British Columbia, Canada, Pers. Obs.). With drying of peatland margins in Burns Bog, British Columbia, drier upland forest trees such as *Betula* spp and *Tsuga heterophylla*, shrubs such as *Gaultheria shallon* and *Rubus* spp., other upland herbs, and dry mosses (e.g., feathermosses) have invaded and laid down a forest humus, and this process is progressing toward the centre of the peatland (Hebda et al., 2000).
8. **Decreases in peat depth at the margins of the peatlands, and decreases of total area of peatlands:** This will happen owing to temperature increase and drying, collapse of the peat matrix owing to water loss out of the matrix, and oxidation and decomposition. In the first decade after forest drainage, subsidence of peat may be in the order of 15 to 40 cm (Päivänen & Paavilainen, 1996). National Forest Inventories in Finland over many decades (Keltikangas et al., 1986) suggest that significant areas of shallow-peated sites have converted to non-peatland with less than 30 cm depth of peat.

9. **Melting of permafrost or palsas in alpine-subalpine zone:** Palsas and palsa melting are key indicators of climate and climate warming (Sollid & Sørbel, 1998; Zuidhoff, 2003). Increases of percent of water cover, expressed as thermokarst and melt lakes, may be an early indicator of permafrost melting (Callaghan et al., 2002). Permafrost thawing is likely to change the flow pathways taken by water as it moves through arctic and subarctic landscapes, and there may be increases in the content of dissolved organic carbon in waterways and lakes in subarctic catchments (e.g., Lyon et al., 2009).
10. **Changes in bog processes: peat decomposition and accumulation, nutrient cycling, and greenhouse gases:** A large amount of research has been done concerning peat accumulation, decomposition, and release of CH_4 and CO_2 greenhouse gases relative to climate change (e.g., Strack et al., 2004). Decomposition rates increase with temperature and increased N availability, and hummock species decompose slower than hollow species (Limpens et al., 2003). There is a great deal of uncertainly, however, in predicting the effect of climate change on the carbon cycling in peatlands (e.g., Moore et al., 1998).

9. Acknowledgements

We are grateful to Erik Cronvall for the help that he provided in the processing of the data and for help with the GIS problems. We also thank the diligent inventory staff of the NILS aerial photography inventory, Per Andersson, Sofia Andreassen, Merit Kindström, Anders Lindblad, Björn Nilsson, Maud Tyboni and Marianne Åkerholm.

10. References

Allard, A., Nilsson, B., Pramborg, K., Ståhl, G. & Sundquist, S. (2005). *Manual for aerial photo interpretation in the National Inventory of Landscapes in Sweden, NILS, year 2003.* Swedish Univ. Agricultural Sciences (SLU), Department of Forest Resource Management and Geomatics.

Allard, A., Löfgren, P. & Sundquist, S. (2004). *Skador på mark och vegetation i de svenska fjällen till följd av barmarkskörning. (Mechanical damage of ground and vegetation in the Swedish Mountains as a consequence of vehicle driving during summer)*, Swedish Univ. Agricultural Sciences, Department of Forest Resource Management, Work Report No. 126,

Allard, A., Cronvall, E., Nilsson, B., Kindström, M., Pramborg, K., Ståhl, G. & Sundquist, S. (2010). *Instruktion för bildtolkningsarbetet vid Nationell Inventering av Landskapet i Sverige, NILS, år 2006. (Manual for aerial photo interpretation in the national inventory of landscapes in Sweden NILS, year 2006)*, Swedish Univ. Agricultural Sciences, Department of Forest Resource Management.

Breeuwer, A., Monique, M.P.D., Heijmans, B.J.M. Robroek, & Berendse, F. (2008). The effect of temperature on growth and competition between Sphagnum species. *Oecologia* 156: 155-167.

Callaghan, T.V., Crawford, R.M.M., Eronen, M., Hofgaard, A., Payette, S., Rees, W.G., Skre, O., Sveinbjörnsson, B., Vlassova, T.K. & Werkman, B.R. (2002). The dynamics of the tundra-taiga boundary: An overview, suggested coordinated and integrated approach to research. *Ambio Special Report,* No. 12:3-5.

Christensen P., Glimskär A., Hedblom M. & Ringvall A. (2008). *Myrarnas areal och vegetation: skattningar från provytedata i NILS 2003-2007 (Mire area and vegetation: estimations from field sample data in NILS 2003-2007).* Swedish Univ. Agricultural Sciences (SLU), Department of Forest Resource Management, Work Report No. 237.

Dale, V. H., Joyce, L.A., McNulty, S., Neilson, R.P., Ayres, M.P., Flannigan, M.D., Hanson, P.J., Irland, L.C., Lugo, A.E., Peterson, C.J., Simberloff, D., Swanson, F.J. Stocks, B.J. & Wotton, B.M. (2001). Climate change and forest disturbances. *BioScience* 51 (9):723–734.

Gorham, E. (1991). Northern peatlands: role in the carbon cycle and probable responses to global warming. *Ecol. Appl.* 1:182.

Gorham, E. (1995). The biogeochemistry of northern peatlands and its possible responses to global warming. In Woodwell, G.M. & MacKenzie, F.T. (eds.), *Biotic feedbacks in the global climatic system: will the warming speed the warming?*, Oxford Univ. Press, pp.169-187.

Gorham, E., Brush, G.S., Graumlich, L.J., & Johnson, A.H. (2001). The value of paleoecology as an aide to monitoring ecosystems and landscapes, chiefly with reference to North America. *Environ. Rev.* 9:99-126.

Government of Sweden. (2009). *Regeringens proposition 2009/10:55. Svenska miljömål, för ett effektivare miljöarbete. (Governemnt Proposition, Swedish Environmental Quality Objectives as a tool for more effective environmental work)*, Available from: http://www.sweden.gov.se/content/1/c6/14/24/56/dca35b38.pdf

Gunnarsson, U. & Löfroth, M. (2009). *Våtmarksinventeringen - resultat från 25 års inventeringar. Nationell slutrapport för våtmarksinventeringen (VMI) i Sverige, (The Wetland Inventory - Results from 25 years of inventory. Final report from the National Wetland Inventory (VMI) in Sweden)*, Swedish Environmental Protection Agency, Report 5225.

Götbrink, E. & Haglund, A, (2010). *Manual för uppföljning i myrar i skyddade områden. (Manual for continued monitoring of mires in protected areas, A field manual for follow-up on European Habitats Directive)*, Swedish Environmental Protection Agency, Typescript. Available from: http://naturvårdsverket.se/upload/04_arbete_med_naturvard/Skydd_och_skotsel_vardefull_natur/Uppfoljning/7_ufmanual_myrar_faststalld20100503.pdf

Hånell, B. (1989). *Skogliga våtmarker I Sverige (Peatlands in Sweden. A description of forest conditions on shallow and deep peatlands and their national and local distribution).* Swedish Univ. Agricultural Sciences, Department of Forest Soils, Rep. 60.

Hebda, R. J., Gustavson, K., Golinski, K. & Calder, A. M. (2000). *Burns Bog Ecosystem Review -Synthesis Report for Burns Bog, Fraser River Delta, South-western British Columbia*, Environmental Assessment Office, Victoria, BC, Canada.

Helmfrid, S. (1996). *Sveriges geografi (Geography of Sweden)*, National Atlas of Sweden Series, SNA, Swedish Land Survey, Vol. 17.

Ihse, M. (2007). Colour infrared aerial photography as a tool for vegetation mapping and change detection in environmental studies of Nordic ecosystems: A review. *Norsk Geografisk Tidsskrift - Norwegian Journal of Geography* 61(4), 170-191.

Joosten, H. & Clarke, D. (2002). *Wise use of mires and peatlands – background and principle including a framework for decision-making.* Internatl. Mire Conserv. Group and Internatl. Peat Soc. Saarijärvi Offset Oy.

Keltikangas, Laine, J., Puttonen, P. & Seppälä, K.M. 1986. Vuosina 1930-1978 metsäojitetut suot: Ojitusalueiden inventoinnin tuloksia. Abstract: Peatlands drained for forestry during 1930-1978: Results from field surveys of drained areas. *Acta For. Fenn.* 193:1-194.

Lennartsson, T. & Stighäll, K. (2005). *Landmiljöer i kust och skärgård* (*Land environment at coasts and archipelagos*), Swedish Environmental Protection Agency, Report 5482.

Limpens, J., Tomassen, H.B. & Berendse, F. (2003) Expansion of Sphagnum fallax in bogs: striking the balance between N and P availability. *Journal of Bryology* 25, pp 83-90.

Lind, P. & Kjellström, E. (2008). *Temperature and precipitation changes in Sweden; a wide range of model-based projections for the 21st century.* Swedish Meteorological and Hydrological Institute, SMHI, RMK, No. 113.

Linderholm, H.W. & Leine, M., (2004). An assessment of Twentieth Century tree-cover changes on a southern Swedish peatland combining dendrochronology and aerial photograph analysis. *Wetlands* 24(2):357-363

Lyon, S.W., Destouni, G., Giesler, R., Humborg, C., Mörth, M., Seibert, J., Karlsson, J. & Troch., P.A. (2009). Estimation of permafrost thawing rates in a sub-arctic catchment using recession flow analysis. *Hydrol. Earth Syst. Sci.* 13:595–604.

Löfroth, M. (1991). *Våtmarkerna och deras betydelse.* (*The wetlands and their importance*), Swedish Environmental Protection Agency, Report 3824.

Mauquoy, D., van Geel, B., Blaauw, M. & van der Plicht, J. (2002). Evidence from northwest European bogs shows 'Little Ice Age' climatic changes driven by variations in solar activity. *Holocene* 12, 1-6.

Mooney, H.A. (1991). Biological response to climate change: An agenda for research. *Ecological Applications* 1, 112-117.

Moore, T.R., Roulet, N.T., & Waddington, J.M. (1998). Uncertainty in predicting the effect of climate change on the carbon cycling of Canadian peatlands. *Climatic Change* 40:229-245.

Morgan, J.L., Gergel, S.E. & Coops, N.C. (2010). Aerial photography: A rapidly evolving tool for ecological management. *Bioscience* 60(1):47–59.

National Wetlands Working Group. (1988). *Wetlands of Canada.* Environment Canada, Ottawa.

Nordic Council of Ministers. 1977. *Naturgeografisk regionindelning av Norden.* (*Natural geographic regions of the Nordic Countries*), NU–serien 1977 (34: 1–137).

Noss, R.F. (1990). Indicators for monitoring biodiversity: a hierarchical approach. *Conservation Biology* 4, 355-364.

Paavilainen, E. & Päivänen, J. (1995). *Peatland forestry: Ecology and principles. Ecological Studies*, Vol. III. Springer-Verlag, Berlin.

Päivänen, J. & Paavilainen, E. (1996). Forestry on peatlands. In Vasander, H. (ed.), *Peatlands in Finland.* Finnish Peatland Society, Helsinki, pp.72-83.

Persson, G., Bärring, L., Kjellström, E., Strandberg, G. & Rummukainen, M. (2007). *Climate indices for vulnerability assessments.* Swedish Meteorological and Hydrological Institute (SMHI), RMK No. 111. Norrköping.

Renman, G., 1989. *Barmarkskörning på fjällen. Effekter av körning med terränghjulingar på mark och vegetation. (Vehicle driving during summer in the Swedish Mountains. Effects of driving with off-road vehicles on ground and vegetation),* Swedish Environmental Protection Agency, Report 3598.

Rydin, H., Sjörs, H., & Löfroth, M. (1999). Mires. *Acta Phytogeographica Suecica* 84: 91–112.

Rydin, H. & Jeglum, J.K. (2006). *The Biology of Peatlands.* Oxford University Press. Oxford.

Seppä, H. (1996). The morphological features of the Finnish peatlands. In Vasander, H. (ed.), *Peatlands in Finland.* Finnish Peatland Society, Helsinki, pp. 27-33.

Sollid, J.L. & L. Sørbel. (1998). Palsa bogs as a climate indicator: examples from Dovrefjell, southern Norway. *Ambio* 27(4:287-291).

Strack, M. (ed.) (2008). *Peatlands and climate change.* International Peat Cong. Tulamoor, Ire.

Strack, M., Waddington, J.M., & Tuittila, E.-S. (2004). Effect of water table drawdown on northern peatland methane dynamics: Implications for climate change. *Global Biogeochem. Cycles* 18, GB4003, doi:10.1029/2003GB002209, 2004.

Ståhl, G., Allard, A., Esseen, P.-A., Glimskär, A., Ringvall, A., Svensson, J., Sundquist, S., Christensen, P., Gallegos Torell, Å., Högström, M., Lagerqvist, K., Marklund, L., Nilsson, B., & Inghe, O. 2011. National Inventory of Landscapes in Sweden (NILS) - scope, design, and experiences from establishing a multiscale biodiversity monitoring system. *Environmental Monitoring and Assessment* 173: 579-595.

Svensson, J. (2002). Succession and dynamics of Norway spruce communities on Gulf of Bothnia rising coastlines. PhD Thesis, *Acta Universitatis Agriculturae Suecicae - Silvestra* 239.

Svensson, J., Allard, A., Christensen, P., Eriksson, Å. & Glimskär, A. (2009), Glimskär, A. & Sandring, S.. Landscape biodivesity monitoring in the Swedish NILS program. *XIII World Forestry Congress, Buenos Aires, Argentina, Conference Proc.* p.8.

Tarnocai, C. (2006). The effect of climate change on carbon in Canadian peatlands. *Global and Planetary Change* 53, pp. 222–232.

Warner, B.G. & Asada, T. (2006). Knowledge gaps and challenges in wetlands under climate change in Canada. In Batti, S., Lal, R., Apps, M.J. & Price, M.A. (eds.), *Climate change and managed ecosystems,* CRC Press, Taylor and Francis Group, Boca Raton FL, pp. 355-371.

Zuidhoff, F.S. (2003). *Palsa growth and decay in Northern Sweden.* PhD Thesis, Uppsala Univ., Uppsala, Sweden. Comprehensive summaries of Uppsala Dissertations from the Faculty of Science and Technology 813.

Zuidhoff, F.S. & Kolstrup, E. (2000). Changes in palsa distribution in relation to climate change in Laivadalen, Northern Sweden, especially 1960-1997. *Permafrost and Periglacial Processes* 11:55-69.

10

Valley Bottom Wetlands Can Serve for Both Biodiversity Conservation and Local Livelihoods Improvements

Pantaleo K.T. Munishi[1], Nice N. Wilfred[1], James S. Nshare[1], Stein R. Moe[2], Deo D. Shirima[1] and Halima H. Kilungu[3]
[1]*Department of Forest Biology, Sokoine University of Agriculture, Morogoro*
[2]*Department of Ecology and Natural Resource Management, Norwegian University of Life Sciences, Aas*
[3]*Open University of Tanzania (OUT), Faculty of Arts and Social Sciences Department of Tourism and Hospitality, Dar es Salaam*
[1,3]*Tanzania*
[2]*Norway*

1. Introduction

Wetlands are among the world's most biologically productive ecosystems and rich in diversity of species and are very important storehouses of plant genetic material some of which are valuable resources for human wellbeing. The Great Ruaha River Basin covering about 6950ha forms one of the major and largest wetland systems composed of numerous ecologically and socio-economically important valley bottom wetlands in Tanzania. This study was conducted in the Ruaha River Basin to assess the contribution of the valley bottom wetlands to livelihoods and biodiversity conservation. Three sites were selected for ecological studies which include Uchindile with Mpombochi River and Isimani stream plus associated swamps, Idete with Idete River and associated swamps/streams and Mapanda with Mkungwe and Kinoga Rivers and associated swamps/streams. The five villages selected for the socio-economic study include Luganga, Matanana, Igowole, Kisada, Njiapanda and Nzivi villages within the little Ruaha sub catchment of the Great Ruaha River. The sites for ecological studies were stratified into broad vegetation types using existing topographic and land cover maps. This stratification gave several major vegetation types in which plant species were then assessed in systematically established temporary nested sample plots measuring 20 x 20 m. Each plant species encountered in each plot was identified and their percent cover estimated. For socio-economic studies the village register was used as a sampling frame, households were then randomly selected and a questionnaire administered to heads of the selected households. The questionnaire sought to get information on whether the household is involved in any kind of wetland utilization, socio-economic activities undertaken in the wetlands, costs and revenues associated with wetlands utilization. Further to questionnaire survey PRA techniques including Focus Group Discussions (FGD) were used to supplement information from household surveys. The ecological data were summarized into tables showing a list of different vegetation types and their species composition, abundance and dominance. The abundance and dominance of each species was determined from their percent cover estimates. Socio-economic data were analyzed and summarized into social economic activities undertaken by a household, agricultural utilization of wetlands and crops grown in dry/wet seasons and the proportional contribution of each wetland related socio-economic activities to household food security (food available for household consumption) and income. The economic benefits were assessed by using gross margin analysis, food available for consumption as indicator of food security was used to assess food security at household level; and the contingent valuation technique was applied to assess the value of wetlands services.

Compared to other habitat types the valley bottom wetlands are the major repository of biodiversity of both flora and fauna, yet they are the most intensively utilized habitats for livelihood enhancement and buffer against drought in the Ruaha River Basin' instead. It was observed that the biodiversity of the valley bottom wetlands is higher than that in any other ecosystem types within the basin in the same locality'. The valley bottom wetlands in these areas are also repositories of some threatened/endangered plant species such as *Prunus africana, Protea sp.* making them important habitats for biodiversity conservation. It was further observed that overall, the total use value of productive activities carried out in upland and valley bottom wetlands was Tanzanian Shillings (Tshs) 3,415,458 (US$ 2,732) per household per year of which 31% of the total economic benefits accrued from utilization of the valleybottom wetlands. Wetland based socio-economic activities included agricultural production (farming) practiced by over 98% of the population followed by livestock grazing and fishing. Wetland based socio-economic activities carried out in valley bottoms commonly known by local people as *vinyungu* contribute about 15% of household food and 55 - 95% of household income annually, equivalent to Tshs 3,234,721 (US$ 2,588). In this respect valleybottom wetlands contribute significantly to household economy and food security. Furthermore over 90% of the dry season agricultural production is associated with valley bottom wetlands. Given the direct benefits of valleybottom wetlands and potential contribution to livelihoods the livelihood potential may override the biodiversity values of these Valleybottom wetlands. If left unattended it is likely that the wetlands will be degraded thus loosing their biodiversity values. The dual value of valley bottom wetlands (biodiversity and household economy) makes them unique habitats requiring an integrated approach to ensure achievement of both without impairing the ecological integrity of these wetlands.

Wetlands are among the world's most biologically rich ecosystems with high species diversity (Mvena et al 1999; Yanda et al., 2005; Munishi and Kilungu, 2004; Munishi et al. 2005). Wetland ecosystems are second only to the rain forests in the number of wildlife and plant species that depend on them for feeding and habitat. Historically, wetlands have been regarded as wastelands but they can also be viewed as being among the last truly wild and untouched places in the world (Maltby, 1986) making them of high repository of biological diversity with most wetlands offering important habitats to a variety of fauna and flora.

It is well documented that In Sub Saharan African countries, wetlands have a considerable importance in provision of innumerable benefits like drinking water, routes for transport, harvestable plants and animals (Ramsar, 1997). Though not well quantified, Tanzania's wetlands contribute in diverse ways to livelihoods of many millions and are chiefly utilized for crop and livestock production (Kashaigili, 2006). Furthermore, wetlands provide income in both dry and wet years for fairly large number of people engaged in agriculture because of their available water and high soil fertility (Munishi, *et al.*, 2003; Mkavidanda and Kaswamila, 2001; Masiyandima *et al.*, 2004; Kashaigili and Mahoo, 2005).

The principal vision of the Government of Tanzania (GoT) is to alleviate widespread poverty by improving several socio-economic opportunities, good governance transparency and by improving public sector performance through the National Strategy for Growth and Reduction of Poverty (NSGRP) or MKUKUTA. The 2025 country's vision overall goal specifically includes references to; 'sustainable development endeavours, on intergeneration equity basis, such that the present generation derives benefits from the rational use of natural resources of the country without compromising the needs of future generations' (ESP, 2003). Further more the National Strategy for Growth and Reduction of Poverty (NSGRP) or MKUKUTA recognize poverty as largely a rural phenomenon and that the rural poor depend solely or to a greater extent on natural resources (Bagachwa, 1994; DPG, 2005; PRSP 2000). Consequently the national environmental policy of 1997 and all the natural resource policies emphasize the clear cause-and-effect relationship between poverty and environmental degradation, and because of this they stress on the need for sectoral policies to address poverty issues by taking into account the need for wise use and sustainable resource exploitation (MNRT 2003).

Wetlands are amongst the most productive ecosystem in Tanzania and have significant economic, social, cultural and biological values (HOORC, 2002). Apart from agricultural use wetlands are considered useful as sources of a variety of natural resources of significance to human welfare. The Government of Tanzania has shown its commitment towards wise use of wetlands as stipulated in the Ramsar Convention and recognizes wetlands as significant natural resources with important ecosystem services and biological values by ratifying the Ramsar Convention on Wetlands in August 2000 (ESP, 2003).

Tanzania is endowed with exceptional wetland resources in which 10% of the country is covered by wetland ecosystems. These ecosystems in Tanzania range from large lake systems to river floodplains, deltaic mangrove formations and associated catchments (Maltby, 1986; Kamukara and Crafter 1993). Of the area covered by wetlands, 5.5% is occupied by four Ramsar sites namely Malagarasi/Moyovosi (32,500 Km^2), Lake Natron Basin (2250 Km^2), Kilombero valley floodplains (7,950 Km^2) and Rufiji-Mafia-Kilwa (5,969.7 Km^2) (MNRT 2003). Broadly the country is divided into nine drainage (river) basins which include Lake Rukwa, Lake Victoria, Lake Tanganyika, Lake Nyasa, Rufiji River, Pangani River, Ruvuma River and Southern Coast, Wami-Ruvu River and Internal Drainage basin. Each of these basins includes a network of rivers and inland valley bottom wetlands scattered throughout the country. These vast biologically rich resources in Tanzania are unique in their biodiversity values as well as support to local livelihoods (MNRT 2003, 2008, Munishi and Kilungu, 2004). The protection of these wetlands against biodiversity loss while maintaining their socio-economic value requires an integrated approach to management though currently protection for biodiversity conservation does not appear to be a serious alternative among majority of the society in Tanzania which jeopardizes future sustainability of these wetlands.

This papers presents results of a study to assess the role of wetland ecosystems in conserving plant diversity and contribution to livelihood enhancement in the Great Ruaha River basin Tanzania

2. Materials and methods

2.1 Study sites

The study site was the Great Ruaha River Basin in Mufindi district, Iringa region (Figure 1). Mufindi District is one of the seven Districts in Iringa Region located in Southern Highland of Tanzania. Mufindi District lies between latitude $8^0.00'$ 9^0 15' South and longitude 34^0 35'-

35⁰ 55' East. The District is situated about 80 km from Iringa Municipality and boarders Iringa rural in the north, Kilolo in north east, Njombe in the south, Kilombero in the south east and Mbarali in the west. Administratively the District is divided into five (5) divisions, 28 wards and 132 villages. The wetlands of Little Ruaha River start from Bumilayinga, Nyololo, Mafinga and Ihalimba wards in Mufindi District (Fig. 1). The River further extends to Kilolo and Iringa Districts crossing the Ipogoro Bridge along the Dar es Salaam - Mbeya road before joining the Great Ruaha River at Mawande village. The entire catchment, lies between latitudes 7° 15' and 8° 35' south and longitudes 35° 00' East covering 4900 km^2 and altitude of 600 m to 2100 m above sea level.

Selection of ecological study sites was based on the presence of and extent of valleybottom wetlands within the sites. For the social economic studies villages were selected for the study based on their proximity to wetland ecosystems. Three sites were selected for the ecological studies which include Uchindile with Mpombochi River and Isimani stream plus associated swamps, Idete with Idete River and associated swamps/streams and Mapanda with Mkungwe and Kinoga Rivers and associated swamps/streams. The five villages selected for the socio-economic study include Luganga, Matanana, Igowole, Kisada, Njiapanda and Nzivi villages within the little Ruaha sub catchment of the Great Ruaha River (Figure 1).

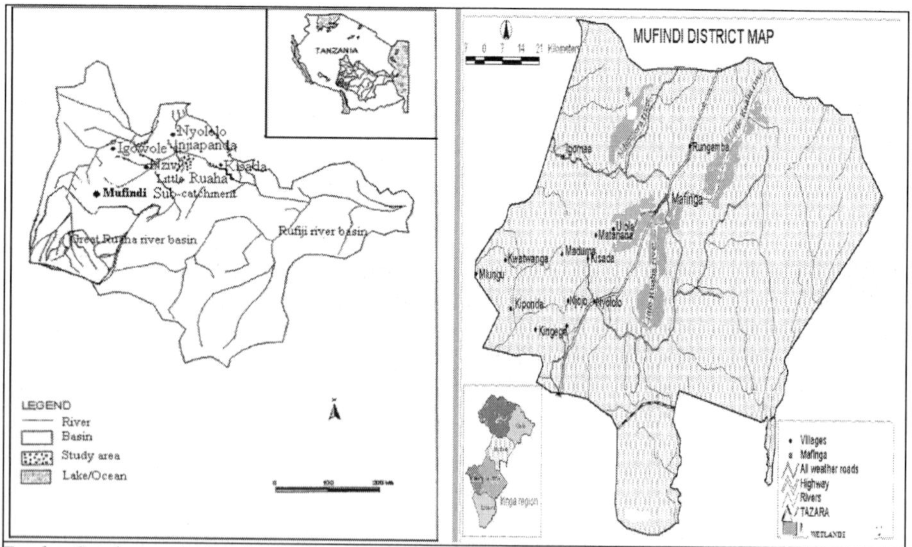

Ruaha Catchment in Mufindi District Tanzania	The Little Ruaha Wetlands in Mufindi

Fig. 1. Map of Mufindi District showing the Little Ruaha Wetlands System

3. Data collection

3.1 Ecological study

Prior to field surveys the areas were stratified into broad vegetation types using existing topographic and land cover maps. This stratification gave several major vegetation types including woodlands, grasslands, valley bottom wetlands/swamps undisturbed natural vegetation and plantations of different types. While in the field the broad types were further sub divided into sub categories depending on field conditions such as wooded grasslands, natural grasslands, pine plantations and eucalyptus plantations.

Temporary nested quadrats (sample plots) measuring 20 x 20 m (with nested subplots of 10m x 10m, 5m x 5m m, 1m x 1m and 0.5m x 0.5 m) were adopted and established systematically in clusters representing different vegetation types in each stratum on transect lines established along a predetermined compass direction. More than 100 plots were established in each site though the number of plots in each site differed depending on area and an adopted minimum sampling intensity of 0.01% ensuring a sufficient coverage and representation of each stratum. The location of each quadrat (plot) was recorded using Global Positioning System (GPS).

At each quadrat all the plant species encountered were identified, recorded and percent cover determined. For species that were difficult to identify in the field voucher specimens were collected and identified at the National Herbarium in Arusha Tanzania.

3.2 Socio-economic study

Multistage sampling was used where two divisions within the Little Ruaha sub catchment were selected randomly. Two wards were then selected randomly from each ward. For each ward two villages were selected randomly for the study. The nature of the distribution of wetlands in the Little Ruaha sub-catchment called for purposeful sampling of villages to be included in the study.The selection of villages was based on close proximity to the wetlands with the assumption that the respective communities were more involved with wetlands cultivation as compared to other villages. In this case, Matanana village (Bumilayinga Ward) being in the upper streams and Luganga village (Ifwagi Ward) on the lower stream of the catchment were sampled. Other villages included in the study are Nyololo Njiapanda (Nyololo Ward), Kisada (Bumilayinga Ward) Igowole and Nzivi (Igowole Ward). In each village the village register was used as a sampling frame, households were then randomly selected and a questionnaire administered to heads of the selected households. Households were taken as sampling units in this study and total of 93 households were sampled (44 and 49 households from Luganga and Matanana respectively). The questionnaire sought to get information on whether the household is involved in any kind of wetland utilization, socio-economic activities undertaken in the wetlands, costs and revenues obtained from wetland utilization. Further to questionnaire survey PRA techniques including Focus Group Discussions (FGD) were used as well as participant field observation to supplement information from household surveys.

4. Data analysis

4.1 Ecological data

All special features of interest in the area e.g. water sources, valley bottom wetlands, natural grassland ecosystem, miombo woodlands and wooded grasslands were identified and compiled into a lists with a descriptions of their species composition. The identified strata and field data were developed into tables showing a list of different vegetation types and their species composition, abundance and dominance. The abundance and dominance of each species was determined from their percent cover estimates.

4.2 Socio-economic data

Socio-economic data were analyzed and summarized into social economic activities undertaken by a household, agricultural utilization of wetlands and crops grown in dry/wet seasons and the proportional contribution of each wetland related socio-economic activities to household food security (food available for household consumption) and income. The economic benefits were assessed by using gross margin analysis, food available for consumption as indicator of food security was used to assess food security at household level; and the contingent valuation technique was applied to assess the value of wetland services. The gross margin analysis was computed as:

$$GM = TR - TC$$

where:

GM	=	Average gross margin (Tshs/kg) or (Tshs/month)
TR	=	Average total revenue (Tshs/kg) or (Tshs/month)
TC	=	Average variable total cost (Tshs/kg) or (Tshs/month)

The food available for consumption at household level was determined to be 300kg of cereal/ person/year. This figure was used to offset post harvest losses (storage loss and handling loss) (FAO, 1985; Ishengoma, 1998). Food available for consumption was obtained by subtracting the amount of food crop that was sold from the total food produced per person per year. Standardization of food available for consumption was done using adult equivalent scale considering age category (Ishengoma, 1998). In this case where the adult above 15 years old has unit equivalent of 1, ages 11 - 15 will have unit equivalent of 0.75 while children with age equal or less than 10 years will have unit equivalent of 0.36 (Ishengoma, 1998).

5. Results and discussions

5.1 Vegetation types and plant species composition

Several vegetation types were identified in the different sites. The major natural vegetation types identified included valleybottom wetlands, natural grasslands, wooded grasslands, and miombo woodlands. In addition planted exotic tree species formed specific vegetation types and included Eucalyptus and Pine Plantation. There was a big variation in plant species composition and richness in all sites studied with the different vegetation types having different species composition and richness within sites. The variation in composition is a reflection of high diversity of plants in the areas. In all three sites studied for plant species composition (Uchindile, Idete and Mapanda landscapes), valleybottom wetlands ranked the highest in plant species richness when compared with other natural vegetation and combined natural vegetation and plantation of exotic species. This general trend shows that valleybottom wetlands in the Ruaha River Basin will likely be the major repositories of biodiversity of plant species. By the fact that vegetation composition may be a reflection of other taxa in an ecosystem there is a high possibility that fauna diversity will follow the same trend. Because plants respond to multiple environmental factors both biotic and abiotic, this richness and diversity of species is a reflection of the heterogeneity of the areas, diverse vegetation types, habitats and landscapes that allow co-existence of species in heterogeneous landscape (Carson & Root, 2000; Franzén, 2004; Tetsuya & Kuniyasu, 2005; Munishi *et al.*, 2007; Tomas & Frantisek 2008).

5.2 Plant species composition by vegetation types in Uchindile landscapes

Three natural vegetation types were identified in Uchindile including (i) Valley bottom Wetlands/Riverine/Riparian, (ii) Natural Grasslands and (iii) Wooded Grasslands. In addition to the natural vegetation there were two types of plantations which are among the major vegetation found on the landscape. These are (i) Pine plantations and (ii) Eucalyptus plantations. The number of plant species varied between vegetation types. In regards to the natural vegetation Valleybottom Wetlands have the highest number of species compared to all other vegetation types (Figure 3a). Further when we consider all the vegetation types Valleybottom Wetlands still rank the highest in plant species richness (Figure 3b). Previous studies have also shown the valley bottom/riverine/riparian vegetation to have higher species richness than the other types of vegetation (Munishi 2006, 2007). The significance of Valleybottom Wetlands in acting as refuge for most species becomes more apparent during the peak of the dry season in which there would definitely be higher species richness for both flora and fauna in valley bottoms because of the wetter conditions compared to the other areas. Most grasslands for example would have higher number of short-lived plant growth forms (annuals) during the rain season which disappear during the dry season.

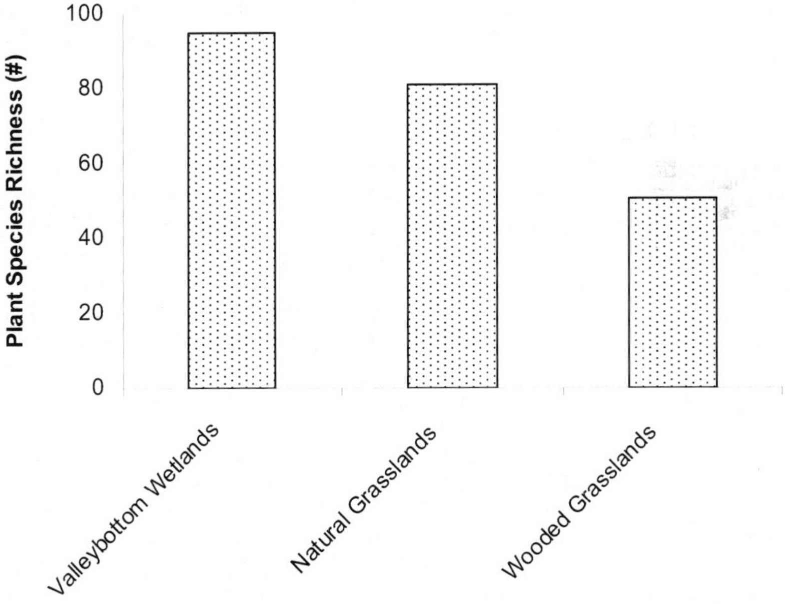

Fig. 3a. Average Number of Plant Species by Natural Vegetation Types in Uchindile

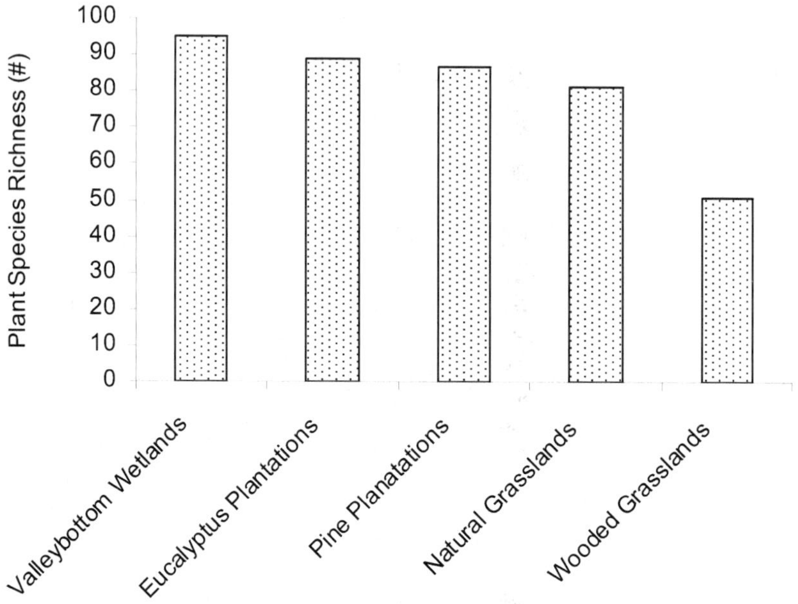

Fig. 3b. Plant Species Richness in all Vegetation Types in Uchindile

5.3 Species dominance in the different vegetation types in Uchindile landscapes

The Valleybottom Wetlands/Riverine Ecosystems were the most diverse compared to other vegetation zones of Uchindile with the most dominant species being *Cyperus papyrus, Cyperus dives, Cyperus corymbetes, Cyperus glaucophyillus, Cyperus articulates and Pteridium aquilinum.* In the natural grasslands the most dominant species were *Hyparrhenia filipendula* (Hochst.) Stapf, *Exotheca abyssinica* Anders, *Parinari curalellifolia* Benth, *Hyparrhenia cymabarica* (L.) Stapf, *Cymbopogon nardus* (Linn.) Rendle, *Protea madiensis* Oliv and *Vernonia sp.*. The wooded grasslands were dominated by *Loudetia simplex* (Nees) C.E.Hub, *Cymbopogon nardus* (Linn.) Rendle, *Protea madiensis* Oliv, *Psorospernum febrifugum* Spach. and *Melinis minutiflora* P.Beauv

5.4 Plant species composition by vegetation types in Idete landscapes

Four different natural vegetation types were identified in the Idete landscapes. These vegetation types include Valleybottom Wetlands/Riparian Areas, Natural Grasslands, Wooded Grasslands and Miombo Woodlands. Valleybottom Wetlands/riparian areas had the highest proportion of species compared to all other vegetation types followed by Wooded Grasslands. Miombo Woodlands and Natural Grasslands had almost the same species richness (Figure 4). These findings indicate that the wetland ecosystems are the most diverse followed by wooded grasslands.

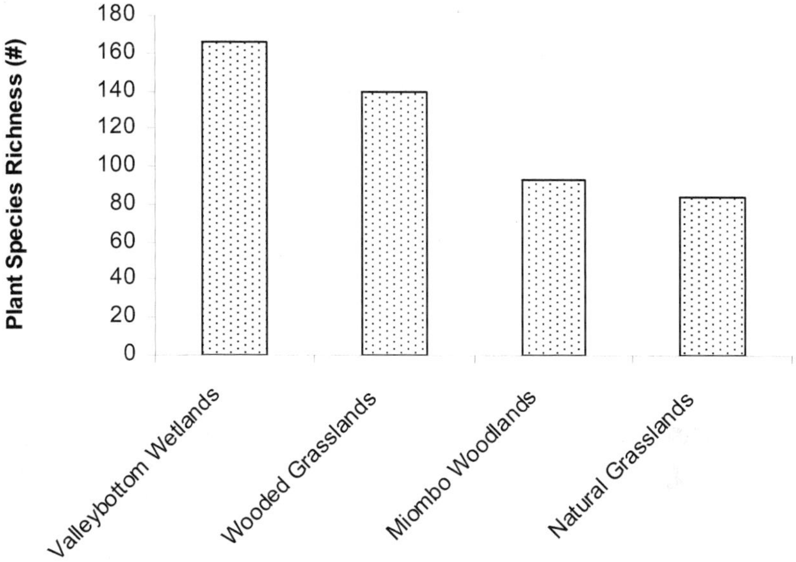

Fig. 4. Plant species richness in different natural vegetation types of Idete Landscapes Iringa Tanzania

5.5 Plant species dominance in the different vegetation types of Idete landscapes

The most dominant species in the Idete landscapes were *Ficus lutea, Loudentia simplex* (Nees) C.E Hubbard, *Nymphaea nguchali, Prunus africana, Themeda triandra* Forssk, *Cymbopogon excavatus, Cynodon sp, Panicum maximum* Jacq, *Faurea saligna* Harvey., *Oxtenanthera abyssinica, Vernonia sp, Dombeya rotundifolia* (Mast) Planch, *Erythrina abyssinica* D.C.*ssp abyssinica, Macaranga capensis, Sorghum bicolor, Annona senegalinsis Pers, Bersama abyssinica, Catha edulis* (M.Vahl.) Forssk, *Cyperus ajax* C.B. Clarke, *Osmunda regalis L.Cynodon sp, Parinari curatelifolia, Setaria sphacelata, Syzygium cordutum* Hochst ex Krauss *and Cyperus ajax* C.B. Clarke. The riverine vegetation closely associated with the Valleybottom wetlands category was mainly dominated by *Prunus africana, Bridelia micrantha, Macaranga capensis, Khaya anthotheca, Cartha edulis, Syzygium cuminii, Syzygium guinensis, Harungana madagascariensis, Myrianthus hostii, Tralepisium madagascariensis, Rauvolfia cafra and Ficalhoa laurifolia.* This vegetation type contains some elements of endangered plant species such as *Prunus africana* which also have medicinal value for the treatment of prostate cancer.

Grasslands were the largest vegetation type in the Idete landscape, dominated by the grass family (Poaceae) with grasses growing to more than 1m tall. Based on species abundance expressed as percent cover the major species in this vegetation type were *Hyparrhenia rufa (Nees) Stapf, Diheteropogon ampelactens, Faurea saligna Harvey, Cymbopogon excavatus, Hyparrhenia sp., Pteridium aquilinum, Londetia simplex (Nees) C.E Hubbard, Allophylus abyssinicus, Protea rupestris R.E.Fr.. Protea welwitschii Engl. Pygmaeothanmnus zeyheri, Syzygium guineense (Willd.) DC, Themeda triandra Forssk* and *Vitex mombassae Vatke.* This vegetation type is a home for rare/threatened species such as *Protea rupestris R.E.Fr.* and *Protea welwitschii Engl.*

The Miombo woodlands in this landscape is dominated by *Brachystegia microphylla Harms., Uapaca kirkiana, Eragrostis sp, Periploca linearifolia* Dill.& A.Rich, *Aristida sp, Combretaum molle, Loudentia simplex* (Nees) C.E Hubbard, *Myrica salicifolia* A.Rich, *Sparmannia ricinocarpa* (Eckl & Zey) Kuntize, *Faurea saligna* Harvey., *Osyris lanceolata* Hochst. & Stend., *Parinari curatelifolia* Planch.ex Beth., *Protea welwitschii* Engl. and *Protea rupestris* R.E.Fr.. The dominant species in the Wooded Grasslands include *Hyparrhenia rufa (Nees) Stapf, Aristida sp, Psorospermum febrifugum* Spach, *Bridelia micrantha, Eragrostis sp, Vernonia lasiopus* O.Hoffm. *Protea rupestris R.E.Fr., Protea welwitschii* Engl., *Cartha edulis, Cymbopogon excavatus, Pteridium aquilinum, Melinis minutiflora* P. Beauv., *Allophylus abyssinicus, Apodytes dimidiata* Arn., *and Vitex mombassae* Vatke.

5.6 Plant species composition by vegetation types in Mapanda landscapes

A total of 320 plant species belonging to 96 families were identified in the Mapanda landscapes which is a reflection of the heterogeneity of the landscape and habitats. Three natural vegetation types were identified in Mapanda which include Valleybottom Wetlands, Wooded Grasslands and Natural grasslands. In addition two vegetation types of planted exotic species were identified as among the major cover types in the landscape. These are Eucalyptus and Pine plantations. Among the natural vegetation types Valleybottom Wetlands had the highest plant species richness (Figure 5a) followed by Wooded grasslands. Of all vegetation types combined Valleybottom Wetlands still rank the highest in species richness (Figure 5b), showing the value of wetland ecosystems as repositories of species in the landscape.

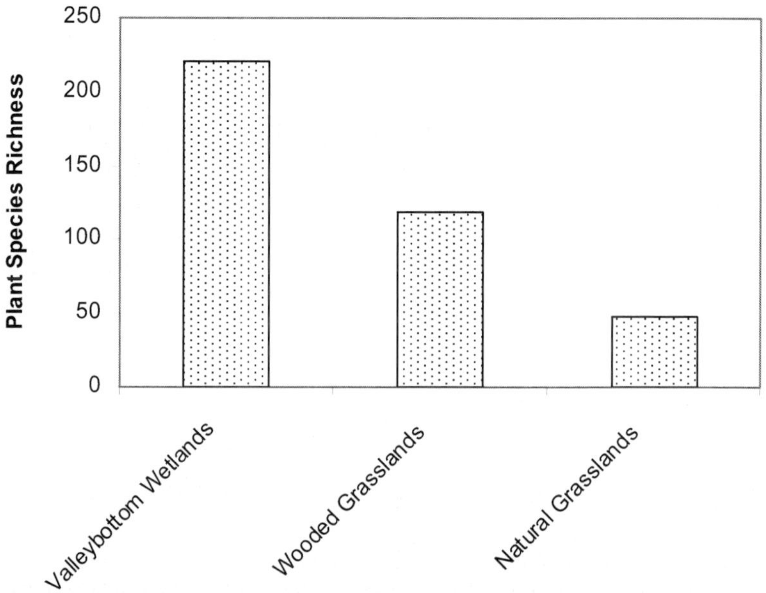

Fig. 5a. Plant species richness in different natural vegetation types of Mapanda Landscapes Iringa Tanzania

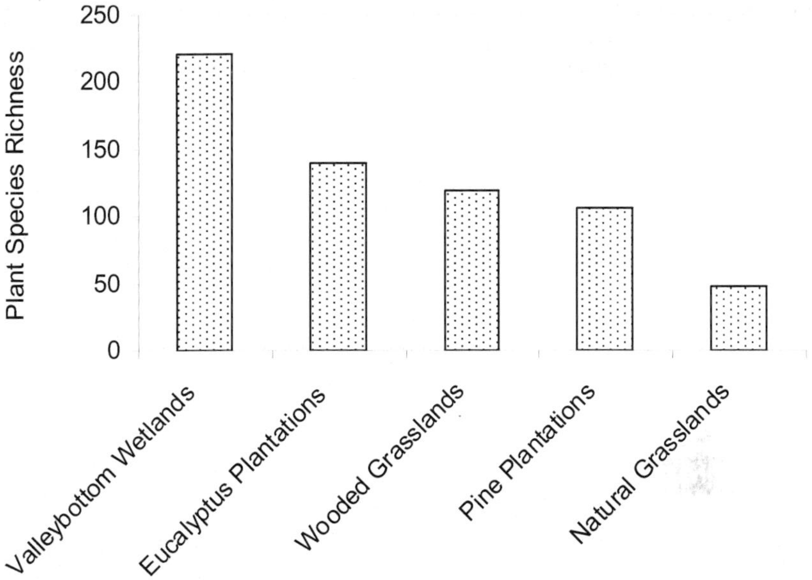

Fig. 5b. Plant Species Richness in all Vegetation Types of Mapanda Landscapes Iringa Tanzania

5.7 Species dominance in the different vegetation types in Mapanda landscapes

The Valleybottom wetlands are dominated by *Hyparrhenia cymabarica* (L.) Stapf, *Dichanthium foveolatum* (Del.) Roberty Clayton & Renvoize, *Syzygium cordatum* Hochst. ex Krauss, *Gnidia glauca* (Fresen.) Gilg. and *Kotschya strigosa* Benth. The most dominant plant species in the natural grasslands include *Cymbopogon nardus* (Linn.) Rendle, *Fadogia odorata* K.Krause and *Hyparrhenia cymabarica* (L.)Stapf,. The Wooded Grasslands were dominated by *Cymbopogon nardus* (Linn.) Rendle, *Loudetia arundinacea* (Hochst. ex A.Rich.) Steud, *Hyparrhenia cymabarica* (L.)Stapf, *Dolichos sericeus* E.Meyer, *Dichanthium foveolatum* (Del.) Roberty Clayton & Renvoize, *Multidentia crassa* (Hiern) Bridson & Verdc and *Protea madiensis* Oliv.

5.8 Socio-economic values of valleybottom wetlands in the Little Ruaha River

Tanzania's wetlands contribute in diverse ways to livelihoods of many millions and wetlands are chiefly utilized for crop production and livestock. An assessment of wetland contribution to livelihoods in 6 villages of the Little Ruaha sub catchment of the Great Ruaha River showed that the total use value of productive activities carried out in upland and valley bottom wetlands was Tanzanian Shillings (Tshs) 3,415,458 (US$ 2,732) per year per household in which 31% of the total economic benefits accrued from utilization of Valleybottom Wetlands. Wetland based socio-economic activities included agricultural production (farming) practiced by over 98% of the population followed by livestock grazing and fishing. Wetland based socio-economic activities carried out in valley bottoms commonly known by local people as *vinyungu* contribute about 15 of household food and 55

- 95% of household income annually, equivalent to Tshs 3,234,721 (US$ 2,588). In this respect Valleybottom Wetlands contribute significantly to household economy and food security. Planning for wetland friendly agricultural activities is pertinent in order to ensure wetlands conservation and sustainable contribution to household economy and food security without impairing the ecological integrity of the wetland ecosystems.

It has been argued that wetlands make appreciable contribution to rural livelihoods in terms of direct cash income and contribution to food security (Mkavidanda and Kaswamila, 2001; Munishi and Halima 2004), and many households that live close to wetland ecosystems in Tanzania and elsewhere utilize wetlands in coping strategies during times of drought and food scarcity. Differences in environmental and socio-economic conditions however result into significant variation in patterns of use between one area and another. The significance of wetlands in agricultural production, poverty reduction and contribution to rural livelihoods have variously been emphasized (Mkavidanda and Kaswamila, 2000; Ngailo et al., 2002; Munishi and Kilungu 2004). The wide range of economic benefits accrued from wetland ecosystems in Tanzania have been iterated (MNRT 2007). It is argued that often overlooked, unappreciated, taken for granted and therefore unmanaged the ecosystem services provided by wetlands in Tanzania include hydro-power production in which 95% of the hydropower production is from wetland related flows. Further 95% of domestic, irrigation, industrial and livestock water is from wetlands, 80% of traditional irrigation schemes depend on wetlands, 95% of rice and vegetable production depends on wetlands, about 850,000 ha of wetlands have potential for future irrigation, 95% of wildlife and wildlife corridors/game migration routes depend on wetlands, 66% of rural animal protein is derived from livestock grazing, game meat or fisheries, 95% of the 25 million livestock is maintained through dry season pastoralism in wetlands, 95% of coastal and wildlife tourism depends wetlands and 33% of the country's GDP depends on wildlife and wetland tourism. Other studies have shown that cultivation of paddy rice in wetlands of Bahi Tanzania contributed significantly to household food security generating 65.4% of total household food crop production compared to other crops grown in drier areas adjacent to the swamp (Rweyemamu, 2009, Munishi et al, - in press). Fishing in this case played a substantial contribution to household food security through household consumption of 10% of fish caught. For household income, sales of paddy rice from the swamp contributed 59.6% while fish sales contributed 36% of the total annual household income. Multiplier activities emerging during fishing season facilitate income to a wider group of communities and on average, 56.2% of the population depend on the swamp for daily socio-economic activities associated with generation of household food and income. The Bahi swamp and related products therefore play a significant role in enhancing local livelihoods for the adjacent communities. Planning for wise use of the swamp in respect of the dominant socioeconomic activities was seen as a means to improve its contribution to livelihoods.

6. Conclusions

Given the direct benefits of Valleybottom wetlands and potential contribution to livelihoods the livelihood potential may override the biodiversity values of these Valleybottom wetlands. If left unattended it is likely that the wetlands will be degraded thus loosing their biodiversity values. The dual value of valley bottom wetlands (biodiversity and household economy) makes them unique habitats requiring an integrated approach to ensure achievement of both without impairing the ecological integrity of these wetlands.

7. Acknowledgements

We acknowledge financial support from the Royal Norwegian Government for the Project NUFU-TZ 2007/10229 without which this study would have been impossible. Thanks to Sokoine University of Agriculture for institutional support in undertaking the NUFU Project and this study.

8. References

Fao/Who/Unu, (1985). *Energy and Protein Requirements*. Report of a joint FAO/WHO/UNU ad hoc Expert Consultation. Geneva, WHO, WHO Tech. Rpt. Ser.724pp.

Franzén, D. (2004). Plant species coexistence and dispersion of seed traits in a grassland Ecography 27(2), 218-224(7) http://botanika.bf.jcu.cz/suspa /ohrazeni/

Ishengoma, C. (1998). The Role of Women in Household Food Security in Morogoro Rural and Kilosa District. Unpublished PhD Thesis. Sokoine University of Agriculture, Morogoro, Tanzania, 337pp.

Maltby, E. (1986). Water Logged Wealth, Why Waste the World's Wet Places? *International Institute for Environment and Development*, London. 19pp.

Mkavidanda, T.A.J. and Kaswamila, A.L. (2001). The Role of Traditional Irrigation Systems in Alleviating Poverty in Iringa Rural District, Tanzania. *REPOA Research Paper No 01.2*.Mkuki na Nyota Publishers, Dar es Salaam. 36pp.

MNRT (2003). *Assessment Needs for Wetlands Inventory and Tools for Assessing, Mapping Wetland Types and their Distribution*. Draft Report, Prepared by Institute of Resource Assessment University of Dar es Salaam. 41pp.

MNRT (2004). *An Issue Paper for the Formulation of the National Wetland Strategy Wildlife Division*. Ministry of Natural Resources and Tourism, Dar es Salaam. 98pp.

MNRT 2009. Wildlife, Wetlands and Climate Change in Tanzania. Wildlife Division 7pp Mtatifikolo, F. and Comoro, C. (1999). Population Dynamics and Poverty. *Case Study of the Tomato Culture Zone Farming System Iringa; Tanzania*. A Final Report Submitted to ENRECA, University of Dar es Salaam.76pp.

Munishi, PKT, and Kilungu, H. (2004). The contribution of wetlands to household income and food security: A case study of "Nyumba Ya Mungu" wetland system in northern Tanzania Paper Presented at the Meeting of the National Wetlands Working Group (NWWG) 30 July 2004, Dar es Salaam

Munishi PKT 2006. Ecological Survey in the Kilombero Forest Project at Uchindile Kilombero District Tanzania

Munishi PKT 2006. Ecological Survey in the Kilombero Forest Project at Uchindile Kilombero District Tanzania

Munishi PKT. 2007. Ecological Survey in the Uchindile and Mapanda Forest Projects of Green Resources (Ltd) Tanzania 48 pp

Munishi PKT. 2008. Assessment of Forest Based Enterprises and Extent of Trade in Non-Timber Forest Products (NTFP) in the Southern Highlands of Tanzania. Report submitted to TAFORI

Munishi, PKT, Rweyemamu, R. Shombe H. and Stein R. Moe (*in press*). The Bahi Swamp Wetlands and Household Food Security and Income of Adjacent Communities

Ngailo, J.A., Kaswamila, A.L. and Senkoro, C.J. (2002). Rice farming system of the Wasukuma and its Contribution to Poverty Alleviation. *Presented at the Fifth REPOA Workshops* held at the White Sands Hotel, Dar es Salaam, Tanzania, 21-22, March, 2002. 23pp.

Rweyemamu R. 2009. The Role of Bahi Swamp Wetlands in Enhancing Household Food Security and Income of Adjacent Communities. Unpublished MSc Thesis Sokoine University of Agriculture Tanzania 74 pp

Tetsuya S. and Kuniyasu M (2005). Organic matter dynamics control plant species coexistence in a tropical peat swamp forest. Proc Biol. Sci. 272(1571):1503–1510

Tomáš H & František K. (2008). Mechanisms of species coexistence in semi naturalmeadow communities; importance of species traits

11

Ecological Studies of Wetland Ecosystem in Manipur Valley from Management Perspectives

S.C. Rai and Adon Raleng
Department of Geography
Delhi School of Economics
University of Delhi, Delhi
India

1. Introduction

Wetlands are complex ecosystems often occupying the interface between land and water. Wetlands function as part of the landscape with or without the presence of humans. They have value because many of their functions have proved to be useful to humans but are at the same time ecologically sensitive and adaptive systems. The increasing demand and pressures on wetlands without understanding their nature and dynamics have often led to their degradation, thereby threatening livelihood of the communities' dependent upon these resources. The world's freshwater lakes of today are vanishing at a much faster rate than they used to be a century ago. In some cases, even the area is lost (Raleng, 2010). Current emphasis on wetland management has two objectives: firstly to conserve biotic diversity and secondly to maximize economic gains. This explains why in recent years much attention has been directed towards the formulation and operation of sustainable management strategies for wetlands[1]. Wetlands comprise lakes, rivers, marshes, bogs and similar areas and they are generally regarded as areas of land that remain waterlogged for a substantial period of the year. Wetlands fulfill a wide range of essential functions include flood proofing, natural sewage treatment, shoreline anchoring and dissipation of erosive forces, sediment trapping, nutrient retention and removal and recharging of aquifers. In addition, many food chains depend upon wetland productivity and form important habitats for fisheries and wildlife. Wetlands are important for maintaining

[1] Wetlands are the only single group of ecosystems to have their own international convention. The call for wetland protection gained momentum in the 1960s, primarily because of their importance as habitat for migratory species. A series of conferences and technical meetings culminated in the "Convention on Wetlands of International Importance Especially as Waterfowl Habitat" (better known as the Ramsar Convention) which came into force in 1975. In 1985 there were 38 signatories, in 1991 this increased to 60, and by 1993 the total number was 75 countries (Dugan, 1993). Currently 153 nations have joined the Ramsar Convention as Contracting Parties and more than1600 wetland sites around the world covering 145 million hectares; have been designated for inclusion in the Ramsar List of wetlands of International Importance (Ramsar Convention Secretariat, 2006). India has designated 25 wetland sites as Ramsar Sites of International Importance.

biodiversity and securing livelihood for local communities. The wetlands are most productive life support system in the world.

Wetland ecosystems are among the most threatened of all environmental resources. These bodies are getting more and more polluted due to anthropogenic activity leading to changes in land-use/cover characteristics and subsequently hydrologic processes, specifically through the large-scale conversion of forests to other land-uses (Jain et al., 2000; Mao & Cherkauer, 2009). After large areas of natural wetlands in all parts of the world were totally lost by drainage and landfills or were highly degraded by other human activities, the values and functions of natural wetlands are now being rediscovered (Gopal, 1991). At present, most of the wetlands of India and elsewhere are threatened to extinction mainly because of cultural eutrophication and conversion into agricultural land. Substantial changes in land-use/cover have occurred over north-eastern part of India in the past few decades with the shortening of *jhum cycle*, spread of settlement and increasing use of land resources for agriculture and economic development. Shifting cultivation, also known as "Slash and Burn" or "Swidden agriculture" and as "Jhum cultivation" in north-east India, where it is a predominant activity for the majority of the population, is often described as an inefficient, destructive practice, which contributes to deforestation and lowland sedimentation (Sillitoe, 1998). Reducing the *jhum cycle* during recent years has put pressure on resources and thus the productivity of land degradation, increased levels of soil erosion, hydrological imbalances and forest degradation all of which have caused reductions in yields and insecurity of food sources (Toky & Ramakrishnan, 1981). Understanding the characteristics of hydrological processes and regime change is important for driving the solutions to rational use of the lake water and limiting the environmental degradation in the region.

In India, studies on wetlands have not yet gained importance though few beautiful fresh water lakes of the Himalaya viz., Dal Lake of Kashmir, Khecheopalari Lake of Sikkim and Nainital Lake of Uttarakhand are losing their charms and reducing their areas at alarming rate. The data on lakes are not scientifically compiled, so it is difficult to provide solutions to lake problems. Several earlier studies have focused on ecological, wetland ecosystem management, rather than on the catchment's scale hydrologic impacts of land-use/cover change. Singh (1989) studies wetland ecosystem management perspective of fish, wildlife and environment in Loktak Lake. Singh & Shymananda (1990) studied management of Loktak Lake. Wetland ecosystems are among the most threatened of all environmental resources. It is important to understand the ecosystem processes in a wetland and the factors responsible for maintaining characteristics which impart particular value to it. There is a broad understanding of land-use change but their associations with the hydro-ecological consequences are not understood properly. So, there is a need to understand the relationship of these changes and those problems. Therefore, in view of this, the present study is focused on ecological studies of wetland ecosystem in Manipur valley from management perspectives, which need immediate conservation.

2. The Loktak Lake ecosystem

Loktak Lake, the largest natural lake in north-eastern India, occupies the southern part of Manipur valley, which runs north-south through the centre of Manipur state. Due to its importance in the socio-economic and cultural life of the people, it is considered as "The life-line of Manipur". A large population living in and around it depends upon the lake resources for their sustenance. Its biological richness, and uniqueness of habitat, has resulted in its designation as a "Wetland of International Importance" under the Ramsar Convention, a distinction it shares with just five other lakes in India. The lake is famous for its floating mats of vegetation locally called *phumdi*, which are heterogeneous mass of soil, vegetation and organic matter at various stages of decomposition and for being the only refuge of the endangered *Sangai* (Manipur brow-antlered deer) which is closer to extinction (LDA & WISA, 1998).

The origin and evolution of Loktak Lake may be ascribed to tectonic activity and neotectonism remarkably influenced by a long history of fluviolacustral processes. The Loktak Lake is situated about 38 km south of Imphal, the capital city of Manipur. It lies between longitudes 93° 46′ to 95° 55′ E and latitudes 24° 25′ to 24° 42′ N (Fig. 1). Overall 53 settlements in and around Lokatak Lake is located having 279935 persons, which accounted for 12% of the total population of Manipur state. The lake, along with its surrounding swamps (locally called *pats*) is an integral part of the floodplain of Imphal River. The oval-shaped Manipur valley (height: 746-798 m asl), bounded by mountains rising 2000-3000 m asl along with the Imphal River and its tributaries (Iril, Thoubal, Heirok, Khunga and Chakpi), and other streams (Nambul, Nambol and Ningthoukhong) that pour their silt-laden waters directly into Loktak Lake. Morphometric data are summarized in Table 1.

Maximum length (L)	26 km
Maximum breadth (B)	13 km
Shore line length (L)	126 km
Orientation	N - S
Surface area	287 km^2
Area covered by *Phumdis*	107 km^2
Area covered by islands	17.2 km^2
Area covered by encroached land of fish farm, paddy field etc.	89.3 km^2
Area of open water	73.5 km^2
Mean breadth	11.04 km
Shoreline development	2.10
Total volume of water	550.21M m^3
Maximum depth (Zm)	4.58 m
Mean depth (V:A)	1.92 m
Mean depth : Maximum depth (Z:Zm)	0.42
Development of volume	1.26
Area of catchment's	980 km^2
Surface area : Catchment's area	0.293
Index of lake permanence	1.92

Table 1. Morphometry of Loktak Lake

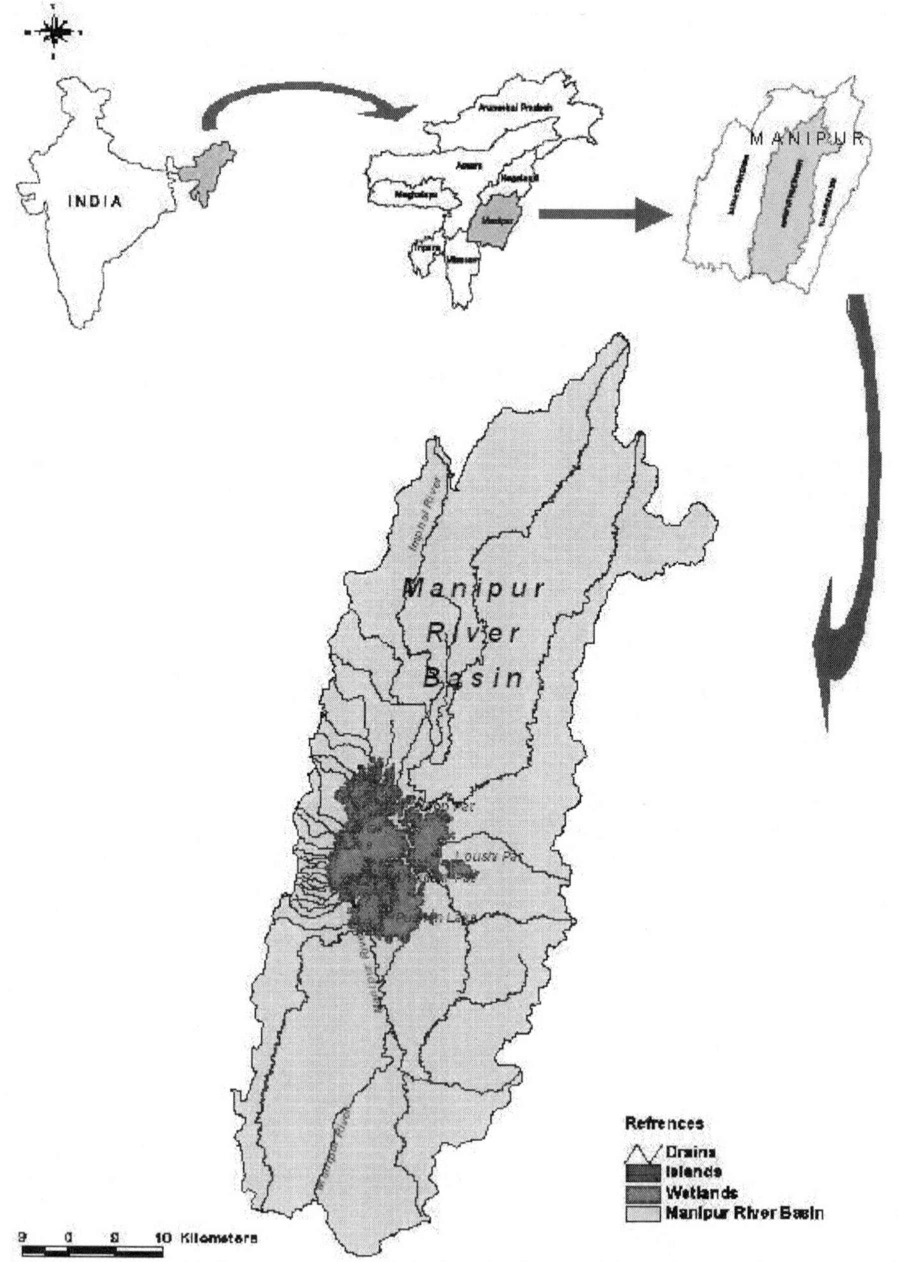

Fig. 1. Location of Loktak Lake

About two-thirds of Manipuri's population lives in the valley (one of the most densely populated areas in India with 415 persons per km^2). These communities are directly or indirectly dependent upon the lake itself. The productivity of the surrounding agricultural fields is on account of the nutrient rich waters of the lake. The wetland's enormous fish resources support large fishing communities in and around the lake area, in particularly villages of Moirang, Ithing, Karang, and Thanga, who completely rely for their livelihood on fish catches. About 60% of the fish catch of Manipur comes from Loktak Lake alone. Besides these vital functions, the lake is also being used for a major lift irrigation and power project.
There are 14 hills appearing as islands in the southern part of the lake. Only four of them, viz., (i) Thanga island, (ii) Karang island, (iii) Ithing island, and (iv) Sendra island are inhabitat. These islands are connected to the lake shore through motorable roads and bridges except Karang island, which is isolated from the other three and can be reached only by boat. The lake can be broadly divided into northern, central and southern zones. The three zones are characteristically different in terms of biodiversity and pressure of human activities. Therefore, these three zones i.e. Northern, Central and Southern were considered for the sampling site. Six sampling sites, two each in every zone, all in the lake periphery except one site (Site IG, which is an island). The three disturbed sites were selected in the study are Moirang (Site MO); Khathinungei (Site KI) and Mayang Imphal (Site MI) whereas the comparatively undisturbed sites are Hayel (Site HA), Khordak (Site KH) and Ithing (Site IG).

3. Land-use/cover change

Land is used to meet a multiplicity and variety of human needs and to serve numerous and diverse purposes. When the users of land decide to employ its resources towards different purposes, land-use change occurs producing both desirable and undesirable impacts. Land-use is the utilization of all developed and vacant lands on a specific space at a given time. The increasing pressure due to population and human activities of land resources to meet the increasing demands are contributing to significant transformation of land for a variety of land-uses.
Shifting cultivation is the characteristic feature of agriculture in the hilly catchments of the Loktak Lake. This form of land-use practices occupies a distinct position in the tribal economy, with about 74% of the tribal population deriving its sustenance from this traditional mode of agriculture. But due to rapid increase in population, the shifting cycle which until a few decades was more than 20 years has reduced to less than five years. This reduction has led to land degradation and increased soil erosion leading to sedimentation of water bodies consequently reducing its water holding and overall carrying capacity. The analysis of land-use/cover change is essential to formulate the suitable plan for lake conservation.
The data for land-use\cover of Loktak Lake catchment have been derived from secondary sources. The temporal land-use/cover data of Loktak Lake catchment of 1990 and 2004 including structural component data of 2004 have been obtained from Loktak Development Authority and Manipur Remote Sensing Application Centre, Imphal. The land-use/cover map has been prepared by using high resolution data like IKONOS satellite, for some area and merge data of 2.5m resolution using LISS-III (Linear Imaging Self Scanner) and PAN (Panchromatic) data. By comparing the temporal land-use data of 1990 and 2004, major land-use/cover changes that occur between the two periods were obtained. Intensive field investigations were carried out for ground verification.

A detailed land-use/cover inventory was developed to assess the current status of the lake catchment and presented in Table 2. The land-use/cover pattern in the lake catchment as a whole showed about 13.99% and 14.85% area under built-up land in 1990 and 2004, respectively. Agriculture is the main occupation of both the people living in the hill and valley area of Loktak Lake. The total agriculture land in the valley accounts for 36.28% and 35.62% in 1990 and 2004, respectively. On the other hand, shifting cultivation or *Jhum* cultivation is a very popular practice in the hill catchments. This type of cultivation is also known as slash and burn cultivation and account for 3.01% and 1.74% in 1990 and 2004, respectively. The total forest land in the lake catchment covered about dense forests have more than 40% canopy and accounts for 1.43% and 0.56% of the total area of the catchment in 1990 and 2004, respectively. The area where canopy cover is less than 10% is considered as degraded forest and its area accounts for 8.95% and 3.35% of the total area of the catchment in 1990 and 2004, respectively. Scrub land is observed mostly in the periphery of dense forest. The total areas under this class accounts for 9.35% and 5.80% in 1990 and 2004, respectively. Open forest represents the forest areas which were earlier deforested and presently natural regeneration is going on. The total areas under this class accounts for 0.37% and 9.76% in 1990 and 2004, respectively. The area under marshy/swampy class accounts about 0.60% and 0.84%, whereas hills/hillocks about 0.72% and 0.67% in 1990 and 2004, respectively. Area under water bodies such as ponds, reservoirs, rivers and aquaculture ponds are clubbed together and accounts for 0.93% and 2.24% in 1990 and 2004, respectively. The total area under wetlands accounts for 24.33% and 24.84% in 1990 and 2004, respectively (Table 2). The detailed descriptions of wetlands are presented in Table 3.

Land use Classes	1990		2004		Variations (%) (1990-2004)
	Area				
	(ha)	(%)	(ha)	(%)	
Built-up land	14558	13.99	15448	14.85	0.86
Agricultural land	37735	36.28	37055	35.62	-0.66
Shifting cultivation	3131	3.01	1811	1.74	-1.27
Dense forest	1493	1.43	586	0.56	-0.89
Open forest	391	0.37	10160	9.76	9.39
Degraded forest	9312	8.95	3514	3.35	-5.58
Scrub forest	9732	9.35	6042	5.80	-3.55
Marshy/swampy land	631	0.60	509	0.48	-0.12
Hill / hillocks	755	0.72	705	0.67	-0.05
Aqua-ponds/water bodies	974	0.93	2331	2.24	1.31
Wetland	25312	24.33	25839	24.84	0.51
TOTAL	104000	100	104000	100	-

Source: Loktak Development Authority (LDA)

Table 2. Area under different land-use/cover in Loktak Lake

Habitat type	Area	
	(ha)	(%)
Open water	2634	10.19
Dense phum	9176	35.52
Moderate phum	1058	4.09
Sparse phum	1744	6.75
Aqua-ponds	6911	26.75
Agriculture	32	0.12
Island with vegetation	288	1.11
Island without vegetation	66	0.25
Settlement	80	0.30
Phum ring area	3841	14.87
Total	25830	100

Source: Loktak Lake Development Authority

Table 3. Structural components of wetland area, 2004

During the 15-years period, the major land-use/cover changes are the expansion of built up area, open forest, aqua-ponds/water bodies and wetland (Table 2). There was increased in built up land up by 0.86 % which was the result of rapid urbanization in the Loktak catchment. The urban population in the catchment has grown at an annual rate of 3.5% during 1991-2001. The increased in open forest area by 9.39% in 2004 and decrease in degraded forest area by 5.58% shows that the activities of afforestation program have been taken up by Lake Development Authority (LDA) for catchment treatment that reduces the shifting cultivation area by 1.27% and ultimately led to the process of natural regeneration in the catchment area. However, there was decreased in dense forest area from 1990 to 2004 by 0.89% which was the effect of past deforestation that need some span of years to regenerate again. The evidence of regeneration is very much clear from the open forest area which increased up to 9.39% in 2004. There has been an increase in the water bodies\aqua-ponds by 1.31%. This is mainly due to conversion of agriculture and marshy/swampy land into aqua-ponds as a result of inundation after the construction of Ithai barrage. This process led to the decreased of agriculture land and Marshy/swampy land by 0.66% and 0.12%, respectively (Table 2). Off the valley's 22 lakes, 9 have silted up and drained for cultivation in the last two or three decades. Loktak Lake itself has shrunk from 495 km^2 in 1971 to just 289 km^2 (Singh, 1989), whereas increase in the number of athapham in lake resulted in prolific growth of *phumdis*. The *phumdis* cover in the lake has increased significantly from 57% in 1989 to 74% in 2002. *Phumdis* in the central zone have increased at an annual rate of 5.7% (Raleng, 2010).

4. Hydrological analyses

The Loktak Lake has been the subject of study since 1950s with the primary objective of flood control and optimal use of water resources for accelerated economic development in the region. Water regime of Loktak Lake is determined by the inflow from various streams and direct precipitation on the lake surface. Overall 34 streams from the western hills and the Manipur river via Ungamel and Khordak channels drain into the lake. The hydrological analysis of the lake is investigated in detail viz., discharge and sediment etc.

4.1 Water Inflow and outflow

The annual average rainfall recorded in the lake watershed was 1392 mm during 2001. Total annual inflow of water into the lake was estimated about 1687 M cusecs. The surface inflow from 34 rivers/streams of the western catchment accounts for 52% of the total inflow into the lake. The total outflow of water from the lake was estimated about 1217 M cusecs. Water abstracted for the hydropower generation by National Hydroelectric Power Corporation (NHPC) accounts for 70% of the total outflow from the lake. Link channels account for 14% of water outflow, while loss of water through evapotranspiration and evaporation was estimated about 9.5% and 6.5%, respectively.

There has been a drastic change in the water exchange pattern between the Manipur river and Loktak Lake after construction of Ithai Barrage. The inflow was reduced to 91 M cusecs and outflow to mere 20M cusecs (Table 4).

Parameters	1958-59 (M cum)	2001-01 (M cum)	Reduction (%)
Inflow	103	91	11
Outflow	315	20	93

Source: Loktak Lake Development Authority, Manipur

Table 4. Reduction in water exchange pattern between Loktak Lake and Manipur river

4.2 Sedimentation

Loktak Lake is gradually silting and the major contributor is sediment from the surrounding catchment. The annual average sediment input into the lake is estimated as 650,000 metric tones. Western catchment accounts for 65% of the total sediment inflow into the lake and the rest 35% from the link channels. The high amount of sediment from the western catchment is mainly due to land-use/cover change and *jhum* cultivation etc. Out of 45 micro-watersheds of the western catchment, Thongjarok yields the maximum sediment load of 58 t/ha and the minimum of 2 t/ha by Merakhong (Table 5). Sedimentation has been observed in the peripheral areas of the lake especially at the mouth of the western streams and link channels, resulting in the reduction in the water holding capacity of the lake, and flooding in the peripheral areas. It has been observed that the silting rates have increased by 34.44% during 1993 to 2003 and by 69.94% during 1967 to 2003. High sediment deposition in the lake turns the lake turbid and deteriorating water quality.

Rivers	Discharge (MCM)	Sediment Load (t)	Area (ha)	Sediment Yield (t/ha)	N (t/yr)	P (t/yr)	K (t/yr)
Potsangbam	22	20752.18	1928.00	10.80	27	6	20
Awang khujairok	9	4229.91	843.40	5.00	54	8	41
Thongjarok	30	133440.83	2302.30	58.00	57	9	45
Merakhong	-	12097.18	5947.20	2.00	-	-	-
Nambol	104	65911.02	9540.00	6.90	243	34	198
Nambul	161	50204.90	19893.00	2.50	2338	270	2081

Source: Loktak Development Authority (LDA), Manipur

Table 5. Sediment Yield during 2000-01

4.3 Flooding
A total area of 63.50 km^2 from different zones of Loktak Lake has been identified to be highly flood prone. The causes of flooding in the wetland system can be mainly attributed to unsystematic operation of Ithai barrage. Silted drainage system, developmental activities, change in land-use practices, loss of forest cover and presence of natural barriers like *Sugnu Hump* and alignment of Chakpi river with Manipur river further compounds the problem. Characteristics of floods in the northern zone are different from that of the southern zone. Many of the drains in the northern zone have heavily silted up and are choked with thick *Phumdi* resulting in reduction of their flushing capacity. This causes flooding in the upstream courses.

5. Water quality analyses

Water quality characteristics of Loktak Lake have been analyzed in detail. Water samples from six sites were collected seasonally i.e., rainy, winter and summer. All analyses were completed within ten days following standard method (Jain et al., 1999). The analysis carried out on various physico-chemical parameters reveals that the lake water of all the sites was circum-neutral to alkaline. The data reveals that pH of lake water ranges from 6.27 to 8.50. The lake water in different sites is severely affected and has also become considerably vulnerable to pollution with a wide range to contaminations. Dissolved oxygen concentration ranges from 6.80 to 9.32 (mg/l) and electrical conductivity in between 98 to 530 (μmoh/cm). Concentration of Chloride ranges from 8 to 30 (mg/l) and Phosphate-phosphorous concentrations fluctuate in between 0.002 to 0.260 (mg/l).
Seasonal variation in water quality characteristics of Loktak Lake have been analyzed in detail (Table 6). Most of the selected parameters reflected the seasonal pattern showing higher values in the rainy season. The pH was recorded highest at Site MI during summer season and lowest at Site KI during winter seasons. Analysis of variance showed that the pH varied significantly at sites and seasons but their interaction was not significant (Table 7). Water temperature ranged from 9.7 °C and 29.4 °C in the temporal cycle, highest during summer and lowest during winter season. Variation in temperature was significant between seasons but not between sites. The turbidity of the lake at various sites ranged from 35 cm to 145 cm. The highest turbidity (145 cm) was recorded in the rainy season at Site MO. Variation in turbidity was significant between sites and seasons, and their interaction was also found to be significant {LSD (0.05) =1.9}. Alkalinity was lowest in the winter season at all sites. The total alkalinity varied significantly among the sites and seasons and their interaction was significant (Table 7) (LSD (0.05) = 2.4}. Electrical conductivity showed marked variation among seasons and sites with higher values ranging from 194 to 530 (μ moh/cm) during rainy season, followed by summer (116 to 428 μ moh/cm) and winter season (98 to 150 μ moh/cm). It varied significantly among the sites and seasons, and their interactions was significant {LSD (0.05) =2.4}. Free carbon dioxide was highest in rainy season and ranged from 4.4 to 63.3 (mg/l) at various sites of the lake across the season. It varied significantly among the sites and seasons, and their interactions was also significant {LSD (0.05) =2.4}. Dissolved oxygen concentration range from 6.8 to 9.32 (mg/l). It varied significantly between sites and seasons and their interaction was also found to significant {LSD (0.05) =0.3}. Chloride concentration was highest in rainy season in all the sites and ranged from 8 to 30 (mg/l). It varied significantly among the sites and seasons and their interaction was significant {LSD (0.05) =4.1}. Phosphate-phosphorus concentrations fluctuate

between 0.21 mg/l to 0.86 (mg/l) at various sites of the lake. Overall high amount of phosphate phosphorous was observed during rainy season. Its concentration increased considerably in rainy season may be attributed to the inflow of nutrients from the catchment area where fertilizers are extremely used for agriculture. It varied significantly among the sites and seasons, and their interaction was also significant {LSD (0.05) =0.03}. Total dissolved solid was recorded higher during rainy season at all sites and low during winter season. It varied from 52.7 to 480 (mg/l) and recorded highest at disturbed sites (360 to 480 mg/l). It varied significantly between sites and seasons, and their interaction was also significant {LSD (0.05) =2.2}. Biological oxygen demand was observed high in rainy season and low during winter season at all the sites. It ranged from 0.4 to 2.98 (mg/l) and varied significantly among the sites and seasons, and their interaction was also significant {LSD (0.05) =0.18} (Table 7).

Loktak Lake suffers from both natural and cultural eutrophication, which is indirectly the result of anthropogenic pressure and land transformation. A large amount of fertilizers residues are washed down the lake from the lake periphery paddy fields during the rainy season and accelerate pollution in the lake. The high silt contents that are brought down into the lake by the feeder streams and from jhumming prone catchment areas have silted up the lake bottom, thereby decreasing its water holding capacity. Changes in the pH of water may be the result of various biological activities (Gupta et al., 1996). The water temperature recorded higher during summer season is basically important for its effects in the biological reactions of the organism. The higher concentration of alkalinity value and free carbon-dioxide in the present study is similar with the report of Pandey & Kumar (1995). The high values of conductivity especially at disturbed sites during rainy season are associated with higher inflow from the surroundings. Total dissolved solids denote mainly the various kinds of materials present in the water. In the polluted waters the concentration of others substances increases depending upon the types of pollution. In present study, total dissolved solids had a cyclic pattern of seasonal changes and maximum during rainy season and minimum in winter. Johnson (1980) observed that total dissolved solid proportionately enhanced the electrical conductance in water and ran parallel to each other. The correlation between conductivity values and total dissolved solids (0.803) is similar with the above views. During the study period, water turbidity was found higher during the rainy season. It might be due to the high silt content of the water carried down into the lake by the feeder rivers and streams from the catchment areas. Such similar results were reported by Zutshi et al., (1980). Chloride concentration was found maximum during rainy season in all the six sites and on the contrary it was low during winter season. High chloride values were observed from the sites where anthropogenic pressure is maximum, which is related with animal wastes from cattle grazing and sewage disposed off by the household residing in the lake catchment. Excess of chloride over 5.5 mg/l in water was associated with contamination from animal organic matter (Khulbe, 1992; Jain et al., 1999). Phosphorous is an important factor in ecological studies and often regards as a limiting element in water ecosystem (Hecky & Kilhan, 1988). Main source of phosphorous in the lake was from domestic sewage, detergents, agricultural runoff. The high values of phosphate-phosphorus was recorded at disturbed site as compared to undisturbed site during rainy season, which might be due to rain draining into the lake with the nutrient rich soil deposited from the catchment areas of lake by its feeder streams and rivers. High level of free carbon-dioxide during rainy season was observed which may be attributed to its influx through rainwater in the form of carbonic acid. This is in conformity with the observation of Chakraborty et al.,

Sites	Season	pH	Temperature (°C)	Turbidity (cm)	Alkalinity (mg/l)	Conductivity (mho/cm)	Free CO_2 (mg/l)	Dissolved oxygen (mg/l)	Chloride (mg/l)	TDS (mg/l)	BOD (mg/)	Total Phosphorus (mg/l)
Mayang Imphal (MI)	W	7.34	10	97	65	150	53	7.1	11.0	52.7	1.43	0.67
	S	8.5	29.4	46	120	428	24.2	8.54	17	160	2.0	0.25
	R	7.8	22.2	140	140	530	63.3	7.4	25	460	2.8	0.84
Hayel (HA)	W	7.81	10	88	105	98	38.9	7.74	12	83	0.7	0.25
	S	8.1	29.3	35	151	116	38	8.89	12	103	1.1	0.21
	R	7.9	22.2	120	205	248	46.2	8.4	15	119	2.1	0.28
Khordak (KH)	W	6.99	9.7	72	98	115	35.7	7.9	8.0	52.7	0.86	0.41
	S	7.4	28.9	54	170	181	26.4	9.32	12.8	75	1.4	0.28
	R	7.2	22.6	118	250	223	40	8.4	13.0	110	1.9	0.51
Khathinungei (KI)	W	6.27	10	96	75	105	28	6.8	9.0	55	1.22	0.7
	S	7.8	29.3	60	128	160	15.8	8.51	12.0	140	2.8	0.40
	R	6.9	22	118	143	350	50	7.3	30.0	360	2.9	0.86
Moirang (MO)	W	7.64	10	95	85	119	12.6	6.9	12.0	60.8	1.49	0.55
	S	8.4	29.2	55	139	186	4.4	8.51	18	170	2.4	0.30
	R	7.9	22.6	145	145	327	59	7.4	28	480	2.98	0.83
Ithing (IG)	W	7.62	10	49	121	121	30.8	7.6	10	61.3	0.4	0.23
	S	7.9	28.2	48	118	120	30	9.2	13.9	99	0.8	0.20
	R	7.8	22	120	182	194	38.5	8.3	15	118	1.4	0.26

Table 6. Seasonal variations in water quality at different sites of Loktak Lake

Source of variation	d. f.	*F*	*P*
pH			
Site	5,36	11.67	0.001
Season	2,36	16.9	0.001
Site × Season	10,36	1.43	NS
Transparecy			
Site	5,36	22.9	0.001
Season	2,36	1990	0.001
Site × Season	10,36	54	0.001
Temperature			
Site	5,36	0.123	NS
Season	2,36	3535	0.001
Site × Season	10,36	1.04	NS
Conductivity			
Site	5,36	5195	0.001
Season	2,36	14598,	0.001
Site × Season	10,36	1064	0.001
Free carbondioxide			
Site	5,36	45.5	0.001
Season	2,36	279	0.001
Site × Season	10,36	31.9	0.001
Alkalinity			
Site	5,36	450	0.001
Season	2,3	2732	0.001
Site × Season	10, 36	116.4	0.001
Disvsoled oxygen			
Site	5,36	15.8	0.001
Season	2,36	49.3	0.001
Site × Season	10,36	9.6	0.001
Phosphate phosphorus			
Site	5,36	154	0.001
Season	2,36	247	0.001
Site × Season	10,36	22.9	0.001
Total Dissolved Solid			
Site	5,36	4335	0.001
Season	2,36	20981	0.001
Site × Season	10,36	2702	0.001
Chloride			
Site	5,36	24.3	0.001
Season	2,36	136.9	0.001
Site × Season	10,36	13.7	0.001
Biological oxygen demand			
Site	5,36	47.9	0.05
Season	2,36	118	0.001
Site × Season	10,36	2.7	0.05

Table 7. ANOVA for seasonal variations in water quality at different sites of Loktak Lake

(1959); Mansoori et al., (1995). The high dissolved oxygen content during summer season is largely attributed to increase in temperature with increase in photosynthetic activity of the aquatic plants and phytoplankton. Our results are consistent with Khulbe (1992). The presence of BOD in the Loktak Lake may be from polluted Nambul river and also from the domestic waste from local areas including from several huts lying inside the lake on the *phumdis*. High value of biological oxygen demand during rainy season might be due to high organic loads along with the rain runoff from the catchment area of the lake. Das & Pandey (1980) have also reported the same view for the lake Nainital of Kumaon Himalaya. The data based on the physico-chemical properties is an indicator of the water quality at the time of sampling. However, regardless of the high variation of the water quality, the results obtained can be representative of the spatial variation of the water quality throughout a wide range.

6. Sustainable management lake ecosystem

The lake has many stakeholders. For the purpose of the present study only some of the important and relevant stakeholders are selected for study. Fishery is the most important occupation of communities in and around the lake, hence fishermen are the most important primary stakeholders of the initiatives. Phum hut dwellers, agricultural farmers of lake shore villages and hill villages, people originally displaced by reservoir flooding and livestock farmers are the other stakeholders. In order to stem and ultimately reverse current pressures on the lake ecosystem, a conservation strategy to promote the sustained use of wetland resources needs to be developed and implemented. The following management strategies warrant immediate attention as suggested by the selected stakeholders:

6.1 Research and monitoring

The most important questions regarding the ecological problems of Loktak Lake remain unanswered viz, the rate of nutrient enrichment, the pollution load, its point and non-point sources, hydrological details such as water retention, water loss or gain through seepage and most importantly, the likely impacts of the proposals considered for implementation both on biota and socio-economy. So, monitoring of lake from time to time is absolutely necessary.

6.2 Siltation control

This will involve soil conservation measures in the catchment, weaning shifting cultivators away from the damaging practice, and involving Sloping Agriculture Land Technology developed by some institutions.

6.3 Encroachment control

It has been pointed out that reclamation of wetland areas is being illegally undertaken by people of means, while others dispossessed of their holding because of submergence due to the damming of the lake are still being taxed. Illegal encroachers should be dispossessed of their holdings, which could then be distributed among the genuinely landless unemployed. Unless these human issues are resolved, all conservation efforts will be futile.

6.4 Control of over fishing

Fishing is the main occupation of island and many lakeshore villages. In recent times fishing has not become sustainable at all. Aquaculture should be encouraged in the catchment to reduce pressure on the fishery resources of the lake. The feasibility of paddy-cum-fish culture as an income generating activity as has been practiced by Apatani communities of Arunachal Pradesh in Apatani valley (Zero valley) could be examined. In addition, the introduction of sewage-fed fisheries would greatly reduce pollution from municipal wastes, and also enhance fish production. At this stage, it is essential that alternative or additional source of livelihood like piggery, duckery and revival of traditional handicraft and handloom industries though adequate incentives and technical and marketing support is necessary.

6.5 Environmental Impact Assessment (EIA)

An EIA of the Loktak Hydro-electric Power Project is must. The damming of the Loktak Lake water seems to have done more damage to the wetland ecosystem than was earlier envisaged; especially in terms of accelerating eutrophication and adversely affecting certain sectors of the traditional economy.

6.6 Water hyacinth control

Facilities to utilize the weed for biogas production, paper pulp and fiber, chemical products, and mulch and compost needs to be introduced. The spread of weevils in the lake for the eradication of hyacinth urgently needs to be investigated. The riotous growth of *Phumdi* also needs to be curtailed after a scientific assessment of the quantity to be removed.

6.7 Fertilizer usage control

Phum compost promises to undo various threats to the lake ecosystem. Removal of *phumdis* lead to better quality water and phum compost are free from the side effect that chemical fertilizers used by farmers, have on the lake ecosystem. Hence, promotion of phum compost at various levels and farm for mass production of phum compost need to be built. Weed infestation is directly related to the large fertilizer inflow into the wetlands. Alternative cropping practices need to be introduced with emphasis on, i.e. horticulture, sericulture, and perhaps apiculture, which could also enhance income levels. Today Loktak Lake has been placed on the informal list of threatened Ramsar Sites of the world. Only a concentrated effort on the part of official agencies, professionals, NGOs, and the local communities themselves, to comprehend the complexities of this delicately balanced, biologically-rich wetland ecosystem, and take appropriate action, can save Loktak Lake from demise.

6.8 Indigenous knowledge

Indigenous knowledge systems which through many years have evolved as a complex practice may be different from scientific conservation management. It would therefore be wise to conduct participatory research and studies to revive and recover traditional knowledge of indigenous management system of both in hill and valley villages as in the case of forest management in Caqueta, Columbia. The villages in the lake area are prone to water borne diseases. Close disease surveillance need to be kept. Mobile medical teams need to visit the villages during peak seasons and provide free medical camp. The hill villages need to given proper awareness regarding the negative effects of deforestation and forest fires not only to the lake ecosystem but also to their own lives. They should encouraged to form joint forest committees, which will control and monitor over exploitation of forest resources.

7. Conclusions

For sustainable development of any natural resources, it is very important that the stakeholders at various levels come together and participate in conserving and managing the resources. After a thorough analysis, it is very much clear that the deforestation particularly jhum cultivation and land-use/cover change has drastically affected the water quality of the lake. The water quality, in general, falls within class C to E as per the CPCB's designated best use criteria. The lake water is not fit for direct drinking without treatment but can be used for irrigation purposes. A comparative analysis of water quality of different sites indicates significant levels of pollution in the densely populated sites as compared to that of less populated sites because dissolved oxygen was low and bio-chemical oxygen demand and other parameters including nutrient levels were higher in the densely populated sites. It also indicates that out of the three zones (i.e. northern, central and southern zones) of the lake, the Northern Zone and Southern Zone shows significant levels of pollution. High intensity of fertilizer usage in the agricultural fields and practice of fish farming contribute significantly to water quality deterioration in the Northern Zone. The highly polluted rivers (like Nambul and Nambol rivers) also finally discharge pollutants in this zone. Southern Zone is polluted due to flow of all the pollutants finally in this zone and their accumulation due to poor flushing. So, while taking up any water management plan of Loktak ecosystem the highly populated sites of the Northern Zone and Southern Zone should be given priority at any cost. The stakeholders who are source of threat to the ecosystem need proper attention and monitoring. Also different types of participation can be practiced as per demand of the situation so that it leads to self management.

8. Acknowledgement

The authors are thankful to the Head, Department of Geography, Delhi School of Economics, University of Delhi, for facilities. We are also thankful to Dr. Alka Jain for her support, critical and constructive comments during the study period and Manipur Pollution Control board for laboratory analysis.

9. References

Chakrabati, P.K., Roy, R.D.P., Singh, S.B., (1959) A quantitative study of the plankton and the physico-chemical conditions of the river Yamuna at Allahabad in 1954-1955. *Indian J. Fish.*, 6(1), 186-203.

Das, S.M., Pande, J., (1980) Pollution, fish mortality and environmental parameters in lake Nainital. *J. Bombay Nat. Hist., Soc.*, 79, 100-109.

Dugan, P.J. (ed.) (1993) *Wetlands in Danger.* IUCN and World Conservation Union, Cambridge.

Gopal, Brij (1991) Wetland (mis) management by keeping people out: two examples from India. *Landscape and Urban Planning*, 20, 53-59.

Gupta, R.K., Sharma, M., Gorai, A.C., Pandey, P.N., (1996) Impact of coal mining effluents on the physico-chemical characteristics of Raja Tank, Jaria (Dhanbad). *J. Freshwater Biol.*, 8(2), 63-73.

Hecky, R.E., Kilhan, P., (1988) Nutrient limitation of phytoplankton in freshwater and marine environment. A review of recent evidence on the effect of enrichment. *Limnol. Oceanogr.*, 33, 796-822.

Jain, Alka, Rai, S.C., Pal, J., Sharma, E., (1999) Hydrology and nutrient dynamics of a sacred lake in Sikkim Himalaya. *Hydrobiologia,* 416, 12-22.

Jain, Alka, Rai, S.C., Sharma, E., (2000) Hydro-ecological analysis of a sacred lake watershed system in relation to land-use/cover change from Sikkim Himalaya. *CATENA*, 40, 263-278.

Johnson, M.E.C., (1980) Total solid content in two freshwater lakes. *Indian J. Bot.*, 11 (2), 188-190.

Khulbe, R.D., (1992) *Studies on Water Pollution in Nainital and Bhimtal Lakes of Kumaun Himalaya.* IInd Annual Progress Report, Ministry of Environment and Forests, Government of India, 187 pp.

King, D.L., (1970) The role of carbon in eutrophication. *J. Water Poll. Contr. Fed.*, 42, 2035-2051.

LDA & WISA (1998) *Project Management Plan, Sustainable Development and Water Resource Management of Loktak Lake.* Loktak Development Authority and Wetland International- South Aisa.

Mansoori, H.A., Lavania, R.K., Tiwari, R.K., (1995) Hydrological study of Lakshmital lake Jhansi with special reference to plankton productivity. *Flora and Fauna,* 1(1), 39-42.

Mao, D., Cherkauer, K.A., (2009) Impacts of land-use change on hydrologic responses in the Great lakes region. *Journal of Hydrology*, 374 (1-2), 71-82.

Pandey, R. and Kumar, A. (1995) Comparative evaluation of potable water quality of tribal and non-tribal villages of Santhal Pargana, Bihar. *Ecology, Environment and Conservation*, 1(1), 71-74.

Ramsar Convention Secretariat (2006) *Participatory Skills. Ramsar Handbooks for the Wise Use of Wetlands.* Ramsar Convention Secretariat, Gland, Switzerland.

Raleng, Adon (2010) *Human Impact on Loktak Lake, Manipur.* Unpublished M. Phil. Dissertation, Department of Geography, University of Delhi, pp.125.

Sillitoe, P., (1998) It's all in the mound: fertility management under stationary shifting cultivation in the Papua New Guinea highlands. *Mountain Research and Development*, 18 (2), 123-134.

Singh, T.H., Shyamanda, R.K., (1990) Management *of Loktak Lake: An Enquiry into the Present Strategy and Possible Alternatives.* Manipur University, Imphal.

Singh, T.H., (1989) *Study of Wetland Ecosystem in Manipur Valley from Management Perspectives of Fish, Wildlife and Environment.* Department of Life Sciences, Manipur University, Imphal.

Toky, O.P., Ramakrishnan, P.S., (1981). Runoff and infiltration losses related to shifting agriculture (jhum) in north-eastern India. *Environmental Conservation*, 8, 313-321.

Zutshi, D.P., Subla, B.A., Khan, M.A., Vanganeo, A., (1980). Comparative limnology of nine lakes of Jammu and Kashmir, Himalayas. *Hydrobiologia*, 72, 101-102.

12

Impact of Domestic Animals on Ecosystem Integrity of Lesotho High Altitude Peatlands

P.J. Du Preez[1] and L.R. Brown[2]
[1]*Department of Plant Sciences, University of the Free State*
[2]*Applied Behavioral Ecology and Ecosystem Research Unit, University of South Africa*
Republic of South Africa

1. Introduction

Wetlands are considered ecologically sensitive ecosystems with unique habitats for a great variety of plants species, birds, small mammals and other aquatic organisms. The term "wetland" is used to describe various habitats where the soil is wet for extended periods of the year but not necessarily permanently waterlogged (Collins, 2005). Wetlands are formed where surface water collects or where ground-water seeps to the surface for long-enough periods to sustain vegetation typically adapted or tolerant of hydric soil conditions. Thus wetlands comprise a large variety of waterlogged habitats such as bogs, estuaries, fens, floodplains, marshes, peatlands, playas (pans), seeps, springs, and swamps located on various altitudes and topographical locations.

In his book on wetlands of southern Africa, Cowan (1995) used the definition of a wetland as it is stated in Article 1.1 of the Ramsar Convention where it has been defined as: *"areas of marsh, fen, peatland or water, whether natural or artificial, permanent or temporary, with water that is static or flowing, fresh, brackish or salt, including areas of marine water the depth of which at low tide does not exceed six metres"*. The South African National Water Act (Act 36 of 1998) defines wetlands as *"land which is transitional between terrestrial and aquatic systems where the water table is usually at or near the surface or the land periodically covered with shallow water, and which in normal circumstances support or would support vegetation typically adapted to life in saturated soil"*.

Peatlands are unique wetland ecosystems in terms of their species composition and age. The process of peat formation is very slow which means that some of these wetlands are ancient as they have a deep peat deposit of over two metres. Peat is organic soil material with a particularly high organic content which, depending on the definition, usually has at least 20% organic carbon by weight. Peatlands are wetlands in which peat (dead plant matter) accumulates due to slow decomposition (Cronk & Fennessy, 2001). These peatlands can be classified into two major types namely bogs and fens. The definition of a **bog** is, a peat-accumulating wetland that has no significant water inflows or outflows and supports acidophilic mosses, particularly *Sphagnum* (usually acidic; pH<7). A **fen** is a peat-accumulating wetland that receives some water from surrounding mineral soil and usually supports marsh-like vegetation (usually alkaline; pH>7)(Cronk & Fennessy, 2001; Mitch & Gosselink, 2000). According to these definitions, the high-altitude wetlands of Lesotho are therefore incorrectly classified as bogs by Grobbelaar & Stegman (1987), Herbst & Roberts

(1974a & b), Jacot-Guillarmod (1962, 1963, 1972), Van Zinderen Bakker (1955), Van Zinderen Bakker & Werger (1974). Intermediate types can also be found which are transitional between a bog and fen and a couple of defining parameters such as pH/alkalinity, hydrology, and nutrient availability as well as plant community structure are needed to distinguish accurately between a bog and fen. In view of this difficulty, Cronk & Fennessy (2001) suggests that the term **peatland** should be used. For this study the Lesotho high-altitude wetlands will therefore be referred to as peatlands.

The functions and values of wetlands are well-known. Collins (2005) classifies these benefits as either direct or indirect benefits to mankind and to the environment. Direct benefits of wetlands are water supply, provision of harvestable resources, socio-cultural significance, tourism and recreation, and education and research. The indirect benefits are hydrological benefits such as water purification, sustained stream flow, flood reduction, groundwater recharge / discharge, and erosion control. Other indirect benefits are biodiversity conservation (integrity and irreplaceablility), and chemical cycling (Kotze, 1999, 2010; Kotze et al., 2008). Direct benefits include grazing, fibre for construction and handcraft production, fisheries, hunting of waterfowl and wildlife, valuable land for cultivation, and a valuable source of water (Collins, 2005). The peatlands of the Lesotho highlands deliver most of these direct and indirect goods and services to the environment.

For many years wetlands throughout the world have been regarded as "cheap land" that could be drained and used for development purposes. In South Africa many slimes dams from mines are located on wetlands or directly adjacent to these sensitive ecosystems. This has led to the destruction of natural vegetation and its associated animal species as well as its total ecosystem functioning. In many cases exotic invasive plant species such as the tree *Eucalyptus* spp. (blue gum), the shrub *Tamarix chinensis,* and the grasses *Cortaderia jubata, C. selloana,* and *Pennisetum clandestinum,* are planted in an attempt to dry up and stabilise the slimes dams. Various other developments such as housing, retail parks, and industrial areas have either completely destroyed wetland ecosystems or hugely impacted on its ecological functioning. Agriculture, damming of water, and invasion of alien plant species have also resulted in the destruction of wetland ecosystems and their functioning. Many crop fields and forest plantations have been established on wetlands thereby displacing the natural vegetation and animal species. The mining of peat for commercial purposes has also degraded a large number of these sensitive ecosystems.

Whereas the wetlands in South Africa are threatened by a number of human activities the wetlands of Lesotho are mostly threatened by agricultural activities such as grazing. The riverheads of the Senqu River System are situated in the Maluti Mountains of the Lesotho Kingdom where they all have their sources in high-altitude peatlands. The functioning of these high-altitude peatlands can be negatively affected by human activities. The peatlands of the Lesotho highlands are under serious threat of human activities. Jacot-Guillarmod already mentioned in 1969 that the Lesotho wetlands are under threat of degradation due to over-utilisation. The land of Lesotho belongs to the people of Lesotho, meaning that all pastoral land is held communally. Everyone has the right to use the land and its products. The land use in the upper catchments is pastoral. The land is subdivided into areas "owned" by a ward chief who may delegate his power to subordinate chiefs or headmen. Control over the numbers of domestic animals and periods they graze the wetlands is almost impossible. The resultant impact on the aquatic and semi-aquatic vegetation is severe.

Overgrazing practices and the making of animal tracks on slopes and through wetlands result in erosion, decline in the biomass production and change in the species composition of the wetlands and surrounding catchments.

Due to the sourveld nature of the high alpine vegetation of Lesotho, the wetlands of Lesotho are considered a critical grazing resource by local Basotho herdsmen, especially in summer when thousands of sheep, goats, cattle and horses can be seen grazing on the open wetlands (Figure 1). Grazing areas are communal and controlled primarily by local chiefs (Letsela et al., 2002). The livestock industry in Lesotho has long been a vital component of the economic and social structure of the country (Mpiti-Shakhane et al., 2002) and due to the open nature (unfenced) of the Lesotho wetlands it will be a challenge to alter this over exploitation in the near future.

Fig. 1. Grazing of domestic animals in the high-altitude wetlands of Lesotho.

The vegetation of the study area falls within the Lesotho Highland Basalt Grassland (Gd 8) as described by Mucina & Rutherford (2006). Although this vegetation type is one of the most endemic-rich vegetation units in the Drakensberg Alpine Centre, only 1% of it is statutorily conserved. Except for the Sehlabathebe and Bokong National Parks, the rest of the conservation areas are situated in South Africa. This implies that in Lesotho, less that 1% of the area is officially conserved. A large transfrontier park, namely the Maloti-Transfrontier Park is planned to protect the high-altitude vegetation of Lesotho and South Africa. In order to determine the effect of grazing on these sensitive ecosystems this investigation was undertaken.

For this study, selected high-altitude peatlands influenced by domestic animals, were studied and compared with near pristine high-altitude peatlands in the nearby eastern Free State of South Africa. This was done by comparing the biomass production, vegetation structure and plant species composition. These comparisons allow for the assessment of the current state of these peatlands and recommendations to manage these human-impacted sensitive ecosystems in a sustainable manner.

2. Study area

The study area is located on the eastern escarpment of South Africa. The South African sites are located on the Platberg plateau, while the Lesotho sites are at Khalong-la-Lithunya (Gunns pass), Koti Sepola (near Sani pass) and the Sugar bowl (Figure 2), an extinct volcano, near the Kao diamond mine. The Platberg sites are located in South Africa and above 2,300 m.a.s.l. while the Lesotho sites are all above 3,200 m.a.s.l.

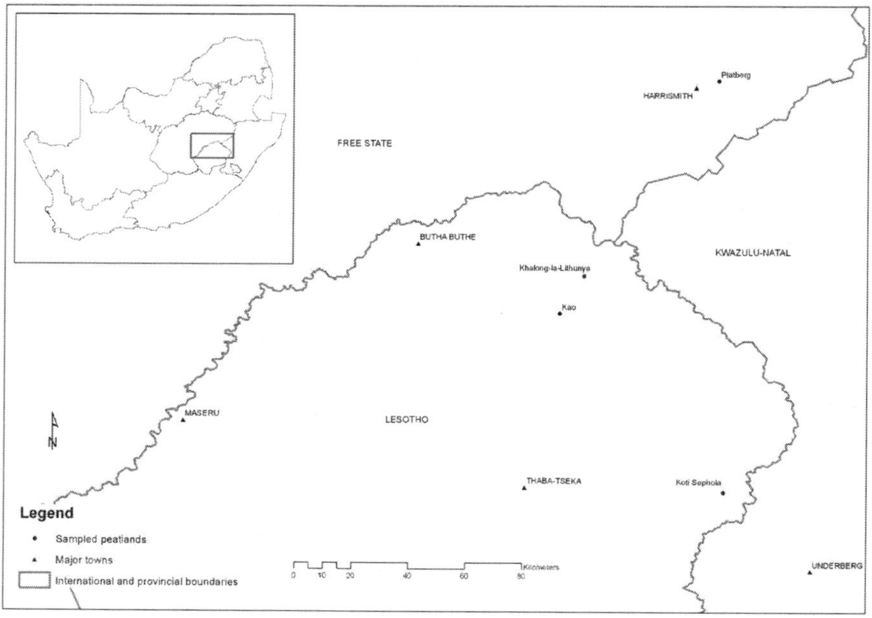

Fig. 2. Location of the study areas.

The climate is temperate cool to cold with dry winters and hot, wet summers. The climate of Lesotho is characterized by warm moist summers, from November to March; and cold dry winters from May to July. The mean monthly maximum measured at Barkley East is 31.4°C. The rainfall is mainly in the form of thunderstorms. The mean annual precipitation (MAP) for Mokhotlong is 575mm. The relatively low MAP can be ascribed to a rain shadow effect. The winters are cold with frost. The mean monthly minimum measured at Barkley East is -10.5°C. Higher elevations above 3,000 m.a.s.l. receive enough snow during winter to cover the ground for several months with sub-freezing temperatures. On rare occasions snowfalls have been recorded in summer as well. Frost-heaving of the soil is a common phenomenon in these areas and occurs especially on barren patches. The ice crystals which are formed during frost-heaving cause the soil to lift. This kind of disturbance prevents seedlings to establish. The barren patches are therefore kept in state by the frost-heaving activities. The frost-heaving phenomenon is probably also the cause of the thufur (hummocks) which occur on peatlands (Van Zinderen Bakker & Werger, 1974).

As far as the geology is concerned the entire Lesotho highland is underlain by basalt of the Lesotho formation and, to a lesser extent, dolerite and gabbro. The basalt is an igneous rock that covered the Clarens Formation about 120 million years ago when Gondwana begun to breakup. Basaltic colluvium and alluvium are the geologic material for the deep soils of the mountain valleys and accumulation glacis (Schmitz & Rooyani, 1987). The soils, at altitude, are mostly shallow and have more or less even proportions of gravel, coarse sand, fine sand, silt and clay particles. The breakdown of organic matter is a slow process due to the long periods of very low temperatures. Especially in the peatlands the accumulation of peat is the result of the decaying roots of the various plants species present on these wetlands. The peat has a high water-retention.

3. Methods

3.1 Vegetation data

A total of 18 sample plots were surveyed in the reference as well as restoration catchments of the four study sites. Plot sizes were fixed at 4 x 4 m to give a total surface area of 16 m^2 (Du Preez, 1991; Malan, 1998). All plant species growing in the sample plots were recorded and their cover abundance assessed using a modified Braun-Blanquet cover abundance scale (Brown, et al., 2005; Mueller-Dombois & Ellenberg, 1974; Westhoff & Van der Maarel, 1980).

Environmental data recorded in each sample plot included aspect, slope, exposure, the size of the rocks, latitude and longitude, altitude, locality, geology, the per cent of area covered by rock, topography, the degree of surface erosion (indicated as low moderate and severe), the degree of trampling, drainage, soil depth and notes on management and utilisation as well as total percentage canopy cover.

The habitat as well as floristic data was processed with the TURBOVEG database (Hennekens & Schaminée, 2001). A first approximation was done with the TWINSPAN (two-way indicator-species analysis) algorithm of Hill (1979). The JUICE program (Lubomír 2002) was used to generate a phytosociological table (Table 1), which was refined by Braun-Blanquet procedures.

3.2 Biomass

From a grazing and resilience point of view, production or yield is one of the most important measures in assessing an ecosystem (Carpenter et al., 2001). Biomass or standing crop usually refers to the weight of organisms present at one time (Pieper, 1978). Most estimates of plant biomass or standing crop include only that above the soil surfaces. This material is commonly available to large herbivores. Direct harvesting is considered the most reliable method of determining aboveground biomass (De Leeuw et al., 1991; Snyman & Fouché, 1993). For this study two exclosure plots of 25 m^2 were erected within each of the Khalong-la-Lithunya and Koti-Sephola peatland sites.

Biomass was determined in each of the exclosure and grazed plots as well in the other peatland systems by randomly placing 0.5 m^2 quadrats in these areas and cutting the herbaceous material down to ground level (Figure 3). The grass and forb species were separated and dried a drying oven at 65°C for 48 hours (Cleaver, 2004).

The Kruskal-Wallis ANOVA by Ranks test (Welman et al., 2007) was used to compare the biomass (kilogram Dry Matter/hectare)(kg DM/ha) of the plots by grazing type while the

the Mann-Whitney U Test (Welman et al., 2007) was used on the ranked data to determine where the pairwise differences lie.

Fig. 3. All rooted herbaceous plants were harvested on a species basis in quadrats ($0.5m^2$), randomly placed in the grazed and exclosure plots in the peatlands.

4. Results

The analysis of the vegetation data resulted in the identification of two major plant communities namely the *Pennisetum sphacelatum-Andropogon appendiculatus* Peatland present on Platberg and the *Haplocarpa nervosa-Isolepis cernua* Peatland present on the Lesotho study sites. The phytosociological classification of the two major wetland communities is presented in Table 1.

4.1 Khalong-la-Lithunya

The vegetation of this peatland is dominated by the grass *Pentachistis oreodoxa* and the forb *Haplocarpa nervosa* that covers up to 75% of the area (Plots 8-11, Species group B - Table 1). The forbs *Isolepis cernua, Ranunculus meyeri* and *Cotula radicalis* are prominent throughout this community. The canopy cover of the wetland ranged between 60% and 100%. A total of 21 different plant species comprising 14% grasses and 86% forbs were identified within this wetland with the average height of the plant species ranging between 2 and 6 cm. Small to medium sized erosion dongas were observed within this system.

In both exclosure plots the biomass was higher than that measured in the grazed plots (Figure 3). The biomass in exclosure plot 1 was 3 868.4 kg DM/ha compared to the 1 020 kg DM/ha in the grazed plot 1 while the biomass of exlosure plot 2 and grazed plot 2 was 2 002.7 kg DM/ha and 759.4 kg DM/ha respectively (Figure 4). The herbaceous layer of the exclosure plots had an average height of 4 cm while the grazed plots had an average height of 2 cm. No difference in species composition was found between the exclosure and grazed plots (Table 1).

		Platberg peatlands							Lesotho peatlands										
Sample plot number		06	01	05	07	04	03	02	18	17	11	14	15	12	16	13	10	08	09
Number of species/plot		08	18	07	12	12	08	15	14	18	20	17	18	18	17	16	19	19	16
Species group A:																			
Andropogon appendiculatus	Grass	5	1	1	1	1	.	4	.	.	.	.	.	.	.	.	.	.	.
Pennisetum sphacelatum	Grass	2	+	2	1	4	.	1	.	.	.	.	.	.	.	.	.	.	.
Denekia capensis	Forb	r	r	r	.	r	r	+	.	.	.	.	.	.	.	.	.	.	.
Senecio isatideus	Forb	+	r	1	+	r	.	+	.	.	.	.	.	.	.	.	.	.	.
Geranium multisectum	Forb	1	1	.	1	.	.	.	.	.	.	.	.	.	.	.	.	.	.
Watsonia lepida	Geophyte	r	+	+	+	.	.	.	.	.	.	.	.	.	.	.	.	.	.
Pentaschistis species	Grass	.	+	5	.	2	.	1	.	.	.	.	.	.	.	.	.	.	.
Fingerhuthia sessleriiformis	Grass	2	.	.	.	3	1	.	.	.	.	.	.	.	.	.	.	.	.
Ficinia stolonifera	Sedge	.	+	.	5	2	5	1	+	r	.	.	.	.	.	.	.	.	.
Agrostis barbuligera	Grass	.	r	.	1	1	1	+	.	.	.	.	.	.	.	.	.	.	.
Helictotrichon longifolium	Grass	.	2	.	+	2	+	4	.	.	.	.	.	.	.	.	.	.	.
Epilobium salignum	Forb	.	+	.	r	r	+	+	.	.	.	.	.	.	.	.	.	.	.
Eriosema salignum	Forb	.	r	.	r	r	.	r	.	.	.	.	.	.	.	.	.	.	.
Lobelia flaccida	Forb	.	+	.	.	r	+	+	.	.	.	.	.	.	.	.	.	.	.
Carex species	Sedge	.	+	.	.	.	.	+	.	.	.	.	.	.	.	.	.	.	.
*Bromus catharticus	Grass	.	+	.	.	.	.	+	.	.	.	.	.	.	.	.	.	.	.
Bromus firmior	Grass	.	4	.	.	.	.	2	.	.	.	.	.	.	.	.	.	.	.
Species group B:																			
Isolepis cernua	Sedge	.	.	.	.	.	.	.	2	2	3	3	3	2	1	2	2	5	5
Haplocarpha nervosa	Forb	.	.	.	.	.	.	.	2	3	5	3	2	1	2	4	5	2	1
Pentaschistis oreodoxa	Grass	.	.	.	.	.	.	.	+	1	4	4	1	5	+	+	4	5	5
Cotula radicalis	Forb	.	.	.	.	.	.	.	4	2	4	1	1	3	1	1	1	1	1
Ranunculus meyeri	Forb	.	.	.	.	.	.	.	1	1	1	1	1	4	+	3	+	+	1
Rhodohypoxis deflexa	Geophyte	.	.	.	.	.	.	.	1	1	+	+	+	+	+	+	+	1	+
Limosella longiflora	Forb	.	.	.	.	.	.	.	+	1	+	1	.	+	+	1	+	+	+
Limosella vesiculosa	Forb	.	.	.	.	.	.	.	+	1	+	+	.	+	+	1	+	+	+
Athrixia fontana	Forb	.	.	.	.	.	.	.	1	+	+	+	r	+	+	1	+	+	+
Cerastium arabidis	Forb	.	.	.	.	.	.	.	.	+	+	+	+	+	r	+	+	+	1
Craterocapsa tarsodes	Forb	.	.	.	.	.	.	.	+	1	+	+	+	.	+	.	+	+	+
Lobelia galpinii	Forb	.	.	.	.	.	.	.	+	+	r	+	r	+	+	+	.	+	+
Dimorphotheca jucunda	Forb	.	.	.	.	.	.	.	r	r	1	.	r	+	+	r	+	+	.
Poa binnata	Grass	.	.	.	.	.	.	.	.	+	+	+	+	+	r	+	+	+	.
Helichrysum bellidiastrum	Forb	.	.	.	.	.	.	.	.	.	+	.	r	+	+	+	1	1	.
Anthoxanthum ecklonianа	Grass	.	.	.	.	.	.	.	.	.	+	.	.	r	.	.	2	1	2
Saniella verna	Geophyte	.	.	.	.	.	.	.	.	.	r	r	r	.	.	.	+	+	+
Wurmbia burttii	Geophyte	.	.	.	.	.	.	.	.	.	r	r	r	.	.	.	+	r	+
Crassula peploides	Succulent	.	.	.	.	.	.	.	.	+	.	+	+	+	+	+	r	.	.
Species group C:																			
Ranunculus multifidus	Forb	.	r	1	+	.	r	r	.	.	r	+	+	+	.	.	.	.	.
Trifolium burchellianum	Forb	r	+	.	+	.	.	.	+	+	+	.	+	+	1	+	+	1	+

Table 1. Phytosociological table of the major high-altitude wetlands of Lesotho and Platberg.

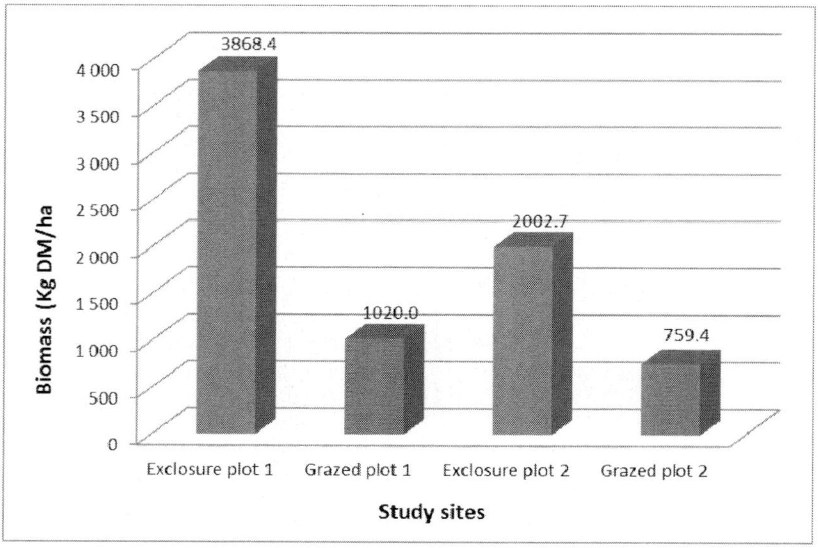

Fig. 4. Biomass of the Khalong-la-Lithunya peatland in Lesotho.

4.2 Koti-Sephola

The grass *Pentachistis oreodoxa* and the forbs *Isolepis cernua, Haplocarpa nervosa* dominate the vegetation each covering between 25% and 65% of the area, while the forbs *Ranunculus meyeri* and *Cotula radicalis* are also prominent (Plots 12-15, Species group B - Table 1). A total

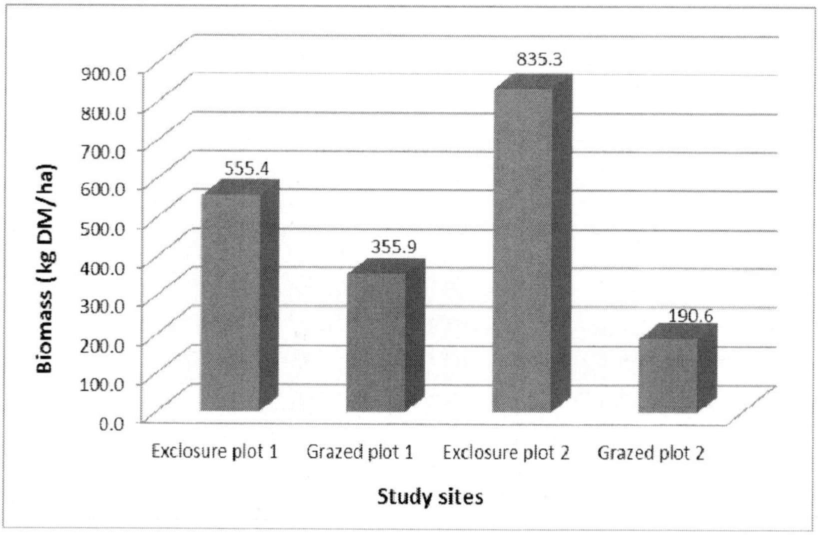

Fig. 5. Biomass of the Koti-Sephola peatland in Lesotho.

number of 24 different species were identified comprising 13% grasses and 87% forbs. The canopy cover of the wetland ranged between 65% and 100% with medium sized erosion dongas present.

Biomass for the exclosure plots was 555.4 kg DM/ha and 835.3 kg DM/ha respectively compared to the low 355.9 kg DM/ha and 190.6 kg DM/ha for grazed plots 3.2 and 4.2 (Figure 5). There was no difference in species composition between the exclosure and grazed plots (Table 1). The average height of the herbaceous layer was 3.8 cm compared to 1.8 cm in the grazed plots.

4.3 Kao

The Kao peatland is characterized by the dominance of the forbs *Cotula radicalis, Isolepis cernua,* and *Haplocarpa nervosa* (Plots 16-18, Species group B - Table 1). A total number of 21 different species were identified comprising 14% grasses and 86% forbs. The canopy cover of the wetland ranged between 60% and 90% with medium to sized erosion dongas present.

Biomass was low at 185.4 kg DM/ha and 145.5 kg DM/ha with the herbaceous layer only 1.5 cm high.

4.4 Platberg peatlands

The Platberg peatlands are characterised by the dominance of the grasses *Andropogon appendiculatus* and *Pennisetum sphacelata* while the grass *Pentaschistis* species and the sedge *Ficinia stolonifera* are prominent locally (Plots 1-7, Species group A - Table 1). A total of 19 different plant species comprising 42% grasses and 58% forbs were present. Canopy cover ranged between 75% and 100% with no signs of erosion observed.

These wetlands had an average biomass of 4302.5 kg DM/ha ranging between 2555.1 kg DM/ha and 5954.4 kg DM/ha (Figure 6). An average height of 51 cm was measured for the herbaceous layer.

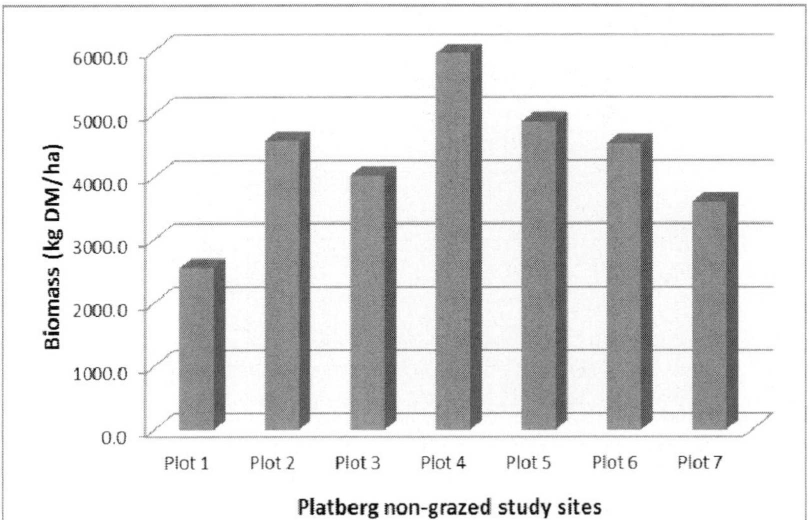

Fig. 6. Biomass of the Platberg peatlands in South Africa.

4.5 Biomass per grazing type

The following boxplot uses the medians and quartiles to compare the distributions of biomass (kg DM/ha) of the plots by grazing type (non-grazed; exclosure plots; grazed) (Figure 7).

As expected there is some overlap in the values of the grazed and exclosure plots, the latter only being fenced off for the past 12 months, while the non-grazed plots have some overlap with the exclosure plots. Thus although there appear to be differences between the grazed and exclosure plots the overlap results in there being a non-significant difference between them.

Although overlaps exist, differences between the different treatments are observed indicating that the non-grazed Platberg peatlands have the highest production followed by the exclosure plots that show variable increases within the short time (12 months) since being fenced off, compared to the open grazed plots.

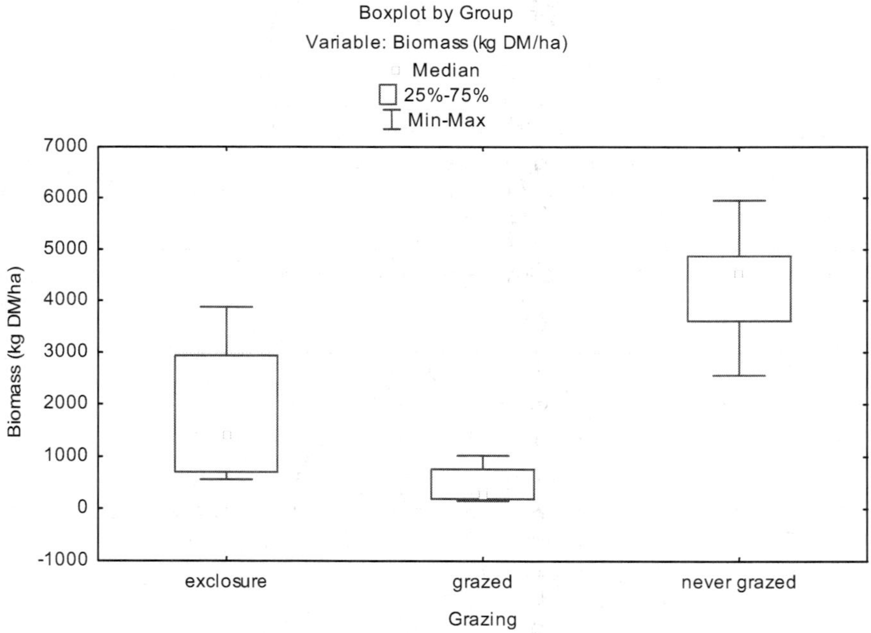

Fig. 7. Boxplots of the medians and quartiles comparing the biomass of the different grazing types.

The Kruskal-Wallis ANOVA by Ranks test was used to compare the Biomass (kg DM/ha) of the plots by grazing type. From the Kruskal-Wallis test [H (2, N=17)=12.071 p =.0024] it shows that there is a significant difference in the ranks of the plots categorised by grazing type (Table 2). Significant differences were found between the grazed, non-grazed and exclosure plots. By using the the Mann-Whitney U Test on the ranked data the significant differences were found to be between the grazed and non-grazed plots ($p<0.01$) and between exclosure and non-grazed plots ($p<0.05$). The Platberg peatlands with no grazing present therefore significantly differs in terms of production from both the grazed and exclosure plots with a much higher biomass as a result of the absence of grazers.

	Code	Valid - N	Sum of - Ranks
Exclosure plots	101	4	33
Grazed plots	102	6	24
Non- grazed plots	103	7	96

Table 2. Kruskal-Wallis ANOVA by Ranks; Biomass Independent (grouping) variable: Grazing Kruskal-Wallis test: H (2, N= 17)

5. Discussion

5.1 Species composition and richness

Two distinctive vegetation communities resulted from the data classification namely the Platberg and Lesotho peatlands respectively (Table 1). Each unit reflects the homogeneity of the communities in terms of plant species composition and dominance. In all wetlands the forbs (including geophytes, succulents and sedges) comprised the largest percentage of species with grasses the lowest. The uniqueness of each unit can be ascribed to various physical factors such as altitude, pH, moisture regime of the peat, and biotic impact.

Studies done by Sieben et al. (2010) revealed that typical wetland plants cannot cope well with the cold conditions during winter months. This opens several niches for forbs, especially species that are not dominant in lowland wetlands to become dominant in high altitude wetlands. The forb *Ranunculus meyeri* (Species group B - Table 1) is such a species, which occur across all altitudes but are dominant in high-altitude wetlands (Sieben et al., 2010). This species was found to be prominent in the Lesotho peatlands.

The peat deposits vary in depth from shallow 10 cm to over 2 m deep (Kuenene, 2008). The peatlands are covered by short mat-like peat forming plants. Dominant species in the Lesotho peatlands include the grass *Pentaschistis oreodoxa,* the sedge *Isolepis cernua,* the geophyte *Rhodohypoxis deflexa,* and the forbs *Haplocarpa nervosa, Cotuls radicalis* and *Ranunculus meyeri* (Table 1). Unlike the Lesotho peatlands, the platberg peatlands are mainly grassland dominated with the grasses *Andropogon appendiculautus* and *Pennisetum sphacelatum* prominent. The sedge *Ficinia stolonifera* and the grass *Helictotrichon longifolium* are locally dominant (Table 1).

Waterlogged conditions are a requirement for peat formation. Peat formation is a very slow process and is mainly formed from the root decay of vascular plants that grow on the peatlands. This process takes place in the absence of oxygen. Anoxia or reduction conditions between and below the roots, decomposes the dead material to form amorphous peat. According to the species composition of the wetlands at Khalong-la-Lithunya, Koti-Sepola, and Kao, sedges such as *Isolepis cernua* are together with forbs such as *Haplocarpha nervosa,* and *Ranunculus meyeri* the main contributors to peat formation. In contrast at Platberg the grasses *Andropogon appendiculatus, Pennisetum sphacelatum* and *Helictotrichon longifolium* are the main contributors to peat formation.

A characteristic of these peatlands is the frost-induced mounds or thufur (Figure 8). Backéus & Grab (1995) and Grab (1994) investigated these thufur and concluded that their morphological changes can be correlated to the changing environmental state of the peatlands. These thufur are present on hill slope seeps and valley head fens. The species composition of these thufur differs from the surrounding wetland vegetation and can be attributed to a relatively lower moisture content. Species that dominate the thufur at the study sites are *Helichrysum bellidiastrum, Athrixia fontana,* and *Sebaea marlothii.*

Fig. 8. Frost-induced mounds or thufur (mounds), with a different species composition (insert) present on the peatlands.

Another characteristic of these peatlands are the open water pools or tarns. The typical vegetation is floating-leaved and submergent aquatic vegetation. Dominant species present in these open water bodies are the grass *Colpodium drakensbergensis*, the forbs *Crassula natans, Lagarosiphon muscoides,* and *Limosella maior* were found at Khalong-la-Lithunya, Kao and Koti-Sephola.

The catchment slopes are characterised by shallow soils and a relatively high percentage of surface rock. The soil has a relatively low organic content due to the high runoff onto the peatlands that has resulted in degraded vegetation and coarse soil texture.

Extensive grazing depletes the root and canopy reserves of perennial grasses (Carpenter et al. 2001) which are then replaced by annual and other forb species. This will have a negative impact on the resilience of the system to maintain its ecosystem functions such as water storage, nutrient cycling and production. In the Lesotho peatlands the forbs comprise the largest percentage (86-87%) of the species composition, while in the Platberg peatlands there is a more even species composition (42% grasses and 58% forbs). Thus, it seems as though the long-term grazing of these systems has affected the species composition. The average species richness for the Platberg peatlands (11.4) is lower than the Lesotho peatlands (17.5). The higher richness in the Lesotho peatlands can be ascribed to greater habitat diversity as a result of various environmental factors resulting in the formation of peat and thurfur on the peatland surface as well as its short structure. In contrast the dense and tall grass tufts of the Platberg peatlands have a shading effect on the lower stratum of the herbaceous layer. This could prevent various forb species from establishing that could explain the lower species richness found. Brown & Bugg (2001) also found that established perennial grasses inhibited the growth of seeded forbs. There are also no thurfur present within the Platberg peatlands with its more homogeneous habitat.

5.2 Effects of grazing on biomass

It is generally believed that overgrazing is one of the most important factors causing vegetation degradation (Danckwerts & Tainton, 1996; David et al., 2000). Furthermore it could influence the hydrological status, stability and productivity of the ecosystem (Kirkman & Moore, 1995; O'Connor & Bredenkamp, 1997; Snyman & Fouché, 1991; Snyman 1997, 1998). Biomass is a good indicator of available forage and the risk of soil erosion (Schino et al., 2003). Biomass is dependent on the root reserves, nutrient and water content of the soil. Overgrazing can negatively affect the species composition and especially biomass production leading to a decrease in resilience of the ecosystem (Carpenter et al., 2001). This is evident from the results of this study where a significant difference was found between the production of the non-grazed Platberg peatlands compared to the grazed and exclosure plots in Lesotho indicating the negative effect of long-term grazing on these peatlands. The grasses in the recently erected exclosures (12 months ago) of the Lesotho peatlands show an increase in production compared to the grazed plots with some overlap with the non-grazed plots of Platberg (Figures 3, 4 & 6). Although not statistically significant at this stage the exclosure plots appear to have a higher production and it is expected that the difference may only become significant with a larger sample size and with continued monitoring over a longer term. This also correlates with the lower height levels of the plants in these plots compared to the exclosure plots. Structurally the Platberg peatlands are much taller with an average height of 51 cm compared to the short Lesotho peatlands with and average height ranging between 1.5 and 6 cm. The near pristine peatlands of Platberg had on average a much higher production and canopy cover with no signs of erosion. The exclosure plots of the Khalong-la-Lithunya peatland had similar biomass values than the Platberg peatlands indicating that if grazing is decreased production will increase.

The grasses in the exclosure plots showed a higher production (598.6 kg DM/ha - 8% of total biomass) than the grasses in the grazed plots (68.6 kg DM/ha - 2.6% of total biomass). It therefore appears that where grazing is absent that the grasses could become more prominent exerting a greater influence on the vegetation thereby causing a future change in species composition.

The Platberg peatlands have a higher amount of energy stored (due to the higher biomass) than the degraded Lesotho peatlands also meaning that it has a higher carbon storage ecosystem functioning.

5.3 Other effects of grazing

Grazing of these peatlands by domestic stock (wild horses, cattle, donkeys, goats and sheep) causes trampling of the vegetation that can result in footpaths (Figure 9) and deep hoof prints in the peatlands. These impacted areas can quicly become devoid of vegetation.

Fig. 9. Footpaths, devoid of vegetation, leading to the formation of erosion gullies in the peatlands as a result of trampling by domestic stock.

However, the heavy animals such as horses and cattle trample holes into the vegetation (Figure 10). These hoof prints form larger holes which eventually start to erode. The

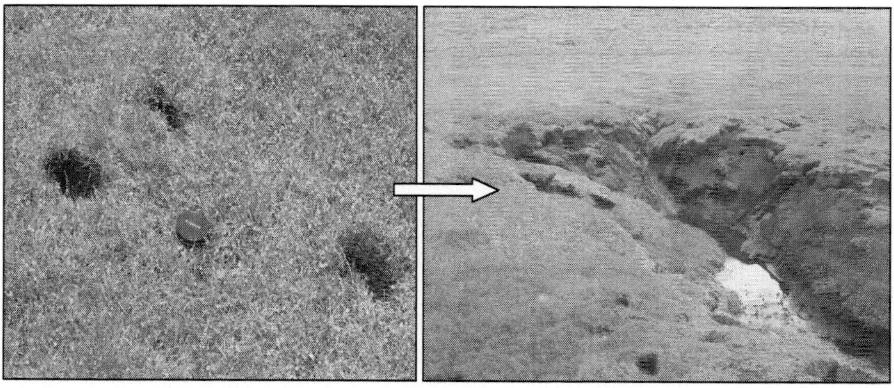

Fig. 10. Deep hoof imprints of animals into the peatlands causing erosion in the long term

vegetation also dies off due to the high urea-content of the urine of animals (Figure 11), while cow dung covers the vegetation for a significant period of time causing the exclusion of sunlight, which put the affected plants under stress leading to die off of the vegetation with other species establishing (Figure 12).

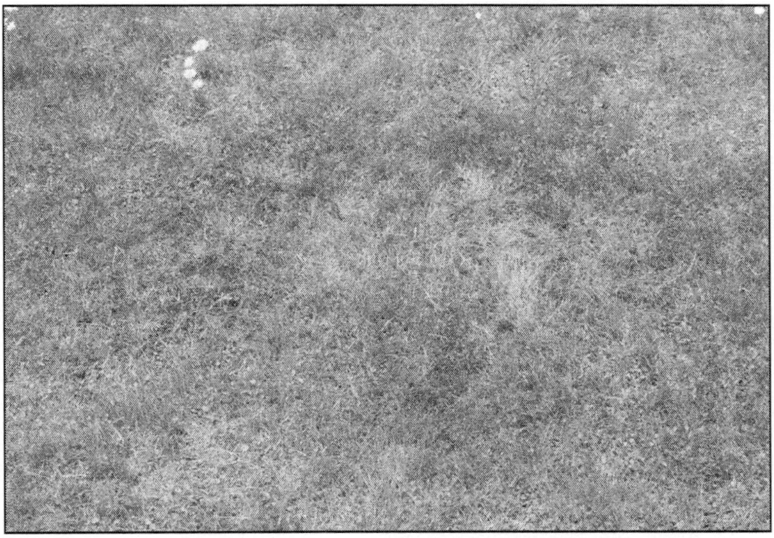

Fig. 11. Vegetation die-off caused by the high urea content of the urine of domestic stock grazing on the peatlands.

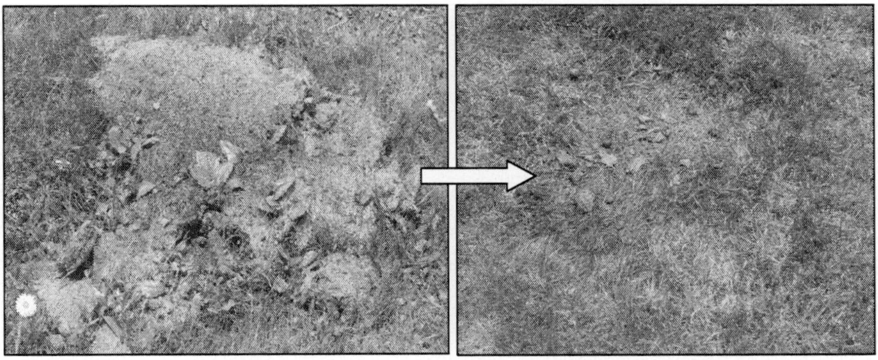

Fig. 12. Cow dung covers the vegetation for a significant period of time also leading to vegetation die-off with other species establishing.

Footpaths (Figure 9) are another major contributor to wetland degradation. On footpaths the animal hooves usually cause destruction of the vegetation cover which in turn results in the formation of large erosion gullies.

Due to the formation of erosion gullies, lateral drainage from the peat takes place at an alarming rate. Desiccation of the peat takes place along the gully which allows the peat to dry out and becomes eroded. A contributing factor is the trampling of the gully banks as animals drink from the streams. Their hooves loosen the dry peat which drains down to the streams (Figure 13).

Fig. 13. Large erosion gullies in the peatlands as a result of trampling and overgrazing.

The ice rat (*Otomys sloggettii*) (Figure 14) is a natural inhabitant of high-altitude vegetation types including peatlands (Schwaibold, 2005). On an undisturbed peatland these rats' dens, runways and tunnels are limited to the peatland fringes and drier areas. The desiccation of the peatlands results in the encroachment of these rodents onto the drier parts, especially around those areas near erosion gullies. Their tunnels contribute to the erosion and agrivating the desiccation the peat (Figure 13) as well as the alteration of the habitat conditions. This leads to various terrestrial plant species establishing thereby displacing the peatland vegetation.

Fig. 14. The ice rat (*Otomys sloggettii*) (insert) with its tunnels draining the peatlands and causing erosion.

6. Conclusion

Due to the low vegetation cover, shallow soils and steep slopes of the valley heads in Lesotho, there is a high water and sediment run off (Figure 15). The peatlands are the only sites where water is being retained and slowly released. These peatlands are very valuable socioecological ecosystems to the highlands of Lesotho because of the goods and services they present. The natural vegetation of these peatlands sustains extensive grazing by domestic livestock. Destruction of these peatlands will have serious implications for the ecosystems and biodiversity of the region. The high runoff sometimes results in the deposition of sediment onto the peatlands. In these instances a sedimentation fan forms on top of the peatland. Subsequently the covered wetland vegetation dies off.

Fig. 15. Sediment runoff from the surrounding catchment areas onto the peatlands.

All the Lesotho peatlands as well as the surrounding upland slopes are heavily affected by grazing and trampling. Species composition of the slopes has been affected significantly. These slopes are all dominated by unpalatable dwarf shrubs such as *Helichrysum trilineatum, Eumorphia sericea, Chrysocoma ciliata, Euryops decurrens* and the absence of grass species is notable.

The degradation of these peatlands cannot really be attributed to the grazing per se of the wetland species. It has been noted that these mat forming plants especially *Isolepis cernua* has a remarkable tolerance of heavy grazing and the peatlands still maintain high species richness. A very dense cover remains despite the heavy grazing pressure.

The canopy cover of the herbaceous plants of both the Lesotho and near pristine Platberg peatland systems was generally high though the latter had a taller and more diverse structure. The more pristine Platberg peatlands had a lower species richness with a larger grass component than the Lesotho peatland areas. Furthermore the Platberg peatlands had a markedly higher biomass production than the Lesotho peatlands. However the exclosure plots showed a noticible increase in production compared to the grazed plots indicating the resilience of these systems if properly managed.

The resilience of a system is dependant on its ability to sustain nutrient cycles and the storage of water. This ability is threatened when degradation of the habitat occurs via soil loss and structural change of the vegetation (Carpenter et al., 2001). Based on the results from this study it is clear that structural changes and soil loss have taken place in the Lesotho high-altitude peatlands (Figure 16) accompanied by a lower production. No improvements in the condition of the wetland and associated upland sites will be possible without a reduction in the grazing pressure in these areas. Degraded vegetation cover already leads to increased overland flow visible as gravel on the soil surface in the hill slopes and gully erosion in the wetlands. Reduced organic matter content in all the soils is expected due to drier soil conditions and short vegetation. Land use change will impact on soil hydrology and that impact will be reflected in the wetland hydrology.

Fig. 16. Large scale erosion as a result of heavy grazing resulting in the degradation of the habitat and a loss of ecosystem functions.

In view of the uniqueness of these peatlands and the high endemicity of the Lesotho Highland Basalt Grassland vegetation it is urgent to identify and proclaim more wildlife conservation areas within the Lesotho highlands. As mentioned above a large transfrontier park namely the Maloti-Transfrontier Park is planned to protect the high-altitude vegetation of Lesotho and South Africa. Mucina & Rutherford (2006) state that this proposed park would increase the conservation status of Lesotho Highland Basalt Grassland. However, the authors of this chapter have their doubts about the effectiveness of such a Transfrontier Park to protect sensitive ecosystems such as peatlands. The reason being that with the proclamation of such as Transfrontier Park the landuse within the transfrontier park would not change significantly.

The authors are therefore proposing smaller conservation areas within the larger Transfrontier Park, which would exclude domestic animals, specifically to protect the peatlands and their catchments. These conservation areas would not only properly protect peatlands, but would also protect the indigenous wildlife populations of grey rheebuck, rock hyrax, and several other species, which are on the brink of extinction in Lesotho. Long term monitoring of the exclosure and grazed plots should be continued on these sensitive ecosystems to provide a basis for decision making.

7. References

Backéus, I. & Grab, S. 1995. Mires in Lesotho. Gunneria 70:243-250. ISSN.0332-8554

Brown, C. S. & Bugg, R. L. (2001), Effects of Established Perennial Grasses on Introduction of Native Forbs in California. *Restoration Ecology*, Vol.9, pp. 38–48, doi: 10.1046/j.1526-100x.2001.009001038.x

Brown, L.R., Marais, H., Henzi, S.P. and Barrett, L. (2005). Vegetation classification as the basis for baboon management in the Bourke's Luck Section of the Blyde Canyon Nature Reserve, Mpumalanga. *Koedoe*, Vol.48, Number2, pp. 71–92, ISSN 0075-6485.

Carpenter, S., Walker, B., Anderies, J.M., & Abel, N. (2001). From Metaphor to Measurement:Resilience of What to What? *Ecosystems* Vol.4, pp. 765–781. DOI: 10.1007/s10021-001-0045-9.

Cleaver G. (2004). Environmental impacts of large-scale groundwater abstraction on ecosystems of the Kammanassie Mountain. M.Tech Dissertaion, University of South Africa, Pretoria.

Collins, N.B. (2005). *Wetlands: The basics and some more.* Free State Department of Tourism, Environmental and Econimic Affairs. ISBN 0-86886-708-X, Bloemfontein.

Cowan,G I (ed.). (1995). *Wetlands of South Africa.* Department of Environmental Affairs and Tourism, ISBN 0-62116817-3, Pretoria.

Cronk, J.K. & Fennessey, M.S. (2001). *Wetland plants: biology and ecology.* Lewis Publishers, ISBN 978-0-415-55426-8, Boca Raton.

Danckwerts, J.E. & Tainton, N.M. (1996). Range management. Optimization of forage production and quality. *Bulletin of the Grassland Society of Southern Africa,* Vol.7, pp. 36-42, ISSN 101-66122.

David, W., Ngairorue, B.T., Apollus, A. & Tjieveze, H. (2000). Perceptions and realities of land degradation in arid Otjimbingwe, Nambia. *Journal of Arid Environments,* Vol.45, pp. 337-356, ISSN 0140-1963.

De Leeuw, P.N., Grandin, B.E. & Solomon, B. (1991). Introduction to the Kenyan rangelands and Kajiado District. In: *Maasai herding: An analysis of the livestock production system of Maasai pastoralists in eastern Kajiado District, Kenya.* Solomon, B., De Leeuw, P.N., Grandin, B.E. & Neate, P.J.H. (Eds.) International Livestock Center for Africa (ILCA), ISBN 92-9053-176-2, Ethiopia.

Du Preez, P.J. (1991). A syntaxonomical and synecological study of the vegetation of the southern and eastern Orange Free State and related areas with special reference to Korannaberg. Ph.D. thesis, Dept of Botany, UFS. Bloemfontein.

Grab, S.W. (1994). Thufur in the Mohlesi valley, Lesotho, southern Africa. *Permafrost and periglacial processes,* Vol. 5, pp. 111-118, ISSN.1099-1530.

Grobbelaar, J.U. & Stegman P. (1987). Limnological characteristics, water quality and conservation measures of a high altitude bog and rivers in the Maluti mountains, Lesotho. *Water South Africa.* Vol. 13, pp. 151-158, ISSN 0378-4738.

Herbst, S.N. & Roberts, B.R. (1974a). Quantitative ecological relationships in the Alpine grassland of Lesotho. *Proceedings of the Grassland Society of South Africa,* Vol. 9, pp. 61-66, ISSN.1551-5028.

Herbst, S.N. & Roberts, B.R. (1974b). The Alpine vegetation of the Lesotho Drakensberg: a study in quantitative floristics at Oxbow. *Journal of South African Botany,* Vol. 40, pp. 257-267, ISSN 0254-6299.

Hennekens, S.M. & Schaminée, J.H.J. 2001. TURBOVEG, a comprehensive database management system for vegetation data. Journal of Vegetation Science 12: 589-591 ISSN.1100-9233

Hill, M.O. (1979). *TWINSPAN: A FORTRAN program for arranging multivariate data in an ordered two-way table by classification of the individuals and attributes.* Cornell University, Ithaca.

Jacot Guillarmod, A. (1962). Bogs and sponges of Basutoland mountains. *Journal of South African Botany,* Vol. 58, pp. 179-182, ISSN 0254-6299.

Jacot Guillarmod, A. (1963). Further observations on bogs of Basutoland mountains. *Journal of South African Science,* Vol. 59, pp. 115-118, ISSN 0038-2353.

Jacot Guillarmod, A. (1972). Bogs and sponges of the Orange River catchment within Lesotho. *Civ. Engr S. Afr.* Vol. 14, pp. 84-85, ISSN. 0305-1 315 .

Kirkman, K.P. & Moore, A. (1995). Towards improved grazing management recommendations for sourveld. *African Journal of Range and Forage Science.* Vol. 12, No. 3, pp. 135-144, ISSN 1022-0119.

Kotze DC, (1999). A system for supporting wetland management decisions. Ph.D. thesis. School of Applied Environmental Sciences, University of Natal, Pietermaritzburg.

Kotze, D.C. 2010. *WET-Sustainable-USE. A system for assessing sustainability of wetland use.* WRC report. WRC Report no TT 438/09

Kotze, D.C., Marnewick,G., Batchelor, A., Lindley, D. & Collins, N.B. 2008. *WET-EcoServices. A technique for rapidly assessing ecosystem services supplied by wetlands.* WRC report TT339/8, Pretoria.

Kuenene, B.T. (2008). Soil morphology as signature of soil hydrology in selected catchments. M.Sc. dissertation. University of Free State, Bloemfontein, South Africa.

Letsela, T., Balkwill, K. & Witkowksi, E.T.F. (2002). Conservation and sustainable livelihoods in Bokong and Tsehlanyane in Lesotho. In: *Contested resources: challenges to governance of natural resources in southern Africa,* T.A. Benjamin, B. Cousins & L. Thomson (Eds.), 192-200. University of the Western Cape, Programs for land and agrarian studies, ISBN 1-86808-537-6, South Africa.

Lubomír, T. (2002). JUICE, software for vegetation classification. *Journal of Vegetation Science* Vol.13, pp. 451-453, ISSN 1100-9233

Malan, P.W. (1998). Vegetation ecology of the Southern Free State. Unpublished PhD thesis, University of the Orange Free State, Bloemfontein.

Mitch, W.J & Gosselink, J.G. (2000). *Wetlands,*(3rd edition). Van Nostrand Reinhold, ISBN 0-471-29232-x, New York.

Mpiti-Shakhane, L.M., Kanyama-Phiri, G.Y., Odenya, W.O. & Banda, J.W. (2002). Sheep response to different stocking rates and seasonal variation under rangeland conditions of Lesotho. *African Journal of Range and Forage Science,* Vol.19, pp. 71-75, ISSN 1022-0119.

Mucina L. & Rutherford M.C. (2006). *The vegetation of South Africa, Lesotho and Swaziland.* South African National Biodiversity Institute, ISBN 978-1-919976-21-1, South Africa.

Mueller-Dombois, D. and Ellenberg, H. (1974). *Aims and methods of vegetation ecology.* J. Wiley & Sons, ISBN 04-1622907, New York.

O`Connor, T.G. & Bredenkamp, G.J. (1997). Grassland. In: *Vegetation of Southern Africa,* Cowling, R.M., Richardson, D.M. & Pierce, S.M. (Eds.), 215-257, Cambridge University Press, ISBN 0-521-57142-1, Cambridge.

Pieper, R.D. (1978). *Measurement techniques for herbaceous and shrubby vegetation.* Mimeograph Series, ISBN 0817639659, New Mexico State University, Las Cruces.

Schino G, Borfecchia F, De Cecco L, Camilla Dibari C, Iannetta M, Sandro Martini S and Franco Pedrott F. (2003). Satellite estimate of grass biomass in a mountainous range in central Italy. *Agroforestry Systems* 59 (2): 157-162, DOI: 10.1023/A:1026308928874

Schmitz, G. & Rooyani, F. (1987). *Lesotho geology, geomorphology, soils.* Morija Printing Works, ISBN 354029144X, The National University of Lesotho, Lesotho.

Schwaibold, U.H. (2005). Foraging biology and habitat use of the southern African ice rat *Otomys sloggettii robertsii.* Unpublished Ph.D. dissertation. University of the Witwatersrand, Johannesburg

Sieben, E.J.J. & Morris, E.D., C.D., Kotze, D.C & Muasya, A.M. (2010). Changes in plant form and function across altitudinal and wetness gradients in the wetlands of the Maloti-Drakensberg, South Africa. *Plant Ecology,* Vol.207, pp.107-139, ISSN 1573-5052.

Snyman, H.A. (1997). The influence of range condition on the hydrological characteristics in semi-arid rangelands. *Proceedings of the XVIII International Grassland congress, Canada,* Vol.2, No.23, pp. 1-2.

Snyman, H.A. (1998). Dynamics and sustainable utilisation of rangeland ecosystems in arid and semi-arid climates of Southern Africa. *Journal of Arid Environments,* Vol.39, pp. 645-666, ISSN 0140-1963.

Snyman, H.A. & Fouche, H.J. (1991). Production and water-use efficiency of semiarid grasslands of South Africa as affected by veld condition and rainfall. *Water South Africa,* Vol17, pp.263-268, ISSN 0378-4738.

Snyman, H.A. & Fouche, H.J. (1993). Estimating seasonal herbage production of a semi-arid grassland based on veld condition, rainfall and evapotranspiration. *African Journal of Range and Forage Science,* Vol.10, No.1, pp. 21-24, ISSN 1022-0119.

Van Zinderen Bakker, E.M, (1955). A preliminary survey of peat bogs of the alpine belt of northern Basutoland. *Acta Geographica.* Vol.14, pp. 413-422.

Van Zinderen Bakker, E.M. & Werger, M.J.A. (1974). Environmental, vegetation and phytogeography of the high-altitude bogs of Lesotho. *Vegetatio,* Vol.29, pp. 37-49,

Welman C, Kruger F & Mitchell B. 2007. *Research Methodology* (3rd edition). Oxford University Press Southern Africa, ISBN 978-0-19-578901-0, Cape Town.

Westhoff, V. and Van der Maarel, E. 1980. The Braun-Blanquet approach, In: *Classification of Plant Communities*. R.H. Whitaker, (Ed.), 287-378, Kluwer Academic Publisher, ISBN 906-193566-0, The Hague.

13

Waterbird Biodiversity and Conservation Threats in Coastal Ecuador and the Galapagos Islands

Juan José Alava[1,2] and Ben Haase[2,3]

[1]*School of Resource & Environmental Management, Faculty of Environment, Simon Fraser University, 8888 University Drive, Burnaby, British Columbia,*

[2]*Fundación Ecuatoriana para el Estudio de Mamíferos Marinos (FEMM), Guayaquil,*

[3]*Museo de Ballenas, Av. Gral. Enríquez Gallo s/n, calles 47 y 50, Salinas, Santa Elena,*

[1]*Canada*

[2,3]*Ecuador*

1. Introduction

Within the context of the Convention on Biological Diversity (CBD) of the United Nations Environment Programme, the conservation of biodiversity is one of the major goals devoted to minimize and mitigate significantly the existing rate of biodiversity loss at the global, regional and national scales. At this level, birds are exemplary sentinels and bioindicators for the conservation and monitoring of biodiversity and ecosystem health. While 1226 bird species are considered globally threatened with extinction due to small and declining populations or ranges (BirdLife International, 2008), a substantial number of seabird populations are declining and threatened with extinction at the global level because of several conservation threats both on land and at sea, including fishery interactions, predation by invasive species and habitat loss due to coastal development (BirdLife International, 2010). The Pacific is an important area for threatened seabirds, where their ranges span multiple Exclusive Economic Zones (EEZs) as well as many areas beyond National Jurisdictions (ABNJs). Although seabirds represent only 3% of the total number of bird species in the world, about 28% (over 130 species) are listed as threatened on the IUCN red list for birds, under which 10% of seabirds are Critically Endangered (BirdLife International , 2010).

Under the CBD, Ecuador is a signatory country (1992) and ratified the Convention in 1993 with the aim to pursue and establish conservation efforts and action plans to preserve the national biodiversity, including birds. Despite of being one of the smallest countries of the world (i.e., 0.19% of the terrestrial surface of the Earth) with 256 370 km^2 and a human population close to 14 millions inhabitants, Ecuador is one of 17 world's megadiverse countries due to its rich biodiversity and high degree of bird endemism (Mittermeier et al., 1997; Stattersfield et al., 1998). Of the 151 wetlands identified as key habitats for Neotropical waterbirds in Ecuador, 40% are present on the continental coast (i.e., 59 wetlands), while 14 exist in the Galapagos (Santander et al., 2006). There, a total of 1640 species of birds are geographically distributed into four well defined geographical zones: the coast (coastal

Ecuador), the Andean region (highlands), the Amazon jungle (eastern region) and the UNESCO Heritage site, the Galapagos Islands. Of the total number of bird species recorded in Ecuador, 13.6 % or 223 species (Granizo et al., 2002; Santander et al., 2006a) are represented by aquatic and seabirds dwelling diverse habitats including oceanic-offshore environments, nearshore habitats, intertidal zones, islands, coastal lagoon, mangroves, shrimp farms, salt ponds and continental freshwater systems.

However, several environmental stressors and human activities threaten the population and survival of waterbirds in both continental Ecuador and the Galapagos. While habitat fragmentation and deforestation, urban sprawl, agriculture, current use pesticides, marine pollution and wetland degradation are the major impacts identified on the Ecuadorian coast, invasive species and pathogens, bycatch (long-line/gillnets) and the regional climate variability are the major threats in the Galapagos Islands. While these conservation threats have been recognized to some degree, most of their impacts have been scarcely identified and assessed. This is critical under the paradigm of conservation biology and preservation of wildlife, depending on science sound data and baseline information intended to support environmental management plans and conservation efforts.

Therefore, the aim of this chapter is to contribute with a review focused on the conservation status of the biodiversity of aquatic birds and an overall environmental impact assessment of current and looming threats in Ecuador, with special emphasis in the Galapagos Islands. To accomplish this goal, a revision of waterbird species and abundance of seabirds, shorebirds and aquatic birds of freshwater systems will be conducted. The chapter will also include a section describing major features of the natural history and conservation status of priority species, including threatened and endemic species (e.g., Waved Albatross, Galapagos Petrel, Galapagos Penguin, Flightless Cormorant, Horned Screamer, Brown-wood Rail, among others), as well as key species for the functioning and health of aquatic ecosystems. This section will be followed by the identification and assessment of current anthropogenic impacts and emerging conservation threats jeopardizing their survival in critical habitats and protected areas. Finally, the chapter will conclude with a section portraying mitigation strategies, recommendations for waterbird conservation and environmental stewardship.

2. Regions and study areas

Under the National Protected Areas System (SNAP, created in 1976), there are currently 40 protected areas, of which 37 are continental (18% of Ecuador); one is insular, The Galapagos National Park (693,700 ha); and, two are exclusively marine, including the Galapagos Marine Reserve (GMR) with 14,110,000 ha. In addition, Ecuador possesses 107 areas designated as Important Bird Areas (IBAs) and 12 Ramsar sites (i.e., wetlands of international importance) for the conservation of biodiversity and protection of threatened species, including critically endangered species (Santander et al., 2009). A review of all sites and areas for waterbirds conservation in Ecuador are out of the scope of this chapter; therefore, for the purpose of this review, only regions and areas in which the authors have been directly involved in censuses, banding, field work and research in southern Ecuador and the Galapagos Islands, described as follows.

2.1 Guayaquil Gulf Estuary Basin

The Gulf of Guayaquil is the largest estuary of the Pacific coast in South America with surface area of approximately 13,701 km and 230 km of length (Fig. 1). The entrance of the

Gulf is located at 3ºS of Ecuador and it goes 204 km from north to south, and enters a distance of 130 km. This estuary is located on the edge of the Guayas River and the city of Guayaquil. It is part of the Guayas Ecosystem, a large tropical area covering the Gulf of Guayaquil, the Guayas River Basin, the Guayas River Estuary and the city of Guayaquil. This ecosystem is home to 45% of the national human population, and articulates 12 provinces and 88 municipalities. Several watersheds drainage into the Guayas River Basin, including a vast geographical area with a hydrological system of 34 000 km^2, which captures the effluents coming from the Daule, Vinces and Babahoyo rivers. The Guayas River Estuary begins on at the Puná Island, across the Jambelí and Del Morro channels, and extends as far as the influence of the tide and salinity, which ends about 100 km within the continent at the confluence of the Daule and Babahoyo rivers. The depth of The Guayas River Estuary ranges from 20 to 180 m.

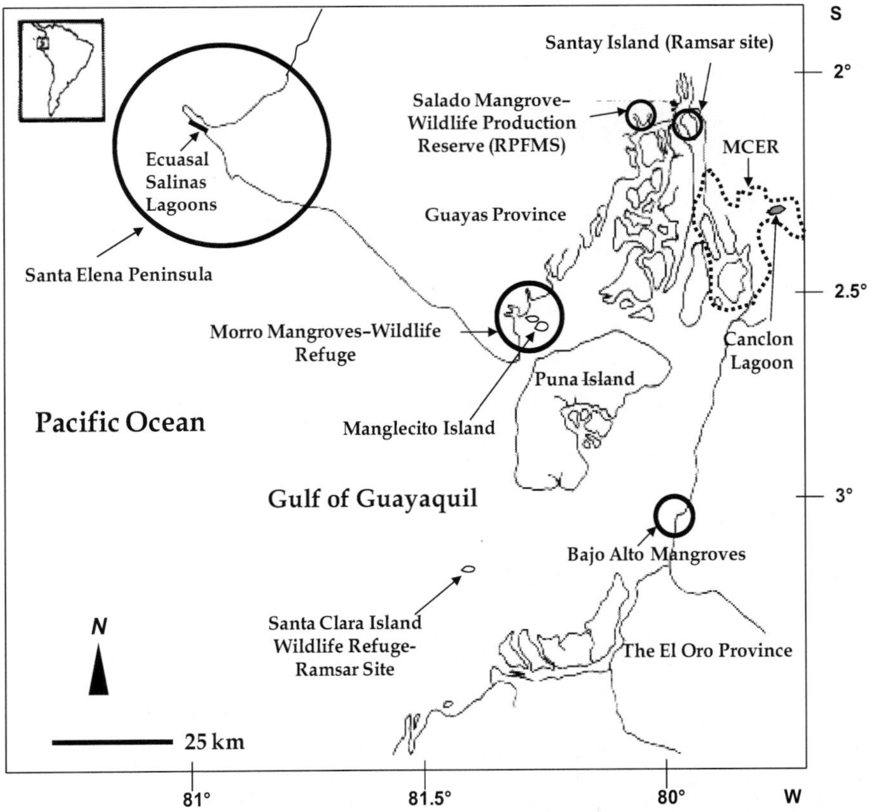

Fig. 1. Study areas on the southern coast of Ecuador, including mangrove-estuarine areas (as indicated by black rings) in the Guayaquil Gulf and Guayas River Basin and several other coastal sites, including the Ecuasal lagoons in the Santa Elena Peninsula (Santa Elena Province), Canclon Lagoon, located in the Manglares Churute Ecological Reserve (MCER); and Santa Clara Island (The El Oro Provi nce).

The Guayaquil Gulf Estuary contains approximately 81% or 121 hectares of the total area of Ecuadorian mangroves, encompassing about 148,000 hectares (CLIRSEN 2007)., and is suitable habitat for vast populations of different species of herons and egrets, shorebirds and frigatebirds, (Alava et al., 2005, Carvajal et al., 2005, Carvajal & Alava, 2007).

2.2 Churute Mangrove Ecological Reserve and El Canclon Lagoon

The Manglares Churute Ecological Reserve (MCER) (02°30'S, 79°42'W; Fig. 1) is located in the Guayaquil Gulf Estuary Basin (INEFAN & Fundación Natura, 1997; Briones et al., 2001) of the Guayas province. El Canclon Lagoon is one of the 32 wetlands identified in the Ecuadorian coastal region and was declared as a Ramsar site in 1996. The surrounding floodplain has an area of 800 ha, and is situated in the northern part of in the northern part of the MCER. Moreover, the El Canclon Lagoon is part of one of the most important endemic bird areas (EBA) in the Neotropics for biodiversity conservation, the Tumbesian region, which extends from southwest Ecuador to northwest Peru (Best et al., 1995; Stattersfield et al., 1998). Its lentic bodies of water and surrounding wetland ecosystem constitute a unique riparian habitat and refuge for Neotropical migrants as well as resident breeding birds. Among the tropical plant species representatives of the area are floating and emergent water plants such as duckweed (*Lemna minima*), water hyacinth (*Eichornia crassipes*), sleeping-beauty waterlily (*Nymphaea blanda*), water lettuce (*Pistia stratoites*), flat sedge (*Cyperus odoratus*) and cattails (*Thypa latifolia* and *T. dominguensis*) (Sierra et al., 1999). In addition, riparian vegetation, lowland bushes, and tropical dry and humid forest remnants border the southeast, east and northeast edges of the lagoon, which lies adjacent to the mountainous watershed of the El Mate, Perequete Chico, Perequete Grande, Cimalon, Pancho Diablo and Masvale hills. The area's unique habitat types, as well as its high level of bird endemism (~ 40%) (Alava et al., 2002), warrant the publication of all ornithological observations of the area, which may prove useful to future conservation efforts. The aquatic species recorded in and around El Canclon Lagoon include mainly waterfowl such as Fulvous Whistling Duck (*Dendrocygna bicolor*) and Black-bellied Whistling Duck (*D. autumnalis*), Neotropical Cormorant (*Phalacrocorax brasilianus*), Purple Gallinule (*Porphyrula martinica*), Wattled Jacanas(*Jacana jacana*), and herons such as Great Egret (*Ardea alba*) and Cattle Egret (*Bubulcus ibis*) egrets (Briones *et al.* 2001; Alava et al., 2002; Alava et al., 2009). A representative threatened species residing year-round in this wetland is the rare Horned Screamer (*Anhima cornuta*), a bird locally known "canclon" (Alava et al., 2007).

2.3 Reserva de Producción de Fauna Manglares el Salado-RPFMS (El Salado Mangrove–Wildlife Production Reserve).

The RPFMS is a wildlife refuge confining 9,748 hectares of remnant mangroves in southeast Guayaquil City (2°10'S, 79°56'W), Guayas Province. About 3% of the total area of this reserve is occupied by shrimp farms (<200 ha), of which 100 ha have been abandoned. Three thermoelectric facilities, a water pump station and sewers also occur within its boundaries. Small-scale clearing of mangrove trees (i.e. timber extraction for coal production) occurs at specific locations, while pollution from solid wastes and water contamination are environmental issues of concern being close to human urbanization and industrial parks (Carvajal et al., 2006). Artisanal fisheries are subsistence activities practiced by communities residing within the RPFMS. Although the RPFMS has a moderate level of fragmentation in some locations, and severe in others, some pristine mangroves in isolated patches are still well conserved and act as biological corridors and provide suitable habitat for local waterbird populations (e.g., heron nesting sites) and mammals. Mangroves around this area, including the Puerto Hondo Mangroves Sanctuary, are roosting habitat for about 214 individuals of the Red-lored Amazon (*A. autumnalis lilacina*) (Berg & Angel, 2006). Five of the six species of mangrove trees recorded for Ecuador are found in this reserve: red mangrove (*Rhizophora mangle*), hybrid red mangrove (*Rhizhopora harrisonii*), black mangrove (*Avicennia germinans*), white mangrove (*Laguncularia racemosa*) and *jeli* or button mangrove (*Conocarpus erectus*).

2.4 Isla Santay (Santay Island - national recreation area)

Isla Santay is a Ramsar site (02°13'S, 79°51'W) with 4,700 ha located in the inner estuary of the Guayaquil Gulf (Guayas River Basin) and also close to the urban periphery of Guayaquil City (Delgado et al., 2000, Jaramillo et al., 2002, Santander & Muñoz, 2004). Isla Santay includes intertidal mud, sand or salt flats and intertidal forested wetlands, encompassing mangrove swamps, and tidal freshwater swamp forests. The total area of mangroves on is 2,224 hectares. The mangroves includes: *R. mangle, R. harrisonii, A. germinans, L. racemosa, C. erectus,* mangrove golden leatherfern (*Acrostichum aureum*), *Salicornia fruticosa* and *Zizhypus thyrsiflora.* During the rainy season (i.e. late December to May) 60% of the island is flooded. The island is also inhabited by humans and several anthropogenic activities such as livestock farming (goats, pigs, and chickens), coal production from burning of trees and rice cultivation (Delgado et al., 2000; Jaramillo et al., 2002). Even though it has a considerable degree of human perturbation, Isla Santay is still a refuge for numerous species of terrestrial birds and waterbirds.

2.5 El Morro Mangroves Wildlife Refuge: Isla Manglecito (Manglecito Island)

The El Morro Mangroves Wildlife Refuge (2°39'S, 80°11'W) is a recent protected area encompassing 10,130 ha of mangrove forest, intertidal mudflats, estuarine channels and creeks in the El Morro Channel (Guayas Province), Gulf of Guayaquil. Manglecito Island, also known as the Frigatebirds' Island, is part of this Reserve and is habitat for a large breeding colony and year-round population of 2,000-6,000 Magnificent Frigatebirds (*Fregata magnificens*), which a large seabird population on the Ecuadorian coast Five species of mangrove trees occur in this island too, including red, hybrid red, black, white and button mangroves.

2.6 Santa Clara Island

Santa Clara Island (3°10′ S, 80° 26′ W) is a Ramsar site and a Wildlife Refuge in the Gulf of Guayaquil and influenced by the highly productive Peruvian upwelling system, delivering nutrient enriched water to the marine ecosystem of the island (Nixon and Thomas, 2001). With 5 hectares and 50-60 meters over the sea level, this refuge harbors the largest colonies of seabirds (15,000-20,000 seabirds), including one of the largest magnificent frigatebird population (5000-8000 frigatebirds) and about 6000−14000 blue-footed boobies and 3000−4000 brown pelicans, in coastal Ecuador (Valle, 1997; Valle, 1998; Suarez & Calle, 2005). The island possesses arid vegetation represented by *Capparis* and *Cordia* shrubs, as well as *Armathocereus* cactuses. Due to the large number of sea birds breeding on the island, the harvesting of "*guano*" has been recorded as an illegal activity, with a total extraction of 120 m^3 equivalent to 2000 sacs per season (Suarez y Calle, 2005). The Santa Clara Island is considered as one of the most important areas used as nesting and resting sites by blue-footed boobies in Ecuador's mainland coast (Miranda et al., 2010).

2.7 Bajo Alto Mangroves

The Bajo Alto mangroves is located in the El Oro Province (3°S, 79.5°W), and encompasses 555 ha, from which 357 ha are mangrove forests and 199 ha is represented by the adjacent estuarine water (Calle, 2003). Bajo Alto mangroves are not yet under any protected area or reserve status even though all the mangroves are protected by the Ecuadorian Mangrove Law. Mangrove remnants are still semipristine, and suitable for wildlife. The El Oro Province is typically characterized by a vast amount of banana plantations and shrimp farms. For instance, shrimp farm ponds are encroaching nearby mangrove habitat where mangrove hawks were sighted. The predominant mangrove tree species in this area is the red mangrove, followed by black and white mangroves (Calle, 2003).

2.8 Santa Elena Peninsula

The Santa Elena Peninsula is located in the southern coast of Ecuador (Fig. 1). This Peninsula is under the influence of the cold Humboldt Current, and, throughout the year, the environmental temperature fluctuates between 22°C in summer (between June and November) and a maximum temperature of 33°C in winter (between December and May). From June, when the southern winter sets in, the temperature of the sea surface decreases. Masses of relatively cold sea air enter the coastal strip, resulting in drizzly weather with very weak rain values, so this is an area of dry weather with a maximum average of 250 mm a year (WHSRN, 2009a). On the south east side of the Santa Elena Peninsula (southwestern tip of the mainland coast) there is a series of salt lakes owned by the ECUASAL Company and categorized as an IBA (see next subsection). In September 2007, a narrow coastal strip of coastal habitat and surrounding maritime stretch of ocean was declared marine protected area. It is located at the westernmost tip of the province of Santa Elena, and includes the beaches and several square miles of water around the Chocolatera. At present, a sea-watching project is being carried out. In order to know about the presence, migration and other behavior of the bird species, regular countings are done from the Chocolatera. The Santa Elena Peninsula is considered an important area and stopover for migratory birds (Haase, 1991; Haase, 2010). In 2010, the results of a total of 250 hours of sea-watching revealed that many poorly known species happen to occur more frequently near Salinas than previously thought. They include threatened species like Waved Albatross (*Phoebastria irrorata*), Galapagos Petrel (*Pterodroma phaeopygia*), Parkinson´s Petrel (*Procellaria parkinsoni*), Pink-footed Shearwater (*Puffinus creatopus*). Other species of concern like Elegant Tern (*Thalasseus elegans*) and Peruvian tern (*Sternula lorata*) are also observed at regular times. The Arctic Tern (*Sterna paradisaea*) and the Sabine's Gull (*Xema sabini*), as well as the six species of boobies occurring in coastal Ecuador (*Sula* spp.) have been also recorded from this area (Haase, 2010). This observation site directly within the Marine Protected area is located near a light house, at 6 meter above the sea level. This panoramic site offers a view of 280 ° of ocean, and is an extraordinary strategic location to watch sea- and coastal bird migration throughout the year. Within one year, 57 species of sea and coastal birds were recorded (Haase, 2010).

2.9 Ecuasal-Salinas Lagoons

The Ecuasal lagoons are located in the southwestern coastal region of Ecuador (Santa Elena Peninsula), including the Salinas (500 ha) and Pacoa (1000 ha) lagoons. The Ecuasal lagoons are man-made lakes situated less than 200 m from the coastline (Salinas: 02° 13´S 80°58´O; Pacoa: 02° 05´S 80°44´O), which were dug out in order to extract sea salt for commercialization by the Ecuadorian Salt and Chemical Products Company (ECUASAL) (WHSRN, 2009a). The Salinas lagoons are located 1 km southeast of the town of the same name and the Pacoa lagoons are 10 Km from Santa Elena (15 Km from Salinas) between San Pablo and Monteverde). The Salinas lagoons face the sea (towards the west) and are surrounded by different types of urban and industrial infrastructure, while the Pacoa lagoons are still mostly surrounded by an arid semi-desert area. Due to the dry, cold weather and scarce precipitation, this is one of the few areas in the country that facilitates salt production at industrial level. The lagoons are a suitable habitat for aquatic birds and currently home to thousands of resident and migratory water birds throughout the year (WHSRN, 2009a,b), and this already prompted its designation as Important Bird Conservation Areas (IBAs) (Santander et al., 2009). The lakes are an important stopover place for migratory birds, particularly during the months of August and September when they gather in large numbers. Systematic census work has been carried out since 1988, and a total of 95 species of aquatic birds have been registered, including 41 species of shorebirds, 9 gull species and 10 tern species (WHSRN, 2009a ; Haase, 2011a). An average of 4000 birds is counted per month. Each year the maximum number of birds is counted in September, when more than 30,000 individuals of the Wilson's phalarope (*Phalaropus tricolor*) have been counted in the area, representing more than 2% of the total population (Haase, 2011a). Additionally, the site is home to over 20,000 waterbirds per year. The Grey-hooded Gull (*Croicocephalus cirrocephalus*) breeds from February until October with over 700 pairs, and 400 pairs of Gull-billed Tern (*Gelochelidon nilotica*), in colonies spread out over the area (Haase, 1991; WHSRN, 2009b; Haase, 2011a). The Kelp Gull (*Larus dominicanus*) and South American Tern (*Sterna hirundinacea*) has also been found to breed and nest in this area for first time in Ecuador (Haase, 1996; WHSRN, 2009b). Other birds, among them some near threatened (NT) species, are regular visitors or permanently resident waterbirds of the site such as Peruvian Tern, (*Sternula lorata*), Chilean Flamingo (*Phoenicopterus chilensis*) and Elegant tern (*Thalasseus elegans*) (WHSRN, 2009b; Haase, 2011a).

2.10 Galapagos islands

The Galapagos comprises an Archipelago with 13 major volcanic islands, situated approximately 1000 km from the Ecuadorian coast, between 01°40´N-01°25´S and 89°15´W-92°00´W (Fig. 2). The roots of their unique nature can be attributed to their remote, oceanic geography. At present, 2,909 marine species have been identified, of which 18.2% are endemic to the Galapagos (Bustamante et al., 2002). Several oceans currents influence the regional climate and drive the population dynamics of native and endemic species. The most important oceanic surface currents are the Panama (El Niño) current, coming from the Northeast and bringing warm, nutrient-poor waters and, and the Peru (Humboldt) current, arriving from the Southern Ocean, and transporting cold, nutrient rich waters. Both current systems merge to form the South Equatorial Current (SEC), which drives surface marine waters to the west of the islands and which has been proposed as the major mean of transportation bringing species from mainland Ecuador to the Galapagos (Banks, 2002;

Bustamante et al., 2002). Remarkably, about 50% of terrestrial and waterbird species are endemic to the Galapagos either at the species or subspecies level (Jimenez-Uzcategui & Wiedenfeld, 2002; Wiedenfeld, 2006). The Galapagos and surrounding waters harbour between 88 and 111 species of seabirds and coastal-shorebirds, including native, breeding and regular migrant species.

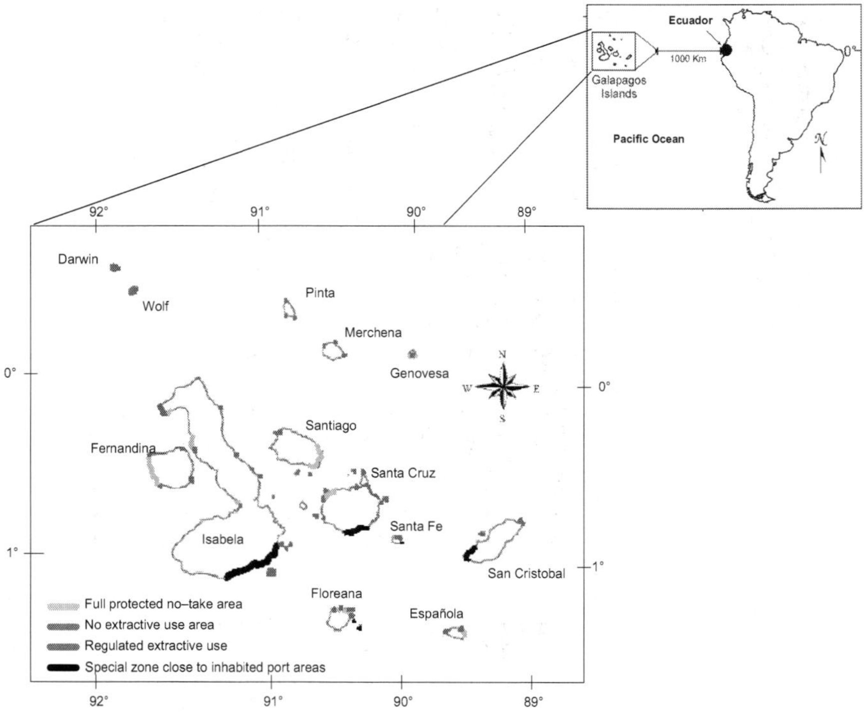

Fig. 2. Location of the Galapagos Islands relative to continental Ecuador, South America. The coastal zoning scheme for the Galapagos Marine Reserve (GMR) is also shown. The zones are fully-protected 'no-take' area, in green; non-extractive use areas, in blue; regulated extractive uses, in red; and, special zones nearby the inhabited port areas, in black. Adapted from Charles Darwin Foundation and World Wildlife Fund (2002).

3. Censuses and field methods

3.1 Ecuador mainland coast

Field observations and censuses were conducted from December 1998 to January 2002 to conduct inventories of waterbirds in mangrove areas of the Gulf of Guayaquil. A total of 22 field trips were deployed, including observations from a boat and an over-flight surveillance trip in an aircraft through mangrove conservation monitoring routes (Carvajal & Alava, 2007; Alava, 2005). Boat trips also followed line transects (n = 9) of 25 and 4 km established in three different areas (northwest, northeast and southeast) of the Gulf of Guayaquil to conduct the censuses (Alava, 2005). From February 2001 to January 2002, weekly bird inventories and 218 alternated field censuses to study aquatic birds were conducted simultaneously or sequentially during a study focused on a Horned Screamer population and habitat along 100m transects around the Canclon lagoon perimeter in five survey areas, showing different degree of perturbation, during the dry and wet seasons during 92 days (Alava et al, 2007; Alava et al., 2009). Daily observations lasted between 15 and 30 minutes at each transect using binoculars (7× 50). These observations were conducted in the morning and at noon (0730-0930 or 1000-1200 hr PDT), and during in the afternoon hours (1300-1500 hr PDT). In order to do the sea-watching from the Chocolatera in Salinas (Santa Elena Peninsula), powerful binoculars (16 x 70) were used to scan the stretch of water in front until the horizon to look for birds, during continued watching periods of 60 minutes (not necessarily on whole clock hours). Additionally, the local atmospheric conditions were also noted down. Most observations were done by one, and sometimes by three people. For the monthly census at the Ecuasal salt lakes one to three observers moved by bicycle or by car on the dikes that divide the pools, noting down the numbers of individuals per recorded species. During most visits a steady route was followed and most visits (90%) were done within the first four hours of daylight. The available optic material included a telescope (Swarovski 25 x60) and binoculars (16 x 70; 10 x 42). Between 1991 and the year 2000, the second author (B. Haase) conducted more than 240 systematic censuses of shorebirds and water birds (Haase, 2011a). As of 2004, two annual water bird census have been conducted (contribution by Wetlands International through the Ecuadorian non governmental organization, Aves & Conservación-Birds & Conservation), as part of the Neotropical Censuses of Water birds. Additionally, more than 2000 hours of shorebird banding has been carried out, with a total number of 6500 birds (smaller waders) captured with mistnets and banded. Basic biometric data were obtained before the birds were equipped with an aluminum band and according to the inter-american standard method, two color flags (for Ecuador: light green over red) were placed on the tibia (Haase, 2011a). Finally, data from Neotropical waterbird censuses conducted in Ecuador were also analyzed in this study and retrieved from published technical reports (Santander & Muñoz, 2005; Santander et al., 2006b; Santander et al., 2007; Santander & Lara, 2008).

3.2 Galapagos islands

Data on field surveys to determine the species and relative abundance and spatial distribution of seabirds in interior and exterior waters of the Galapagos Marine Reserve (GMR) were retrieved from the information and datasets collected by oceanographic expeditions aboard the R/V BAE Orion (INOCAR-Ecuadorian Navy) on August 2000 and April 2009, as well as on the R/V Roger Revelle (SCRIPPS Institution of Oceanography, University of California, San Diego, UCSD) from August to September 2001 (Cruz-Delgado, 2001; Alava, 2002; Jimenez et al., 2010; Alava et al., 2010). This study also includes data on censuses of Galapagos penguins and flightless cormorants reported elsewhere (Vargas et al., 2005; Jiménez-Uzcátegui et al., 2006; Jiménez-Uzcátegui & Vargas, 2007; Jiménez-Uzcátegui and Devineau 2009) and compared to time series of sea surface temperature (SST) anomalies to asses the impact of the El Niño-Southern Oscillation (ENSO), used here as a proxy to explore the effect of regional climate change. Data on SST anomalies were retrieved from the National Weather Service of the National Oceanic and Atmospheric Administration (NOAA, 2011):
(http://www.cpc.ncep.noaa.gov/products/analysis_monitoring/ensostuff/ONI_change.shtml)

3.3 Richness and abundance of waterbird species

Several censuses of Neotropical waterbids have been conducted in Ecuador since 1995 (Santander & Lara, 2008). However, the censuses were not carried out each year, and census data is only available for 1995, 2004, 2005, 2006 and 2007, with a gap between 1995 and 2004, as reported elsewhere (Santander & Muñoz, 2005; Santander et al., 2006b; Santander et al., 2007; Santander & Lara, 2008). Censuses for 2008, 2009 and 2010 were not yet available at the time the present review was conducted. For the purpose of this study, the 2007-Neotropical waterbird census, including the coastal zone, highlands and Amazon region (Santander & Lara, 2008; retrieved online, http://lac.wetlands.org/), was used to measure waterbird biodiversity. The rationale to select this census is based on the fact that it represents the census with the highest records of both species and number of individuals in the history of Neotropical waterbird censuses undertaken in Ecuador since 1995. Under this premise, the total annual biodiversity of waterbirds for mainland Ecuador was estimated using the Shannon-Weaver and the Simpson biodiversity indices. The former was calculated using the equation (Krebs, 1999):

$$H' = -\sum pi\ln(pi) \quad (1)$$

Where H' is the Shannon-Weaver diversity index, and pi is the relative abundance of each group of species. The Shannon-Weaver index is usually expressed as $e^{H'}$. Typically the value of the index ranges from 1.5 (i.e., low species richness and evenness) to 3.5 (i.e., high species evenness and richness), but values beyond this range can be found (McDonald, 2003). The latter was calculated using the Simpson's index (D) as follows:

$$D = \sum (pi)^2 \quad (2)$$

Where equation (2) is subtracted from 1to yield 1-D (i.e., the Simpson's index of diversity), and pi is the fraction of all organisms which belong to the i-th species. The value of this index ranges between 0 and 1, with 1 representing infinite diversity and 0 representing no diversity (e.g., the greater the value, the greater the sample diversity). In this case, the index represents the probability that two individuals randomly selected from a sample will belong to different species. Additionally, seabird observations conducted in the Galapagos were sorted by species and relative abundances. For each threatened species identified in the present review, both the national (Ridgely & Greenfield 2001, Granizo et al., 2002) and global assigned categories by the IUCN Red List for birds (BirdLife International, 2011) are noted. Scientific names and English common names for waterbirds follow the South American Classification Committee (SACC) Classification (Freile, 2010).

4. Waterbird biodiversity

4.1 Richness and evenness

Both the number of species and abundances have increased since the first Neotropical census was deployed in 1995 in Ecuador (Table 1). This can be explained due to an increase in the observation efforts and number of sites visited for censuses. The maximum number of species was registered in February 2007, with a total abundance close to 40000 individuals.
At the Ecuasal lagoons, large numbers of shorebirds were recorded, with maximum counts for common and abundant species shown in Table 2. The most abundant shorebird species is the Wilson's Phalarope, followed by the Semipalmeated Sandpiper, Western Sandpiper

and Least Sandpiper. The environmental conditions of the Ecuasal lagoons offer a suitable stopover and habitat for a substantial number of shorebird and seabird species.

	1995	2004	2005		2006		2007	
	February	July	February	July	February	July	February	July
Sites censused	11	22	23	25	29	26	27	21
Number of species	45	68	67	74	59	62	80	77
Number of individuals	3750	13759	21201	15533	17600	21509	39764	24704

Table 1. Data of Neotropical waterbird censuses conducted in Ecuador from 1995 to 2007. The censuses are generally conducted twice per year (February and July). In 1995 and 2004, the census was carried out only in February and July, respectively. Adapted from Santander & Lara (2008)

Abundant species	Maximum Counts
Snowy Plover (*Charadrius alexandrinus*)	221
Sanderling (*Calidris alba*)	> 1,000
Black-necked Stilt (*Himantopus mexicanus*)	1,500
Western Sandpiper (*Calidris mauri*)	1,500
Least Sandpiper (*Calidris minutilla*)	1,500
Semipalmated Sandpiper (*Calidris pusilla*)	2,000
Wilson's Phalarope (*Phalaropus tricolor)*	32,000

Source: WHSRN, (2009b). http://www.whsrn.org/site-profile/lagunas-de-ecuasal

Table 2. Major shorebird species and maximum abundaces recorded in Ecuasal lagoons.

During the oceanographic cruises conducted around the Galapagos Islands in 2000, 2001 and 2009, a total of 24 species of seabirds were recorded (Fig. 3), with abundances ranging from 830 individuals in 2001 to 2242 individuals in 2002. The Nazca Booby (*S. granti*) was the most abundant seabird accounting for 51% of the total abundance of species recorded (1560 seabirds) in 2009 (Fig. 3). Most of the sightings were aggregated in places southwestern Galapagos (0°-2°S; 94-91°W), where generally nutrient-enriched, upwelling areas are found (Jimenez et al. 2010; Alava et al., 2010). A high abundance of sea birds was also observed in areas exhibiting the highest values of primary production (0.5-1.7 mg/m^3 in 2000 and 0.46-0.50 mg/m^3 in 2009), southeast of the Galapagos (2-3°S; 88°W) (Alava et al., 2010). These observations underline the role of seabirds as eco-markers of primary productivity in a highly stochastic marine environment. Several seabirds, including critically endangered species such as the Waved Albatross (*Phoebastria irrorata*) and Galapagos Petrel (*Pterodroma phaeopygia*), were recorded in foraging areas (87°-84° W) off the GMR boundaries, implying the risk of bycatch in these unprotected areas. A reduction of 82% and 87% in the abundance of Waved Albatross and Galapagos Petrel recorded at sea (using similar tracks and cruises with the INOCAR Ecuadorian Navy for both years) is observed from 2001 to 2009, respectively (Fig. 3). In addition, large numbers of the Red-necked Phalarope (*Phalaropus lobatus*) were observed in several oceanographic cruises off Ecuador and around Galapagos waters (B. Haase, pers. obs.). On the contrary, a negative local population trend of this species has been reported at coastal Ecuador from 1991 to 2011 (Haase, 2011b). This species seems to be more oceanic, showing offshore habits compared to the Wilson's Phalarope, which is a more coastal species.

Several rare species of seabirds such as the Wandering Albatross (*Diomedea exulans*), Black-footed Albatross (*Phoebastria nigripes*), Gould's Petrel (*Pterodroma leucoptera*), Buller's Shearwater (*Puffinus bulleri*) and White-faced Storm-Petrel (*Pelagodroma marina*) have been registered for the Galapagos, but never at the Continent's mainland coast (Annex I).

Interestingly, the Parkinson's Petrel (*Procellaria parkinsoni*) is a locally common species found offshore of Ecuador's southern coast, where more than 100 individuals have been registered (B. Haase, pers. obs.), underscoring that this petrel is more frequent than previously thought.

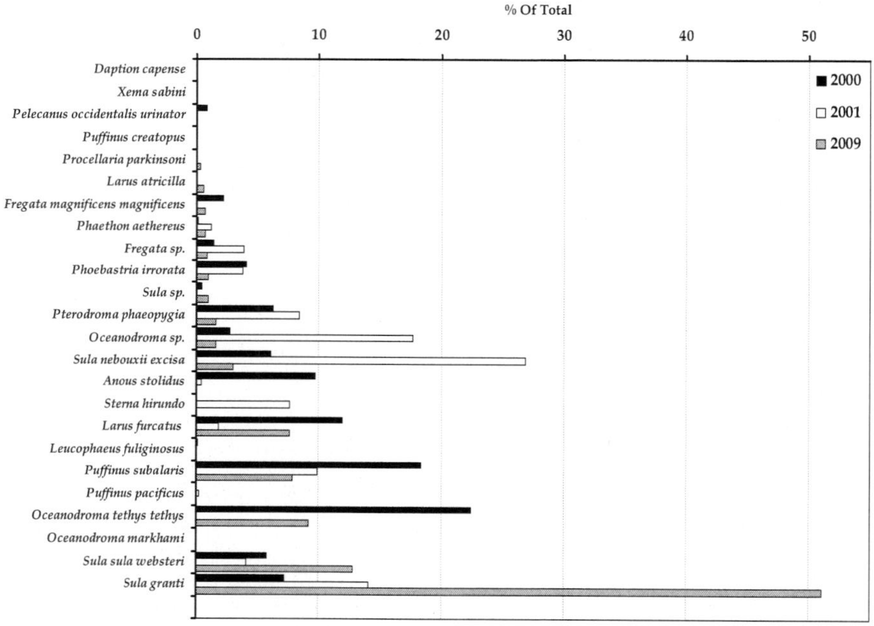

Fig. 3. Proportion of seabirds by species as a fraction of totals recorded around Galapagos waters aboard oceanographic cruises in 200, 2001 and 2009 (adapted from Alava et al., 2010).

The Masked Booby (*Sula dactylatra*) and the Brown Booby (*Sula leucogaster*) have recently been confirmed as booby species inhabiting the marine-coastal zone, including La Plata Island and Salinas, respectively (Haase, 2011a).

Based on the 2007 Neotropical census in Ecuador (Santander & Lara, 2008) and using both the Shannon-Weaver and Simpson indexes, a high biodiversity of waterbirds is found in Ecuador. The Shannon-Weaver index value was 3.50, while the Simpson biodiversity index value was close to 1.0 (Table 3). Although data for the Galapagos Islands was not present for the 2007 census, this still corroborates the high degree of richness and abundance of waterbird species in Ecuador, as previously reported (Santander et al., 2006a). A complete list of the waterbird species found and recorded for Ecuador and the Galapagos Islands, including the newest records, is available in Annex I.

Species	February	July	Total	(*pi*)	ln(*pi*)	(*pi*)*ln(*pi*)	$(pi)^2$
Podiceps occipitalis	69	134	203	0.00315	-5.7607	-0.0181	0.00001
Podilymbus podiceps	91	1093	1184	0.01837	-3.9973	-0.0734	0.00034
Tachybaptus dominicus	19	33	52	0.00081	-7.1227	-0.0057	0.00000
Pelecanus occidentalis	871	4970	5841	0.09060	-2.4013	-0.2176	0.00821
Pelecanus thagus	1546	1175	2721	0.04221	-3.1652	-0.1336	0.00178
Phalacrocorax brasiliensis	2272	1576	3848	0.05969	-2.8186	-0.1682	0.00356
Anhinga anhinga		34	34	0.00053	-7.5476	-0.0040	0.00000
Ardea alba	1374	887	2261	0.03507	-3.3504	-0.1175	0.00123
Tigrisoma fasciatum	4		4	0.00006	-9.6876	-0.0006	0.00000
Ardea cocoi	259	46	305	0.00473	-5.3536	-0.0253	0.00002
Ardea herodias	1		1	0.00002	-11.0739	-0.0002	0.00000
Butorides striatus	169	90	259	0.00402	-5.5171	-0.0222	0.00002
Egretta caerulea	630	390	1020	0.01582	-4.1464	-0.0656	0.00025
Egretta thula	1997	1539	3536	0.05485	-2.9032	-0.1592	0.00301
Egretta tricolor	43	94	137	0.00213	-6.1539	-0.0131	0.00000
Ardea ibis	1094	473	1567	0.02431	-3.7170	-0.0903	0.00059
Botaurus pinnatus	2	3	5	0.00008	-9.4645	-0.0007	0.00000
Botaurus spp	3		3	0.00005	-9.9753	-0.0005	0.00000
Nyctanassa violacea	50	128	178	0.00276	-5.8921	-0.0163	0.00001
Nycticorax nycticorax	94	96	190	0.00295	-5.8269	-0.0172	0.00001
Ixobrychus exilis	2		2	0.00003	-10.3808	-0.0003	0.00000
Eudocimus albus	27	52	79	0.00123	-6.7045	-0.0082	0.00000
Plegadis falcinellus	7		7	0.00011	-9.1280	-0.0010	0.00000
Theristicus melanopis	24	4	28	0.00043	-7.7417	-0.0034	0.00000
Mycteria americana	4		4	0.00006	-9.6876	-0.0006	0.00000
Phoenicopterus chilensis	738	264	1002	0.01554	-4.1642	-0.0647	0.00024
Anhima cornuta		24	24	0.00037	-7.8959	-0.0029	0.00000
Anas andium	426	111	537	0.00833	-4.7879	-0.0399	0.00007
Anas bahamensis	2103	511	2614	0.04055	-3.2053	-0.1300	0.00164
Anas discors	2896		2896	0.04492	-3.1028	-0.1394	0.00202
Anas clypeata	1		1	0.00002	-11.0739	-0.0002	0.00000
Anas georgica	888	371	1259	0.01953	-3.9359	-0.0769	0.00038
Dendrocygma bicolor	43	21	64	0.00099	-6.9150	-0.0069	0.00000
Dendrocygna autumnalis	389	60	449	0.00696	-4.9669	-0.0346	0.00005
Netta erythrophthalma	4		4	0.00006	-9.6876	-0.0006	0.00000
Cairina moschata		2	2	0.00003	-10.3808	-0.0003	0.00000
Oxyura ferruginea	177	176	353	0.00548	-5.2075	-0.0285	0.00003
Aramus guarauna		2	2	0.00003	-10.3808	-0.0003	0.00000
Laterallus albogularis	1	9	10	0.00016	-8.7713	-0.0014	0.00000
Rallus longirostris	5	4	9	0.00014	-8.8767	-0.0012	0.00000
Aramides axillaris		12	12	0.00019	-8.5890	-0.0016	0.00000
Neocrex erythrops		4	4	0.00006	-9.6876	-0.0006	0.00000
Porphyrio martinicus	40	35	75	0.00116	-6.7564	-0.0079	0.00000
Gallinula chloropus	454	98	552	0.00856	-4.7604	-0.0408	0.00007
Fulica ardesiaca	1157	1099	2256	0.03499	-3.3526	-0.1173	0.00122
Jacana jacana	126	87	213	0.00330	-5.7126	-0.0189	0.00001
Haematopus palliatus	31	32	63	0.00098	-6.9308	-0.0068	0.00000
Himantopus mexicanus	3379	1821	5200	0.08066	-2.5175	-0.2031	0.00651

Vanellus chilensis	2		2	0.00003	-10.3808	-0.0003	0.00000
Vanellus resplendens	153	176	329	0.00510	-5.2779	-0.0269	0.00003
Pluvialis squatarola	378	49	427	0.00662	-5.0171	-0.0332	0.00004
Pluvialis dominica	1		1	0.00002	-11.0739	-0.0002	0.00000
Charadrius alexandrinus	43	37	80	0.00124	-6.6919	-0.0083	0.00000
Charadrius collaris	169	35	204	0.00316	-5.7558	-0.0182	0.00001
Charadrius semipalmatus	244	181	425	0.00659	-5.0218	-0.0331	0.00004
Charadrius vociferus		4	4	0.00006	-9.6876	-0.0006	0.00000
Charadrius wilsonia	59	25	84	0.00130	-6.6431	-0.0087	0.00000
Charadrius spp.	227		227	0.00352	-5.6490	-0.0199	0.00001
Actitis macularius	213	53	266	0.00413	-5.4904	-0.0227	0.00002
Aphriza virgata		6	6	0.00009	-9.2822	-0.0009	0.00000
Arenaria interpres	238	123	361	0.00560	-5.1850	-0.0290	0.00003
Calidris alba	1005	31	1036	0.01607	-4.1308	-0.0664	0.00026
Calidris bairdii	365	6	371	0.00575	-5.1577	-0.0297	0.00003
Calidris mauri	1533	281	1814	0.02814	-3.5706	-0.1005	0.00079
Calidris melanotos	6	5	11	0.00017	-8.6760	-0.0015	0.00000
Calidris minutilla	786	552	1338	0.02075	-3.8750	-0.0804	0.00043
Calidris pusilla	1139	278	1417	0.02198	-3.8176	-0.0839	0.00048
Calidris spp.	253	47	300	0.00465	-5.3701	-0.0250	0.00002
Tringa semipalmata	292	140	432	0.00670	-5.0055	-0.0335	0.00004
Limnodromus griseus	239	160	399	0.00619	-5.0850	-0.0315	0.00004
Limosa haemastica	2	1	3	0.00005	-9.9753	-0.0005	0.00000
Micropalama himantopus	322	220	542	0.00841	-4.7787	-0.0402	0.00007
Numenius phaeopus	158	97	255	0.00396	-5.5327	-0.0219	0.00002
Phalaropus tricolor	1529	1642	3171	0.04919	-3.0121	-0.1482	0.00242
Tringa flavipes	472	56	528	0.00819	-4.8048	-0.0394	0.00007
Tringa melanoleuca	260	51	311	0.00482	-5.3341	-0.0257	0.00002
Tringa solitaria	4		4	0.00006	-9.6876	-0.0006	0.00000
Leucophaeus atricilla	2591	612	3203	0.04968	-3.0021	-0.1492	0.00247
Chroicocephalus cirrocephalus	1466	1143	2609	0.04047	-3.2072	-0.1298	0.00164
Larus dominicanus		3	3	0.00005	-9.9753	-0.0005	0.00000
Leucophaeus modestus	12	217	229	0.00355	-5.6402	-0.0200	0.00001
Leucophaeus pipixcan	44		44	0.00068	-7.2897	-0.0050	0.00000
Chroicocephalus serranus	179	74	253	0.00392	-5.5405	-0.0217	0.00002
Thalasseus elegans	67	121	188	0.00292	-5.8375	-0.0170	0.00001
Sterna hirundinacea		53	53	0.00082	-7.1036	-0.0058	0.00000
Sterna hirundo	4	19	23	0.00036	-7.9384	-0.0028	0.00000
Thalasseus maximus	1479	500	1979	0.03070	-3.4836	-0.1069	0.00094
Gelochelidon nilotica	37	140	177	0.00275	-5.8978	-0.0162	0.00001
Thalasseus sandvicensis	132	1	133	0.00206	-6.1836	-0.0128	0.00000
Sterna spp.	150	4	154	0.00239	-6.0370	-0.0144	0.00001
Rynchops niger	1	1	2	0.00003	-10.3808	-0.0003	0.00000
Total abundance	**39764**	**24704**	**64468**	**1.00**		**-3.5206**	**0.04128**
						***H'* = 3.52**	**1-*D* = 0.960**

Table 3. Richness and evenness for waterbirds accounted during 2007 Neotropical census in Ecuador and measurements of biodiversity indices, including the Shannon-Weaver (*H'*) and Simpson (1-*D*) indices of biodiversity. The 2007 census also included water systems (e.g., lakes, lagoons and wetlands) from the highland and Amazon regions.

4.2 Accounts of priority species

4.2.1 Piping Plover (*Charadrius melodus*)

The Pipping Plover is a globally Near Threatened species and its population appears to be increasing (BirdLife International, 2011). However, this shorebird species is Critically Endangered in Ecuador (Granizo et al., 2002; Santander et al., 2006a). At present, the only record of this accidental visitor for South America has occurred in the Ecuasal lagoons (Santander et al. 2006a; WHSRN, 2009b).

4.2.2 Buff-breasted Sandpiper (*Tryngites subruficollis*)

A Near Threatened species also recorded in the Ecuasal lagoons (WHSRN, 2009b; B. Haase, pers. obs). The population of this species is globally declining due to overhunting in the past, habitat degradation and environmental contaminants (BirdLife International, 2011)

4.2.3 Waved Albatross (*Phoebastria irrorata*)

The Waved Albatross is an endemic, Critically Endangered species with almost the entire population breeding on a single island, Española Island, (Galapagos). Very few pairs breed in La Plata Island (Machalilla National Park) at coastal Ecuador. While the total population (i.e., breeders and non breeders) is currently estimated to be 15475 albatrosses, a population decline of 47% was observed from 1994 to 2007 (Anderson et al., 2008). It accounted for only 1% of the total abundance of seabirds observed in the 2009-INOCAR/Navy oceanographic cruise conducted in the Galapagos. Longline fisheries and targeted-direct fishing are the major threats in marine waters off Ecuador and Peru (Wiedenfeld and Jiménez-Uzcátegui 2008; Hardesty et al., 2010). In contrast to the serious problems that plastic ingestion causes in some other species of albatrosses, this pollution threat appears to pose a minor impact to this species.

4.2.4 Galapagos Petrel (*Pterodroma phaeopygia*)

The Galapagos petrel is an endemic, Critically Endangered species breeding on Santa Cruz, Floreana, Santiago, San Cristóbal, Isabela and possibly other islands in the archipelago, where the global population ranges from 10,000 to 19,999 individuals, based on an estimate of 4,500-5,000 active nests in 2008 (BirdLife International 2011; F. Cruz-Delgado, pers. comm.). Similar to the Waved Albatross, this species represented about 1.50% of the total number of seabirds recorded in the 2009-INOCAR/Navy oceanographic cruise. Long-line fishing in the southeastern Pacific is a major threat; however, long-lining within the GMR limits is particularly likely to affect foraging birds, while introduced species such as cats, pigs and rats are a major threat at the breeding grounds (Wiedenfeld and Jiménez-Uzcátegui 2008; BirdLife International 2011b).

4.2.5 Lava Gull (*Leucophaeus fuliginosus*)

This species is one of the rarest gulls in the world and widespread throughout the Galapagos Islands. With a very small endemic population of about 600-800 individuals, this gull only breeds in Galapagos and is considered Vulnerable by the IUCN (BirdLife International, 2011), which might understimate the conservation status risk for this species (Wiedenfeld and Jiménez-Uzcátegui 2008). Introduced predators such as cats and dogs as well as fishing activities (e.g., hooks and nets) are major conservation threats for this particular species (Cepeda & Cruz, 1994; Wiedenfeld and Jiménez-Uzcátegui, 2008).

4.2.6 Galapagos Penguin (*Spheniscus mendiculus*)

The tropical Galapagos penguin is an endemic and Endangered species breeding in the Galapagos Islands, where its decreasing population is estimated in 1,800 mature individuals, with approximately 95% of the population restricted to Isabela and Fernandina islands in the western part of the Galapagos (Vargas et al., 2005; Jiménez-Uzcátegui & Vargas, 2007; Vargas et al., 2007). The El Niño events driven by climate change is the primary threat, while oil spills, fishing activities, alien predators and emerging infectious diseases are categorized as looming threats impacting this unique population of penguins (Vargas et al., 2006; Vargas et al., 2007; Wiedenfeld and Jiménez-Uzcátegui, 2008; BirdLife International, 2011).

4.2.7 Flightless Cormorant (*Phalocrocorax harrisi*)

Similar to the Galapagos penguin, the flightless cormorant is an endemic and Endangered seabird breeding on Isabela and Fernandina in the Galapagos Islands, where its population has been estimated close to 1680 individuals (BirdLife International, 2011). The threats for this species are similar to that identified for Galapagos penguins, mainly the El Niño events, and oils spills, as well as flooding (Wiedenfeld and Jiménez-Uzcátegui, 2008; BirdLife International, 2011).

4.2.8 Horned Screamer (*Anhima cornuta*)

The Horned Screamer is threatened (i.e., Endangered) in Ecuador (Granizo et al., 2002) and categorized as a species of Least Concern at the global level (BirdLife International, 2011). The population (i.e. the El Canclon Lagoon population) is confined to the Manglares Churute Ecological Reserve (MCER). Studies on the Horned Screamer (*Anhima cornuta*) population, conservation status and habitat deterioration in the El Canclon Lagoon wetland have been documented elsewhere (Alava et al., 2002; Alava et al., 2007a; Alava et al., 2009). The most recent data indicate that the estimated overall mean number of screamers in the lagoon was 68 ± 48 birds (Alava et al., 2009). The extrapolated, absolute mean density is estimated in approximately less than 1 bird/ha for the whole lagoon wetland (68 birds /800ha) or a relative density of 0.7 individuals per transect. Recently, 24 individuals were recorded in the 2007-Neotropical census (Table 3; Santander & Lara, 2008). The abundance of screamers was lower (six-ten individuals) in more disturbed areas containing farms, agriculture fields and cattle ranching. Abundance was inversely associated with the presence of cattle and was not seasonally dependent during the dry and wet seasons. Screamers were significantly associated with vegetation coverage. Because the Horned Screamer strongly relies on the wetland vegetation, cattle overgrazing jeopardizes its habitat and survival in this Ramsar wetland.

4.2.9 Comb Duck (*Sakidiornis melanotos*)

This is a Vulnerable and rare species of waterfowl scarcely observed in coastal Ecuador (Granizo et al., 2002; Santander et al., 2006a). In December and January 2002, a total of three records of this species were made at the southeastern edge of the El Canclón Lagoon, where individuals were sighted on the water surface during normal flood conditions (Alava et al., 2007a). The sightings of Comb Ducks in Ecuador's costal regions have been sporadic and scattered, suggesting a marked rareness of this species in lowlands of southwest Ecuador. The first two specimens of this species were recorded on the slopes of the Cayambe Volcano in the Ecuadorian highlands during 1951-1952 (Norton *et al.* 1972). Additionally, flocks of about 15-25 individuals were observed in the early morning (06:30) along a mangrove channel of the MCER near the El Canclón Lagoon in November 1987 (Ortiz-Crespo 1988). At the extreme border of southern Loja and west Macara provinces, a few individuals, with a potential resident population, have been sighted along Rio Sabiango and Zapotillo area (Best *et al.* 1993, Ridgely & Greenfield 2001). Likewise, a total of 12 individuals were recently reported at the El Azúcar Dam, Santa Elena Peninsula (Guayas Province) in July 2004 during a Neotropical waterbird census (Santander & Muñoz 2004). Hunting and use of pesticides in rice fields are likely to be major threats for this species (Santander et al., 2006).

4.2.10 Muscovy Duck (*Cairina moschata*)

The Muscovy Duck is currently rare and locally uncommon in lotic, lentic, and wetland (e.g., marshes) ecosystems in both east and southwest Ecuadorian lowlands, even though it was probably an abundant waterfowl in the past (Ridgely & Greenfield 2001). The species is Endangered in Ecuador (Granizo et al., 2002). In the El Canclon laggon, this species was seldom recorded, with 7 sightings at the beginning of the rainy season from December 2001 to January 2002 (Alava et al., 2007a). All observed individuals showed the typical feather coloration of wild birds. These are the second most recent sightings of this species in the MCER after more than a decade since about 25 birds were recorded in MCER in March 1998 by J. C. Matheus (as cited by Ridgely & Greenfield 2001). This species has also been sighted in the Yaguachi marshes since 1980 (Ridgely & Greenfield 2001). Recently, one individual of this species was recorded at the Santay Island during the 2004-Neotropical waterbird census (Santander & Muñoz 2004), while two individuals were recorded in the 2007- Neotropical census (Table 3; Santander & Lara, 2008). As with the Comb Duck, sightings of this species are extremely rare along the Ecuadorian coastal, and has been suggested that hunting pressure is the major cause of its population decline in recent years (Ridgely & Greenfield 2001).

4.2.11 Southern Pochard (*Netta erythrophthalma*)

This is a very rare and Critically Endangered species in Ecuador (Granizo et al., 2002), but considered as a species of Least Concern globally (BirdLife International, 2011). Only four individuals were recorded in the 2007 Neotropical census (Santander & Lara, 2008). Major threats include hunting, aquatic habitat degradation and wetland transformation due to agricultural encroachment (Santander et al., 2006a; BirdLife International, 2011). Conservation efforts for this species need to be focus on habitat protection.

4.2.12 Great Blue Heron (*Ardea herodias*)

The great heron blue is found in the Ecuadorian coast and Galapagos inhabiting mangroves and coastal wetlands. The great blue heron recorded for the Galapagos is an endemic subspecies (*A. h. cognata*), breeding in the islands (Jackson, 2001; Jiménez-Uzcátegui et al., 2007). This heron is not as abundant as other species of herons and its population number is probably declining due to habitat loss.

4.2.13 Roseate Spoonbill (*Platalea ajaja*)

Research and conservation aspects on the Roseate Spoonbill population of the Gulf of Guayaquil were documented by Alava (2005). The mean number of recorded birds was 40.5

birds (95% CI: 16.0–64.8), ranging from 1 to 100 individuals. Censusing of birds based on the Hayne model yield a number of 662 individuals in an area of 48,000 ha (i.e., 2000 birds/1350 km^2), with absolute and relative abundances of 0.7 birds/km and 0.014 birds/ha, respectively (Alava, 2005). Individuals were mainly aggregated in the northwestern and northeastern areas of the Guayaquil Gulf Estuary, and followed by the southeastern part. Wetlands destruction such as mangrove areas, agriculture, urbanization and non-controlled hunting have negative impacts on Roseate spoonbills. On the short term, the Roseate Spoonbill population may become threatened because of habitat loss.

4.2.14 Greater Flamingo (*Phoenicopterus ruber*)

This charismatic species is mainly found in saline and brackish coastal lagoons of the Galapagos Archipelago, mainly in Isabela, Santiago, Rábida, Floreana and San Cristobal islands. Although the flamingo population found in the Galapagos is not endemic to the islands and it is not threatened at the global level, the native population breeding in the islands is considered as threatened with approximately 320-520 individuals (Granizo et al., 2002; Wiedenfel & Jiménez-Uzcátegui, 2008). Sea level rise due to climate change has been identified as a potential conservation threat for flamingos in the Galapagos (Granizo et al., 2002; Jiménez-Uzcátegui, 2006; Jiménez-Uzcátegui et al., 2007). Flamingos are very sensitive to human perturbations and predation of nests by introduced species such as pigs (Jackson, 2001; Wiedenfel & Jiménez-Uzcátegui, 2008).

4.2.15 Brown Wood-Rail (*Aramides wolfi*)

The Brown Wood-rail is categorized as Vulnerable globally and as Endangered in Ecuador (Granizo et al., 2002; BirdLife International, 2011), where it is one of the rarest birds and, during the last two decades, seldom recorded, mainly in mangrove (Taylor, 1996; Ridgely & Greenfield, 2001). Most sightings have been in either the north-west or south-west, from: north of Quinindé; pristine humid forest at Paraíso de Papagayos ranch; secondary forest and disturbed wetlands at Jatun Sacha-Bilsa Biological Reserve (within Mache Chindul Ecological Reserve) (Esmeraldas province); forest remnants at Río Palenque Research Station, (Los Ríos province; and mangroves in Manglares Churute Ecological Reserve (MCER), (Guayas province) (Alava et al., 2007b). On 29 June 2001, a bird was seen on a dirt road bordering disturbed riparian vegetation and secondary foothill forest on the east side of El Canclon lagoon (02°30'S 79°42'W) (Alava et al., 2007b). Coloration of the head, neck and body identified it as *A. wolfi* (i.e., ashgrey head and cinnamon-rufous neck, with the rest of the upperparts and underparts pale olivaceous-brown). This observation is the first at El Canclon (a Ramsar site) and in the entire MCER for over a decade, following that of two birds in mangrove on 28 December 1989 (Ridgely & Greenfield, 2001). Unconfirmed sightings were made in 2005–2006 at Puerto Hondo, a mangrove relatively close to Cerro Blanco Protected Forest. Attempts are underway to verify these observations as Rufous necked Wood-rail *Aramides axillaris* is frequent there and juveniles have similar plumage to adult *A. wolfi* (Alava et al., 2007b). The two are locally sympatric in mangrove, but *A. wolfi* seems to be more frequent inland (Ridgely & Greenfield, 2001). It has been suggested that *A. wolfi* is capable of surviving in fragmented or deteriorated habitats (Taylor, 1996; Ridgely & Greenfield, 2001). On the other hand, it is perhaps less tolerant of habitat perturbation than Grey-necked Wood-rail *A. cajanea* (Ridgely & Greenfield, 2001), and is definitely much more sensitive than *A. axillaris* (Vulnerable in Ecuador). The latter is commonly found close to disturbed or human modified mangroves, including in Guayas (Puerto Hondo) and El Oro (Bajo Alto) provinces (J.J. Alava, pers. obs.). Additional studies of *A. wolfi* are urgently required to increase knowledge of its natural history.

4.2.16 Galapagos Rail (*Laterallus spilonotus*)

As an endemic bird to the Galapagos, this a rare species of rail occurring in several islands, mainly those offering suitable habitat with humid zone vegetation as that found in Pinta, Fernandina, Isabela, San Cristobal, Santa Cruz, Floreana, and Santiago (Rosenberg, 1990), as well as Wolf, Drawin and Alcedo (BirdLife International, 2011). This rail is threatened (i.e., Vulnerable) with a decreasing population, estimated on 5000-10000 mature individuals (BirdLife International, 2011). The threats for its conservation include continuing conversion of highland habitat to agriculture, invasive plants (quinine tress, *Cinchona pubescens*, and black berries, *Rubus niveus*), overgrazing by exotic mammalian herbivores (e.g., feral goats, cattle and horses) and predation by introduced cats, pigs and rats (Rosenberg, 1990; Gibbs et al., 2003; Wiedenfel & Jiménez-Uzcátegui, 2008; BirdLife International, 2011).

5. Environmental impacts and conservation threats

5.1 Habitat loss, degradation and fragmentation

Birds are closely associated with forests, and approximately 30% of the world's species of birds are highly restricted to tropical forests used as either winter grounds or year-round habitats (Myers 1992). In Ecuador, Western and Tumbesian forests of Ecuador are being cleared by farming and ranching and are highly threatened by browsing and trampling of domestic livestock, with about 4% of the original forest coverage remaining by 1998 (Dodson & Gentry, 1991; Best & Kessler, 1995). For instance, uncontrolled cattle grazing of the native vegetation, deforestation, and agricultural sprawl (rice crops and farms) have negatively impacted the El Canclon Lagoon at the MCER, jeopardizing its conservation and affecting the local population of the Horned Screamer (*A. cornuta*) and several other waterbirds depending on this wetland (Alava et al., 2007; Alava et al., 2009). Likewise, it is estimated that about 55,400 hectares (27% of the original total area: 203,625 hectares) of mangrove forests has been lost in coastal Ecuador from 1969 to 2006 due to uncontrolled clear-cutting of mangroves (Fig. 4), not only for construction of illegal shrimp farms (aquaculture), but for agriculture, illegal extraction of timber and urban sprawl (CLIRSEN, 2007; Carvajal & Alava, 2007). The decrease of slat flat areas is also of concern with a reduction of 93% since 1969 (Fig. 4). Extensive banana plantations are found in southwest Ecuador, and are primarily located in coastal provinces such as Los Rios, El Guayas and the El Oro (INEC 2007). A total area averaging 232,235 ha is dedicated to the production of bananas at the national level. About 79% of this total are located on the coastal zone, mainly in the El Guayas and the El Oro provinces (an average of 51,183 and 44,607 ha, respectively), which are relatively close to mangrove areas. Presence of solid wastes (i.e., plastic bags and bottles) and illegal camp fires are signs of human activity in mangrove areas, as well. There, it has been suggested that deforestation and fragmentation in mangrove habitats have affected the local and nesting population of Roseate Spoonbills (*P. ajaja*) in the Guayaquil Gulf Estuary (Alava, 2005), as well as the declining population of the Brown Wood Rail (*A. wolfi*), which is endangered and less tolerant to habitat deterioration in Ecuador (Alava et al., 2007).

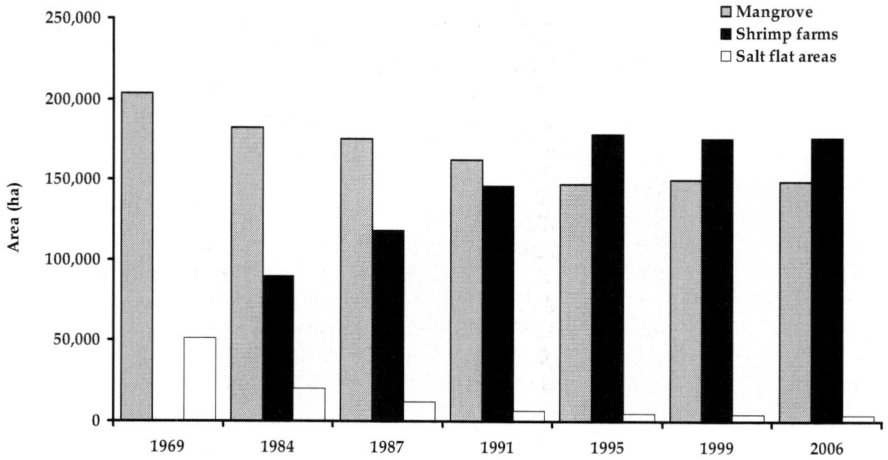

Fig. 4. Temporal and spatial evolution of mangroves, shrimp farming and salt flat areas (ha) on the continental coast of Ecuador from 1969 to 2006 (CLIRSEN, 2007).

5.2 Fishery interactions

Fishery bycatch, including longline fisheries, is categorized as the single major threat affecting many seabird populations and on the order of hundreds of thousands of seabirds, especially albatrosses, are caught and killed each year (BirdLife International, 2008; BirdLife International, 2009; Brothers et al., 2010). Industrial and artisanal fisheries, including commercial longline, gillnet and trawl fisheries, cause a significant mortality of seabirds (i.e., hundreds of thousands) each year around the global ocean, and in some cases, the effort of fishing activities (e.g., longline) overlaps with foraging grounds for seabirds (BirdLife International, 2008). Fishing activities outside of the boundaries (unprotected areas) of the Galapagos Marine Reserve possess a looming threat for seabirds such as albatrosses, petrels and shearwaters foraging frequently in these areas close to the continent (Jiménez-Uzcátegui & Wiedenfeld, 2002; Wiedenfeld & Jiménez-Uzcátegui 2008; Anderson et al., 2008). Among these, the waved albatross is probably the most affected seabird by fisheries interactions (e.g., longline fisheries) in the region. For instance, about 9-13 waved albatrosses were observed bycaught in longlines (i.e., 155 longline sets; 350 hooks per set) during a field study conducted with the artisanal fishing community of Santa Rosa in coastal Ecuador (Hardesty et al., 2010; J. Hardesty, pers. comm.). The bycatch incidence has been preliminary estimated in 0.11 albatrosses/1000 hooks, and most of the interactions are associated with artisanal longline fishing gears to capture hake (*Merluccius gayi*) (Arteaga et al., 2010). The bycatch assessment of pelagic longlining (High Seas Experimental Pilot Plan) conducted around water of the Galapagos Marine Reserve (GMR) in 2003 (Murillo et al., 2004) to evaluate the impact on epipelagic species and top predators did not report seabirds (e.g. albatrosses) as victims of bycatch, although it underscored the potential risk of bycatch for seabirds (Murillo et al., 2004). In contrast, a local artisanal tuna fishery, using single-hook lines with live sardines as bait within the GMR, reported catches up to five waved albatrosses per boat per day, indicating that a serious bycatch risk exists within the GMR if longlining occurs there (Anderson et al., 2003). In marine waters off northern Peru, thousands of waved albatrosses were estimated to be bycaught in small-scale longline fisheries (Jahncke et al., 2001). Mortality of adult albatrosses due to incidental and intentional (i.e., targeted fisheries for human consumption) in the artisanal Peruvian fishery possesses a serious threat for its conservation and remains as one of the stressors influencing the population dynamics of this species in recent years (Awkerman et al., 2006; Anderson et al., 2008).

5.3 Invasive species and emerging diseases: The Galapagos case

Biological invasions are considered a leading cause of extinctions in terrestrial and marine ecosystem of marine protected areas (Boersma & Parrish, 1999; Bax et al., 2003). The introduction of exotic marine species and pathogens (viruses, bacteria and parasites) represents major threats for biodiversity and ecosystem functions, with potentially serious implications for fisheries resources, tourism, human health in marine protected areas and biosphere reserves (Carlton, 1989; Carlton & Geller, 1993; Carlton, 1996; Bax et al., 2003). Furthermore, emerging marine diseases in marine organisms have been linked to anthropogenic factors (Harvell et al., 1999). The Hawaiian Islands represents an extraordinary example of the negative effects of the biological invasion on endemic and native species (Vitousek et al., 1987). This is supported by the fact that Hawaii contains a large proportion of the imperilled USA endemic birds (43%) and plants (40%) threatened by alien species (Gurevitch & Padilla, 2004). The Galapagos Islands are facing a similar fate unless control and conservation strategies take place to mitigate biological invasion. Terrestrial invasive species, including mammalian predators and plants, significantly jeopardize native and endemic species inhabiting these remote islands (Snell et al., 2002).The number of registered introduced species in the archipelago has increased 10 times from 112 species in 1900 to 1321 in 2007 (Watkins & Cruz, 2007). Yet, this does not include introduced pathogens. Among the invasive pathogens, viruses, bacteria and parasites are the ones possessing serious risk to the endemic fauna.

Introduced plants including berries (*Rubus* spp.; black berry *Rubus niveus*) and quinine trees (*Cinchona pubescens*) have caused habitat loss and alteration for endemic species of birds such as the Galapagos Petrel and Galapagos Rail (Wiedenfeld & Jiménez-Uzcátegui, 2008). Introduced vertebrates are mainly predators affecting bird populations by killing many species of adult birds (cats *Felis catus*) and flightless or nesting species (dogs *Canis familiaris* and pigs *Sus scrofa*); and by destroying nests and young (cats, dogs, black rat, *Rattus rattus*) (Jiménez-Uzcátegui & Wiedenfeld, 2002; Wiedenfeld & Jiménez-Uzcátegui, 2008). Some introduced viral diseases from domestic animals such as avian virus or avipoxvirus by domestic birds, fowlpox virus infecting chicken have threatened endemic species of birds (e.g., Darwin's finches) in the Galapagos (Wikelski et al., 2004). Thiel et al., (2005) has recently found presence of canarypox-like viruses in pox-like lesions of endemic passerine birds (Yellow Warblers, *Dendroica petechia*; finches, *Geospiza* spp.; and Galápagos mockingbirds, *Nesomimus parvulus*) from the inhabited islands of Santa Cruz and Isabela. A seroprevalence of 66% (29/44) to adenovirus group 1 has been found in waved albatrosses (*P. irrorata*) inhabiting Espanola Island (Padilla et al., 2003). Newcastle disease, Marek's disease virus (herpes) and mycoplasmosis detected in domestic chickens farmed on the islands (Vargas & Snell, 1997), has the potential to cause declines of the Flightless Cormorant (*P. harrisi*), Lava Gull (*L. fuliginosus*), and Galapagos Penguin (*S. mendiculus*), species with small population sizes. West Nile Virus (WNV) is expected to reach Ecuador anytime and there is a high probability risk of its introduction into Galapagos unless strict control and preventive strategies are implemented prior to the arrival of the disease (GGEPL, 2004). The incidental transport of mosquitoes by boat or of infected vertebrate hosts is also significant risks for WNV invasion. If WNV is introduced in to Galapagos it is likely to cause catastrophic mortality of endemic birds, reptiles and mammals, leading to irreparable ecological and economic damage to the islands (GGEPL, 2004). The introduction of this disease is most likely to occur through the human transport of infectious mosquitoes, particularly via inadvertent transport in airplanes. Recently, several kinds of bacteria have already been detected in endemic sea bird and pinnipeds of the Galapagos. For example, while antibodies to avian adenovirus type 1 and *C. psittaci* were found in 31% (21/68) and 11% (7/65) of flightless cormorants, respectively, seventy-five of 84 (89%) Galapagos penguins had antibodies to *Chlamydophila psittaci,* but chlamydial DNA was not detected via polymerase chain reaction in samples from 30 birds (Travis et al., 2006a; Travis et al., 2006b). Waved albatrosses showed a seroprevalence of 9% (4/44) to avian encephalomyelitis; however, cloacal swabs were negative for *C. psittaci*-DNA. (Padilla et al., 2003). *Salmonella* sp. was reported in domestic pigeons (introduced rock doves, *Columba livia*) on San Cristóbal and may cause severe disease in species such as Galapagos doves (*Zenaida galapagoensis*) and other native birds (Harmon et al., 1987; Wikelski et al., 2004; Padilla et al., 2004). Among parasites, *Haemoproteus* sp., the only hemoparasite identified, was found in 89% of the Galapagos doves sampled but not in the rock doves (Padilla et al., 2004).

Currently, the major parasitic disease that could cause widespread mortality of native, endemic birds is the avian malaria, if it is introduced into Galapagos ecosystems. This parasite has caused severe mortality and decimation of a significant proportion of Hawaiian's endemic birds since it was introduced at beginning of 20th century (Wikelski et al., 2004). At present, despite its vector, the mosquito *Culex quinquefasciatus* (Diptera: Culicidae), is already established on the Galapagos Islands (Peck et al., 1998; Whiteman et al., 2005), there has been no report or detection of *Plasmodium relictum* (Wikelski *et al.* 2004; Thiel *et al.* 2005). A protozoan, *Trichomonas gallinae,* was reported in domestic pigeons (introduced rock doves, *Columba livia*) on San Cristóbal and may cause severe disease in species such as Galapagos doves (*Zenaida galapagoensis*) and other native birds (Harmon et al., 1987; Wikelski et al., 2004; Padilla et al., 2004). Because endemic species of birds of the Galapagos were not exposed to alien parasites transmitted by invasive species prior human occupation of the islands, they are more susceptible to the pathogenesis generated by parasitic diseases with potential risk at the population health level.

5.4 Anthropogenic pollution

Pollution coming from agriculture, forestry and industry significantly affects birds' population. Marine oil spills and persistent organic pollutants (POPs) can have a significant impact on population of seabirds (BirdLife International, 2008a). In Ecuador, the Guayaquil Gulf Estuary Basin has become the sink receiving point and non point sources of contamination over the last 80 years. As agriculture is the fundamental base for the Ecuador economic activity, the predominant crops are banana plantation, rice fields, sugar cane and cocoa in the Gulf of Guayaquil. In 2005, the total land used/harvest area for banana, coffee, rice, maize and cocoa ranged from 1,269,775 to 1,652,600 ha (INEC, 2007; FAO, 2007). In the past, farmers conducted both extensive and intensive use and application of fertilizers, herbicides and pesticides, including some organochlorine pollutants banned in others countries such as DDTs, chlordanes, heptachlor, dieldrin, aldrin, mirex, and lindane

(Solórzano, 1989). For example, banana plantations and agricultural lands use a broad spectrum of synthetic pesticides transported via run off and aerial dispersion to the estuaries and mangrove forests. However, the negative effects of chemical pollution on the coastal-estuarine environment have been scarcely characterized. The demand of pesticide usage for agricultural area is reflected by the importations of fungicides, insecticides and herbicides from January to May for both 2002 and 2003, with a total of 2,494 and 3,254 metric tonnes, respectively (SICA-MAG, 2003)

The Salado Estuary, harboring the El Salado Mangrove–Wildlife Production Reserve, has been receiving about 60% from domestic use and 40% from industrial use, causing degradation of the water and sediment conditions of this estuary. Several studies from the Municipality of Guayaquil, National Fisheries Institute, and the Polytechnic School of the Litoral have reported low dissolved oxygen (DO) levels at the Salado Estuary, ranging from 0.74 mg/L to 2.4 mg/L, and pH as low as 5.7 over the surface sediment (Calle & Alava, 2009). A recent study on pollution by pesticides on the Taura River Basin, Gulf of Guayaquil, revealed the presence of several organochlorine (OC) and organophosphate (OP) and pyrethroid pesticides in samples of water, sediment and aquatic organisms (Montaño & Resabala, 2005). Some industrial and agricultural POPs such as PCBs and DDT were used in Ecuador after they were banned in the 1970s in developed countries, and therefore released to soil and water bodies. In continental Ecuador, DDT was applied inside houses (intra-domiciliary applications) between 1957 and 1999 (Ministerio del Ambiente & ESPOL-ICQ , 2004), and a massive use of DDT was carried out during the 1980s to control the malaria vector-mosquito (Dr. Hugo Jurado, pers. comm.). The huge scale use of DDT culminated in 1988. At that time, however, DDT was also distributed without any control and used illegally for the agricultural sector to control crop pests (Dr. Hugo Jurado, pers. comm.). DDT was used, overused or misused, and therefore released to the soil and water bodies. To date, it has been pointed out that the only country still using DDT during the mid-1990 in South America was Ecuador; ironically, it was also the only country that experienced a significant decline in malaria (Mangu-Ward, 1997). DDT concentrations were reported on the Taura River Basin, Gulf of Guayaquil, in sediment (1.36 ug/kg wet weight) and aquatic organisms (2.87 ug/kg wet weight). The DDE/DDT ratio for these samples indicate relatively recent contamination by DDT-parental compound in sediment (ratio DDE/DDTs = 0.66) and fish (ratio DDE/DDTs=0.14) from the Taura River. The environmental implications and health effects of DDT use in aquatic birds and raptors is poorly understood and assessed in this country. The current levels, distribution, fate and effects of these POPs in environmental matrices (e.g., water, sediments, soil, fish and birds) have received scant attention.

Relatively high metal concentrations in sediment were reported for Hg (2.89 mg/kg dw), Pb (112 mg/kg dw), Cu (250 mg/kg dw), and Zn (550 mg/kg dw) exceeding the Effects Range Low (ERL) and the Effects Range Medium (ERM) sediment quality guidelines for Hg (0.71) and for Zn (410) (Calle & Alava, 2009). Organic (i.e., pesticides) and inorganic (metals) chemicals contaminants are a major problem not only for waterbirds, but for raptors associated to aquatic environments and several other species of wildlife. It is likely that individuals of Mangrove Black Hawk inhabiting mangroves close to commercial banana cultivation (i.e. the El Oro Province) might be facing exposures to chemicals and lethal effects both in the long and short terms (Alava et al., 2011), similar to that suggested for the Snail Kite (*Rosthramus sociabilis*) inhabiting and foraging in zones of vast rice fields and flooded areas of coastal Ecuador (e.g., Guayas, Los Rios and the El Oro provinces) where pesticides are broadly used (Alava et al., 2007). Ecotoxicological research is strongly encouraged in Ecuador to determine the levels, food web bioaccumulation and effects of insecticides and herbicides (e.g., organochlorines, organophosphates, carbamates, bipiridyls) in top predator birds, including water birds and raptors.

Marine pollution by debris in Galapagos waters is emerging as a significant concern for biota. A beach-shoreline cleanup program around the Galapagos in 1999 retrieved 22,140 kg of debris, with plastics and metals being the predominant objects at 25 and 28% of the total (Fundación Natura & WWF 2000; Alava, 2011). At sea, the accidental or deliberate disposal of solid waste (e. g., plastic, fishery gear) from both tourism and fishing vessels represent a threat for marine vertebrates such as large pelagic fish, sea turtles, cetaceans, sea lions, fur seals and sea birds (Alava, 2011). Likewise, both intentional (operational) and unintentional (accidental) fuel and oil releases occur around the islands from ships, with the former occurring in the long-term causing chronic degradation and latter resulting in acute impacts to the marine environment (Lessmann, 2004). Oil spills offer perhaps the most visible example of pollutant impacts on sea life. During the last two decades, several oil spills have taken place in the Galapagos (Table 4). A major oil spill that threatened a significant part of the GMR was the *MV Jessica* spill on 16 January 2001 at the entrance of Naufragio Bay (89 37′15″W, 053′40″S), San Cristóbal Island. The oil tanker released almost 100% of its total cargo consisting of 302,824 L of IFO 120-bunker fuel (Fuel Oil 120) and 605,648 L of Diesel oil # 2 (DO#2) (Lougheed et al., 2002; Edgar et al., 2003). Although no oiled seabirds were recorded at the time of this oil spill (Lougheed et al., 2002), researcher doing field work in Española Island found five oiled Nazca boobies (*Sula granti*) in January 2001, one oiled waved albatross in June 2001, and two oiled Nazca boobies in November 2001, confirming that these birds were polluted by spilled oil (Anderson et al., 2003). In early July 2002, a second oil spill took place in the Galapagos, when a small tanker (*BAE/Taurus*) sank and spilled diesel fuel in waters off the coast of Puerto Villamil, Isabela Island. Fortunately, no sign of fuel was found on the beaches or on marine animals (including sea birds), due to mitigation efforts conducted by the GNPS and Charles Darwin Foundation. Other low magnitude oil spill events have also occurred (Lessmann, 2004). The chronic toxic effects of the 2001–Jessica oil spill's residues on unique vulnerable population of Galapagos marine iguanas (*Amblyrhynchus cristatus*) has been well documented elsewhere (Wikelski et al., 2001; Wikelski et al., 2002). Less visible and more insidious global toxicants of concern involve POPs (i.e., PCBs, DDTs and several other organochlorine pesticides), which have recently been detected and assessed in fish collected from Galapagos waters and in Galapagos sea lions (*Zalophus wollebaeki*) (Alava et al., 2009; Alava et al., 2011; Alava, 2011), but these contaminants still need to be investigated in seabirds endemic to the Galapagos.

Boat/Tanker	Date	Site	Quantity (L)
Motor Yacht Iguana	June 1988	Santa Cruz Island	189,265
MV/Jessica	16 January 2001	Naufragio Bay, San Cristóbal	908,472
BAE/Taurus	4-7 July 2002	Puerto Villamil, Isabela Island	7571
MV/Galapagos-Explorer	13-14 September 2005	Academia Bay, Puerto Ayora, Santa Cruz Island	Not reported*

Table 4. Inventory of oil and diesel spills in the Galapagos from 2001 to 2006. *151,412 L of fuel were estimated to be contained in the boat, but actual volume spilled was not reported (adapted from Alava, 2001).

5.5 Regional climate variability

Increasing emissions of greenhouse gases, including carbon dioxide (CO_2) due to fossil fuel, and increases in global average air and ocean temperatures are the major forces driving global warming in the last century and in recent times (IPCC, 2007). A warming (global surface temperature) of about 0.2 °C per decade is projected for the next two decades according to scenarios of the Intergovernmental Panel on Climate Change (IPPC, 2007). Warming is larger in the Western Equatorial Pacific than in the Eastern Equatorial Pacific over the past century, suggesting that the increased West-East temperature gradient may have increased the likelihood of strong El Niños, such as those of 1983 and 1998 (Timmermann et al., 1999; Hansen et al., 2006). It has been predicted that anthropogenic warming and sea level rise would continue for centuries due to the time scales associated with climate processes and feedbacks, even if greenhouse gas concentrations were to be stabilized (IPPC, 2007). Global warming of more than ≈1°C, relative to 2000, will constitute dangerous climate change due to likely effects on sea level and extinction of species (Hansen et al., 2006).

Seabirds are key indicators of the impact of climate change on the global ocean (BirdLife International, 2008b). Although the impact of climate change on several large-scale oceano-climatic fluctuations, including the El Niño-Southern Oscillation, (ENSO) is difficult to predict, it has been suggested that global warming may result in more frequent and intense

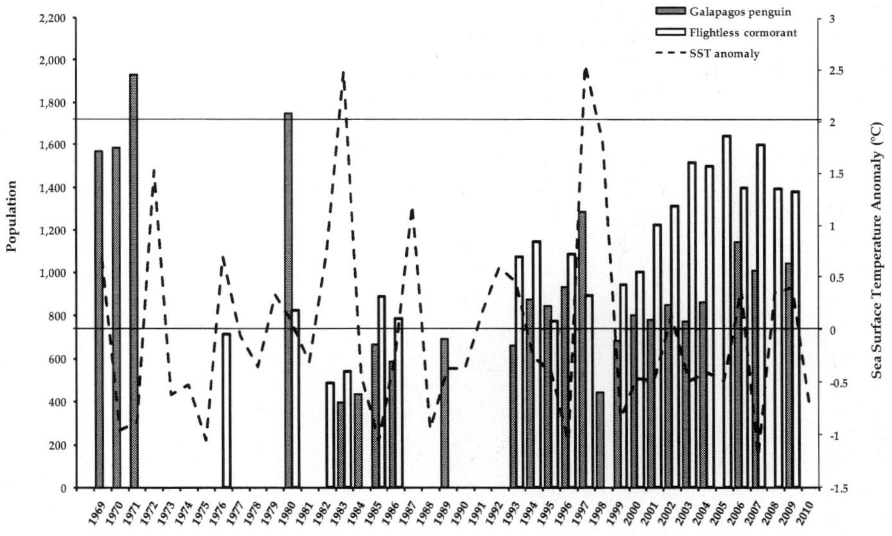

Fig. 5. Time series data of Galapagos penguin (grey bars) and Flightless cormorant (white bars) populations and sea surface temperature (SST) anomalies (dashed line) for El Niño regions 1 and 2, which engulf the Galapagos Archipelago in the Southeastern Tropical Pacific Ocean region. The positive temperature anomalies exceeding 2°C (solid, black line) indicate strong El Niño events (i.e., 1982-1983; and, 1997-1998). SST anomalies are good indicator of El Niño events; thus, SST anomaly data was collected between 1969 and 2009. SST anomalies were available from the NOAA National Weather Service (NOAA, 2011): (http://www.cpc.ncep.noaa.gov/products/analysis_monitoring/ensostuff/ONI_change.shtml).

El Niño events (Timmermann et al., 1999). Therefore, it is likely that the most significant threat from global climate change is its potential to affect the frequency and severity of ENSO events and the associated to lack of primary productivity, impacting endemic Galapagos seabirds and coastal waterbirds (Vargas et al., 2006; Wiedenfeld & Jiménez-Uzcátegui, 2008). Increases in sea surface temperature deplete primary production disrupting the bottom of marine food webs, and therefore top predators. El Niño may severely affect marine species especially small population of seabird such Galapagos penguins and Flightless cormorants (Wiedenfeld & Jiménez-Uzcátegui, 2008; Vargas et al., 2006). For instance, the 2004 penguin population (858 penguins) was estimated to be less than 50% of that prior to the strongest 1982–1983 El Niño event, including the population counted in the early 1970s, when the total number was 1931 penguins (Vargas et al., 2005; Vargas et al., 2006; Vargas et al., 2007), as shown in Fig 5. This underlines that the strong El Niño events of 1982-1983 and 1997-1998 were followed by population declines of more than 60% from which the species has yet to recover (Vargas et al., 2007). The censuses for Galapagos penguin and Flightless cormorants conducted in the last decade (2000-2010) appear to show a moderate positive or stable trend for both species (Jiménez-Uzcátegui et al., 2006; Jiménez-Uzcátegui & Vargas, 2007; Jiménez-Uzcátegui & Devineau, 2009), underscoring the potential recover during cold La Niña episodes (Vargas et al., 2006), but above all the ecological resilience of these endemic seabirds to overcome the environmental change and. In addition, sea level rise and shift in suitable climatic conditions attributable to global warming may damage coastal habitats such as mangroves and lagoons (Wiedenfeld & Jiménez-Uzcátegui, 2008; BirdLife International, 2008b).

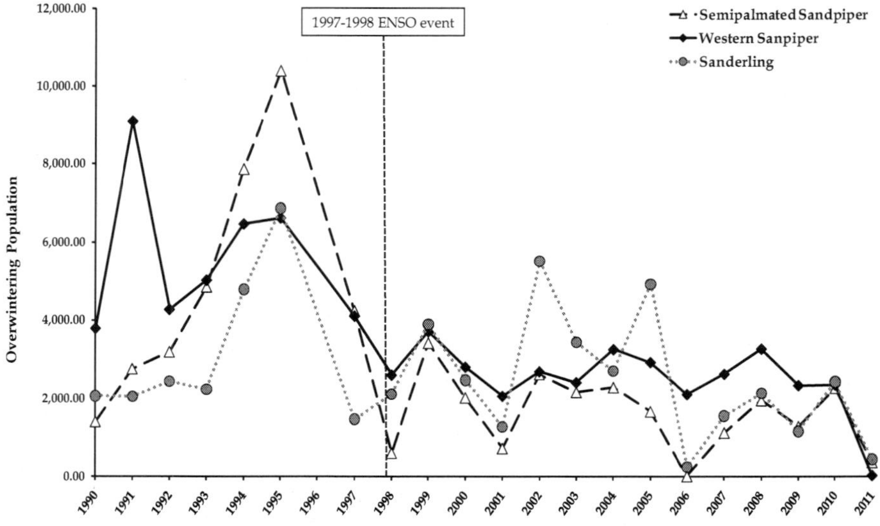

Fig. 6. Local population trends of migratory sandpipers observed at Ecuasal lagoons on coastal Ecuador from 1990 to 2011 (data adapted from Haase, 2011b).

Similarly, negative trends have recently been observed for local populations of several migratory shorebird species overwintering at coastal Ecuador from 1991 to 2011 (Haase,

2011b). Some of the species include the Western Sandpiper (*C. mauri*), Semipalmated Sandpiper (*C. pusilla*) and Sanderling (*C. alba*). The decline is more evident after the El Niño event of 1997-1998 (Haase, 2011b), as shown in Fig. 6.

6. Conservation and management implications

Several international conventions aimed to conserve and protect the biodiversity and environment as well as cultural and natural heritage within the country have been signed and ratified by the Ecuadorian government. These include the Convention on Biological Diversity (CBD, ratified in 1993), World Heritage Convention (signed in 1973), Convention on Migratory species (ratified in 2004), Convention on International Trade in Endangered Species of Wild Fauna and Flora (signed in 1974), Ramsar Convention on Wetlands (ratified in 1990). Ecuador has also signed bilateral environmental agreements with Peru and Colombia. Despite of Ecuador's commitment to internalize and pursue the goals of these agreements and the existence of several environmental laws, regulations and acts at the National level, lack of law empowering is observed at the local and regional levels and violations are scarcely sanctioned. Empowerment and enforcement of regulations and laws are necessary to accomplish legal protection of threatened species and conservation of critical habitats for waterbirds. Best management practices and effective land-use zoning, utilizing buffer zones between agricultural (plantations and farms) areas and the mangrove wetlands would ensure protection of local biodiversity. Best framing practices and the establishment of buffer zones (i.e., 100m) are needed to mitigate the agriculture expansion and cattle ranching in some areas such as those found around the El Canclon lagoon and Santay Island. Community-based conservation and environmental awareness might be undertaken by building capacity of local stakeholders (e.g. farmers and ranchers) in sustainable aquaculture/agriculture and nature tourism in areas that have received scant attention. At the Ecuasal lagoons, for instance, the owner and manager (Ecuasal Company) of the two production plants (Salinas and Pacoa lagoons) has lent its facilities to local ornithologists for the study of the birds and has shown great interest in supporting bird conservation and ecotourism during the last 10 years. Likewise, the Control and Surveillance System of the Mangrove Clear-Cutting Project conducted by Fundacion Natura and the National Chamber of Aquaculture hampered significantly the deforestation of mangrove forests on the Ecuadorian continental coast between 1998 and 2002 (Carvajal & Alava, 2007). Yet, field research is necessary for studying the relationship between the abundance waterbird species, including common and rare species, and disturbed and undisturbed mangrove areas.

Introduced plants and animals represent one of the greatest threats to the ecosystems of Galapagos. The invasive species eradication and control program of the Galapagos National Park Service and the Galapagos Inspection and Quarantine System (SICGAL) promise the avoidance and restriction of alien species to the islands. A recent triumph in this arena was the successful eradication of introduced and feral goats in Santiago Island (Cruz et al., 2009); yet, goats from other island still need to be removed. Similarly, industrial longline fishing activities have continued within Galapagos waters and several illegal vessels have been confiscated since the GMR was established. It has been suggested that the reduction of adult mortality of albatrosses in the coastal fishery of Ecuador and Peru appears to be the most effective means to stabilize this threatened species (Anderson et al., 2008). Currently, the ongoing institutional cooperation and surveillance system involving the Galapagos National Park Service, The Ecuadorian Navy Forces and non-governmental organizations assure in somehow the control and enforcement of fishing prohibitions to mitigate the bycatch problem within the GMR.

7. Acknowledgements

We acknowledge the field work by volunteers and our assistants, M. Constantino, E. Astudillo, X. Arosemena, A. Martinez, M. Peñafiel and C. Bohórquez, C. Gutierrez, G. Nazareno, M. Bruno, G. Baez, P. J. Jimenez, R. Carvajal and C. Cajas as well as the volunteers of the Ecuadorian Foundation for the Study of Marine Mammals. Part of the information and data generated for this contribution was possible under the project 'Conservation of the El Canclon Lake to protect *Anhima cornuta* in the Manglares Churute National Ecological Reserve (ECU/00/009), awarded by the GEF–Small Grant Programme (Operational programme OP2: Coastal, Marine and Freshwater Ecosystems) of the United Nations Development Programm, as well as the Project, Control and Surveillance System of the Mangrove Clear-Cutting on the Ecuadorian Continental Coast sponsored by the National Chamber of Aquaculture (Camara Nacional de Acuacultura). Both projects were administered by Fundación Natura Capítulo. We are greatful to the Ecuasal Company, for giving permission to do the census and field work within their property. The census work and the banding project were carried out with financial help from the US Fish and Wildlife Service, the Canadian Wildlife service and the Manomet Centre for conservation Sciences. Special thanks to Brian Harrington, Robert Elner and Ron Ydenberg for stimulating our conservation efforts in Ecuador.

Order: Family/common name	Species	Remarks/**Global conservation status**
ANSERIFORMES: Anhimidae		
Horned Screamer	*Anhima cornuta*	**Least Concern**
ANSERIFORMES: Anatidae		
Black-bellied Whistling-Duck	*Dendrocygna autumnalis*	
Fulvous Whistling-Duck	*Dendrocygna bicolor*	
Comb Duck	*Sarkidiornis melanotos*	
Orinoco Goose	*Neochen jubata*	**Near-threatened**
Muscovy Duck	*Cairina moschata*	**Near-threatened**
Torrent Duck	*Merganetta armata*	
Blue-winged Teal	*Anas discors*	
Cinnamon Teal	*Anas cyanoptera*	Extirpated
Northern Shoveler	*Anas clypeata*	Rare/Accidental
White-cheeked Pintail	*Anas bahamensis*	
Yellow-billed Pintail	*Anas georgica*	
Andean Teal	*Anas andium*	
Southern Pochard	*Netta erythrophthalma*	**Least Concern**
Lesser Scaup	*Aythya affinis*	Rare/Accidental
Masked Duck	*Nomonyx dominicus*	
Ruddy Duck	*Oxyura jamaicensis*	
PODICIPEDIFORMES: Podicipedidae		
Least Grebe	*Tachybaptus dominicus*	

Order: Family/common name	Species	Remarks/**Global conservation status**
Pied-billed Grebe	*Podilymbus podiceps*	
Great Grebe	*Podiceps major*	Rare/Accidental
Silvery Grebe	*Podiceps occipitalis*	**Least Concern**
PHOENICOPTERIFORMES: Phoenicopteridae		
American Flamingo	*Phoenicopterus ruber*	
Chilean Flamingo	*Phoenicopterus chilensis*	**Near-threatened**
SPHENISCIFORMES: Spheniscidae		
Humboldt Penguin	*Spheniscus humboldti*	Rare/Accidental/**Vulnerable**
Galapagos Penguin	*Spheniscus mendiculus*	Endemic/ **Endangered**
PROCELLARIIFORMES: Diomedeidae		
Black-browed Albatross	*Thalassarche melanophris*	Rare/Accidental
Wandering Albatross	*Diomedea exulans*	Rare/Accidental
Waved Albatross	*Phoebastria irrorata*	**Critically endangered**
Black-footed Albatross	*Phoebastria nigripes*	Rare/Accidental/**Endangered**
PROCELLARIIFORMES: Procellariidae		
Southern Giant-Petrel	*Macronectes giganteus*	Rare/Accidental/**Near-threatened**
Southern Fulmar	*Fulmarus glacialoides*	Rare/Accidental
Cape Petrel	*Daption capense*	Rare/Accidental
Galapagos Petrel	*Pterodroma phaeopygia*	Breeding endemic/**Critically endangered**
Gould's Petrel	*Pterodroma leucoptera*	Rare/Accidental
White-chinned Petrel	*Procellaria aequinoctialis*	Rare/Accidental/**Vulnerable**
Parkinson's Petrel	*Procellaria parkinsoni*	**Common/Vulnerable**
Pink-footed Shearwater	*Puffinus creatopus*	Accidental/**Vulnerable**
Wedge-tailed Shearwater	*Puffinus pacificus*	Rare
Buller's Shearwater	*Puffinus bulleri*	Rare/Accidental/**Vulnerable**
Sooty Shearwater	*Puffinus griseus*	Near-threatened; Common
Galapagos Shearwater	*Puffinus subalaris*	Endemic/Rare coastal
PROCELLARIIFORMES: Hydrobatidae		
Wilson's Storm-Petrel	*Oceanites oceanicus*	Rare/Accidental
Elliot's Storm-Petrel	*Oceanites gracilis*	Data deficient
White-faced Storm-Petrel	*Pelagodroma marina*	Rare/Accidental
White-bellied Storm-Petrel	*Fregetta grallaria*	Rare/Accidental
Polynesian Storm-Petrel	*Nesofregetta fuliginosa*	Vulnerable
Ringed Storm-Petrel	*Oceanodroma hornbyi*	Accidental/**Data deficient**
Leach's Storm-Petrel	*Oceanodroma leucorhoa*	Rare/Accidental

Order: Family/common name	Species	Remarks/**Global conservation status**
Ashy Storm-Petrel	*Oceanodroma homochroa*	Rare/Accidental **Endangered**
Band-rumped Storm-Petrel	*Oceanodroma castro*	Accidental
Wedge-rumped Storm-Petrel	*Oceanodroma tethys*	Common
Black Storm-Petrel	*Oceanodroma melania*	Common
Markham's Storm-Petrel	*Oceanodroma markhami*	Rare/Accidental/**Data deficient**
Least Storm-Petrel	*Oceanodroma microsoma*	Accidental
PHAETHONTIFORMES: Phaethontidae		
Red-billed Tropicbird	*Phaethon aethereus*	Coastal Accidental
CICONIIFORMES: Ciconiidae		
Jabiru	*Jabiru mycteria*	Rare/Accidental
Wood Stork	*Mycteria americana*	
SULIFORMES: Fregatidae		
Magnificent Frigatebird	*Fregata magnificens*	Common
Great Frigatebird	*Fregata minor*	Coastal rare
SULIFORMES: Sulidae		
Nazca Booby	*Sula granti*	Common
Blue-footed Booby	*Sula nebouxii*	Common
Peruvian Booby	*Sula variegata*	Common
Red-footed Booby	*Sula sula*	Coastal Rare
Brown Booby	*Sula leucogaster*	Hypothetical
SULIFORMES: Phalacrocoracidae		
Flightless Cormorant	*Phalacrocorax harrisi*	Endemic/**Endangered**
Neotropic Cormorant	*Phalacrocorax brasilianus*	Common
Guanay Cormorant	*Phalacrocoraxbougainvillii*	Rare/**Near-threatened**
SULIFORMES: Anhingidae		
Anhinga	*Anhinga anhinga*	
PELECANIFORMES: Pelecanidae		
Brown Pelican	*Pelecanus occidentalis*	Common
Peruvian Pelican	*Pelecanus thagus*	Common/**Near-threatened**
PELECANIFORMES: Ardeidae		
Pinnated Bittern	*Botaurus pinnatus*	**Least Concern**
Zigzag Heron	*Zebrilus undulatus*	
Least Bittern	*Ixobrychus exilis*	
Rufescent Tiger-Heron	*Tigrisoma lineatum*	
Fasciated Tiger-Heron	*Tigrisoma fasciatum*	
Great Blue Heron	*Ardea herodias*	
Cocoi Heron	*Ardea cocoi*	Coastal common
Great Egret	*Ardea alba*	Common

Order: Family/common name	Species	Remarks/**Global conservation status**
Little Egret	*Egretta garzetta*	
Snowy Egret	*Egretta thula*	Common
Little Blue Heron	*Egretta caerulea*	Locally common
Tricolored Heron	*Egretta tricolor*	Common
Cattle Egret	*Bubulcus ibis*	Common
Green Heron	*Butorides virescens*	
Striated Heron	*Butorides striata*	Locally Common
Agami Heron	*Agamia agami*	
Whistling Heron	*Syrigma sibilatrix*	Rare/Accidental
Capped Heron	*Pilherodius pileatus*	
Black-crowned Night-Heron	*Nycticorax nycticorax*	Common
Yellow-crowned Night-Heron	*Nyctanassa violacea*	Common
Boat-billed Heron	*Cochlearius cochlearius*	
PELECANIFORMES: Threskiornithidae		
White Ibis	*Eudocimus albus*	Locally Common
Scarlet Ibis	*Eudocimus ruber*	Rare/Accidental
Glossy Ibis	*Plegadis falcinellus*	
Green Ibis	*Mesembrinibis cayennensis*	
Bare-faced Ibis	*Phimosus infuscatus*	Rare/Accidental
Black-faced Ibis	*Theristicus melanopis*	**Least Concern**
Roseate Spoonbill	*Platalea ajaja*	Very locally common
GRUIFORMES: Rallidae		
Rufous-sided Crake	*Laterallus melanophaius*	
White-throated Crake	*Laterallus albigularis*	
Gray-breasted Crake	*Laterallus exilis*	
Galapagos Rail	*Laterallus spilonotus*	Endemic/**Vulnerable**
Clapper Rail	*Rallus longirostris*	**Least Concern**
Virginia Rail	*Rallus limicola*	
Brown Wood-Rail	*Aramides wolfi*	**Vulnerable**
Rufous-necked Wood-Rail	*Aramides axillaris*	**Least Concern**
Gray-necked Wood-Rail	*Aramides cajanea*	
Red-winged Wood-Rail	*Aramides calopterus*	
Uniform Crake	*Amaurolimnas concolor*	
Chestnut-headed Crake	*Anurolimnas castaneiceps*	
Russet-crowned Crake	*Anurolimnas viridis*	
Black-banded Crake	*Anurolimnas fasciatus*	
Sora	*Porzana carolina*	
Colombian Crake	*Neocrex colombiana*	**Data deficient**
Paint-billed Crake	*Neocrex erythrops*	
Spotted Rail	*Pardirallus maculatus*	
Blackish Rail	*Pardirallus nigricans*	

Order: Family/common name	Species	Remarks/**Global conservation status**
Plumbeous Rail	*Pardirallus sanguinolentus*	
Purple Gallinule	*Porphyrio martinica*	Locally Common
Azure Gallinule	*Porphyrio flavirostris*	
Common Moorhen	*Gallinula chloropus*	Locally Common
American Coot	*Fulica americana*	**Extirpated**
Slate-colored Coot	*Fulica ardesiaca*	Coastal Accidental
GRUIFORMES: Heliornithidae		
Sungrebe	*Heliornis fulica*	
EURYPYGIFORMES: Eurypygidae		
Sunbittern	*Eurypyga helias*	
GRUIFORMES: Aramidae		
Limpkin	*Aramus guarauna*	
GRUIFORMES: Psophiidae		
Gray-winged Trumpeter	*Psophia crepitans*	
CHARADRIIFORMES: Burhinidae		
Peruvian Thick-knee	*Burhinus superciliaris*	Locally common
CHARADRIIFORMES: Charadriidae		
Pied Lapwing	*Vanellus cayanus*	
Southern Lapwing	*Vanellus chilensis*	
Andean Lapwing	*Vanellus resplendens*	Rare
Black-bellied Plover	*Pluvialis squatarola*	Common
American Golden-Plover	*Pluvialis dominica*	Accidental
Pacific Golden-Plover	*Pluvialis fulva*	Rare
Collared Plover	*Charadrius collaris*	Locally common
Snowy Plover	*Charadrius alexandrinus*	Locally Common
Wilson's Plover	*Charadrius wilsonia*	Locally Common
Semipalmated Plover	*Charadrius semipalmatus*	Common
Piping Plover	*Charadrius melodus*	Rare/Accidental/**Near-threatened**
Killdeer	*Charadrius vociferus*	Locally Common
Tawny-throated Dotterel	*Oreopholus ruficollis*	Extirpated
CHARADRIIFORMES: Haematopodidae		
American Oystercatcher	*Haematopus palliatus*	Locally Common
CHARADRIIFORMES: Recurvirostridae		
Black-necked Stilt	*Himantopus mexicanus*	Common
American Avocet	*Recurvirostra americana*	Rare

Order: Family/common name	Species	Remarks/**Global conservation status**
CHARADRIIFORMES: Jacanidae		
Wattled Jacana	*Jacana jacana*	Locally Common
CHARADRIIFORMES: Scolopacidae		
Spotted Sandpiper	*Actitis macularius*	Common
Solitary Sandpiper	*Tringa solitaria*	
Wandering Tattler	*Tringa incana*	Locally Common
Greater Yellowlegs	*Tringa melanoleuca*	Common
Willet	*Tringa semipalmata*	Common
Lesser Yellowlegs	*Tringa flavipes*	Common
Upland Sandpiper	*Bartramia longicauda*	Coastal accidental
Whimbrel	*Numenius phaeopus*	Common
Hudsonian Godwit	*Limosa haemastica*	Accidental
Marbled Godwit	*Limosa fedoa*	Rare/Accidental
Ruddy Turnstone	*Arenaria interpres*	Common
Black Turnstone	*Arenaria melanocephala*	Rare/
Surfbird	*Aphriza virgata*	Locally common
Red Knot	*Calidris canutus*	Accidental
Sanderling	*Calidris alba*	Common
Semipalmated Sandpiper	*Calidris pusilla*	Common
Western Sandpiper	*Calidris mauri*	Locally Common
Least Sandpiper	*Calidris minutilla*	Common
White-rumped Sandpiper	*Calidris fuscicollis*	Coastal accidental
Baird's Sandpiper	*Calidris bairdii*	Accidental
Pectoral Sandpiper	*Calidris melanotos*	Accidental
Curlew Sandpiper	*Calidris ferruginea*	Rare/Accidental
Stilt Sandpiper	*Calidris himantopus*	Locally Common
Buff-breasted Sandpiper	*Tryngites subruficollis*	**Near-threatened**
Short-billed Dowitcher	*Limnodromus griseus*	Locally Common
Long-billed Dowitcher	*Limnodromus scolopaceus*	Rare
South American Snipe	*Gallinago paraguaiae*	
Noble Snipe	*Gallinago nobilis*	
Andean Snipe	*Gallinago jamesoni*	
Imperial Snipe	*Gallinago imperialis*	**Near-threatened**
Wilson's Phalarope	*Phalaropus tricolor*	Locally Common
Red-necked Phalarope	*Phalaropus lobatus*	Common
Red Phalarope	*Phalaropus fulicarius*	Accidental
CHARADRIIFORMES: Thinocoridae		
Rufous-bellied Seedsnipe	*Attagis gayi*	
Least Seedsnipe	*Thinocorus rumicivorus*	Rare/Accidental

Order: Family/common name	Species	Remarks/**Global conservation status**
CHARADRIIFORMES: Laridae		
Swallow-tailed Gull	*Creagrus furcatus*	Coastal accidental
Sabine's Gull	*Xema sabini*	Locally Common
Bonaparte's Gull	*Chroicocephalus philadelphia*	Rare/Accidental
Andean Gull	*Chroicocephalus serranus*	Coastal Accidental
Gray-hooded Gull	*Chroicocephalus cirrocephalus*	Common
Gray Gull	*Leucophaeus modestus*	Locally Common
Laughing Gull	*Leucophaeus atricilla*	Common
Franklin's Gull	*Leucophaeus pipixcan*	Common
Lava Gull	*Leucophaeus fuliginosus*	Endemic/**Vulnerable**
Belcher's Gull	*Larus belcheri*	Rare
Ring-billed Gull	*Larus delawarensis*	Rare
Lesser Black-backed Gull	*Larus fuscus*	Rare
Kelp Gull	*Larus dominicanus*	Common
Glaucous-winged Gull	*Larus glaucescens*	Rare
Brown Noddy	*Anous stolidus*	Rare
Sooty Tern	*Onychoprion fuscatus*	Not coastal/**Data deficient**
Bridled Tern	*Onychoprion anaethetus*	Locally Common
Yellow-billed Tern	*Sternula superciliaris*	Not Coastal
Peruvian Tern	*Sternula lorata*	**Endangered**
Large-billed Tern	*Phaetusa simplex*	Coastal Rare
Gull-billed Tern	*Gelochelidon nilotica*	Locally Common
Inca Tern	*Larosterna inca*	Near-threatened/Rare
Black Tern	*Chlidonias niger*	Accidental
Common Tern	*Sterna hirundo*	Accidental
Arctic Tern	*Sterna paradisaea*	Locally uncommon
South American Tern	*Sterna hirundinacea*	Locally Common
Royal Tern	*Thalasseus maximus*	Common
Sandwich Tern	*Thalasseus sandvicensis*	Common
Elegant Tern	*Thalasseus elegans*	Common /**Near-threatened**
Black Skimmer	*Rynchops niger*	Accidental/ Locally Common/Least Concern
CHARADRIIFORMES: Stercorariidae		
Pomarine Jaeger	*Stercorarius pomarinus*	Accidental
Parasitic Jaeger	*Stercorarius parasiticus*	Locally Common
Long-Tailed Jaeger	*Stercorarius longicaudus*	Rare
South Polar Skua	*Stercorarius maccormicki*	Rare

Annex I. List of waterbird species of Ecuador, including species of the Galapagos Islands Scientific and English names follow the South American Classification Committee (SACC) Classification, Version 31 March 2011 (Freile, 2010).

8. References

Alava, J.J. (2002). Registros y abundancia relativa de mamíferos marinos durante el crucero oceanográfico insular B/ I ORION (CO-II-2000) en las Islas Galápagos y sus alrededores. *Acta Oceanográfica del Pacífico* Vol.11, No. 1, pp. 165-172

Alava, J.J.; Costantino, M.; Astudillo, E. & Arosemena, X. (2002). *Estudio de la Población del canclón, Anhima cornuta y su habita en la Reserva Ecológica Manglares Churute, Ecuador.* PPD-PNUMA, Fundación Natura Capitulo Guayaquil, REMCH, Ministerio del Ambiente, Club Observadores de Aves, Guayaquil, Ecuador.

Alava, J.J. (2005). Censo, distribución y estado de conservación de la Garza pico de espatula (*Ajaia ajaja*) en el Estuario del Golfo de Guayaquil, Ecuador. *Ornitologia Neotropical,* Vol.16, No.2, (May 2005) pp. 175–185, ISSN 1075-4377

Alava, J.J.; Arosemena, X. & Angel, R. (2007). Brown Wood-rail *Aramides wolfi* at El Canclon Lagoon, Manglares-Churute Ecological Reserve, Ecuador. *Cotinga,* Vol. 27, pp. 81-82, ISSN 1353-985X

Alava, J.J.; Arosemena, X.; Astudillo, E.; Costantino, M.; Peñafiel, M. & Bohorquez, C. (2007). Occurrence, abundance, and notes on some threatened Ecuadorian birds observed in the El Canclon Lagoon, Manglares Churute Ecological Reserve. *Ornitologia Neotropical* Vol.18, No.2, (May 2007), pp. 223–232, ISSN 1075-4377

Alava, J.J., Costantino, M.; Astudillo, E.; Arosemena, X. & M. Peñafiel. (2009). Population, seasonality and conservation threats of the Horned Screamer (*Anhima cornuta*) in Southwestern Ecuador. *Waterbirds* Vol.32, No. 1, (March 2009), pp.81-86, ISNN 1524-4695

Alava, J.J., Saavedra, M.; Arosemena, X.; Calle, M.; Vinueza, C.; Carvajal, R. & Vargas, F. H. (2010). Observations, distribution and potential threats to the Mangrove Black Hawk *Buteogallus anthracinus subtilis* in mangrove habitat of southwestern Ecuador. *Boletín de la Sociedad Antioqueña de Ornitología (Boletin SAO)* Vol. 20, No.2, *In press,* Available from http://www.sao.org.co/boletinsao.html

Alava, J.J.; Jiménez, P.; Benner, J.; Jimenez-Uzcategui, G.; Cruz-Delgado, F.; Torres, S. & Vargas, F. H. (2010). At sea distribution and abundance of seabirds around the Galapagos Islands-Ecuador: Conservation implications *Abstracts of the 1st World Seabird Conference: Seabirds: Linking the Global Oceans,* pp. 36–37 , Victoria Canada, September 7-11, 2010

Alava, J.J.; Ikonomou, M. G.; Ross, P. S.; Costa, D. P.; Salazar, S.; Aurioles-Gamboa, D.; & Gobas, F.A.P.C. (2009). Polychlorinated biphenyls and polybrominated diphenyl ethers in Galapagos sea lions (*Zalophus wollebaeki*). *Environmental Toxicology and Chemistry* Vol. 28, No.11, (May 2009), pp. 2271-2282, ISSN 0730-7268

Alava, J.J.; Ross, P. S.; Ikonomou, M. G.; Cruz, M.; Jimenez-Uzcategui, G.; Salazar, S.; Costa, D. P.; Villegas-Amtmann, S.; Howorth, P.; & Gobas, F.A.P.C. (2011). DDT in endangered Galapagos Sea Lions (*Zalophus wollebaeki*) from eight rookeries in the Galapagos Islands. *Marine Pollution Bulletin* Vol. 62, No. 4, (April 2011), pp. 660-671, ISSN 0025-326X

Alava, J.J. (2011). *Bioaccumulation of pollutants in Galapagos sea lions and marine mammals from British Columbia, Canada.* PhD Thesis, School of Resource and Environmental Management, Faculty of Environment, Simon Fraser University, BC, Canada.

Anderson, D. J.; Huyvaert, K. P.; Wood, D. R.; Gillikin, C. L.; Frost, B. J. & Mouritsen, H. (2003). At-sea distribution of waved albatrosses and the Galapagos Marine Reserve. *Biological Conservation* Vol. 110, No.3, (April 2003), pp. 367-373, ISSN 0006-3207

Anderson, D.J.; Huyvaert, K. P.; Awkerman, J. A.; Proaño, C. B.; Milstead, W. B.; Jiménez-Uzcátegui, G.; Cruz, S. & Grace, J. K. (2008). Population status of the critically endangered Waved Albatross *Phoebastria irrorata*, 1999-2007.*Endangered Species Research* Vol. 5, No. (December 2008) pp. 185–192 doi: 10.3354/esr00089

Awkerman, J.A.; Huyvaert, K.P.; Mangel, J.; Alfaro Shigueto, J. & Anderson, D.J. (2006). Incidental and intentional catch threatens Galápagos waved albatross. *Biological Conservation* Vol. 133, No. 4, (December 2006), pp. 483–489, ISSN: 0006-3207

Bax, N.; Williamson, A.; Aguero, M.; Gonzalez, E. & Geeves, W. (2003). Marine invasive alien species: a threat to global biodiversity. *Marine Policy* Vol. 27, No. 4, (July 2003), pp. 313–323, ISSN: 0308-597X

Best, B. J. & Kessler, M. (1995). *Biodiversity and conservation in Tumbesian Ecuador and Peru*, BirdLife International, ISBN 0-946888-26-4, Cambridge, United Kingdom.

BirdLife International. (2008a). *State of the world's birds: indicators for our changing world*, BirdLife International, ISBN 978-0-946888-63-4, Cambridge, UK.

BirdLife International. (2008b) Seabirds are key indicators of the impact of climate change on the worlds oceans. Presented as part of the BirdLife State of the world's birds website, Available from http://www.birdlife.org/datazone/sowb/casestudy/279

BirdLife International. (2009). Seabird bycatch mitigation measures, *Western and Central Pacific Fisheries Commission BirdLife International Global Seabird Programme*, WCPFC-SC5-2009/EB-WP-03., Port Vila, Vanuatu, August 10-21, 2009.

BirdLife International. (2010). *Marine Important Bird Areas: priority sites for the conservation of biodiversity*, BirdLife International, ISBN 978-0-946888-74-0, Cambridge, UK.

BirdLife International (2011). IUCN Red List for birds, 14.04.2011, Available from http://www.birdlife.org

Boersma, P.D. & Parrish, J.K.(1999). Limiting abuse: marine protected areas, a limited solution. *Ecological Economics* Vol. 31, No. 2, (November 1999), pp. 287–304, ISSN: 0921-8009

Briones, E.; Gómez, J.; Hidalgo, A.; Tirira, D. & Flacher, A. (2001). *Inventario de Humedales del Ecuador. Segunda Parte: Humedales Interiores de la Provincia del Guayas*. Tomo II, Convención de Ramsar, Ministerio del Ambiente and EcoCiencia, Quito, Ecuador.

Brothers, N.; Duckworth, A.R.; Safina, C. & Gilman, E.L. (2010). Seabird bycatch in pelagic longline fisheries is grossly underestimated when using only haul data. PLoS ONE Vol. 5, No. 8, e12491 doi:10.1371/journal.pone.0012491

Calle, M. (2003). Diagnostico y Propuesta de Manejo Ambiental del Remanente de Manglar de Bajo Alto-La Puntilla, Provincia El Oro. Tesis de Maestria, Universidad Agraria del Ecuador, Guayaquil, Ecuador.

Calle, D. P. & Alava, J. J. (2009). Coastal chemical pollution at continental Ecuador: The case of the Gulf of Guayaquil, *9th Congress of the Society of Environmental Toxicology and Chemistry (SETAC LA) and the II Peruvian Congress of Ecotoxicology and Environmental Chemistry (SETAC PERU)*– International Meeting "Ecotoxicology and Social

Responsibility". SETAC PERU. The Biologist (Lima) Vol. 7, No. 1-2, pp. 13, Lima, Perú, October 05-09, 2009

Carlton, J.T. (1989). Man's role in changing the face of the ocean: biological invasions and implications for conservation of near-shore environments. *Conservation Biology* Vol. 3, No. 3, (September 1989), pp. 265–273. ISSN: 0888-8892

Carlton, J.T. & Geller, J.B.(1993). Ecological roulette: the global transport of nonindigenous marine organisms. *Science* Vol. 261, No. 5117, (July 1993), pp. 78–82, ISSN 0036-8075

Carlton, J.T. (1996). Pattern, process, and predictions in marine invasion ecology. *Biological Conservation* Vol. 78, No. 1-2, (October-November 1996), pp. 97–106, ISSN: 0006-3207

Carvajal-M., R.; Jiménez, P.; Saavedra, M. & Iturralde-Muñoz, G. (2006). *Zonificación de la Reserva de Producción de Fauna Manglares El Salado,* Fundación Natura Capitulo Guayaquil, Ministerio del Ambiente y M. I. Municipalidad de Guayaquil, Guayaquil, Ecuador.

Carvajal, R. & Alava, J.J. (2007).Mangrove Wetlands Conservation Project and the Shrimp Farming Industry in Ecuador:

Lessons Learned. *World Aquaculture* Vol. 38, No. 3, pp. 14-17

Cepeda, F. & Cruz, J. B. (1994). Status and management of seabirds on the Galápagos Islands, Ecuador, In: *Seabirds on islands: threats, case studies and action plans* D.N., Nettleship, J. Burger & M., Gochfeld, (Eds.), pp. 268-278 BirdLife International (BirdLife Conservation Series), Cambridge, U.K.

CLIRSEN. (1999). *Estudio multitemporal de manglares, camaroneras y salinas al año 1999.* Centro de Levantamiento Integrado de Recursos Naturales por Sensores Remotos/Ministerio de Defensa & PATRA, Guayaquil, Ecuador.

CLIRSEN. (2007). *Estudio multitemporal de manglares, camaroneras y salinas al año 2006.* Centro de Levantamiento Integrado de Recursos Naturales por Sensores Remotos/Ministerio de Defensa, Ministerio del Ambiente, Programa de Recursos Pesqueros Guayaquil, Ecuador.

Cruz-Delgado, F. (2001). Distribución de aves marinas al norte y oeste del Archipiélago de Galápagos. Área de Ecología y Monitoreo de Vertebrados, Estación Científica Charles Darwin, Unpublished Report, Septiembre 2001, Puerto Ayora, Galapagos Islands, Ecuador.

Cruz, F.; Carrion, V.; Campbell, K. J.; Lavoie, C. & Donlan, C. J. (2009). Bio-economics of large-scale eradication of feral goats from Santiago Island, Galápagos. *The Journal of Wildlife Management* Vol. 73, No. 2, (February 2009), pp.

191–200, ISSN: 1937-2817

Delgado, J. D.; Mejia, A.; Bohorquez, S.; Marechal, C. & J. Orellana. (2000). Ficha Informativa de los Humedales Ramsar: Isla Santay. Convención Ramsar sobre los Humedales. Recomendación 4.7 Conferencia de las Partes (COP), Guayaquil, Ecuador

Darquea-Arteaga, J.; Baquero, A.; Hardesty, J. & Vallejo, F. (2010). Primear evaluación de captura incidental de aves marinas en la costa de Ecuador, *Primer Simposio de Biodiversidad Marina y Costera de Latinoamérica y el Caribe, Segundo Simposio Ecuatoriano de Biodiversidad Marina y Costera,* pp. 1, Grupo de Trabajo sobre

Biodiversidad Marina y Costera del Ecuador, Ministerio del Ambiente del Ecuador. Manta, Ecuador, Diciembre 9-12, 2010

Dodson, C. H. & A. H. Gentry. (1991). Biological extinction in western Ecuador. *Annals of the Missouri Botanical Gardens* Vol. 78, No.2, pp.273-295, ISSN: 00266493

Edgar, G.J.; Snell, H. L. & Lougheed, L. W. (2003). Impacts of the Jessica oil spill: an introduction. *Marine Pollution Bulletin* Vol. 47, No. 7-8, (July-August 2003), pp. 273-275, ISSN 0025-326X

Food Agriculture Organization (2007). FAOSTAT. In: *FAO Statistics Division*, 18.05.2007, Available from http://faostat.fao.org/site/567

Freile, J. (2010). Species lists of birds for South American countries and territories: Ecuador. Version 31 August 2010. 27.04.2011, Available from http://www.museum.lsu.edu/~Remsen/SACCCountryLists.html

Fundación Natura & WWF. (2000). Anexos 9. In: *Informe Galápagos 1999–2000,* Fundación Natura - World Wildlife Fund, 119-120, Quito, Ecuador.

GGEPL. (2004). Proceedings of the Galapagos West Nile Virus Workshop. In: *Galapagos Genetics, Epidemiology and Pathology Laboratory*. Galapagos National Park Headquarters, Puerto Ayora, Galapagos Islands, Ecuador, April 29, 2004. 02.05.2007, Available from http://www.fbs.leeds.ac.uk/ggepl/English/ggepl_publications.htm

Gibbs, J. P.; Shriver, W. G. & Vargas, H. (2003). An assessment of a Galapagos Rail population over thirteen years (1986–2000). *Journal of Field Ornithology* Vol. 74, No. 2, (Spring 2003), pp. 136–140, ISNN 0273-8570

Granizo, T.; Pacheco, C.; Ribadeneira, M. B.; Guerrero, M. & Suárez, L. (Eds.). (2002). *Libro Rojo de las aves del Ecuador*. SIMBIOE, Conservación Internacional, EcoCiencia, Ministerio del Ambiente & UICN, Quito, Ecuador.

Gurevitch, J. & Padilla, D. K. (2004). Are invasive species a major cause of extinctions? *Trends in Ecology and Evolution* Vol. 19, No. 9, (September 2004), pp. 470-474, ISSN 0169-5347

Haase, B. (1991a). The Santa Elena Peninsula: an important stopover for migratory birds, *Abstracts of the 4th Neotropical Ornithology Congress*, pp. 24, Ecuadorian Corporation of Ornithology (CECIA), Quito, Ecuador, November 3-9, 1991

Haase, B. (1991b). Presence and distribution of gulls in Ecuador. *Abstracts of the 4th Neotropical Ornithology Congress*, pp. 154, Ecuadorian Corporation of Ornithology (CECIA), Quito, Ecuador, November 3-9, 1991

Haase, B. (1991c). New marine bird and coastal bird species observed in Ecuador. *Abstracts of the 4th Neotropical Ornithology Congress*, pp. 22, Ecuadorian Corporation of Ornithology (CECIA), Quito, Ecuador, November 3-9, 1991

Haase, B. (1996). Kelp Gull *Larus dominicanus*: a new breeding species for Ecuador. *Cotinga* Vol. 5, pp. 73-74, ISSN 1353-985X

Haase, B. (2010). La migración de aves marinas observada desde la Chocolatera en Salinas, *Primer Simposio de Biodiversidad Marina y Costera de Latinoamérica y el Caribe, Segundo Simposio Ecuatoriano de Biodiversidad Marina y Costera*, pp.39, Grupo de Trabajo sobre Biodiversidad Marina y Costera del Ecuador, Ministerio del Ambiente del Ecuador, Manta, Ecuador, Diciembre 9-12, 2010

Haase, B.J.M. (2011a). *Las Aves Marinas de Ecuador Continental y Acuáticas de las Piscinas Artificiales de Ecuasal*, Gráficos Hernández, Guayaquil, Ecuador.

Haase, B. (2011b). Fluctuations in shorebirds populations along the Ecuadorean southwest coast (1990-2010), pp. 73. *Abstracts of the Fourth Western Hemisphere Shorebird Group Meeting*, Simon Fraser University, Burnaby, BC, Canada, August 11-15, 2011

Hansen, J.; Sato, Mki.; Ruedy, R.; Lo, K.; Lea, D. W. & Medina-Elizade, M. (2006). Global temperature change. *Proceedings of the National Academy of Sciences* Vol. 103, No. 39, (September 2006), pp. 14288-14293, ISSN 0027-8424

Harmon, W. M.; Clark, W. A.; Hawbecker, A. C. & Stafford, M. (1987). *Trichomonas gallinae* in columbiform birds from the Galápagos Islands. *Journal of Wildlife Diseases* Vol. 23, No. 3, (July 1), pp. 492-494, ISSN 0090-3558

Harvell, C. D.; Kim, K.; Burkholder, J. M.; Colwell, R. R.; Epstein, P. R.; Grimes, D. J.; Hofmann, E. E.; Lipp, E.K.; Osterhaus, A.D.M.E.; Overstreet, R.M.; Porter, J.W.; Smith, G.W. & Vasta, G.R. (1999). Emerging marine diseases-climate links and anthropogenic factors. *Science* Vol. 285, No. 5433, (September 1999), pp. 1505-1510, ISSN 0036-8075

INEC. (2007). VI Censo de Población y de Vivienda, In: *Instituto Nacional de Estadística y Censos (INEC)*, Quito, Ecuador. Available from http://www.inec.gov.ec/default.asp

INEFAN & Fundación Natura. (1996). *Plan de manejo – Reserva Ecológica Manglares Churute*. Fase I, Diagnóstico. Tomo I. Instituto Ecuatoriano Forestal y de Áreas Naturales & Fundación Natura, Guayaquil, Ecuador

IPCC. (2007). Summary for Policymakers, In: *Climate Change 2007: The Physical Science Basis*, S. Solomon, D. Qin, M. Manning, Z. Chen, M. Marquis, K.B. Averyt, M.Tignor & H.L. Miller, (Eds.), Contribution of Working Group I to the Fourth Assessment Report of the Intergovernmental Panel on Climate Change. Cambridge University Press, Cambridge, United Kingdom and New York, NY, USA

Jackson, M., H. (2001). *Galápagos, A Natural History*, University of Calgary Press, Alberta, Canada

Jahncke, J.; Goya, E. & Guillen, A. (2001). Seabird by-catch in smallscale longline .sheries in northern Peru. *Waterbirds* Vol. 24, No. 1 (April 2001) pp. 137–141, ISNN 1524-4695

Jaramillo, A.; Porozo, N.; Molina, R.; Naranjo, J. & A. Pacalla. (2002). Plan de Manejo del Humedal Isla Santay. Comité Ecológico del Litoral con la colaboración del Ministerio del Ambiente/DED Servicio Técnico de Cooperación Alemana. Guayaquil, Ecuador

Jiménez-Uzcátegui, G. & Wiedenfeld, D. A. (2002). Aves marinas, In: *Reserva Marina de Galápagos, Línea Base de laBiodiversidad*, E. Danulat & G.J. Edgar, (Eds.), pp. 343-372, Fundación Charles Darwin/Servicio Parque Nacional Galápagos, Santa Cruz, Galapagos, Ecuador

Jiménez Uzcátegui, G. (2006). Censo parcial de Flamencos *Phoenicopterus ruber* 2006. Unpublished Report, Fundación Charles Darwin & Servicio Parque Nacional Galápagos, Puerto Ayora, Galapagos, Ecuador

Jiménez-Uzcátegui, G.; Hernán Vargas, F.; Larrea, C.; Milstead, B. & Llerena, W. (2006). Galapagos Penguin and Flightless Cormorant survey 2006. Report for the Charles

Darwin Foundation, The Galapagos National Park and the Seaworld & Busch Gardens Conservation Fund, Puerto Ayora, Galapagos, Ecuador

Jiménez-Uzcátegui, G. & Vargas, F. H. (2007). Censo del pingüino de Galápagos y comoran no volador 2007. Charles Darwin Foundation and Galapagos National Park, Puerto Ayora, Galapagos, Ecuador

Jiménez-Uzcátegui, G.; Milstead, B.; Márquez, C.; Zabala, J.; Buitrón, P.; Llerena, A.; Salazar, S. & Fessl, B. (2007). Vertebrados de Galápagos: estado de amenaza y acciones de conservación, In: *Informe Galápagos 2006–2007*, Fundación Charles Darwin, Servicio Parque Nacional Galápagos & INGALA, Puerto Ayora, Galapagos, Ecuador

Jiménez-Uzcátegui, G. & Devineau, O. (2009). Censo del pingüino de Galápagos y comoran no volador 2009. Charles Darwin Foundation and Galapagos National Park, Puerto Ayora, Galapagos, Ecuador

Jiménez, P., Torres, S. & Alava, J.J. (2010). Inventario, Abundancia y Distribución Espacial de Mamíferos Marinos y Aves en las Islas Galápagos Durante el Crucero Oceanográfico (BAE Orión) en abril 2009. Implicaciones para Conservación. *Acta Oceanográfica del Pacifico (under review)*

Krebs, C. J. (1999). *Ecological methodology,* (2nd edition), Addison-Wesley Longman, Inc., Menlo Park, California.

Lessmann, R.P. (2004). Current protections on the Galapagos Islands are inadequate: the International Maritime Organization should declare the islands a Particularly Sensitive Sea Area. *Colorado Journal of International Environmental Law and Policy* Vol. 15, No. 1, (Winter 2004), pp. 117-151, ISSN 1050-0391

Lougheed, L.W., Edgar, G.J., & Snell, H.L. (Eds.). (2002). *Biological Impacts of the Jessica Oil Spill on the Galápagos Environment.* Final Report, Version No.1.10, Charles Darwin Foundation, Puerto Ayora, Galápagos, Ecuador

Mangu-Ward, K. (1997). Suffering in Silence, *The Wall Street Journal.* (April 20), W23

McDonald, G. (2003) *Biogeography: Space, Time and Life,* John Wiley & Sons Inc, New York, USA.

Ministerio del Ambiente & ESPOL-ICQ. (2004). Inventario de Plaguicidas COPs en el Ecuador. Informe Técnico Final. Proyecto GEF/2732-02-4456, Global Environmental Facility (GEF), Ministerio del ambiente del Ecuador, Programa Nacional Integrado para la Gestión Racional de las Sustancias Químicas, Guayaquil, Ecuador

Miranda, C., Valle, C. & Baquero, A. (2010). Isla Santa Clara: Zona critica de importancia para las aves marinas, *Primer Simposio de Biodiversidad Marina y Costera de Latinoamérica y el Caribe, Segundo Simposio Ecuatoriano de Biodiversidad Marina y Costera,* pp.39-40, Grupo de Trabajo sobre Biodiversidad Marina y Costera del Ecuador, Ministerio del Ambiente del Ecuador, Manta, Ecuador, Diciembre 9-12, 2010

Mittermeier, R.A.; Robles-Gil, P. & Mittermeier, C. G. (1997). *Megadiversity. Earth's biologically wealthiest nations.* Conservation International and Cemex S.A., ISBN 978-9686397505, Washington, USA

Montaño, M. & Resabala, C. (2005). Pesticidas en sedimentos, aguas, y organismos de la Cuenca del Rio Taura. *Revista Cientifica Ciencias Naturales y Ambientales* Vol. 1, pp. 93-98, ISSN 1019-6161

Myers, N. (1992). *The primary source: tropical forests and our future*. Norton, ISBN 978-0393017953, New York, USA

Murillo, J.C., Reyes, H., Zarate, P., Banks, S., & Danulat, E. (2004). Evaluación de la captura incidental durante el Plan Piloto de Pesca de Altura con Palangre en la Reserva Marina de Galápagos, Fundación Charles Darwin y Dirección Parque Nacional Galápagos, Santa Cruz, Galápagos, Ecuador

NOAA. (2011). Climate prediction center, In: *National Weather Service*, 27.02. 2011, Available from http://www.cpc.ncep.noaa.gov/products/analysis_monitoring/ensostuff/ONI_change.shtml.

Nixon, S. & Thomas, A. (2001). On the size of the Peru Upwelling Ecosystem. Research Note. *Deep-Sea Research I: Oceanographic Research Papers* Vol. 48, No. 11, (November 2001), pp. 2521-2528, ISSN 0967-0637

Padilla, L.R.; Huyvaert, K.P.; Merkel, J.; Miller R.E. & Parker, P.G. (2003). Hematology, plasma chemistry, serology, and *Chlamydophila* status of the waved albatross (*Phoebastria irrorata*) on the Galapagos Islands. *Journal of Zoo and Wildlife Medicine* Vol. 34, No. 3, (September 2003), pp. 278-283, ISSN 1042-7260

Padilla, L.R.; Santiago-Alarcon, D.; Merkel, J.; Miller R.E. & Parker, P.G. (2004). Survey for *Haemoproteus* spp., *Trichomonas gallinae, Chlamydophila psittaci*, and *Salmonella* spp. in Galapagos Islands columbiformes. *Journal of Zoo and Wildlife Medicine* Vol. 35, No. 1, (March 2004), pp. 60-64, ISSN 1042-7260

Peck, S.B.; Heraty, J.; Landry, B. & Sinclair, B.J. (1998). Introduced insect fauna of an oceanic archipelago: The Galápagos Islands, Ecuador. *American Entomologist* Vol. 44, No. 4, (January 1998), pp. 218- 237, ISSN 1046-2821 Ridgely, R. S. & Greenfield, P. J. (2001). *The birds of Ecuador: Status, Taxonomy, and Distribution*, Volume I, Cornell University Press, ISBN 0-8014-8720-X, Ithaca, NY, USA

Rosenberg, D. K. (1990). The impact of introduced herbivores on the Galápagos Rail (*Laterallus spilonotus*). *Monographs in Systematic Botany from the Missouri Botanical Garden*. Vol. 32, pp. 169-178, ISSN 0161-1542.

Santander, T.; Freile, J. F. & Loor-Vela, S. (2009). Ecuador. In: *Important Bird Areas Americas - Priority sites for biodiversity conservation*, C. Devenish, D. F. Díaz Fernández, R. P. Clay, I. Davidson & I. Yépez Zabala (Eds.), pp. 187 –196, BirdLife International (BirdLife Conservation Series No. 16), Quito, Ecuador

Santander, T.; & Muñoz, I. (2005). Ecuador-Informe anual 2004. In: *El Censo Neotropical de Aves Acuáticas 2004*, B. López-Lanus, & D. E. Blanco, (Eds.). Wetlands International, Global Series No. 17, Buenos Aires, Argentina

Santander, T.; Hidalgo, J. R. & Haase B. (2006a). Aves Acuáticas en Ecuador. Reporte final preparado para Waterbird Conservation for the Americas. Aves&Conservación (Corporación Ornitológica del Ecuador), Quito, Ecuador

Santander T.; Muñoz, I. & Lara, A. (2006b). Ecuador: informe anual. Censo Neotropical de Aves Acuáticas 2005. In: *El Censo Neotropical de Aves Acuáticas 2005: Una herramienta para la conservación,* B. López-Lanús & D.E. Blanco (Eds.), Wetlands International, Buenos Aires, Argentina. Available from http://lac.wetlands.org/

Santander, T., A. Lara & I. Muñoz. (2007). Ecuador: informe anual. Censo Neotropical de Aves Acuáticas 2006, In: *El Censo Neotropical de Aves Acuáticas 2006: Una herramienta para la conservación,* A.J. Lesterhuis & D.E. Blanco (Eds.), Wetlands International, Buenos Aires, Argentina. Available from http://lac.wetlands.org/

Santander, T.G. & Lara, A. (2008). Ecuador: informe anual. Censo Neotropical de Aves Acuáticas 2007, In: *El Censo Neotropical de Aves Acuáticas 2007: Una herramienta para la conservación,* D.A. Unterkofler & D.E. Blanco (Eds.), Wetlands International, Buenos Aires, Argentina. Available from http://lac.wetlands.org/

SICA-MAG. (2003). Informe de fertilizantes y agroquímicos. In: *Proyecto Servicio de Información y Censo Agropecuario (SICA) Ministerio de Agricultura, Ganadería, Acuacultura y Pesca del Ecuador (MAG),* Quito, Ecuador. 15.05.2007. Available from http://www.sica.gov.ec/agro/insumos/inf_08_03

Sierra, R. (Ed.). (1999). *Propuesta preliminar de un sistema de clasificación de vegetación para el Ecuador continental,* Proyecto INEFAN/GEFBIRF, Wildlife Conservation Society and EcoCiencia, Quito, Ecuador.

Snell, H.L.; Tye, A.; Causton, C.E. & Bensted-Smith, R. (2002). Current status of and threats to the terrestrial biodiversity of Galapagos. In: *A biodiversity vision for the Galapagos Islands.* R. Bensted-Smith (Ed.), pp. 30-47, Charles Darwin Foundation and World Wildlife Fund, Puerto Ayora, Galapagos.

Solórzano, L. (1989). Status of Coastal Water Quality in Ecuador, In: *A Sustainable Shrimp Mariculture Industry for Ecuador,* S. Olsen, and L. Arriaga, E. (Eds.), pp. 163-177, Technical Report Series TR-E-6. Coastal Resources Center, University of Rhode Island, Narragansett, RI.

Stattersfield, A. J.; Crosby, M. J.; Long, A. J. & Wege, D. C. (1998). *Endemic Birds Areas of the World,* BirdLife International, ISBN 0-0946888-33-7, Cambridge, U.K.

Suárez, H. & Calle, M. (2005). Monitoreo del estado poblacional y reproductive de aves marinas en el refugio de vida Silvestre Isla Santa Clara (RVS-ISCAL), Julio 2005

Taylor, P. B. (1996). Family Rallidae (rails, gallinules and coots), In: *Handbook of the birds of the world, Hoatzin to Auks,* Volume 3, J. del Hoyo, A. Elliott & J.Sargatal, (Eds.), Lynx Edicions, ISBN 978-8487334207, Barcelona, Spain

Thiel, T.; Whiteman, N.K.; Tirape, A.; Baquero, M.I.; Cedeño, V.; Walsh, T.; Jimenez-Uzcategui, G. & Parker, P.G. (2005). Characterization of canarypox-like viruses infecting Endemic birds in the Galapagos Islands. *Journal of Wildlife Diseases* Vol. 41, No. 2, (April 1), pp. 342–353, ISSN 0090-3558

Timmermann, A.; Oberhuber, J.; Bacher, A.; Esch, M.; Latif, M. & Roeckner, E. (1999). Increased El Niño frequency in a climate model forced by future greenhouse warming. *Nature* Vol. 398, No.6729, (April 1999), pp. 694–697, ISSN 0028-0836

Travis, E. K.; Vargas, F. H.; Merkel, J.; Gottdenker, N.; Miller, R. E. & Parker, P. C. (2006a). Hematology, plasma chemistry, and serology of the flightless cormorant (*Phalacrocorax harrisi*) in the Galapagos Islands, Ecuador. *Journal of Wildlife Diseases* Vol. 42, No. 1, (January 1), pp. 133-141, ISSN 0090-3558

Travis, E. K.; Vargas, F. H.; Merkel, J.; Gottdenker, N.; Miller, R. E. & Parker, P. G. (2006b). Hematology, serum chemistry, and serology of Galapagos penguins (*Spheniscus mendiculus*) in the Galapagos Islands, Ecuador. *Journal of Wildlife Diseases* Vol. 42, No.3, (July 1), pp. 625-632, ISSN 0090-3558

Valle, C. (1997). Isla Santa Clara: estudio del ambiente terrestre y establecimiento de un sistema de monitoreo de las colonias de aves marinas. Informe de Consultoría para el PATRA/MMA, Guayaquil, Ecuador

Valle, C. (1998). Monitoreo de las Poblaciones de Aves Marinas en Isla Santa Clara, Golfo de Guayaquil, In: Informe de Consultoría preparado para EDC- Ecuador Ltda, Vol. I., M. Hurtado et al., (Eds.). Guayaquil, Ecuador

Vargas, H. & H. L. Snell. (1997). The arrival of Marek's disease to Galápagos. *Noticias de Galápagos* Vol. 58, No. 58, (May 1997), pp. 4-5, ISSN 0550-1067

Vargas, H.; Lougheed, C. & Snell, H. (2005). Population size and trends of the Galapagos penguin *Spheniscus mendiculus*. *Ibis* Vol. 147, No. 2, (April 2005), pp. 367–374, ISSN 1474-919X

Vargas, F.H.; Harrison, S.; Rea, S. & Macdonald, D.W. (2006). Biological effects of El Niño on the Galapagos penguin. *Biological Conservation* Vol. 127, No. 1, (January 2006), pp. 107–114, ISSN 0006-3207

Vargas, F.H.; Lacy, R. C.; Johnson, P. J.; Steinfurth, A.; Crawford, R. J.M.; Boersma, P. D. & Macdonald, D. W. (2007). Modelling the effect of El Niño on the persistence of small populations: The Galápagos penguin as a case study. *Biological Conservation* Vol. 137, No. 1, (June 2007), pp. 138-148, ISSN 0006-3207

Vitousek, P. M.; Loope, L. L. & Stone, C. P. (1987). Stone Introduced species in Hawaii. Biological effects and opportunities for ecological research. *Trends in Ecology and Evolution* Vol. 2, No. 7, (July 1987), pp. 224-227, ISSN 0169-5347

Watkins, G. & Cruz, F. (2007). *Galapagos at Risk: A Socioeconomic Analysis of the Situation in the Archipelago*, Charles Darwin Foundation, Puerto Ayora, Province of Galapagos, Ecuador

WHSRN (2009a). Lagunas de ECUASAL: Description. In: *Western Hemisphere Shorebird Reserve Network,* 20.04.2011 Available from http://www.whsrn.org/site-profile/lagunas-de-ecuasal

WHSRN (2009b). Lagunas de ECUASAL: Ecology and Conservation. In: *Western Hemisphere Shorebird Reserve Network,* 20.04.2011, Available from http://www.whsrn.org/site-profile/lagunas-de-ecuasal

Whiteman N. K.; Goodman, S. J.; Sinclair, B. J.; Walsh, T.; Cunningham, A.A.; Kramer, L. D. & Parker, P.G. (2005). Establishment of the avian disease vector *Culex quinquefasciatus* Say 1823 (Diptera: Culicidae) on the Galápagos Islands, Ecuador. *Ibis* Vol. 147, No. 4, (October 2005), pp. 844–847, ISSN 1474-919X

Wiedenfeld, D. A. (2006). Aves, The Galapagos Islands, Ecuador. *Checklist* Vol. 2, No. 2, (July 2006), pp. 1–27. ISSN 1809-127X

Wiedenfled, D.A. & Jiménez-Uzcategui, G. (2008). Critical problems for bird conservation in the Galapagos Islands. *Cotinga* Vol. 29, pp. 22-27, ISSN 1353-985X

Wikelski, M.; Foufopoulos, J.; Vargas, H. & Snell, H. (2004). Galápagos Birds and Diseases: Invasive Pathogens as Threats for Island Species. *Ecology and Society* Vol. 9, No.1 (January 2004), Art. 5, ISSN: 1708-3087, Available from http://www.ecologyandsociety.org/vol9/iss1/art5/

Wikelsky, M.; Romero, L.M. & Snell., H.L. (2001). Marine iguanas oiled in the Galapagos. *Science* Vol. 292, No. 5516, (April 2001), pp. 437-438, ISSN 0036-8075

Wikelsky, M.; Wong, V.; Chevalier, B.; Rattenborg, N. & Snell, H.L. (2002). Marine iguanas die from trace oil pollution. *Nature* Vol. 417, No. 6889, (June 2002), pp. 607-608, ISSN 0090-3558

14

Biodiversity Drifts in Agricultural Landscapes

Olga Ameixa[1,2] and Pavel Kindlmann[1,3]
[1]*Department of Biodiversity Research, Global Change Research Centre AS CR, České Budějovice*
[2]*Faculty of Science, University of South Bohemia, České Budějovice*
[3]*Charles University, Prague*
Czech Republic

1. Introduction

We are in the midst of the sixth global mass extinction event (McNeely & Scherr, 2002; Thomas et al., 2004). Around the globe, biological communities that took millions of years to develop—including tropical rain forests, coral reefs, old-growth forests, prairies and coastal wetlands—have been devastated as a result of human actions. Biologists predict that tens of thousands of species and millions of unique populations will go extinct in the coming decades (Brown & Laband, 2006; Millennium Ecosystem Assessment, 2005a). If the current predictions are correct, the rates of environmental changes may outpace the capacities of organisms to adapt to the changes.

There are seven major threats to biodiversity: habitat destruction; habitat fragmentation; habitat degradation (including pollution); global climate change; the overexploitation of species for human use; the invasion of exotic species; and the increased spread of disease. Most threatened species and ecosystems face at least two or more of these threats, which can interact synergistically to speed the way to extinction and hinder efforts at protecting biodiversity (Burgman et al., 2007; Millennium Ecosystem Assessment, 2005b). All seven threats are the result of an expanding human population's ever increasing use of the world's natural resources (Primack, 2008).

Agroecosystems include a large proportion of the world's biodiversity (Pimentel et al., 1992). Over the past two decades, research has demonstrated the value of agricultural biodiversity in all its forms, including crop and livestock genetic diversity, and associated species important for production, for example, pollinators, soil microorganisms, beneficial insects, and predators of pests and wild species that occur in agricultural landscapes (Uphoff et al., 2006). Some species are almost completely dependent on agricultural habitats for survival, e.g. Great Bustard *Otis tarda*, Grey Partridge *Perdix perdix* or the Black-tailed Godwit *Limosa limosa* (Kleijn et al., 2006).

Since the 1960's both industrial agriculture in developed countries and the original green revolution in developing countries have depended on improved seeds, chemical fertilizers, pesticides and irrigation. This production model involved a small number of crops, generally in monoculture (to increase efficiency in use of inputs and mechanization), increased pesticide and fertilizer use and short crop-rotations (Benton et al., 2003). Wild flora and fauna were considered direct competitors for resources or harvested products, while water was diverted from wetlands and natural habitats for irrigation (Uphoff et al., 2006), and intensification has reduced the suitability of agricultural fields for a wide range of organisms (Benton et al., 2003). The cultivation of annual crops has expanded at the cost of non-crop habitats such as extensive grasslands, fallow, hedges and field margins (Benton et al., 2003; Tilman et al., 2001b). Non-crop habitats provide dispersal corridors for wildlife and habitat islands required by many species as refuges and feeding areas (Öckinger & Smith, 2007; Stoate et al., 2001). Non-crop habitats can also act as biodiversity reservoirs for natural enemies, which can potentially improve natural pest control in agricultural landscapes (Ives et al., 2000; Wilby & Thomas, 2002), however, they can also act as reservoirs for pest species, which can colonize the crops (van Emden, 1965).

The expansion of agricultural intensification (AI) is often considered to be an important factor that has contributed to a rapid decline in biodiversity in agroecosystems (Benton et al., 2003; Mattison & Norris, 2005) and negatively affected the production of ecosystem services, e.g., maintenance of fertile soils, biotic regulation, nutrient recycling, assimilation of wastes, sequestration of carbon dioxide, and maintenance of genetic information (Benton et al., 2003; Chamberlain et al., 2000; Hooper et al., 2005; Robinson & Sutherland, 2002; Tilman et al., 2002). Wilcove et al. (1998) estimated that 38% of the endangered species in the United States are negatively affected by agricultural practices. Changes in landscape composition and intensive management practices are believed to be the main factors causing this decline. Also many species of raptor have been negatively affected by prey declines, probably associated with AI (Tucker & Heath, 1994). Furthermore, the potential of biodiversity for providing ecological resilience, i.e., the capacity to recover from disruption of functions, and the mitigation of risks caused by disturbance (Holling, 1996; Swift, 2004) is poorly documented. A better knowledge of which goods and services are provided by agroecosystems is urgently needed since we live on the brink of no return.

At the present time, 10% of the global land area is under intensive agricultural use, 17% is under extensive use associated with the use of far fewer artificial inputs, and 40% is grazed by domestic livestock (Mooney et al., 2005; Wood et al., 2000). The world's population of 6.3 billion people is projected to grow to 7.2 billion by the year 2015, 8.3 billion by 2030 and to 9.3 billion by 2050 (FAO, 2003). By 2050, food production must double to meet human needs. In order to meet this increasing demand for food and fibre, production systems are expected to become increasingly dependent on synthetic inputs of fertilizers and pesticides (Clay, 2004). Since the world's population will continue to increase, we will increase agricultural output by 30–50% over the next 30 years; thus, the need to protect biodiversity will compete directly against the need for new agricultural land (Tilman et al., 2001a).

Not only biodiversity is at risk, lately there has been an increase in public awareness of the possible effects of agro-chemicals. Many studies document increased risk of cancer among children and adults associated with exposure to an array of pesticides (Alavanja et al., 2007; Dich et al., 1997; Zahm & Ward, 1998). Sometimes the dangers are ignored by the responsible entities, for example, the fungicide vinclozolin, which is widely used in vineyards, was registered for use in 2000, despite laboratory tests indicating that it causes testicular cancer and disrupts normal androgen activity in laboratory animals (U.S. Environmental Protection Agency, 2000). Pesticide poisoning is also a daily hazard for the majority of the world's rural population (Dinham & Malik, 2003). The World Health Organization (WHO, 1990) has indicated that 20,000 women, men and children die of accidental pesticide poisoning each year, three million are poisoned, and nearly three fourths of a million new people each year will suffer from chronic effects of exposure. For all these reasons, new solutions are necessary for producing more food and fibre, protecting the resource base upon which agriculture depends and promoting social well-being (Millennium Ecosystem Assessment, 2005b).

2. Agriculture intensification and Agri-environmental schemes

In Europe, the Common Agricultural Policy (CAP), born 50 years ago, began by subsidizing production of basic foodstuffs in the interests of self-sufficiency, after the difficult period of the war. Currently, CAP, give farmers an important role in improving quality, preserving biodiversity and traditional landscapes and keeping rural economies alive. Furthermore, more informed consumers are entitled to food that is safe and of high quality; this induced the creation of regulations defining organic foods and also what can be considered an organic farm. More extensive systems, such as organic farming, aim to mitigate the negative effects of modern agriculture and enhance biodiversity (Krebs et al., 1999; Reganold et al., 2001; Tybirk et al., 2004). Agri-environmental schemes (AES) were introduced into the European Common Agricultural Policy (CAP) in the early 1990s to reduce biodiversity loss in agricultural landscapes and mitigate other harmful effects of modern agriculture. AES are considered the most important policy instruments for protecting biodiversity in agricultural landscapes (European Environment Agency report, 2004) as they provide financial incentives to farmers for adopting environmentally friendly practices mostly at the field scale (i.e., reduction in pesticide and fertiliser applications or delaying harvesting).

With the increasing number of organic farms, several studies and meta-analyses have been conducted, with the sole purpose of finding a correspondence between the decline in biodiversity and the AI in conventional versus organic farms. Nevertheless, sometimes these studies are inconclusive, contradictory and sometimes positive results are found. Recent European-wide studies have questioned the effectiveness of AES for biodiversity conservation. Over half the studies showed significant positive effects of AES on the diversity or abundance of target groups such as plants, birds or arthropods, but the remaining studies showed non-significant or even negative effects (Kleijn et al., 2006; Kleijn & Sutherland, 2003). Usually the positive effects of organic farming relative to conventional agriculture are in terms of botanic diversity (Bengtsson et al., 2005; Hald, 1999; Hyvönen et al., 2003) whereas arthropods appear to respond ambiguously to organic cropping (reviewed in Hole et al., 2005). There are also other studies on other measures of agriculture intensification, for example, grazing intensification, extensive vs. intensive farming, etc.

One, however, should not expect immediate results from the introduction of AES. For example, Ameixa & Kindlman (2008) did not find any relation between agricultural practices and the diversity and abundance of carabids in several agricultural fields, which was probably because the species that live in agricultural fields have already undergone some kind of selection and are for this reason adapted to the constant changes. For example, in many parts of Europe, agricultural landscapes are well over 2000 years old (Groppali, 1993; Williamson, 1986), so organisms must be adapted to this environment. Thus, studies that compare organic vs. conventional fields should not aim to see an immediate change in biodiversity patterns in agricultural landscapes after years of intense land use, but find other methods to access this problem.

Another expectation is that even if AES are applied and therefore agriculture becomes less intensive, diversity will increase only until a certain maximum in agricultural fields above which no more species will be found (Figure 1). This is because the number of species that can live in a particular habitat is always finite, defined by local climatic and soil conditions and this maximum number is not affected by the way people are handling this habitat: lion will never be found in arctic tundra. On the other hand, even heavily exploited habitats will still harbour some species: the carabid, *Pterostichus melanarius*, is a good example of a species well adapted to intensively managed agroecosystems and was found to be even more numerous in these, compared with more natural habitats (Ameixa & Kindlmann, 2008).

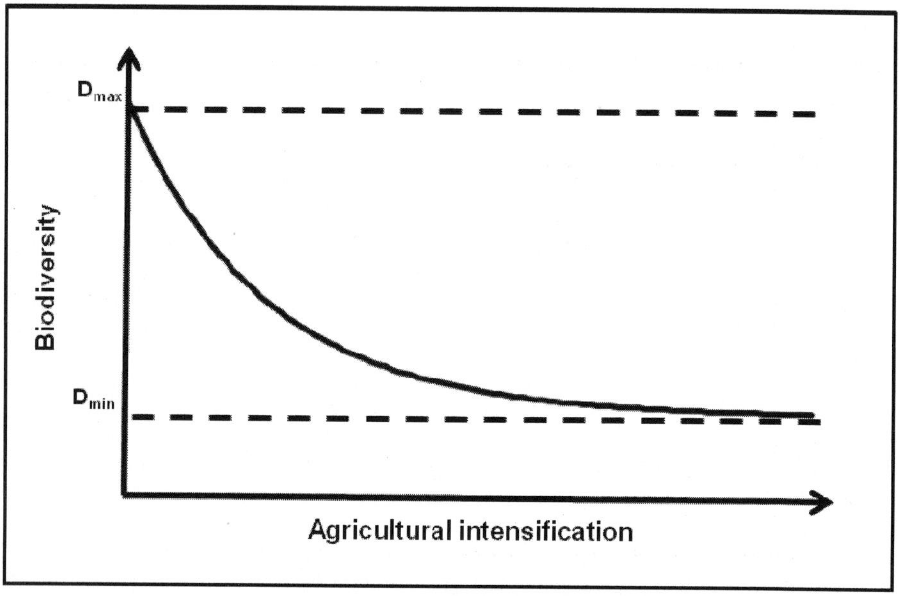

Fig. 1. Hypothetical representation of the diversity expected to be found in agricultural fields. D_{max}: Maximum diversity that can be found in agricultural lands; D_{min}: minimum diversity that can be found in agricultural lands.

3. Landscape composition

Krebs et al. (1999) suggest that biodiversity in agroecosystems depends on both farm management and landscape heterogeneity. Landscape context can modify the influence of organic farming on plants (Roschewitz et al., 2005a) or may be even more important for the diversity of bees, butterflies, carabids and spiders than the local farming system (Kremen et al., 2002; Schmidt et al., 2005; Weibull et al., 2000; Weibull et al., 2003). The contrasting results between organic and conventional fields maybe larger when these fields are isolated in homogeneous landscapes and the species pool may be too small to allow a response in terms of biodiversity to organic farming (Tscharntke et al., 2005).

The landscape context of an agricultural field may make a difference in compensating field isolation or agricultural practices that reduce diversity. Field boundaries, hedges and fallows satisfy a set of wildlife requirements (refuge, food, breeding sites, etc.) that promote species persistence in agricultural landscapes (Benton et al., 2003) facilitating both recolonization and maintenance of populations in agricultural landscapes (Duelli & Obrist, 2003). Duelli & Obrist (2003) attribute the lack of effectiveness of AES to the simplification of agricultural landscapes.

However, again we have to take in to account that diversity is expected to increase with complexity of the landscape only above a minimum threshold (Figure 2), as landscape will always harbour some species. Positive effects of landscape complexity will eventually level-off after a given level of complexity is reached, as the number of species that can live in a particular habitat is always finite, defined by local climatic and soil conditions (Concepción et al., 2008).

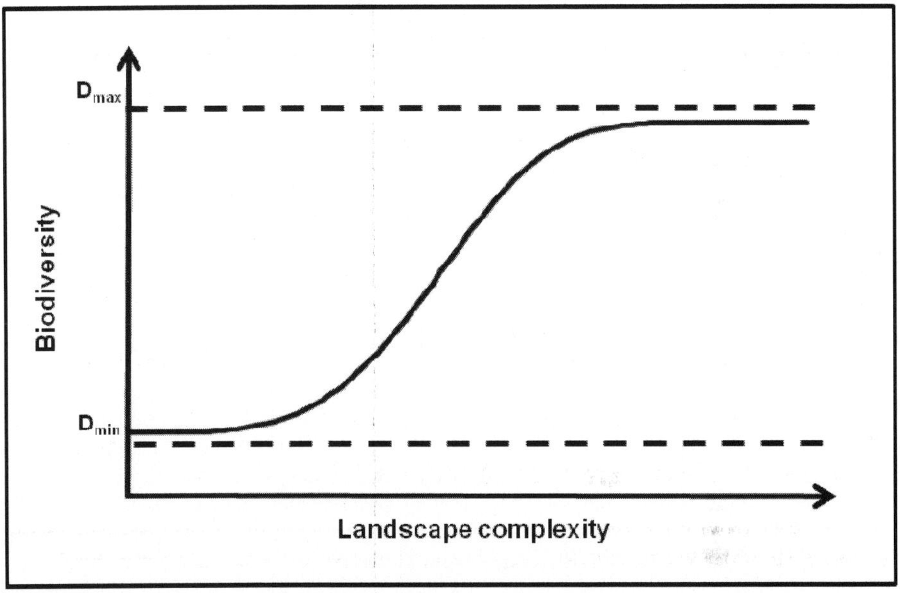

Fig. 2. Hypothetical non-linear effects of landscape complexity around cultivated fields on the biological diversity in such fields. D_{max}: saturation point of complexity, above which landscapes are so complex that no further effects of complexity are expected; D_{min}: minimum threshold of complexity below which landscapes are too simple for maintaining biodiversity (adapted from Conception 2008).

4. Meta-analysis on different taxa

The above indicates there is enough evidence that agriculture has become much more intensive during recent decades and simultaneously there has been a drastic decline in biodiversity in agroecosystems. This means that biodiversity in agroecosystems is negatively correlated with AI. However, correlation does not necessarily mean causation, and therefore - in theory - the decline in biodiversity in agroecosystems might have been caused by other factors and from a practical point of view, the effects of these should be minimized. This doubt provoked abundant case studies on how exactly AI can affect biodiversity of particular groups of organisms. The results of such studies, however, are contradictory and many are inconclusive. Because there is now a great number of these studies the results need to be summarized and patterns revealed. Although there are several such reviews, none of them analyze the situation in its entirety, which is the task of this section.

We searched the Web of Science using the following key-words: agriculture intensification; organic agriculture; agro-environmental schemes; effects of agriculture; landscape composition; land use; biodiversity. The search was restricted to studies on invertebrates, birds, plants and mammals. We then categorized them according to their conclusion regarding the relationship between AI or Landscape composition and biodiversity as having a positive, negative, none or mixed effect. We used 54 studies for determining the relationships between AI and biodiversity and 36 for those between landscape composition and biodiversity. If a study considered more than one taxon, more than one parameter (e.g. diversity and abundance) or more than one measure of comparison (e.g., AI and landscape composition) we treated them as independent studies. During this procedure, all reviews and studies containing only models or did not provide a clear statement allowing us to categorize them into one the four categories, were excluded.

When categorizing, we have always respected the conclusions formulated by the authors, even if we did not always agree with them, because introducing our personal views could have affected the outcome of our analyses. We accepted all measures of diversity used in the papers studied, which includes the number of species, diversity indexes, and even number of individuals in the group studied, even though we do not consider the latter as valid, because the number of individuals can be affected by one or a few dominant species, adapted to the particular conditions.

There was a wide variety of measures of AI in the papers. These include usage/absence of pesticides and/or artificial fertilizers, amounts of pesticides/fertilizers used, intensity of tillage, comparison of organic vs. conventional farms, grazing intensity or comparison of extensive vs. intensive cropping. Measures of landscape composition, include, more or less heterogeneity, land use, average size of fields, percentage and size of arable land and/or non-crop habitats in the landscape. In table 1 are the reference to the studies used in the meta-analysis of the relationship between AI and Biodiversity.

Relationship between AI and biodiversity	**Birds**	**Invertebrates**	**Mammals**	**Plants**
Ameixa & Kindlmann, 2008		x		
Batáry et al., 2007		x		
Bates & Harris, 2009			x	
Benton et al., 2002	x			
Blackburn & Wallace, 2001		x		
Bradbury et al., 2004	x			
Brittain et al., 2010		x		
Burel et al., 1998	x	x	x	x
Clough et al., 2007a		x		x
Clough et al., 2007b		x		
Cole et al., 2005		x		
Davey et al., 2010	x			
Davy et al., 2007			x	

Delgado & Moreira, 2010	x			
Di Giulio et al., 2001		x		
Dıaz & Telleria, 1994	x			
Doxa et al., 2010	x			
Duelli et al., 1999		x		
Feehan et al., 2005		x		x
Geiger et al., 2010	x	x		x
Genghini et al., 2006	x			
Gibson et al., 2007				x
Hald, 1999				x
Hasken & Poehling, 1995		x		
Hendrickx et al., 2007		x		
Hodgson et al., 2010		x		
Holzschuh et al., 2007		x		
Hutton & Giller, 2003		x		
Hyvönen et al., 2003				x
Kleijn & van Zuijlen, 2004	x			
Kleijn et al., 2001	x	x		x
Kleijn et al., 2004	x	x		x
Kleijn et al., 2006	x	x		x
Knop et al., 2006		x		x
Kremen et al., 2002		x		
Melnychuk et al., 2003		x		
Östman, 2002		x		
Ottvall & Smith, 2006	x			
Peach et al., 2001	x			
Pfiffner & Luka, 2003		x		
Pocock & Jennings, 2008		x	x	
Roschewitz et al., 2005a		x		
Roschewitz et al., 2005b		x		
Rundlof & Smith, 2006		x		
Rundlöf et al., 2008		x		
Schmidt et al., 2005		x		
Schmitzberger et al., 2005				x
Shah et al., 2003		x		
Thorbek & Bilde, 2004		x		
Weibull et al., 2000		x		
Weibull et al., 2003		x		x
Wickramasinghe et al., 2003			x	
Wickramasinghe et al., 2004			x	
Wilson et al., 2007	x			

Table 1. References to studies used in the meta-analysis of the relationships between AI and biodiversity for the four groups of organisms studied.

Figure 3 shows the frequencies of papers listed in Table 1, claiming different effects (positive, negative, no effect or mixed) of AI on biodiversity. There is no clear pattern in the results.

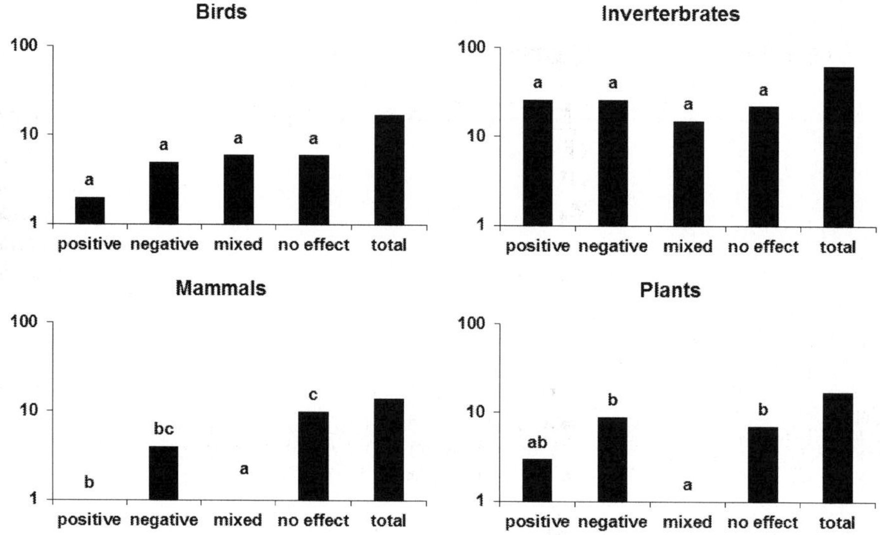

Fig. 3. Frequencies of papers claiming various types of effect of AI on biodiversity for the four taxa studied. G-test, significance 5%

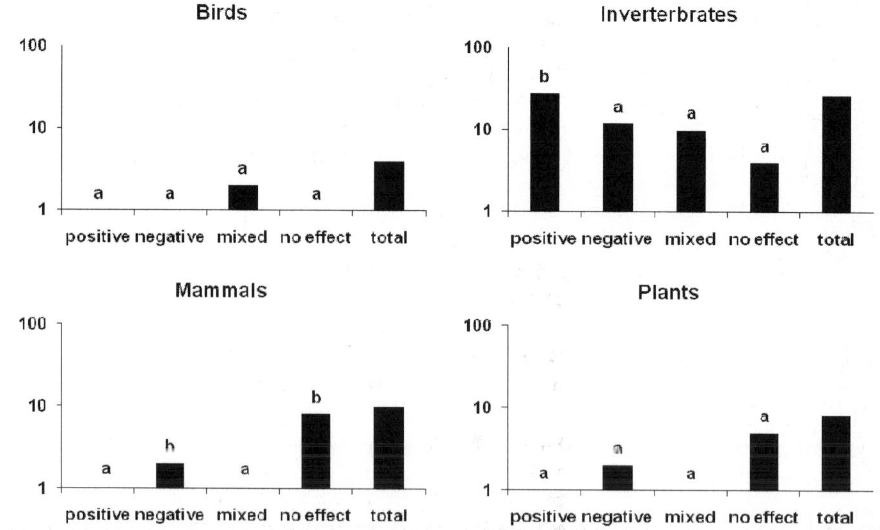

Fig. 4. Frequencies of papers claiming various types of effect of landscape structure on biodiversity for the four taxa studied. G-test, significance 5%

Figure 4 shows the frequencies of papers listed in Table 2, claiming different effects (positive, negative, no effect or mixed) of landscape structure on biodiversity. Similarly, there is no clear pattern in the results, except for the significantly larger number of papers claiming a positive effect of landscape structure on biodiversity of invertebrates, compared with the other groups of organisms studied.

Landscape structure and biodiversity	Birds	Invertebrates	Mammals	Plants
Asteraki et al., 1995		x		
Aviron et al., 2005		x		
Banks & Stark, 2004		x		
Batáry et al., 2007		x		
Bates & Harris, 2009			x	
Bradbury et al., 2004	x			
Brittain et al., 2010		x		
Burel et al., 1998	x	x	x	x
Burel et al., 2004		x	x	
Clough et al., 2007b		x		
Cole et al., 2005		x		
Dıaz & Telleria, 1994	x			
Doxa et al., 2010	x			
Duelli & Obrist, 2003		x		
Genghini et al., 2006	x			
Gibson et al., 2007				x
Hendrickx et al., 2007		x		
Holzschuh et al., 2007		x		
Kitahara & Sei, 2001		x		
Kleijn et al., 2004	x	x		x
Kremen et al., 2002		x		
Kremen et al., 2004		x		
Öckinger & Smith, 2007		x		
Peach et al., 2001	x			
Petit & Burel, 1998		x		
Pocock & Jennings, 2008		x	x	
Roschewitz et al., 2005b		x		
Rundlöf et al., 2008		x		
Schmidt et al., 2005		x		
Schmitzberger et al., 2005				x
Thies & Tscharntke, 1999		x		
Thorbek & Bilde, 2004		x		
Vollhardt et al., 2008		x		
Weibull et al., 2000		x		
Weibull et al., 2003		x		x
Wickramasinghe et al., 2003			x	

Table 2. References to studies used in the meta-analysis dealing with the relationship between landscape structure and biodiversity of the four groups of organisms studied.

5. Final remarks

The studies on the relationship between biodiversity and AI conducted so far do not always indicate a negative relationship between AI and biodiversity. Despite this, the number of studies showing this relationship is worrying. In addition, toxicological studies should be undertaken as they are unlikely to support the idea that AI should be promoted.

There is an inevitable conflict between the increasing need for higher agricultural production and the need to preserve biodiversity. Even though agroecosystems are increasingly subjected to human disturbance, they are still able to sustain some diversity and undoubtedly important for some aspects of life of certain species, like, farmland birds or predators of pests. For example, *Otis tarda* usually lives in areas traditionally cultivated using a cereal - fallow rotation.

Without any doubt, current production models seem to fail in the maintenance of biodiversity, and the more sustainable traditional systems are not so appealing, because they are less productive. However, these comparisons are mostly based only on yield and ignore the cost of agrochemicals or fuel. Producers should change their ways of thinking and realize the importance of their role in preserving diversity for future generations as a source of revenue. This change is especially important in developing countries, which are starting their "green revolution" and should learn from the mistakes of others.

Small steps are being undertaken to change intensive production models into more sustainable ones. One should not expect an immediate response of biodiversity to environmentally-friendly changes in agricultural practices, which by intensive land use over many years have systematically selected plant and animal species, even if the environmentally-friendly changes might lead eventually to an increase in biodiversity. Landscape composition also plays an important role. For example, the existence of sources of plants and animals close to agricultural fields like, boundaries, hedges and fallows, (heterogeneous landscape) are extremely important refuges, food sources and/or overwintering places. Heterogeneous landscape can also help some species to cope with the disturbances common in agricultural fields.

The enormous variation among taxa suggests that some species are more sensitive to disturbance than others. It is not surprising that insects that are *R* – selected species, with a short maturation time, breeding at a young age, short lifespan, producing many small offspring quickly, high mortality rates of young and no parental care, can more easily adapt to agroecosystem disturbances than birds and bats that are considered to be *K* – selected species, with a long maturation time, breeding relatively late in life, a long lifespan, producing relatively few large offspring, low mortality rates of young and extensive parental care. Large species tend to have long life-cycles and consequently require a degree of stability of resources over time (Blake et al., 1994).

Agroecosystems will always be linked with human activities, and in this way the future of biodiversity in these systems will always be dependent on human actions. Mankind should realize, however, that not only biodiversity suffers from the consequences of agriculture intensification but so does public health.

6. Acknowledgments

The research was supported by the grants No. LC06073 and CZ.1.05/1.1.00/02.0073 (Czechglobe) of the MSMT and grant No. GD206/08/H044 of the GA CR.

7. References

Alavanja, M. C. R.; Ward, M. H. & Reynolds, P. (2007). Carcinogenicity of agricultural pesticides in adults and children, Journal of Agromedicine, 12, 39-56.

Ameixa, O. M. C. C. & Kindlmann, P. (2008). Agricultural policy-induced landscape changes: effects on carabid abundance and their biocontrol potential, European Journal of Entomology, 105, 467-476.

Asteraki, E. J.; Hanks, C. B. & Clements, R. O. (1995). The influence of different types of grassland field margin on carabid beetle (Coleoptera, Carabidae) communities, Agriculture, Ecosystems & Environment, 54, 195-202.

Aviron, S.; Burel, F.; Baudry, J. & Schermann, N. (2005). Carabid assemblages in agricultural landscapes: impacts of habitat features, landscape context at different spatial scales and farming intensity, Agriculture, Ecosystems & Environment, 108, 205-217.

Banks, J. E. & Stark, J. D. (2004). Aphid response to vegetation diversity and insecticide applications, Agriculture, Ecosystems & Environment, 103, 595-599.

Batáry, P.; Báldi, A.; Szél, G.; Podlussány, A.; Rozner, I. & Erdős, S. (2007). Responses of grassland specialist and generalist beetles to management and landscape complexity, Diversity and Distributions, 13, 196-202.

Bates, F. S. & Harris, S. (2009). Does hedgerow management on organic farms benefit small mammal populations?, Agriculture Ecosystems & Environment, 129, 124-130.

Bengtsson, J.; Ahnström, J. & Weibull, A.-C. (2005). The effects of organic agriculture on biodiversity and abundance: a meta-analysis, Journal of Applied Ecology, 42, 261-269.

Benton, T. G.; Bryant, D. M.; Cole, L. & Crick, H. Q. P. (2002). Linking agricultural practice to insect and bird populations: A historical study over three decades, Journal of Applied Ecology, 39, 673-687.

Benton, T. G.; Vickery, J. A. & Wilson, J. D. (2003). Farmland biodiversity: is habitat heterogeneity the key?, Trends in Ecology & Evolution, 18, 182-188.

Blackburn, J. & Wallace, A. (2001). Comparative abundance of centipedes on organic and conventional farms, and its possible relation to declines in farmland bird populations, Basic and Applied Ecology, 2, 373-381.

Blake, S.; Foster, G. N.; Eyre, M. D. & Luff, M. L. (1994). Effects of habitat type and grassland management practices on the body size distribution of carabid beetles, Pedobiologia, 38, 502–512.

Bradbury, R. B.; Browne, S. J.; Stevens, D. K. & Aebischer, N. J. (2004). Five-year evaluation of the impact of the Arable Stewardship Pilot Scheme on birds, Ibis, 146, 171-180.

Brittain, C.; Bommarco, R.; Vighi, M.; Settele, J. & Potts, S. G. (2010). Organic farming in isolated landscapes does not benefit flower-visiting insects and pollination, Biological Conservation, 143, 1860-1867.

Brown, R. M. & Laband, D. N. (2006). Species imperilment and spatial patterns of development in the United States, Conservation Biology, 20, 239-244.

Burel, F.; Baudry, J.; Butet, A.; Clergeau, P.; Delettre, Y.; Le Coeur, D.; Dubs, F.; Morvan, N.; Paillat, G.; Petit, S.; Thenail, C.; Brunel, E. & Lefeuvre, J.-C. (1998). Comparative biodiversity along a gradient of agricultural landscapes, Acta Oecologica, 19, 47-60.

Burel, F.; Butet, A.; Delettre, Y. R. & Millàn de la Peña, N. (2004). Differential response of selected taxa to landscape context and agricultural intensification, Landscape and Urban Planning, 67, 195-204.

Burgman, M. A.; Keith, D.; Hopper, S. D.; Widyatmoko, D. & Drill, C. (2007). Threat syndromes and conservation of the Australian flora, Biological Conservation, 134, 73-82.

Chamberlain, D. E.; Fuller, R. J.; Bunce, R. G. H.; Duckworth, J. C. & Shrubb, M. (2000). Changes in the abundance of farmland birds in relation to the timing of agricultural intensification in England and Wales, Journal of Applied Ecology, 37, 771-788.

Clay, J. W. (2004). World Agriculture and the Environment: A Commodity-by-Commodity Guide to Impacts and Practices, Island Press, Washington DC, USA.

Clough, Y.; Holzschuh, A.; Gabriel, D.; Purtauf, T.; Kleijn, D.; Kruess, A.; Steffan-Dewenter, I. & Tscharntke, T. (2007a). Alpha and beta diversity of arthropods and plants in organically and conventionally managed wheat fields, Journal of Applied Ecology, 44, 804-812.

Clough, Y.; Kruess, A. & Tscharntke, T. (2007b). Local and landscape factors in differently managed arable fields affect the insect herbivore community of a non-crop plant species, Journal of Applied Ecology, 44, 22-28.

Cole, L.; McCracken, D.; Downie, I.; Dennis, P.; Foster, G.; Waterhouse, T.; Murphy, K.; Griffin, A. & Kennedy, M. (2005). Comparing the effects of farming practices on ground beetle (Coleoptera: Carabidae) and spider (Araneae) assemblages of Scottish farmland, Biodiversity and Conservation, 14, 441-460.

Concepción, E.; Díaz, M. & Baquero, R. (2008). Effects of landscape complexity on the ecological effectiveness of agri-environment schemes, Landscape Ecology, 23, 135-148.

Davey, C. M.; Vickery, J. A.; Boatman, N. D.; Chamberlain, D. E.; Parry, H. R. & Siriwardena, G. M. (2010). Assessing the impact of Entry Level Stewardship on lowland farmland birds in England, Ibis, 152, 459-474.

Davy, C. M.; Russo, D. & Fenton, M. B. (2007). Use of native woodlands and traditional olive groves by foraging bats on a Mediterranean island: consequences for conservation, Journal of Zoology, 273, 397-405.

Delgado, A. & Moreira, F. (2010). Between-year variations in Little Bustard Tetrax tetrax population densities are influenced by agricultural intensification and rainfall, Ibis, 152, 633-642.

Di Giulio, M.; Edwards, P. J. & Meister, E. (2001). Enhancing insect diversity in agricultural grasslands: the roles of management and landscape structure, Journal of Applied Ecology, 38, 310-319.

Dıaz, M. & Telleria, J. L. (1994). Predicting the effects of agricultural changes in central Spain croplands on seed-eating overwintering birds, Agriculture Ecosystems & Environment, 49, 289-298.

Dich, J.; Zahm, S. H.; Hanberg, A. & Adami, H.-O. (1997). Pesticides and Cancer, Cancer Causes & Control, 8, 420-443.

Dinham, B. & Malik, S. L. L. M. (2003). Pesticides and Human Rights, International Journal of Occupational and Environmental Health, 9, 40-52.

Doxa, A.; Bas, Y.; Paracchini, M. L.; Pointereau, P.; Terres, J. M. & Jiguet, F. (2010). Low-intensity agriculture increases farmland bird abundances in France, Journal of Applied Ecology, 47, 1348-1356.

Duelli, P. & Obrist, M. K. (2003). Regional biodiversity in an agricultural landscape: the contribution of seminatural habitat islands, Basic and Applied Ecology, 4, 129-138.

Duelli, P.; Obrist, M. K. & Schmatz, D. R. (1999). Biodiversity evaluation in agricultural landscapes: above-ground insects, Agriculture, Ecosystems & Environment, 74, 33-64.

European Environment Agency report (2004). High Nature Value Farmland - Characteristics, Trends and Policy Challenges., Copenhagen.

FAO (Food and Agriculture Organization) (2003). Jelle Bruinsma, Rome.

Feehan, J.; Gillmor, D. A. & Culleton, N. (2005). Effects of an agri-environment scheme on farmland biodiversity in Ireland, Agriculture, Ecosystems & Environment, 107, 275-286.

Geiger, F.; Bengtsson, J.; Berendse, F.; Weisser, W. W.; Emmerson, M.; Morales, M. B.; Ceryngier, P.; Liira, J.; Tscharntke, T.; Winqvist, C.; Eggers, S.; Bommarco, R.; Pärt, T.; Bretagnolle, V.; Plantegenest, M.; Clement, L. W.; Dennis, C.; Palmer, C.; Oñate, J. J.; Guerrero, I.; Hawro, V.; Aavik, T.; Thies, C.; Flohre, A.; Hänke, S.; Fischer, C.; Goedhart, P. W. & Inchausti, P. (2010). Persistent negative effects of pesticides on biodiversity and biological control potential on European farmland, Basic and Applied Ecology, 11, 97-105.

Genghini, M.; Gellini, S. & Gustin, M. (2006). Organic and integrated agriculture: the effects on bird communities in orchard farms in northern Italy, Biodiversity and Conservation, 15, 3077-3094.

Gibson, R. H.; Pearce, S.; Morris, R. J.; Symondson, W. O. C. & Memmott, J. (2007). Plant diversity and land use under organic and conventional agriculture: a whole-farm approach, Journal of Applied Ecology, 44, 792-803.

Groppali, R. (1993). Breeding birds in traditional tree rows and Hedges in the central Po Valley (Province of Cremona, Northern Italy), In Landscape Ecology and Agroecosystems (Eds, Bunce, R. G. H., Ryszkowski, L. and Paoletti, M. G.) Lewis Publishers, Boca Raton, pp. 153-158.

Hald, A. B. (1999). Weed vegetation (wild flora) of long established organic versus conventional cereal fields in Denmark, Annals of Applied Biology, 134, 307-314.

Hasken, K.-H. & Poehling, H.-M. (1995). Effects of different intensities of fertilisers and pesticides on aphids and aphid predators in winter wheat, Agriculture, Ecosystems & Environment, 52, 45-50.

Hendrickx, F.; Maelfait, J.-P.; Van Wingerden, W.; Schweiger, O.; Speelmans, M.; Aviron, S.; Augenstein, I.; Billeter, R.; Bailey, D.; Bukacek, R.; Burel, F. O.; Diekötter, T. I. M.; Dirksen, J.; Herzog, F.; Liira, J.; Roubalova, M.; Vandomme, V. & Bugter, R. O. B. (2007). How landscape structure, land-use intensity and habitat diversity affect components of total arthropod diversity in agricultural landscapes, Journal of Applied Ecology, 44, 340-351.

Hodgson, J. A.; Kunin, W. E.; Thomas, C. D.; Benton, T. G. & Gabriel, D. (2010). Comparing organic farming and land sparing: optimizing yield and butterfly populations at a landscape scale, Ecology Letters, 13, 1358-1367.

Hole, D. G.; Perkins, A. J.; Wilson, J. D.; Alexander, I. H.; Grice, F. & Evans, A. D. (2005). Does organic farming benefit biodiversity?, Biological Conservation, 122, 113-130.

Holling, C. S. (1996). Engineering resilience versus ecological resilience, In Engineering Within Ecological Constraints (Ed, Schulze, P.) National Academy Press, Washington, DC, USA, pp. 31-44.

Holzschuh, A.; Steffan-Dewenter, I.; Kleijn, D. & Tscharntke, T. (2007). Diversity of flower-visiting bees in cereal fields: Effects of farming system, landscape composition and regional context, Journal of Applied Ecology, 44, 41 - 49.

Hooper, D. U.; Chapin, F. S.; Ewel, J. J.; Hector, A.; Inchausti, P.; Lavorel, S.; Lawton, J. H.; Lodge, D. M.; Loreau, M.; Naeem, S.; Schmid, B.; Setälä, H.; Symstad, A. J.; Vandermeer, J. & Wardle, D. A. (2005). Effects of biodiversity on ecosystem functioning: a consensus of current knowledge, Ecological Monographs, 75, 3-35.

Hutton, S. A. & Giller, P. S. (2003). The effects of the intensification of agriculture on northern temperate dung beetle communities, Journal of Applied Ecology, 40, 994-1007.

Hyvönen, T.; Ketoja, E.; Salonen, J.; Jalli, H. & Tiainen, J. (2003). Weed species diversity and community composition in organic and conventional cropping of spring cereals, Agriculture, Ecosystems & Environment, 97, 131-149.

Ives, A. R.; Klug, J. L. & Gross, K. (2000). Stability and species richness in complex communities, Ecology Letters, 3, 399-411.

Kitahara, M. & Sei, K. (2001). A comparison of the diversity and structure of butterfly communities in semi-natural and human-modified grassland habitats at the foot of Mt. Fuji, central Japan, Biodiversity and Conservation, 10, 331-351.

Kleijn, D.; Baquero, R. A.; Clough, Y.; Díaz, M.; De Esteban, J.; Fernández, F.; Gabriel, D.; Herzog, F.; Holzschuh, A.; Jöhl, R.; Knop, E.; Kruess, A.; Marshall, E. J. P.; Steffan-Dewenter, I.; Tscharntke, T.; Verhulst, J.; West, T. M. & Yela, J. L. (2006). Mixed biodiversity benefits of agri-environment schemes in five European countries, Ecology Letters, 9, 243-254.

Kleijn, D.; Berendse, F.; Smit, R. & Gilissen, N. (2001). Agri-environment schemes do not effectively protect biodiversity in Dutch agricultural landscapes, Nature, 413, 723-725.

Kleijn, D.; Berendse, F.; Smit, R.; Gilissen, N.; Smit, J.; Brak, B. & Groeneveld, R. (2004). Ecological Effectiveness of Agri-Environment Schemes in Different Agricultural Landscapes in The Netherlands, Conservation Biology, 18, 775-786.

Kleijn, D. & Sutherland, W. J. (2003). How effective are European agri-environment schemes in conserving and promoting biodiversity?, Journal of Applied Ecology, 40, 947-969

Kleijn, D. & van Zuijlen, G. J. C. (2004). The conservation effects of meadow bird agreements on farmland in Zeeland, The Netherlands, in the period 1989-1995, Biological Conservation, 117, 443-451.

Knop, E. V. A.; Kleijn, D.; Herzog, F. & Schmid, B. (2006). Effectiveness of the Swiss agri-environment scheme in promoting biodiversity, Journal of Applied Ecology, 43, 120-127.

Krebs, J. R.; Wilson, J. D.; Bradbury, R. B. & Siriwardena, G. M. (1999). The second Silent Spring?, Nature, 400, 611-612.

Kremen, C.; Williams, N. M.; Bugg, R. L.; Fay, J. P. & Thorp, R. W. (2004). The area requirements of an ecosystem service: crop pollination by native bee communities in California, Ecology Letters, 7, 1109-1119.

Kremen, C.; Williams, N. M. & Thorp, R. W. (2002). Crop pollination from native bees at risk from agricultural intensification, Proceedings of the National Academy of Sciences of the United States of America, 99, 16812-16816.

Mattison, E. H. A. & Norris, K. (2005). Bridging the gaps between agricultural policy, land-use and biodiversity, Trends in Ecology & Evolution, 20, 610-616.

McNeely, J. A. & Scherr, S. J. (2002). Ecoagriculture: Strategies to Feed the World and Save Wild Biodiversity, Island Press, Washington, DC.

Melnychuk, N. A.; Olfert, O.; Youngs, B. & Gillott, C. (2003). Abundance and diversity of Carabidae (Coleoptera) in different farming systems, Agriculture, Ecosystems & Environment, 95, 69-72.

Millennium Ecosystem Assessment (2005a). Ecosystems and Human Well-Being, In Millennium Ecosystem Assessment Island Press, Covelo, CA.

Millennium Ecosystem Assessment (2005b). Ecosystems and Human Well-Being: Biodiversity Synthesis, In Millennium Ecosystem Assessment World Resources Institute, Washington, D.C.

Mooney, H. A.; Cooper, A. & Reid, W. (2005). Confronting the human dilemma, Science, 434, 561-562.

Öckinger, E. & Smith, H. G. (2007). Semi-natural grasslands as population sources for pollinating insects in agricultural landscapes, Journal of Applied Ecology, 44, 50-59.

Östman, Ö. (2002). Distribution of bird cherry-oat aphids Rhopalosiphum padi (L.) in relation to landscape and farming practices, Agriculture, Ecosystems & Environment, 93, 67-71.

Ottvall, R. & Smith, H. G. (2006). Effects of an agri-environment scheme on wader populations of coastal meadows of southern Sweden, Agriculture, Ecosystems & Environment, 113, 264-271.

Peach, W. J.; Lovett, L. J.; Wotton, S. R. & Jeffs, C. (2001). Countryside stewardship delivers cirl buntings Emberiza cirlus in Devon, UK, Biological Conservation, 101, 361-373.

Petit, S. & Burel, F. (1998). Effects of landscape dynamics on the metapopulation of a ground beetle (Coleoptera, Carabidae) in a hedgerow network, Agriculture, Ecosystems & Environment, 69, 243-252.

Pfiffner, L. & Luka, H. (2003). Effects of low-input farming systems on carabids and epigeal spiders - a paired farm approach, Basic and Applied Ecology, 4, 117-127.

Pimentel, D.; Stachow, U.; Takacs, D. A.; Brubaker, H. W.; Dumas, A. R.; Meaney, J. J.; O'Neill, J. A. S.; Onsi, D. E. & Corzelius, D. B. (1992). Conserving biological diversity in agricultural/forestry systems, Bioscience, 432, 354-362.

Pocock, M. J. O. & Jennings, N. (2008). Testing biotic indicator taxa: the sensitivity of insectivorous mammals and their prey to the intensification of lowland agriculture, Journal of Applied Ecology, 45, 151-160.

Primack, R. B. (2008). A Primer of Conservation Biology, Sinauer Associates, Sunderland , MA.

Reganold, J. P.; Glover, J. D.; Andrews, P. K. & Hinman, H. R. (2001). Sustainability of three apple production systems, Nature, 410, 926-930.

Robinson, R. A. & Sutherland, W. J. (2002). Post-war changes in arable farming and biodiversity in Great Britain, Journal of Applied Ecology, 39, 157-176.

Roschewitz, I.; Gabriel, D.; Tscharntke, T. & Thies, C. (2005a). The effects of landscape complexity on arable weed species diversity in organic and conventional farming, Journal of Applied Ecology, 42, 873-882.

Roschewitz, I.; Hücker, M.; Tscharntke, T. & Thies, C. (2005b). The influence of landscape context and farming practices on parasitism of cereal aphids, Agriculture, Ecosystems & Environment, 108, 218-227.

Rundlöf, M.; Bengtsson, J. & Smith, H. G. (2008). Local and landscape effects of organic farming on butterfly species richness and abundance, Journal of Applied Ecology, 45, 813-820.

Rundlof, M. & Smith, H. G. (2006). The effect of organic farming on butterfly diversity depends on landscape context, Journal of Applied Ecology, 43, 1121-1127.

Schmidt, M. H.; Roschewitz, I.; Thies, C. & Tscharntke, T. (2005). Differential effects of landscape and management on diversity and density of ground-dwelling farmland spiders, Journal of Applied Ecology, 42, 281-287.

Schmitzberger, I.; Wrbka, T.; Steurer, B.; Aschenbrenner, G.; Peterseil, J. & Zechmeister, H. G. (2005). How farming styles influence biodiversity maintenance in Austrian agricultural landscapes, Agriculture, Ecosystems & Environment, 108, 274-290.

Shah, P. A.; Brooks, D. R.; Ashby, J. E.; Perry, J. N. & Woiwod, I. P. (2003). Diversity and abundance of the coleopteran fauna from organic and conventional management systems in southern England, Agricultural and Forest Entomology, 5, 51-60.

Stoate, C.; Boatman, N. D.; Borralho, R. J.; Carvalho, C. R.; Snoo, G. R. d. & Eden, P. (2001). Ecological impacts of arable intensification in Europe, Journal of Environmental Management, 63, 337-365.

Swift, M. (2004). Biodiversity and ecosystem services in agricultural landscapes - Are we asking the right questions?, Agriculture, Ecosystems & Environment, 104, 113-134.

Thies, C. & Tscharntke, T. (1999). Landscape structure and biological control in agroecosystems, Science, 285, 893-895.

Thomas, J. A.; Telfer, M. G.; Roy, D. B.; Preston, C. D.; Greenwood, J. J. D.; Asher, J.; Fox, R.; Clarke, R. T. & Lawton, J. H. (2004). Comparative losses of British butterflies, Birds, and Plants and the Global Extinction Crisis, Science, 303, 1879-1881.

Thorbek, P. & Bilde, T. (2004). Reduced numbers of generalist arthropod predators after crop management, Journal of Applied Ecology, 41, 526-538.

Tilman, D.; Cassman, K. G.; Matson, P. A.; Naylor, R. & Polasky, S. (2002). Agricultural sustainability and intensive production practices, Nature, 418, 671-677.

Tilman, D.; Fargione, J.; Wolff, B.; D'Antonio, C.; Dobson, A.; Howarth, R.; Schindler, D.; Schlesinger, W. H.; Simberloff, D. & Swackhamer, D. (2001a). Forecasting agriculturally driven global environmental change, Science 292 (5515), 281-284.

Tilman, D.; Fargione, J.; Wolff, B.; D'Antonio, C.; Dobson, A.; Howarth, R.; Schindler, D.; Schlesinger, W. H.; Simberloff, D. & Swackhamer, D. (2001b). Forecasting agriculturally driven global environmental change, Science, 292, 281-284.

Tscharntke, T.; Klein, A. M.; Kruess, A.; Steffan-Dewenter, I. & Thies, C. (2005). Landscape perspectives on agricultural intensification and biodiversity - ecosystem service management, Ecology Letters, 8, 857-874.

Tucker, G. M. & Heath, M. F. (1994). Birds in Europe: Their conservation status, Birdlife International, Cambridge.

Tybirk, K.; Alrøe, H. F. & Frederiksen, P. (2004). Nature quality in organic farming: A conceptual analysis of considerations and criteria in a european context, Journal of Agricultural and Environmental Ethics, 17, 249-274.

U.S. Environmental Protection Agency (2000). Washington DC.

Uphoff, N.; Ball, A. S.; Fernandes, E.; Herren, H.; Husson, O.; Laing, M.; Palm, C.; Pretty, J. & Sanchez, P. (Eds.) (2006) Biological approaches to sustainable soil systems, CRC Press, Boca Raton.

van Emden, H. F. (1965). The role of uncultivated land in the biology of crop pests and beneficial insects, Scientific Horticulture, 17, 121-136.

Vollhardt, I. M. G.; Tscharntke, T.; Wäckers, F. L.; Bianchi, F. J. J. A. & Thies, C. (2008). Diversity of cereal aphid parasitoids in simple and complex landscapes, Agriculture, Ecosystems & Environment, 126, 289-292.

Weibull, A.-C.; Bengtsson, J. & Nohlgren, E. (2000). Diversity of butterflies in the agricultural landscape: the role of farming system and landscape heterogeneity, Ecography, 23, 743-750.

Weibull, A.-C.; Östman, Ö. & Granqvist, Å. (2003). Species richness in agroecosystems: the effect of landscape, habitat and farm management, Biodiversity and Conservation, 12, 1335-1355.

WHO and UN Environment Program (1990). Geneva, Switzerland, pp. 85-9.

Wickramasinghe, L. P.; Harris, S.; Jones, G. & Vaughan Jennings, N. (2004). Abundance and Species Richness of Nocturnal Insects on Organic and Conventional Farms: Effects of Agricultural Intensification on Bat Foraging, Conservation Biology, 18, 1283-1292.

Wickramasinghe, L. P.; Harris, S.; Jones, G. & Vaughan, N. (2003). Bat activity and species richness on organic and conventional farms: impact of agricultural intensification, Journal of Applied Ecology, 40, 984-993.

Wilby, A. & Thomas, M. B. (2002). Natural enemy diversity and pest control: patterns of pest emergence with agricultural intensification, Ecology Letters, 5, 353-360.

Wilcove, D. S.; Rothstein, D.; Dubow, J.; Phillips, A. & Losos, E. (1998). Quantifying threats to imperiled species in the United States, Bioscience, 48, 607-615.

Williamson, T. (1986). Parish boundaries and early fields: Continuity and discontinuity, Journal of Historical Geography, 12, 241-248.

Wilson, A.; Vickery, J. & Pendlebury, C. (2007). Agri-environment schemes as a tool for reversing declining populations of grassland waders: Mixed benefits from Environmentally Sensitive Areas in England, Biological Conservation, 136, 128-135.

Wood, S.; Sebastian, K. & Scherr, S. J. (2000). In International Food Policy Research Institute and World Resources InstituteWashington, DC, USA.

Zahm, S. H. & Ward, M. H. (1998). Pesticides and Childhood Cancer, Environmental Health Perspectives, 106, 893-908.

15

Environmental Evaluation and Monitoring of Agro-Ecosystems Biodiversity

Fabio Taffetani, Michele Rismondo and Andrea Lancioni
Dipartimento di Scienze Agrarie, Alimentari e Ambientali (3A)
Università Politecnica delle Marche
Via Brecce Bianche, Ancona
Italy

1. Introduction

Agroecosystems are open systems, whereby continual human intervention is needed for the maintenance of an equilibrium created by man himself with the aim of maximising the production provided by a few domesticated plant and animal species. In these systems, the equilibrium is maintained through energy input from outside that makes the operations possible, which are designed to create optimal conditions for the growth of the cultivated plants and the animals raised.

The individual agricultural systems can be very different depending on the pedoclimatic characteristics of the territories, the types of cultivation practiced, and the intensive practices used, although in all cases, to maintain specific productive characteristics there is the need for human intervention. If they are abandoned to their natural evolution, these systems tend to recover their natural state, with increases in the areas occupied by spontaneous vegetation and with substantial changes in the population dynamics of the various species that make up the wild fauna, obviously at the expense of the species that benefit from the human actions.

Despite being characterised by a specific level of structural simplicity and by a decrease in their natural state that is caused by the modifications to the original environment that had the aim of obtaining animal and plant production for mainly food and energy uses, these agroecosystems cover a large part of the Community area. For this reason, the conservation of the biodiversity on a large scale needs interventions that are aimed at the preservation of the contexts that are most hard hit by human activities.

As well as having less value with respect to the areas characterised by a higher natural state, in various situations, the agricultural systems can provide great heterogeneity of the environment, and in cases in which the actions of man have not had a great impact, they can contain and maintain a considerable biodiversity patrimony.

For extensive agricultural systems, which are generally more common in the medium-high hilly and low mountain areas and are associated with agricultural systems of low impact, with good diversity of cultivation and the presence of semi-natural elements, these can provide a reasonable variety of habitats and large areas of contact between environments with different ecological characteristics (the ecotones). In these cases, the animal and plant biodiversity levels are high and their maintenance is strictly linked to the management practices adopted by man, whereby the traditional activities favour such biodiversity.

On the other hand, the agricultural systems that are subjected to intensive cultivation that are generically represented in the plain or low hilly areas are poorer overall in landscape elements that are useful for the protection, survival and reproduction of the wild animal and plant species.

Over the last ten years, a greater environmental awareness has guided public opinion towards a re-evaluation of the role of man in the management of the agroecosystems on which the maintenance of a satisfactory level of biological complexity depends. From the agricultural point of view, we have therefore passed from a largely productive approach to management strategies that are aimed at an increase in the quality of the products and the rural environment.

Today, the recovery of ancient cultivated varieties or autochthonous races is encouraged and has taken on crucial importance from the politico-economic and social points of view within specific rural contexts. The appreciation and commercialisation of these products, together with the official recognition that guarantees their quality, can return the original identity to the rural territories of high landscape value. Given their specific characteristics, such territories would have been progressively excluded from the commercial networks of intensive and industrialised agriculture.

The return of the original production and local traditions is allowed by the measures introduced with the regulatory practices that are a part of agriculture today. The Common Agricultural Policy that is applied at a regional level through the Regional Plans for Rural Development, and the regulation of the Cross Compliance favour the modernisation of the sector and of the individual farms, the improvement of the socio-economic conditions of the rural populations, and the protection of the environment within which these populations live and work.

In this sense, the farms take on a multifunctional role: as well as providing quality production, and in some cases services such as refection, or educational or recreational activities (e.g., holiday and educational farms), are especially required to manage and maintain their environment. This guarantees periodic maintenance of the non-productive infrastructure, the presence of which has fundamental importance in the correct ecological functioning of the agroecosystem.

This non-productive infrastructure of the agricultural landscape serves many functions, among which there is the conservation of biodiversity, the management and filtering of the water, the protection from pollution, erosion, and hydrogeological problems, and the fixation of CO_2. Moreover, these constitute an excellent indirect opportunity for the farms, because they allow the farms themselves to identify with a varied productive context that is rich in life and well maintained.

It is obvious that the actions of those who operate in agriculture must be evaluated periodically in order to verify the effective validity and efficacy of their practices. For this, there is the need for indicators or indexes that provide a sufficiently detailed measure of the quality of the rural environment.

With the present study, following on from a description of the main vegetation typologies of an agroecosystem context, we will illustrate a method for the evaluation of the functionality of the rural contexts that we have applied in past years to various territories examined. The system that we have adopted is based on vegetation analysis and on the application of

indexes specifically designed for this. This system allows precise measures to be carried out, which can be integrated with other fauna bioindicators. It has been applied at different scales, which go from hydrographic basins of individual farms to small herbaceous areas.
The analytical instrument that we present is particularly useful for the monitoring of the conservation state of agroecosystems, and in general it provides a tool for the better management of rural territories.

2. Vegetation of the agroecosystem

2.1 Study of the plant landscape

The study of the vegetation represents the best-suited method for interpretation of the ecological potential of a territorial context. The arrival and establishing of one specific vegetational typology rather than another are phenomena that are mainly linked to the particularities of the climate, pedology and use of a specific area. The phytosociological method is a rapid and efficient analytical procedure based on floristic and statistical measures that is widely used throughout the World (Westhoff & Van der Maarel, 1978; Gèhu & Rivas-Martinez, 1981).
It is based on phytosociological sampling that is carried out in the field following the individuation of an ecologically homogeneous area. It consists in the forming of a list of all of the plants that are found in the chosen site and the assignment of coverage and interactive values to all of these following specifically designed scales (range, + → 5 for the coverage, and 1 → 5 for the interactions). A re-evaluation of the relevés of the area then follows, through comparisons with Tables of previously defined relevés or through statistical analyses, with the aim of defining the plant associations.
The association is the basic unit upon which phytosociology is based, which is found within a unit of territory that is ecologically homogeneous and is represented by significantly repeated floristic components. The phytocoenoses are organised according to a hierarchical system. The upper level of the association is represented by the alliance, which in turn is included in an order. At the apex of the hierarchical organisation there is the vegetation class, which expresses the structure, evolution and phytogeographical significance of the various typologies that are referred to it.
The identification and description of the plant communities represents the first level of investigation, which goes under the name of classical phytosociology analysis. The study of the vegetation is, however, expressed according to two further levels: synphytosociological and geosynphytosociological.
The synphytosociological investigation is based on the identification of vegetation series. This phase studies the dynamic relationships between the various associations within the same sigmetum, which consists of a portion of the territory characterised by the same pedoclimatic particularities and within which there is therefore space for a single typology of potential vegetation (the climax). This level allows identification of the interactions between the different phytocoeneses, as evolutive or regressive, that occur within a "tessella" of ecologically homogeneous territory (Biondi, 1996). The tessella therefore represents the biogeographical environmental unit of a hypothetical mosaic and it is defined on the basis of various factors: the nature of the substrate, the altitude, the exposure, and the slope.
Within each tessella, the associations are organised in serial interactions and the areas occupied by the same plant communities are not constant with time, but can vary according to the management practices that are followed.

The third level of analysis is known as geosynphytosociology, and it studies the interactions between typologies of the vegetation across different geographical environmental units. In this case, these interactions are defined as chain interactions, and the evolution of a coenosis towards another in chain contact is never expected to occur, because two such coenoses occupy landscape tessellae with different vegetational potentials.
The method has an important predictive role in the ecological analysis of the landscape and has particular use in the planning and management of the territory. It can be used on different scales and allows generalisable results to be obtained. Depending on historic data, it allows the performing of diachronic analyses. It has been adopted by the European Community as the official means for identification and description of the habitats that warrant protection according to Directive 92/43/EEC.
This investigation system constitutes, moreover, the basis of our analysis of the functionality of the agroecosystems. This has been possible following the application of the floristic-vegetational indexes that are now presented (Taffetani & Rismondo, 2009; Rismondo *et al.*, 2011).

2.2 The rural ecomosaic

As already discussed in the introduction, the agroecosystem normally appears within a context that has been modelled and strongly influenced by man. Although recognising the existence of agricultural realities that are greatly different both from the point of view of the vegetational potentiality and that relative to the productive uses and the impact of man, it is still possible to say that generally the rural ecomosaic is formed by an alternation between various tesserae that can be differentiated one from the other as a function of their ecological significance.
In the agricultural landscape, a succession of areas can be seen that are cultivated with techniques that are more or less intensive, as the plots, alternating with diffuse semi-natural elements that gradually tend to occupy more space in passing on to contexts with less human impact. These are usually found in the zones that are less accessible and where mechanical tillage is less possible, and they are more frequent in the agricultural areas that are less intensively cultivated or are subjected to particular protection regulations.
The semi-natural elements are identified by formations such as rows of trees, hedges, small woods (either spontaneous or planted by man) and their ecotones, grasslands, abandoned fields, herbaceous strips along ditches, slopes, small farm tracks, edges of fields and ditches, and riparian formations.
The agricultural areas that are more variegated and richer from a naturalistic point of view usually occupy the zones in which the local population knew how to maintain and value traditional techniques and production in equilibrium with the potentiality of the surrounding environment. However, more and more often the areas that are disadvantaged from a productive point of view tend to be progressively abandoned; this allows evolution of grass coenoses, an increase in the areas of the shrub and tree communities, and the consequent loss of ecosystem variety. All of the units making up the agroecosystem mosaic take on their own functional significance within the context analysed. This significance is the fruit of all of the previous progressive productive uses, and it can be photographed in a clear and precise way, with the identification of the floristic-structural characteristics and with the subsequent syntaxonomic classification of the investigated areas.
In this section of the present study, the vegetation classes will be presented that are more represented in the rural systems that are under analysis. These last refer to the Adriatic

sector of the Italian peninsula, from the region of Fruili Venezia Giulia to Molise (Fig. 1). The geographical localisation and the pedoclimatic and morphological characteristics of the areas under study are summarised later. Brief descriptions of the ecological characteristics and the evolutive significance of the syntaxonomic classes are given. Moreover, the main orders are illustrated, and some of the vegetational typologies are described that can be typically seen in the agricultural ecosystems.

The classes are ordered according to their physiognomic and ecological characteristics: the serial typologies (therophyte vegetation, perennial grasslands, forest edges, shrub layers and mantles, woods) have been differeniated from the category relative to the acquatic azonal vegetation (both herbaceous and arboreal).

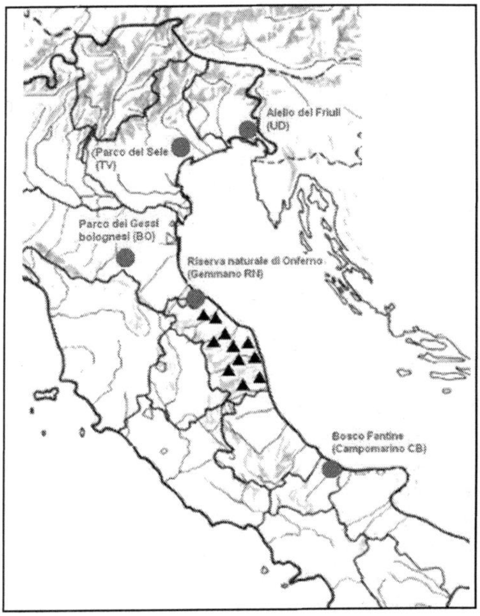

Fig. 1. Map of the study areas.

2.3 The therophyte vegetation

2.3.1 Class *Stellarietea*

The plant communities that have mainly therophyte species are referred to the class *Stellarietea*, and they are characterised by an annual cycle and reproduction by seed (Rivas Martinez et al., 2002).

This syntaxon includes phytocoenoses of commensals of annual-seeded crops, as both autumn to winter and spring to summer, and initial and immature grasslands that are seen under conditions of recurring disturbance, mainly by man. This disturbance can be caused by herbicides or by movement or removal of the surface layer of the soil.

In a serial context, the class *Stellarietea* occupies the first evolutionary step. The phytocoenoses of this syntaxonomic unit take on pioneering characters and tend to be the first to grow on bare substrata. In these communities, exotic species or species with a large distribution that were accidentally introduced by man are often strongly represented, and these are distinguished by their high invasive capacity; in some cases, these same species have a dominant physiognomy and end up by taking over from the other species.

Stellarietea has a ruderal, nitrophilous or semi-nitrophilous character, and it is diffuse throughout the World, except in the tropical zones. Relative to the temperate zones and to part of the Mediterranean bioclimate distribution area, five different orders can be identified, as described below.

- *Centaureetalia cyani*: This syntaxon includes the infesting coenoses of the autumn to winter crops (cereals) without irrigation. This is well represented on marly-arenaceous or calcareous soils more or less rich in nutrients and alkaline or subalkaline, in the hilly and montane zones of the temperate bioclimate.

  **Biforo testiculatae-Adonidetum cupanianae* (Fig. 2): This is the reference association for the commensals of the autumn to winter seeded crops of the marly-arenaceous hilly sectors of central Italy. It can be found in the Mesomediterranean and temperate hilly bioclimate belts, and it belongs to the alliance *Caucalidion lappulae*. The species belonging to this association have relatively early flowering, and the characteristic species are *Legousia hybrida, Adonis annua* ssp. *cupaniana, Papaver hybridum, Bifora testiculata, Gladiolus italicus* and *Valerianella eriocarpa*.
- *Aperetalia spicae-venti*: This order groups the coenoses that are mainly found on the alluvial plains of the temperate zones, as commensals of autumn-winter annual-seeded crops. It is well represented on subacid, sandy or sandy-silty soils that are well supplied with water.
- *Solano nigri-Polygonetalia convolvoli*: The associations coming from this syntaxonomic unit are composed of infesting species that can colonise the sown spring-summer crops (both irrigated and not) or the rows between grapevines that have not been seeded. This syntaxon is distributed mainly in the temperate zones and in part in the zones with a Mediterranean bioclimate.

  **Linario spuriae-Stachyetum annuae*: This belongs to the alliance *Polygono-Chenopodion polyspermi* and it represents the infesting vegetation of spring-summer crops in the hilly clayey and marly zones that are not irrigated. It is diffuse in many sectors of the Italian peninsula south of the River Po, and it is mainly present in situations of strong edaphic aridity, interrupted only by the contribution of the limited summer rain. The characteristic species are *Linaria spuria, Stachys annua* and *Picris echioides* (Baldoni, 1995).
- *Chenopodietalia muralis*: This order represents the nitrophilous-ruderal vegetation of the Mediterranean area that is relatively mesophilous. It includes herbaceous pioneering coenoses, often dominated by exotic species, which grow on soils heavily enriched in nitrogen nutrients.
- *Sisymbrietalia officinalis*: This syntaxon includes the pioneer and ruderal plant communities of temperate Europe and the Mediterranean that grow along the edges of farm tracks and paths, which is also a habitat that is often subjected to compaction (and therefore on substrata that are particularly compact) and near to human habitation.

  **Sinapidetum albae*: This coenosis belongs to the alliance *Sisymbrion officinalis* (of the temperate zones) and overall it is dominated by the early flowering Crucifera *Sinapis alba*, which characterises the edges of the tracks and fields in March and April with its garish flowers. It has been described for the hilly Marche territories (Allegrezza Ballelli and Biondi, 1987) and prefers soils rich in organic deposits that are frequently disturbed or removed.

Aveno barbatae-Brometum diandri: This is an association of herbaceous margins that is relatively diffuse in the central Adriatic sector of Italy and it belongs to the alliance *Hordeion murini*. It is characterised by the species *Avena barbata* and *Bromus gussonei*. It has a spring development and it is easy to recognise in the month of April, while it tends to completely dry out with the advancing season; it is often found in situations where the soil is removed periodically to avoid invasion of the farm tracks by the vegetation and to keep the rain drains free (Biondi & Baldoni, 1991).

- *Thero-Brometalia*: This includes the nitrophilous-ruderal associations of a xero-thermophilous type that are limited to Mediterranean and Submediterranean territories. These coenoses are usually found where there are high nutrient concentrations in the soil.

Fig. 2. *Biforo testiculatae-Adonidetum cupanianae*.

2.3.2 Class *Polygono-Poetea*

The class *Polygono-Poetea* is diffuse throughout the World and includes the vegetation typologies that are characteristic of habitats that are compacted. It is linked to humanised and strongly compacted environments, as it has the ability to grow in environments such as the edges or centres of dirt tracks; some communities can colonise the spaces between the bricks of paved tracks or paths (Rivas-Martinez et al., 2002). The phytocoenoses in this class are ephemeral and are made up of annual plants, and in some cases, small perennials, which are anyway always characterised by the ability to grow in edaphic contexts with relatively high nitrate contents. In the agroecosystem, the communities that are included in this syntaxon can frequently be found where the tyres of farm vehicles have compacted the ground.

Although this is a vegetation class that has a strong human footprint, *Polygono-Poetea* does not tolerate excessive disturbance and reacts negatively to practices such as weeding or soil tillage, which, as already indicated, more favours species from the class *Stellarietea*. From an evolutive point of view, *Polygono-Poetea* is positioned a step higher than *Stellarietea*.

The only order recognised for this class is *Polygono arenastri-Poetalia annuae*.

- *Polygono arenastri-Poetalia annuae*: This order includes communities of grass, whether or not creeping, that can be found in temperate and Mediterranean climates, in sandy, silty and clayey edaphic contexts, and in some cases, rich in limestone. This order has a large ecological valence: some of the coenoses included in the order favour more wet and shady locations, others instead grow in more arid contexts.

Poetum annuae: These communities are dominated by the small Graminacea *Poa annua*, and they are part of the alliance *Matricario-Polygonion avicularis* and are found in contexts subjected to strong compaction and with soils that are not subjected to water retention. The dominant species is accompanied by many other entities of the same class that are particularly resistant to soil compaction. This coenosis is spread diffusely through the agroecosystems analysed, although it had not been previously noted for the territories investigated. Instead, it had been found in various locations on the Balcan peninsula Carni et al., 2002; Carni, 2005; Silc & Kosir, 2006).

Coronopodo procumbentis-Sclerochloetum durae (Fig. 3): This coenosis grows in situations under more intense impaction and it usually has a very low coverage index (50%-60%) and is dominated by the Graminacea *Sclerochloa dura*. It belongs to the alliance *Sclerochloo durae-Coronopodion squamati*, reaching its full development between the end of April and the beginning of May, and it is found along farm tracks. It has been described for the subcoastal territory of Marche (Biondi et al., 2002).

Fig. 3. *Coronopodo procumbentis-Sclerochloetum durae.*

2.4 The perennial grasslands

2.4.1 Class *Artemisietea*

This syntaxonomic unit includes pioneer associations that are mainly made up of biennial and perennial species, usually medium to tall (Rivas Martinez et al., 2002), that can be found in contexts affected by human activities, and that can rapidly colonise locations with uncovered soil or occupied by less evolved coenoses (of the class *Stellarietea mediae*).

Typical examples are represented by the plant communities of abandoned fields, of the calanchi, of the edges of fields and tracks, in the ruderal zones near to houses, and also in the urban environments or on the outskirts of cities. These coenoses grow under conditions of recurrent disturbance, although not excessive or systematic, and they represent the first evolutive stage towards perennial grass formations that are particularly stable. These last can grow in better conserved environments and are usually managed by periodic removal of the biomass or with pasturing, and they are included in the classes *Molinio-Arrhenatheretea* and *Festuco-Brometea*.

The class *Artemisietea* is found in territories with a Mediterranean-pluviseasonal and holartic-temperate climate and it definitely prefers substrata enriched in nutrients, and especially in nitrates. The syntaxa referred to this class grow on various types of substrata, mainly clayey and alkaline. From the evolutive point of view, they represent a higher level than the other two classes presented above. In the areas of temperate Europe and the Mesomediterranean, three orders are found.

- *Agropyretalia repentis*: This syntaxon includes the post-cultivation, ruderal grassland formations that are medium nitrophilous and are found in the Eurosiberian temperate and Submediterranean areas, on substrata that are mainly of the clayey type. The coenoses here are particularly diffuse in agricultural areas that have been abandoned, as they mainly characterise plots that have not been used for 3 or 4 years or the calanchi areas, with low or higher salt concentrations in the soil, that have been subjected to strong erosive morphological modifications.
  **Senecio erucifolii-Inuletum viscosae* (Fig. 4): This phytocoenosis is found in the abandoned fields of the central Adriatic hilly sectors with temperate or Submediterranean bioclimate (Biondi & Allegrezza, 1996). The diagnostic species are *Senecio erucifolium, Bellevalia romana* and *Cirsium italicum*. This community is included in the alliance *Inulo-Agropyrion* and it is usually physiognomically dominated by the Composita with a Euromediterranean distribution area, *Inula viscosa*. It prefers marly and clayey terrain and in situations that are more mesophilous it is enriched in species of the class *Molinio-Arrhenatheretea*, such as *Holcus lanatus* and *Agrostis stolonifera.*
- *Artemisietalia vulgaris*: This represents the central order of the class, and it includes ruderal and moderately mesophilous communities of the temperate areas. These coenoses often grow on soils that are sufficiently deep and wet, as well as being well supplied with nitrogenous nutrients, and they can also be found along the edges of water courses.
- *Onopordetalia acantii*: This order is represented by the ruderal and pioneering phytocoenoses of the meso-orotemperate or thermo-oromediterranean areas. These communities are usually dominated by species of thistles, and in particular of the genus *Onopordum*, and on average, they are more thermophilous than the formations included in the other two orders, They grow on soils of various natures: sandy, clayey and more or less calcicole.

Fig. 4. *Senecio erucifolii-Inuletum viscosae*

2.4.2 Class *Molinio-Arrhenatheretea*

Molinio-Arrhenatheretea is the vegetation class to which the regularly cut grasslands are referred, or in some cases pastured, that develop on deep, wet, neutral to subacid soils that are well supplied with nutrients, in territories with a prevalently flat or slightly sloped morphology (Rivas Martinez et al., 2002). The species of this class are mesophilous, nitrophilous, and well used as forage, and they are favoured by human practices, with the cutting and removal of the biomass.

In the agroecosystem, the communities of this syntaxonomic unit are spread in particular in the valley bottoms, along with grasslands or in strips of grassed terrain at the edges of ditches, fields or farm tracks. In some cases, they collonise contexts subjected to great human disturbance, such as the compacted environments or the grass areas by farm houses. The phytocoenoses of *Molinio-Arrhenatheretea* can survive short periods of flooding, which frequently occur during spring, but they cannot tolerate periods of drought. This class has a Mediterranean, thermo-boreal and temperate distribution area, although it finds its optimum in the Eurasiatic area. From an evolutive point of view, the coenoses of this syntaxon can be considered among those herbaceous with intermediate maturity and stability levels. In the temperate and Submediterranean contexts, the hierarchical subdivision reveals the presence of six orders.

- *Molinietalia caeruleae*: This order has a central-European distribution area and includes temperate and mesophilous grasslands and pastures that are often cut and are linked to the presence of terrain that is almost always wet. They include both the communities of the non-fertilised and oligo-mesotrophilous grasslands and the fertilised and often flooded habitats that are found on the low-montane to montane zones.
- *Arrhenatheretalia eliatioris*: This includes fertilised grasslands and pastures with a Eurosiberian and Submediterranean distribution area. The communities are mainly dominated by *Arrhenatherum eliatius* or *Cynosurus cristatus*, and grow on deep and wet soils that have clayey or calcareous origins and good levels of mineral fertility, and that are well drained.
- *Plantaginetalia majoris*: This syntaxonomic unit includes the herbaceous communities that grow on wet eutrophic soils, which are sometimes flooded and are pastured and compacted, with a temperate and Medierranean distribution area. They are particularly well represented in the agroecosystem, on substrata that are mainly clayey or on alluvial plains, at the edges of ditches or at the sides of tracks, in edaphic contexts generally wet also in summer.

  **Lolio perennis-Plantaginetum majoris*: This association is dominated by *Lolium perenne* and *Plantago major*. It is included in the alliance *Lolio-Plantaginion*, and is linked to conditions of compact soil that is rich in nutrients and is wet. It is well enough spread in the investigation area, and in particular at the sides or along the centres of farm tracks, or in herbaceous areas subjected to intense impaction and enriched in nitrates due to the pasturing of animals.

  **Festuco fenas-Caricetum hirtae* (Fig. 5): The grasslands of *Carex hirta* of the alliance *Mentho-Juncion* find their ecological optimum in small depressions that are flooded and disturbed due to compacting, pasturing and ruderalisation (Ninot et al., 2000). Although they were not previously found in the area investigated, the association was found in some locations in Marche, along the banks of rivers and ditches, where it tends to grow on deep and wet substrata that are rich in nutrients and are periodically flooded.

  **Ranunculetum repentis*: Overall, these formations are dominated by *Ranunculus repens* and are characterised by their hygrophilous condition, as they can colonise the valley bottoms that are adjacent to water courses and often flooded, or plain areas that have slight depressions and with surface water. They belong to the alliance *Mentho-Juncion* and in the agroecosystem their presence is limited to the zones that are under good conservation.

- *Paspalo-Heleochloetalia*: This order includes grassland coenoses that are characeristic of the river banks that are sometimes flooded or submerged, or of the edges of small stagnant ponds or pools of water. These grow on muddy terrains that are rich in minerals and nitrogenous substances, and they are often characterised by being physiognomically dominated by stolonate exotic plants
- *Trifolio-Hordeetalia*: This syntaxon includes the permanent grasslands that grow on the previously lacustrine basins or on the alluvial plains. Its distribution area is limited to the Balcan peninsula and the central-southern Apennines, in which environment it is well represented by various associations that can be found on the karst high-plains, mainly in depressions that are particularly subjected to occasional flooding.
- *Holoschoenetalia vulgaris*: This includes grasslands with a dominance of rushes with a Mediterranean or Submediterranean distribution area and summer development. These coenoses are found on deep and muddy soils with a mainly silty and clayey fabric, although in other cases also on gravely or sandy sediments. They are well represented mainly corresponding to or near to bends in rivers with weak water currents.

Fig. 5. *Festuco fenas-Caricetum hirtae*

2.4.3 Class *Festuco-Brometea*

The class *Festuco-Brometea* includes the grasslands of human origins that in general are used for animal pasture. The communities included in this syntaxon are mainly composed of hemicryptophyte species characterised by good forage value; of the grasslands, these represent those with greater levels of stability and maturity. These formations are often particularly rich from a floristic point of view and they play host to numerous entities of great interest, among which there are various species of orchids.

The distribution area of this syntaxonomic unit includes the Eurosiberian and the western Mediterranean regions. From an edaphic point of view, the grasslands of *Festuco-Brometea* are baso-neutrophilous or, in some case, slightly acidophilous, and they generally prefer deep and rich soils, and anyway well drained, although they sometimes also grow on relatively superficial substrata (Rivas Martinez et al., 2002).

In the agroecosystem, the coenoses of this class are relatively diffuse in well conserved contexts that are under techniques of extensive cultivation, while their diffusion is lower in the zones that are more subjected to human influence. From the management point of view, the abandonment of grasslands of *Festuco-Brometea* favours the recollonisaton of the pastures by shrub species, with the consequent floristic impoverishment and drastic reduction in the plant biodiversity. The associations that grow on calcareous substrata are protected according to the specific Community Directive 92/43/EEC, as habitat 6210, and therefore they need to be managed in such a way as to guarantee their conservation.

In the territories that were included in our analysis, the grasslands of this class can be included in the order *Brometalia erecti*.

- *Brometalia erecti*: This order includes the grasslands that substitute the forest vegetation and that are found on calcareous or marly-arenaceous substrata, which is more or less deep, in the temperate bioclimate areas, including the zones that fall within the Submediterranean variant. Its distribution area includes central and western Europe (Royer, 1991). The coenoses of this syntaxon are diffusely represented throughout the Italian peninsula.

  Centaureo bracteatae-Brometum erecti (Fig. 6): The grasslands attributed to *Centaureo-Brometum* are mesophilous continuous formations that to grow often as a post-cultivation stage and that are present on marly-arenaceous, arenaceous-clayey, and marly-calcareous substrata of the hilly slopes. These are widely diffuse in the northern and central Apennines (Biondi et al., 2006). They are included in the alliance *Bromion erecti* and their characteristic species are *Brachypodium rupestre*, *Galium album*, *Carex flacca* ssp. *flacca* and *Centaurea bracteata*.

Fig. 6. *Centaureo bracteatae-Brometum erecti*.

2.5 The vegetation of the forest edges

2.5.1 Class *Galio-Urticetea*

The class *Galio-Urticetea* is represented by the high-grass communities that grow along neutrophilous margins that are in contact with pre-forest or forest formations, and in some cases also near to artificial tree plantations, as for example the reforestation of conifers. Its distribution area includes territories with a thermoboreal, pluviseasonal Mediterranean and holarctic temperate bioclimate (Rivas Martinez et al., 2002).

The sites where it is possible to find vegetation of the class *Galio-Urticetea* are still today, or were in the past, under modifications attributable to human activities, which results in particular in enrichment of the soil with nitrogenous substances (Poldini, 1989). The plants

of the groups referred to this class are generally characterised by a good ability to live under shady conditions. They prefer deep soils with a good water content and rich in nutrients.
In the agroecosystem the plant associations of this syntaxon can be seen in sites such as the margins of ditches or the edges of copses, rows of trees, or hedges, and they can often be found also in human contexts, like in and around urban areas. They can grow on acidic substrata as well as on alkaline terrain.
From the serial point of view, the vegetation of this syntaxon can be identified as a pre-forest herbaceous stage that grows in the contact zones beween the herbaceous hemicryptophytic and shrub-arboreal vegetation. From the evolutive aspect, as it is mainly linked to the influence of man, and it is considered less mature with respect to the edge vegetation of the class *Trifolio-Geranietea,* which instead is found in contact with arboreal formations that are better conserved and structured.
Galio-Urticetea can be subdivided into two orders according to their ecological characteristics.

- *Galio-Alliarietalia*: This order is represented by the coenoses of the nitrophilous pre-forest or forest margins (Font et al., 1988). These can also be seen in small clearings, and they are linked to temperate and Mediterranean bioclimates, and require conditions of deep soil. The formations of this syntaxon can be more or less sciaphilous and their survival depends on good water availability of the soil also in summer; they are. However. never linked to situations characterised by stagnant water.
  Galio aparines-Smyrnietum olusatri: This association was described by Allegrezza et al. (1987). It is dominated by the *Smyrnium olusatrum* species, and it can be found at the edges of semi-natural and artificial arboreal formations, often along the sides of roads in shady and wet areas in and around urban zones. This coenosis is included in the alliance *Galio-Alliarion* and is particularly nitrophilous and distributed both on marly-calcareous substrata and on arenaceous and clayey hills. It is well enough represented in the rural landscape.
  Alliario petiolatae-Chaerophylletum temuli (Fig. 7): This nitrophilous edge association with *Alliaria petiolata* and *Chaerophyllum temulum* is also included in the alliance *Galio-Alliarion,* and has a sciaphilous character and grows in locations with soil rich in nutrients, e.g. in the zones next to the banks of rivers and edges of ditches, where the fine and rich soils are deposited (Hruska, 1988). Moreover, it is present at the edges of forest formations and in dense coenoses along the sides of roads. It is very well represented in the egroecosystem environment.
  Petasitetum hybridi: This association is characterised physiognomically by the large leaves of the species *Petasites hybridus,* and it is usually found on the external parts of the banks of the water courses, sometimes at the edges of riparian formations of *Salix alba,* generally under conditions of high atmospheric humidity (Baldoni & Biondi, 1993). The soil on which this grows is rich in organic matter and is constantly wet. This belongs to the alliance *Aegopodion podagrariae* and is well represented in the terriory analysed.
- *Calystegietalia sepium*: The phytocoenoses of this syntaxonomic unit are instead closely linked to the presence of water. Indeed, they can be found mainly on the banks of ditches and small pools of water, on soils with a high content of nitrates and water. Often, these associations can survive also under conditions of temporary stagnant water. The plants that form a part of these communities are climbers and moderately sciaphyle.

**Arundini donacis-Convolvuletum sepium*: The reed beds of *Arundo donax* are included in the alliance *Calystegion sepium* and are often associated with the fluvial margins that are degraded following the effects of remodelling of the banks and of herbicide use. The dominant species, which is almost always accompanied by the climbing Convulvulacea *Calystegia sepium*, was introduced for agricultural purposes and has become naturalised since. Today it is a constant element of the rural landscape, where it grows mainly on wet clayey terrain.

**Convolvulo sepii-Epilobietum hirsuti*: The formations dominated by *Epilobium hyrsutum*, which can also themselves be included in the alliance *Calystegion sepium*, grow on constantly wet soil, on various types of substrata. In the agricultural environment they can be found both on the borders of water courses with a permanent flow, in contact with more hygrophilous vegetation, and in the internal parts of ditches with periodic water flow, which are situated along the hilly slopes.

Fig. 7. *Alliario petiolatae-Chearophylletum temuli*.

2.5.2 Class *Trifolio-Geranietea*

This syntaxonomic class includes the plant communities of the margins of the pre-forest or forest formations of the Eurosiberian and Submediterranean regions (Rivas Martinez et al., 2002). The associations of this syntaxon are found in the hilly and montane bioclimate belts, both along the borders of the thermophilous oak and hornbeam woods as well as along the margins of the more mesophilous beech woods.

The phytocoenoses *Trifolio-Geranietea* can be seen on various terrains of calcareous, arenaceous and clayey origins. They usually prefer wet, deep-soiled sites with a good humus layer and constantly in the shade; in some cases they can even grow on particularly poor substrata. Some coenoses are more adapted to growing on acidic terrain, others in edaphic contexts that are more alkaline (Biondi et al., 2001).

The communities of this class are linked in contact with well-conserved forest environments and often show high floristic richness due to their growth in situations that are ecologically transitory; this last factor increases the plant biodiversity.

In the agroecosystem, this class is not widespread, both for its scarcity and for the often not optimal conservation state of the residual woods of the rural territory. Indeed, with these residual woods, the ecotone space of the transition towards contexts used by man, e.g. cultivated fields, is often missing; frequently the terrain in contact with the small wooded areas found in the countryside are tilled right up to the margins of the wood, with the consequent destruction of the edge coenoses.

From the synphytosociological point of view, the class *Trifolio-Geranietea* can be considered as the herbaceous state with greater evolutive significance. The reference order for the coenoses found is *Origanetalia vulgaris* (Biondi et al., 2006).

- *Origanetalia vulgaris*: This order includes the coenoses of the edges of the woods that grow on mature soils that have a lot of humus and are mainly calcareous or marly-arenaceous. The coenoses are diffuse in the temperate and Submediterranean bioclimates, and the order is diffusely spread throughout the Italian peninsula, and especially in the Apennines, at the margins of both thermophilous and mesophilous woods

  **Buglossoido purpureocaeruleae-Glechometum hirsutae* (Fig. 8): This coenosis is the forest edge formations of downy oak or hop hornbeam of the areas with a temperate or Submediterranean climate. It has been described for a forest area of the Marche hinterlands ("Monaci Bianchi"Wood; Taffetani et al., 2009). It is included in the alliance *Trifolion medii* and it is not infrequent to see it at the margins of the residual woods of rural territories, in ecotone zones that are not greatly disturbed and in situations with deep and humus-rich soil. It is dominated by the species *Glechoma hyrsuta* and in the better preserved zones it can be noted for its good floristic richness and the presence of species of the upper heirachical levels.

Fig. 8. *Buglossoido purpureocaeruleae-Glechometum hirsutae*

2.6 The shrub and arboreal vegetation

2.6.1 Class *Rhamno-Prunetea*

The class *Rhamno-Prunetea* includes all of the coenoses of the forest and shrub mantle, which are composed mainly of nanophanerophyte or phanerophyte species that normally grow in contact with forest communities of the classes *Querco-Fagetea* and *Salici-Populetea*.

This syntaxon is mainly diffuse in the Eurosiberian and Mediterranean regions and the coenoses that belong to it can be more or less mesophilous or thermophilous (Rivas-Martinez et al. 2002). From the edaphic point of view, the class shows good ability for colonisation of various substrata. Among the communities that are a part of this class, some prefer fresh and deep soils, while others have adapted to also live on very poor and superficial substrata. In this last case, sometimes the shrub vegetation can take on the role as head of the series, due to a lack of the right ecological conditions for the development of mature forest formations.

In the rural landscape, the vegetation of *Rhamno-Prunetea* is well enough represented, both in the margins of woods or of riparian arboreal formations, and along rows of trees or in hedges. The shrubs colonise abandoned fields and pastures, and they represent the serial stage that precedes the return of the woods.

For the syntaxonomical classification, relative to the temperate and Submediterranean contexts analysed, the main reference order is *Prunetalia spinosae*.

- *Prunetalia spinosae*: This order collects the substitution phytocoenoses (mantles, shrubs, hedges) of the woods of the class *Querco-Fagetea* that grow mainly on well-structured soils that are often calcareous or calcareous-marly. Within this order, there are mesophilous coenoses that can be found on the hilly and montane zones of the temperate bioclimate, and others that are subthermophilous and thermophilous, which are typical of the Submediterranean areas.

  Clematido vitalbae-Rubetum ulmifolii: These coenoses with a dominance of *Rubus ulmifolius* are usually found as extended shrub layers that are almost impenetrable, among which there are brambles, with their ability to quickly expand due to their intense production of suckers, and which are associated with the Ranuncolacea vines of *Clematis vitalba* (Poldini, 1989). These shrub coenoses of the alliance *Berberidion vulgaris* are usually characterised by a floristic poverty, and they are indifferent to soils of different origins due to the wide ecological valance of brambles. In the rural landscape, they are very widespread, especially in the zones abandoned for longer times; they are frequenly found on post-cultivation grasslands (*Senecio-Inuletum*) and often show high stability, slowing the return of the wood, especially in situations particularly degraded and with a scarcity or absence of the propagules of arboreal forest species.

  Symphyto bulbosi-Sambucetum nigrae (Fig. 9): This mesophilous pre-forest formation of *Sambucus nigra* has been described for the hilly territory of Marche (Biondi & Allegrezza, 2004), and it prefers deep and fresh soils and makes up the pre-forest stage of elm coppices, both on sandy substrata and on alluvial substrata. It is ascribable to the alliance *Pruno-Rubion* and it is differentiated from analogous formations with elder of the northern Apennines and the eastern Alps by the presence of some Mediterranean and south European species, such as *Arum italicum*, *Symphytum bulbosum* and *Rubus ulmifolius*.

Fig. 9. *Symphyto bulbosi-Sambucetum nigrae*

2.6.2 Class *Querco-Fagetea*

This class includes the mesophilous woods that are spread over the hilly and montane zones of the areas with a temperate macrobioclimate, with penetration into zones with a Mediterranean influence (Rivas-Martinez et al. 2002).

The formations of the class *Querco-Fagetea* can be more or less structured depending on their previous management, and as with all woods, these are characterised generally by a multi-layered structure; indeed, they are composed of arboreal species, which characterise them physiognomically, as well low arboreal, shrub and herbaceous species.

The coenoses that can be included in this sytaxonomic unit can be found on various types of substrata, from those of calcareous origins to those marly, clayey and arenaceous, and they generally prefer soils with a certain depth, even if in some situations they can grow in relatively superficial edaphic contexts.

The vegetation of *Querco-Fagetea* represents the climax, i.e. the maximum evolutive expression found in a landscape context. In the agroecosysems, the woods have now become rare, but precious, elements, because of their richness in life forms and their complexity, and because of the possibility to provide autochthonous germplasm of nemoral arboreal, shrub and herbaceous species. These therefore need to be carefully managed according to their potential.

With regard to the sector of the Italian peninsula, as the object of this analysis, this class is represented by two orders.

- *Fagetalia sylvaticae*: This syntaxon has a central Europe and Caucasian distribution area and includes mesophilous forest phytocoenoses of deciduous trees of the montane and hilly zones of the areas with a temperate macrobioclimate. They have their optimum in fresh locations and with particularly deep and wet soils that have a good humus content. They include, in particular, beech, European hornbeam, lime and mountain ash woods.
- *Quercetalia pubescentis*: This order groups the wood formations that are essentially constituted of thermophilous broad-leaved trees, among which the most widespread are downy oak, turkey oak, bay oak, hop hornbeam and flowering ash. The spread of these coenoses that can be seen in the temperate and Submediterranean areas is limited to hilly zones.

  **Roso sempervirentis-Quercetum pubescentis* (Fig. 10): The woods of downy oak of this association are diffuse across the meso-Submediterranean thermoclimate belt of the hilly subcoastal territory of the central Adriatic region (Biondi & Allegrezza, 1996). These can be found on substrata of various origins, as arenaceous, clayey, and more or less calcareous, and they usually grow in the zones that are more exposed to the sun or with more superficial soil. They are characterised by the presence of various thermophilous species, like *Rosa sempervirens, Smilax aspera, Rubia peregrine* and *Laurus nobilis*. They are included in the alliance *Carpinion orientalis* and can usually be seen on the summit ridges of the hills.

  **Asparago acutifolii-Ostryetum carpinifoliae*: This neutrophilous hop hornbeam wood is included in the alliance *Carpinion orientalis* and it occupies soils derived from calcareous, marly and pelitic-arenaceous substrata of the lower mesotemperate bioclimate belt (Biondi et al., 2006). In the rural landscape of the territories investigated, it mainly occupies the fresh and usually north-facing slopes. Among the more representive species, as well as the hop hornbeam, there are *Asparagus acutifolius, Buglossoides purpureocaerulea, Smilax aspera* and *Acer obtusatum*.

**Lonicero xylostei-Quercetum cerridis*: The turkey oak woods of this association are included in the alliance *Carpinion orientalis* and they can be found in the central-southern sectors of the Italian Adriatic aspects. In particular, they occupy slopes with arenaceous-pelitic outcrops with a mainly sandy texture (Biondi & Allegrezza, 2004; Taffetani et al., 2005). Among the diagnostic species, as well as the hop hornbeam, there are *Lonicera caprifolium, Lonicera xylosteum, Sorbus domestica, Sorbus torminalis* and *Cyclamen hederifolium*.

Fig. 10. *Roso sempervirentis-Quercetum pubescentis.*

2.7 The herbaceous and arboreal hygrophilous vegetation
2.7.1 Class *Phragmito-Magnocaricetea*

The hygrophilous vegetation of the class *Phragmito-Magnocaricetea* grows in the marshy zones, at the edges of water courses, or ponds or lakes, in zones more-or-less constantly flooded. The class is well represented in territories with the temperate and Mediterranean macrobioclimate of continental Europe.

The communities that can be included in his syntaxonomic unit are often poor in species and composed mainly of helophyte plants, i.e. those characterised by their growth outside of the water although they are rooted in submerged soils. Among these, there are in particular the reeds, rushes, Carex sedges, and graminoid perennial species (Rivas-Martinez et al., 2002).

All of the coenoses of this class well tolerate long periods of submersion and soils that are very heavy or muddy, and deep. They grow both on clayey and silty soils, as well as on gravely alluvial plains that are typical of the valley bottoms. They cannot survive periods of drought, which can cause a decisive reduction in the water content of their soil. Some formations show from moderate to high resistance to the force of the water current.

In the rural contexts, *Phragmito-Margnocaricetea* characterises the more internal margins of ditches, channels, and small ponds or hill lakes, The presence of coenoses of this syntaxonomic unit within the rural territories contributes to the stabilisation and consolidation of the banks, and is an important index of good conservation: indeed, some of the species that characterise these communities can disappear if the level of human disturbance is high or in cases in which the concentration of nitrogen compounds in the water or in the soil are above specific levels.

From the syntaxonomic point of view, in the area under investigation, the coenoses seen can be referred to three main orders.

- *Phragmitetalia*: This syntaxonomic unit mainy concerns the marsh reed-beds and the *Typha* formations. These communities are characterised by the fact that the greater part of the biomass is produced by one, or a few, species (Poldini, 1989); usually these grow at the edges of water pools or along the banks of rivers and ditches where the water flows slowly (Rivas-Martinez et al., 2002).
  **Typhetum latifoliae*: The communities of *Typha latifolia* are included in the order *Phragmition communis;* these are poor in species and in the countryside they find the optimal conditions for their development in proximity to channels or ditches characterised by stagnant water that is present throughout the year. In general, they cannot survive conditions of high eutrofixation (Baldoni & Biondi, 1993) and they indicate a conservation level of the riparian environment that is sufficiently good.
- *Magnocaricetalia*: This order gathers the communities comprising the species of Carex of medium to high size that can be found on the edges of pools, ponds, lakes and water courses. These formations are often found interpositioned beween the coenoses of the order *Phragmitetalia* and riparian arboreal formations (Poldini, 1989).
  **Cypero longi-Caricetum otrubae*: These formations are physiognomically dominated by *Carex otrubae* and *Cyperus longus* and belong to the alliance *Magnocaricion elatae,* and they are found on heavy and constantly wet soils on the inside edges of channels and ditches. In central Italy, this has been reported for the wet environments of Lakeo Trasimeno (Umbria; Venanzoni & Gigante, 2000) .In the rural territory, these are becoming more rare because of intensive cultivation practices.
- *Nasturtio-Glycerietalia*: This syntaxon instead includes the plant communities at the edges of rivers or of small water courses that include the small rhizomatose pioneer plants and characterise a good level of resistance to the force of the water current. These pioneer plants grow in contexts where there are frequent variations in water flow (Rivas-Martinez, 2002).
  **Helosciadetum nodiflori* (Fig. 11): The formations of *Apium nodiflorum* are relatively diffuse in the the rural territory investigated, both in the high hilly areas and in the better conserved subcoastal zones. They form a part of the alliance *Glycerio-Sparganion* and they colonise the more internal parts of the banks of the rivers and ditches with sufficiently oxygenated water flow. They resist the impact of the current and periodic submersion well.

Fig. 11. *Helosciadetum nodiflori.*

2.7.2 Class *Salici-Populetea*

The class *Salici-Populetea* includes the riparian arboreal phytocoenoses of deciduous trees of the Eurosiberian and Mediterranean regions (Rivas Martinez et al., 2002). These plant formations are made up of species of willow and poplar, and they prefer soils of alluvial origin subjected to periodic flooding as a result of river overflow; some of these are located in close contact with the water, with others at greater distances, as a function of their decreasing ability to resist the force of the current. These same coenoses are frequently positioned in close contact with communities of the class *Phragmito-Magnocaricetea* and often it is possible to find various species of this last syntaxon within them.

The presence of coenoses of the class *Salici-Populetea* is relatively common along the banks of the more important rivers; instead, at the margins of small water courses, these communities are often very rare, if not completely absent.

In the agroecosystems analysed, the associations of this syntaxon are widely present, but in general little structured and reduced to thin strips positioned in direct contact with the water courses. This occurs because of the use of the major part of the terrain of the plains for the cultivation of annually seeded crops and as a function of the use of the soil that is determined by the development of commercial and industrial areas on river plains near to inhabited centres. Moreover, the arboreal riparian formations are always more often damaged extensively and heavily by badly designed reshaping and remodelling of the river banks, carried out to try to regulate the water flow of the main water courses.

Within this class, two orders can be distinguished, which include coenoses with different abilities to resist river flooding.

- *Populetalia albae*: This order includes the hygrophilous arboreal coenoses with a dominance of black poplar or elm that grow in zones not subjected to the continual influence of the water. These are usually located on alluvial terraces, on soils with a generally sandy texture and with good permeability, or in the bottom of the clayey gullies with deep and constantly wet soil.

  **Symphyto bulbosi-Ulmetum minoris* (Fig. 12): The elm coppices are found in the gullies with deep and fresh soil of clayey origins, or on the alluvial terraces with a superficial water table. These are included in the alliance *Alnion incanae*. The association illustrated here has been described for the hilly territory of a hilly sector of the central Adriatic (Biondi & Allegrezza, 1996), and it is a constant enough element of the hilly agricultural territory studied. In the herbaceous layer, the most represented species are usually *Symphytum bulbosum* and *Arum italicum*.
- *Salicetalia purpureae*: This syntaxon includes the riparian coenoses with a dominance of bushy willows that can be found in particular in the internal parts of the fluvial beds. The species that are part of these communities show high resistance to the impact of the water flow and they manage to rapidly recover their vegetative ability after the flooding of the river.

  **Rubo ulmifolii-Salicetum albae*: The riparian arboreal formations with a dominance of *Salix alba* are characteristic of the edges of the water courses which have a constant water flow. These are often seen as linear formations in direct contact with the adjacent cultivated fields. Those relative to the subcoastal hilly territory of the central-southern part of the Italian Adriatic sector are included in the association *Rubo ulmifolii-Salicetum albae* (Allegrezza et al., 2006). This coenosis belongs to the alliance *Salicion albae* and it is differentiated from analogous coenoses of white willow with a central European distribution area by the presence of species with a Mediterranean chorotype, such as *Rubus ulmifolius, Hedera helix* and *Laurus nobilis.*

Fig. 12. *Symphyto bulbosi-Ulmetum minoris.*

3. Evaluation of the quality and functionality of the agroecosystem

3.1 A system of floristic-vegetational indexes for the analysis of ecosystem functionality

Aspects of the ecological functionality relative to the agricultural systems can be evaluated and understood using the sysem of bioindicators introduced Taffetani & Rismondo (2009) and then updated (Rismondo *et al.*, 2011). It is based on the floristic-syntaxonomic particularities of the areas examined. In this type of analysis, the study of the plant landscape is examined with the aim of collecting and interpreting the effects caused by the diverse types of previous management.

The model of the study in question is based on two specifically formed databases:

- The floristic-syntaxonomic database collects the ecological information relative to each single species present in the Flora of Italy (Pignatti, 1982; Conti et al., 2005). Various indicators are associated with each taxon, among which there is the one relative to the syntaxonomic class to which it belongs (following consultation with various literature contributions: Guinochet et al., 1973; Rameau et al., 1989; Oberdorfer, 1990; Royer, 1991; Biondi et al., 1995; Biondi et al. 2005; Rivas Martinez et al., 2002), and those regarding the biological form, chorological type and the possible influences of the physical conditions of the terrain (edaphism).
- The database of the syntaxonomic classes includes the main classes of vascular vegetation found on national soil. A numerical value is given to every class (with a range of 0 → 9; Fig. 13) on the basis of the maturity level determined:

1. Commensal vegetation of the annually seeded crops;
2. Herbaceous pioneer vegetation;
3. Herbaceous ruderal nitrophilous vegetation of the margins and the abandoned fields;
4. Herbaceous perennial vegetation of the margins and the grasslands with cutting;
5. Herbaceous perennial vegetation of pastures and mature grasslands;
6. Mesonitrophilous herbaceous perennial vegetation of the ecotones of the arboreal formations;

7. Herbaceous perennial vegetation of the edges of the forest formations;
8. Vegetation of the forest mantle and the shrubs;
9. Arboreal forest vegetation.

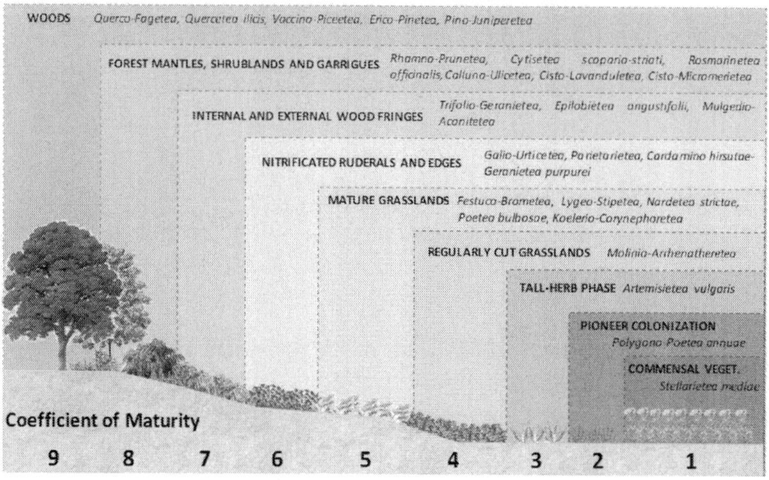

Fig. 13. Scheme which illustrates the maturity significance of each single community.

The numeric values introduced therefore increase bit by bit passing from pioneering vegetation typologies, like those relative to the commensals of annually seeded crops (*Stellarietea*), or to terophytic herbaceous formations (*Stellarietea, Polygono-Poetea*), to the herbaceous formations of the grasslands (*Artemisietea, Molino-Arrhenatheretea, Phragmito-Magnocaricetea, Festuco-Brometea*) and of the edges (*Galio-Urticetea, Trifolio-Geranietea*), and finally to the typologies identifiable with the pre-forest (*Rhamno-Prunetea*) or forest (*Salici-Populetea, Querco-Fagetea*) stages. This is in agreement with that which is illustrated in the section dedicated to the description of the agroecosysem landscape.

The classes characterised by a cerain type of edaphic determinism (adaptability of the vegetation to grow in contexts with soils always rich in water, or arid, or with high concentrations of salts) are distinguished by specific hygrophilous, xerophilous and alophilous coefficients; these last are functional in the calculation of the edaphic indexes, as well as for the calculation of the maturity indexes (see paragraph 3.1.1.).

The results of the vegetational investigations (relating to the presence and abundance of each taxon found) and the information contained in the two databases (ecological characteristics of every species) represent the applicative basis of the indexes used, through which numerical values are obtained that measure the conservation level of each individual coenosis (coenotic indexes) or of entire territorial contexts (cartographic indexes).

The system was specifically designed for the rural areas and as a consequence, its set-up is for areas mainly characterised by agricultural use of the territory.

The method is seen to be particularly efficacious in the analysis of the grassland formations, because a good part of the scale used for the quantification of the levels of maturity refer to the classes of herbaceous vegetation. This last is by far the most represented in the rural contexts than the mature elements of the landscape, such as shrubs, pre-woods and woods,

as the herbaceous vegeation is characterised by being particularly susceptible and responsive to the ways in which it is used, which can also be observed even over short periods of time.
This makes the method particularly efficacious for the monitoring of environmental quality of contexts within which the ecomosaic is characterised by being influenced by human activities in an evident way. In the following paragraphs, the structure and significance of the indexes used in the analytical and applicative part of this study are described

3.1.1 Coenotic indexes

The indexes relative to the phytocoenoses allow the measuring of some ecological characteristics of the plant communities found in the field.
This information essentially regards the maturity, i.e. the evolutive level reached (Index of Maturity; *IM*), the biodiversity, expressed as the species number for each phytocoenosis (Index of Floristic Biodiversity; *IFB*), and various other parameters, like the edaphism (Indexes of Hygrophylia, Xerophylia and Alophylia; *IW, IX, IA*), the relative presence of the various biological forms (indexes of the terophytic, hemicryptophytic and non-hemicryptophytic perennial components; *IT, IH, IF*), and the relative presence of some of the chorological types (indexes of the endemic, wide distribution and exotic components; *IL, ID, IE*). The index of maturity is expressed in a range that goes from 0 to 9 (according to the maturity scale described in the previous paragraph), the index of floristic biodiversity is a simple number (of species per relevé), and all the others are as percentages.
In most cases, these indexes are calculated as a function of the coverage that can be attributed to each single species. The coverage can be considered within each single relevé or as the mean for a Table of relevés that are all attributable to one phytocoenosis.
The indexes that we have used with the greatest frequency for the functional characterisation of the herbaceous communities are the Index of Maturity, the Index of Floristic Biodiversity, and the edaphic indexes, with the Index of Hygrophylia in particular. Of these, the following provides brief descriptions of the formulas for their calculation.

- Index of Maturity: This allows the evaluation of the level of evolution of the coenoses analysed, which will be higher when there is a greater presence and coverage of species of the more evolved vegetation classes (e.g. shrubs, pre-woods, woods). For the herbaceous communities, the value will be higher as a function of the presence of perennial entities that are typical of mature grasslands or of ecotone coenoses.

$$IM = \frac{\sum_{i=1}^{n}(c_i \times y)}{C_{(tot)}}$$

where *IM* is the Index of Maturity; c_i is the coverage value for each single species, as an absolute value for single relevés or as a mean for groups of relevés in a Table; y is the value corresponding to m ($y = m$) for each single species and assigned on the basis of the information contained in the databases described above; and $C_{(tot)}$ is the value of the total coverage obtained from the sum of the coverage values for all of the species.

- Index of Floristic Biodiversity: This measure is the number of species relative to each of the coenoses, which are given as groups of floristic species that can develop in an ecologically homogeneous context. This does not depend on the coverage value of each single taxon (Puppi, 2008).

$$IFB = sp/ril$$

where *IFB* is the Index of Floristic Biodiversity; *sp* is the number of species of a given coenosis; and *ril* is the single relevé.

- Edaphic Indexes: These allow the qualtification of the weight of the hygrophilous, xerophilous and alophilous components of each coenosis and they are used to gather and measure the adaptability of each coenosis to develop under particular edaphic conditions regarding the availability of water and salts in the soil.

$$IW = \frac{\sum_{i=1}^{n}[c_{(sw)}]\ i}{C_{(tot)}} x100$$

$$IX = \frac{\sum_{i=1}^{n}[c_{(sx)}]\ i}{C_{(tot)}} x100$$

$$IA = \frac{\sum_{i=1}^{n}[c_{(sa)}]\ i}{C_{(tot)}} x100$$

where *IW* is the Index of Hygrophylia; *IX* is the Index of Xerophylia; and *IA* is the Index of Alophylia; $[c_{(sw)}]_i$; $[c_{(sx)}]_i$; $[c_{(sa)}]_i$ are the coverage values of each single species that belongs to the hygrophilous (*sw*), xerophilous (*sx*) and alophilous (*sa*) classes, as an absolute value for single relevé or as a mean for groups of relevés in a Table; and $C_{(tot)}$ is the value of the total coverage obtained from the sum of the coverage values for all of the species.

3.1.2 Cartographic indexes

The cartographic indexes come from the integration of the content of the maturity index with the data derived from the cartographic representations. This last is realised through the use of GIS (geographic information system). The cartographic analyses are carried out at a number of levels and allow thematic maps of different contents to be obtained.

The first phase is the photointerpretation of the territory under study, with the consequent attribution of each single patch to a specific category (previously defined). This allows the drawing up of the map of the use of the soil. Then, the cartographic investigation is deepened and the single polygons are assigned to defined vegetation typologies, inserted in their specific serial and landscape context. In this way, this provides the maps of the vegetation and the plant landscape, which illustrate the study area in a qualitative way.

The final level of the cartographic analysis is the assigning of the maturity values calculated for every coenosis to each of the corresponding spaces reported on the map. In this way it is possible to construct the maturity map, which provides a qualitative-quantitative vision related to the evolutive state and grade of conservation of the area investigated.

The ecological functionality of the hydrographic network of the agricultural areas can be measured with an analysis system based on the reconstruction of the graph of the fluvial segments and on the intersection of this with the vegetation map. This method, which will be considered more deeply in a following publication, allows a map of the conservation state of the hydrographic network to be obtained, as a predictive instrument that is useful for the identification of the part of territories that are potentially susceptible to erosion phenomena and hydrogeological problems, as characterised by high numbers of closed

down ditches or with profoundly altered riverbank vegetation. In contrast to the previous example, in this case the map does not come from an analysis of the surface, but from a linear investigation, as the ditches are divided into segments to which a numerical value is associated, which is relative to the vegetational typology of the corresponding margin. The interpretation of the qualitative characteristics and of the management dynamics that can be revealed in the various contexts investigated is made possible by the superpositioning and the reading of the various cartographic representations presented.
The map of the maturity provides the foundation for the calculation of the so-called cartographic indexes, the index of synthetic maturity (*ISM*) and the index of the unproductive areas (*IUA*), expressed as a 1 → 9 scale and as a percentage, respectively, and as described in the following.

- Index of Synthetic Maturity: This is based on the area occupied on the map by each vegetational typology reported in the cartography and on the maturity value attributed to the same phytosociological typology.

$$ISM = \frac{\sum_{i=1}^{n}(IM_i \times \Omega_i)}{\Omega_{(tot)}}$$

where *ISM* is the Index of Synthetic Maturity; IM_i is the index of maturity relative to the ith vegetational typology from the cartography; $\boldsymbol{\Omega}_I$ is the area of the ith vegetational typology from the cartography; and $\boldsymbol{\Omega}_{(tot)}$ is the total area of the carography.

$$IUA = \frac{\sum_{i=1}^{n}[\Omega_{(u)}]\ i}{\Omega_{(tot)}} x100$$

- Index of Unproductive Areas: This index is based on the distinction between cultivated or disturbed areas, characerised by $IM \leq 2$, and unproductive areas with $IM > 2$, and it is calculated as the relative presence of the cultivated or disturbed areas included in the cartography.

where *IUA* is the Index of the Unproductive Areas; $[\boldsymbol{\Omega}_{(u)}]_I$ is the area of the ith vegetation typology with $IM > 2$; and $\boldsymbol{\Omega}_{(tot)}$ is the total area on the cartography.

3.2 Applying the indexes

In the following, there are some examples of the application of the system of the bioindicators just described. These refer to some of the territorial contexts described over the last few years and they are aimed at helping the reader in their understanding of the method presented here and in their appreciation of its potential use. The applications presented refer to both the coenosis indexes and the cartographic indexes.

3.2.1 Analysis of the ecological gradient in a vegetation transect

The application is based on the reconstruction of transects of the vegetation typologies that are revealed under particular conditions of morphology and use. These refer in particular to the edges of the water courses and of the dirt tracks. The schemes and the graphics that support the analysis allow the demonstration of how the method of bioindication on a floristic-vegetational basis provides precise and easily understood results regarding the influence of particular ecological or management conditions on specific phytocoenoses.

- Vegetation of the diches edges: Considering agroecosystems, the plant coenoses of the edges of the water courses often provide a high floral richness and an elevated naturalistic interest (as for the better-conserved edges of the fields and of the farm tracks) and perform multiple functions: they contribute to the protection of the soil from erosive phenomena, they act as connecting ecological corridors between semi-natural habitats, they host numerous forms of life and favour interactions between living and non-living organisms, and they absorb CO_2.

Unfortunately, the coming of intensive agriculture has favoured the practice of management techniques of rural territories that are particularly aggressive, such as the stripping of vegetation on a large scale and the tillage of areas previously not used, which have been performed with the aim of obtaining the maximum areas possible for cultivation. This has often resulted in the removal or alteration of a large part of the riverbank plant component.

In our studies into the understanding of the state of the biodiversity of the rural territories, following investigations of the countryside, we have been able to reconstruct the plant successions that can be potentially found along the edges of the primary and secondary water courses. The coenoses found can be referred to classes that are particularly hygrophilous, such as *Phragmito-Magnocaricetea* and *Salici-Populetea*, or to mesophilous classes, such as *Galio-Urticetea* and *Molinio-Arrhenatheretea*. To the reconstruction of the vegetation transects, we added the application of two of the floristic-vegetational indexes, as the maturity index (*IM*) and the hygrophilous index (*IW*), which allowed us to reveal interactions between the morphology of the soil, the gradient of available water, and the vegetation.

Scheme A of Figure 14 shows the succession of the coenoses seen at the edges of the main ditches with permanent water-flow throughout the seasons of the year (Rismondo *et al.*, 2011). The gradient of available water is shown by the progression of the hygrophilous index along the transect: it is higher for the coenoses that are in direct contact with the water, such as *Helosciadetum nodiflori* (*Phragmito-Magnocaricetea*), the *Carex pendula* communities (*Phragmito-Magnocaricetea*) and the association *Rubo ulmifolii-Salicetum albae* (*Salici-Populetea*), and lower for the ecotone formations, such as *Petasitetum hybridi* (*Galio-Urticetea*), and for the mesophilous grasslands of the external margins of the ditch, attributed to *Ranunculetum repentis*, *Festuco fenas-Caricetum hirtae* and *Lolio multiflori-Plantaginetum majoris Juncus bufonius* variant (all three of which are included in the class *Molinio-Arrhenatheretea*).

The grade of maturity is greatest for the riparian arboreal formation of willow, while it decreases for the herbaceous coenoses: it is lower for the vegetation with *Juncus bufonius* of the compacted and slightly depressed zones, as a function of the greater human disturbance (passing of agricultural machines, contact with the cultivated fields).

Scheme B of Figure 14 shows instead the sequence of coenoses of the ditches of the slopes with water flowing only during and immediately after rain events. The indexes of Maturity and Hygrophylia are higher for the formations within the ditch, such as *Convolvulo sepii-Epilobietum hirsuti*, and lower for the mesophilous coenoses of the external edges of the ditch, such as *Agropyron repens* and *Galium album* communities. In the agroecosystems the presence of these two plant formations (both of the class *Galio-Urticetea*) are seriously endangered by the profound alterations of the segments of the hydrographic network positioned in the hilly zones with the greater slopes.

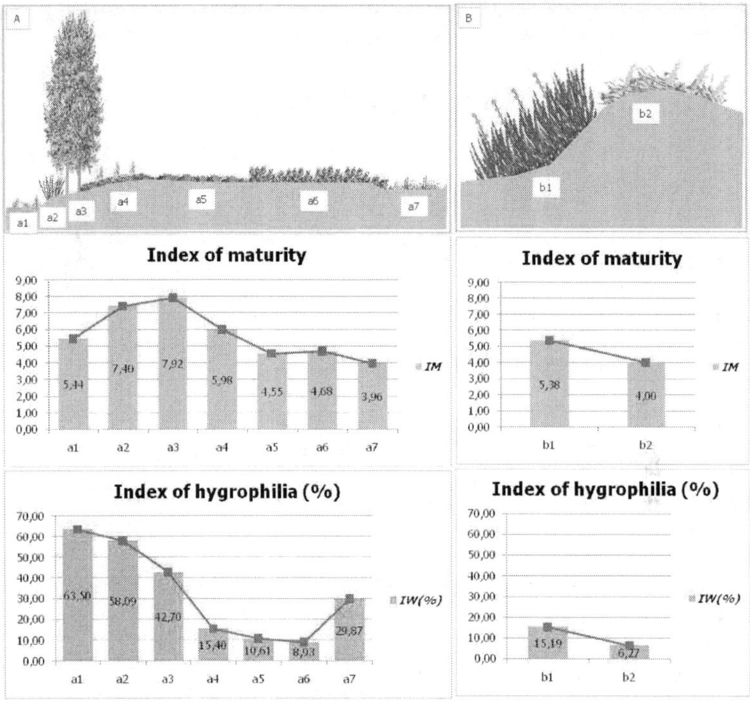

Fig. 14. Indexes of Maturity and Hygrophylia of the vegetation of the principal and minor ditches (Scheme B). Legend: a1=*Helosciadetum nodiflori;* a2= *Carex pendula* communities; a3=*Rubo ulmifolii-Salicetum albae;* a4=*Petasitetum hybridi;* a5=*Ranunculetum repentis;* a6=*Festuco fenas-Caricetum hirtae;* a7=*Lolio perennis-Plantaginetum majoris Juncus bufonius* variant; b1=*Convolvulo sepii-Epilobietum hirsuti;* b2= *Agropyron repens and Galium album* communities.

- Vegetation of the dirt tracks: The farm roads made of beaten earth provide the connections between the farms and their fields and they are used mainly for the passage of agricultural machinery. These have little traffic, but are anyway subjected to high compaction of the terrain. The particular ecological conditions that are created corresponding to these secondary communication tracks allow the development of vegetation that is well adapted to this mechanical disturbance and the alterations to the structure of the terrain, which is little aerated. The coenoses that can be found mainly belong to the class *Polygono-Poetea*, although in some cases also to the classes *Stellarietea* or *Molinio-Arrhenatheretea*. These habitats are also today decreasing, mainly due to the vegetation stripping that also affects the environments outside of the cultivated fields, and to the extreme simplification of the network of tracks between the fields and houses in the countryside (as a function of the depopulation of the countryside, which has resulted in the abandoning of many houses in the countryside, and the incorporation of small plots in the larger cultivated fields). In some situations, however, the state of conservation is still sufficiently good and our study allowed the reconstruction of an example that illustrates the vegetation variation and the relative maturity level as a function of the level of compaction of the soil.

In Scheme C of Figure 15, it can be seen how the coenoses of the class *Polygono-Poetea*, which are *Coronopodo procumbentis-Sclerochloetum durae* and *Poetum annuae*, are positioned. The first of these is found in the individual tracks left by the passage of the agricultural machinery, and the second is in the centre and at the margins of the dirt track; *Lolio perennis-Plantaginetum majoris*, of the class *Molinio-Arrhenatheretea*, is instead found at the more external margins. In the transect, it can also be seen that there is an inverse relationship between the level of compaction of the soil and the evolutive level of the phytocoenoses: indeed, this last decreases going from the margin to the centre of the dirt track (in particular under the tracks of the tyres of the tractors).

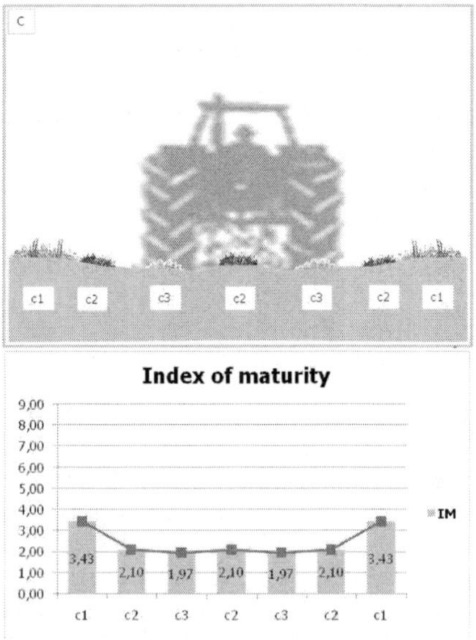

Fig. 15. Indexes of Maturity of the vegetation of the dirt tracks (Scheme C). Legend: c1=*Lolio perennis-Plantaginetum majoris*; c2=*Poetum annuae*; c3=*Coronopodo procumbentis-Sclerochloetum durae*.

3.2.2 Evaluation of the environmental quality of the territory at different scales of enquiry

The example reported is based on the calculation of the maturity index of the coenoses found within determined territory contexts, on the construction of the maturity maps (derived from the vegetation maps), and on the calculation of the consequent cartographic indiexes. The analysis system is shown to be an excellent means for comparison of the areas characterised by different management modalities that can influence the state of conservation of the habitats. The following presents applications relative to the different scales of investigations.

- Analyses of the state of conservation of small herbaceous areas: This example concerns the characterisation of the threshing grounds surrounding an abandoned farmhouse (named as C2) located in the rural territory in a medium hilly area of Marche. The study has involved various threshing grounds of country houses located beween the provinces of Ancona and Macerata (inside Musone River Basin), and it involved the definition of all of the vegetation typologies present for every location analysed, the drawing up of the vegetation map, and the subsequent calculations of the coenosis indexes. Following this, the maturity map was constructed and the cartographic indexes were calculated on the basis of the area occupied by each coenosis and their maturity values.

The investigations carried out allowed us to see how these herbaceous areas represent habitats rich in biodiversity and definitely worth further investigation from the point of view of the relationships that exist beween the flora and the historic usage. The threshing grounds are mainly characterised by coenoses of the class *Molinio-Arrhenatheretea*, grasslands that are regularly cut and that develop, as illustrated, on soils well supplied with nutrients and water. These communities are however on the road to extinction in the agroecosystem environment, due to the abandoning of the traditional management methods.

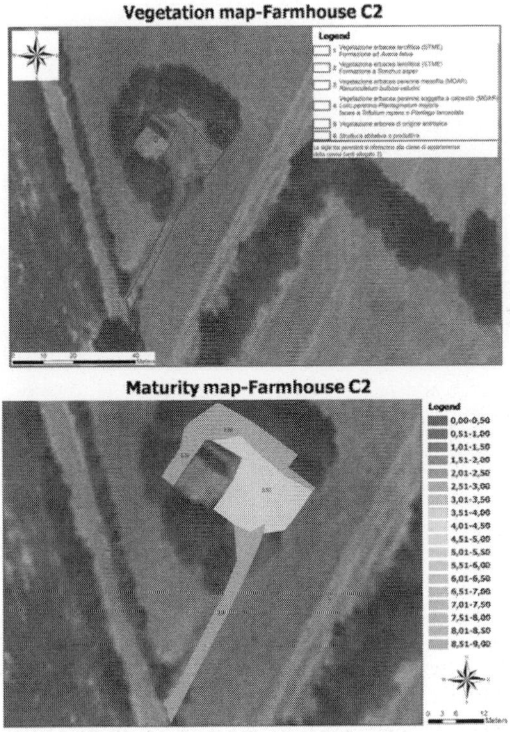

Fig. 16. Vegetation and maturity maps of Farmhouse C2.

The maturity map shown in Figure 16 reveals how the method allows the obtaining of high-precision data regarding the conservation state of the plant communities and of the entire area (as a function of the values of the indexes; Table 1) and provides the possibility of monitoring possible subsequent variations in the floral composition and the level of maturity due to changes in the type of management of the areas involved.

Vegetation community	**Area (m^2)**	***IM***	***IFB***	***IW(%)***	***IX(%)***	***IA (%)***
Avena fatua community	47,78	2,52	20,00	0,00	0,00	0,00
Sonchus asper community	136,41	2,98	18,00	0,00	0,00	0,00
Ranunculetum bulbosi-velutini	245,76	3,52	23,00	0,00	0,00	0,00
Lolio perennis-Plantaginetum majoris Trifolium repens e *Plantago lanceolata* facies	141,52	2,80	33,00	0,00	0,55	0,00
	Area (m^2)	***IMS***				
Farmhouse 2	571,47	3,13				

Table 1. Coenosis indexes and IMS of Farmhouse C2.

- Comparison of the functionality of the agroecosystems at the territory scale: This example refers to two agricultural territories of medium hilly areas of Marche, located in the province of Ancona: the Spescia and Bottiglie sub-basins (municipality of Serra de' Conti; Taffetani & Rismondo, 2009). The two areas are located a few kilometres from each other on soils that originated from sedimentary rock and on clayey, silty-clayey, marly, silty and sandy substrata of various grades of compaction. The reference bioclimate is of the Submediterranean variant of the temperate region.

The Spescia sub-basin shows extreme simplification of the landscape and a lack of semi-natural structures like herbaceous strips, small woods, or rows of trees and shrubs. It is cultivated in an intensive way according to conventional agricultural techniques. The whole of the territory is divided into only two properties and as a consequence the cultural diversification is minimal, also because the plots cultivated with seeded crops are very large and there is no tree growth. The study of the vegetation, the application of the indexes, and the obtaining of the relative maturity map (Fig. 17) reveal a very low level of conservation, as can be seen by the large area covered by coenoses of a very low grade of evolution and by the high homogeneity of use of the areas.

The Bottigle sub-basin instead presents a more diverse plant landscape and its area is subdivided into many small properties. The agricultural use of the terrain includes annual seeded crops, forage crops and polyphytic grasslands; some of the owners operate following the regulations of organic production. The variety of the environments is also favoured by a good level of semi-natural environments, such as herbaceous strips at the edges of the farm tracks, rows of trees and shrubs located at the limits of the fields, riparian vegetation and a small, well structured, turkey oak wood. These residual environments are characterised by vegetation with a good level of maturity; e.g. along the edges of the access tracks there are grasslands of the class *Festuco-Brometea,* the conservation of which is favoured by the periodic cutting. When the maturity map is compared to that of the Spescia sub-basin (Fig. 17), it clearly reveals higher environmental diversification and a good presence of semi-natural marginal coenoses, with *IM* ≥2.

What can be easily noted graphically from the maturity map can be translated into numerical data through the use of the cartographic indexes. As can be seen from Table 2, both the Index of Syntetic Maturity (*ISM*) and the Index of the Unproductive Areas (*IUA*) are a lot higher for Bottiglie sub-basin with respect to the agricultural territory of Spescia.

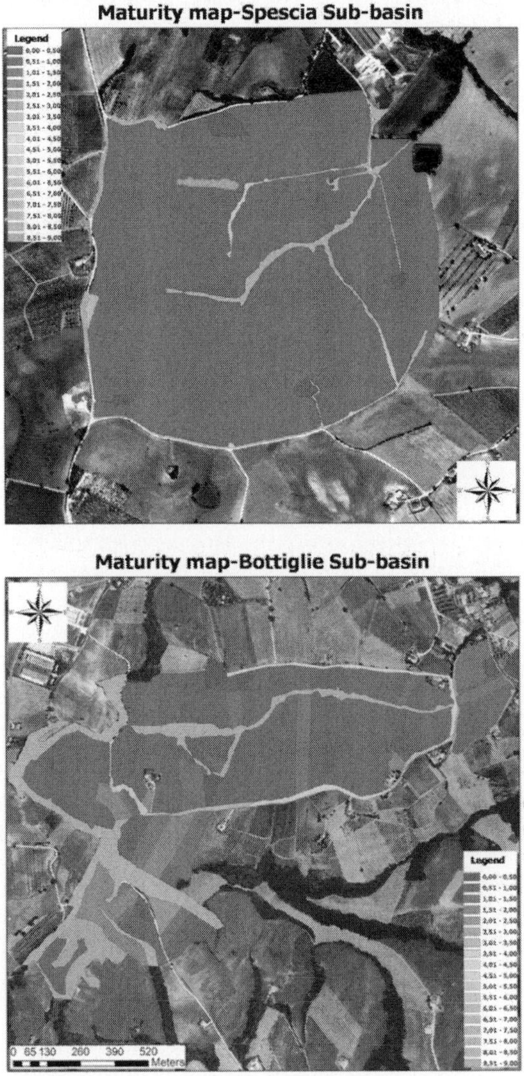

Fig. 17. Maturity maps of Spescia and Bottiglie Sub-basins.

Area	Area (Ha)	*ISM*	*IUA*
Spescia Sub-basin(AN)	84,74	0,96	5,49%
Bottiglie Sub-basin (AN)	107,32	2,31	22,12%

Table 2. Cartographic indexes of Spescia and Bottiglie Sub-basins.

4. Conclusions

4.1 Directives and Regulations for the conservation of biodiversity in agricultural environments

Today, the normative instruments to promote eco-sustainable management of rural territories have a fundamental role for the realisation of conservation strategies for the biodiversity of agricultural environments. The ecosystem services produced by the correct use of the rural territory cannot indeed be given a direct monetary value and therefore cannot be paid to the farmers. Due to this, the introduction of norms and of incentives and financing applicable to specific forms of management represent the only way to give an effective and concrete value to these positive ecosystem benefits.
In the following paragraphs, we consider the main Community legislative references connected with the planning and conservation management of agrobiodiversity.

4.1.1 Rural Development Programme and cross compliance

The Rural Development Programme (RDP) is an economic and policy tool of the European Union for non-urbanised areas. The normative framework to which it refers and its financial funding are known as FEARS. Every region had to take on Regulation (EC) 1698/05, to prepare an RDP for the period of 2007-2013 (the preceding one referred to 2000-2006), and to send it to the European Commision, which needed to evaluate and approve it. The main aims that need to be followed with the RDP regard the improvement of the competitivity in the agricultural and forestry sectors, the improvement in the quality of the rural environment, and the improvement in the quality of life and in the economic opportunities connected to the rural territories.
In exchange for the high levels of public resources directed to the agricultural sector, the intervention priorities according to the second of the objectives just introduced are dictated by the opportunity to guarantee the attribution to agriculture of a role that goes beyond the simple production of foods. This should extend to the function of providing environmental services directed at the protection of the territory and the regeneration of the basic elements, like air, water and soil. Every single farm can have access to funds provided by the RDP following their participation in specific funding applications. Among the measures considered by the RDP, there is, for example, the possibility of access to agro-environmental payments for those who take on the implementation of farm management models that focus on the conservation of the environmental quality (regulations for organic production with low environmental impact).
The Cross Compliance involves all of the farmers who from the start of 2005 intended to benefit from funding made available by the European Union through the Common Agricultural Policy itself, obviously including the payments planned by the RDP. Indeed, from that date, the farmers had to assure that they would respect a series of obligations for the correct agronomic management of the terrain, the protection of the environment, public health and animal health, and animal welfare. Non-conformity with these obligations results in the activation of a mechanism for the reduction of the direct payments to which each of farmers would have the right.
The obligations to which every farmer must make reference are subdivided into two large categories.

- Obligatory Management Criteria, as provisions of law, indicated as Acts already in force and deriving from the national implementation of the corresponding Community arrangements.
- Good Agronomic and Environmental Conditions, indicated as norms set up at a national level to guarantee reaching five priority objectives fixed by the European Union, as: protection of the soil through the use of appropriate measures; maintenance of the levels of organic matter in the soil through the use of appropriate practices; protection of the structure of the soil through the use of adequate measures; assurance of a minimum level of maintenance of the ecosystem and avoidance of deterioration of the habitats; and protection and management of the water resources.

The Cross Compliance is based on the Regulation (EC) 1782/03, subsequently abrogated by Regulation (EC) 73/09.

4.1.2 HNV Farmland Areas

The politics of rural development at the European level proposed the specific objective of the conservation of the agricultural areas of High Natural Value (HNV Farmland Areas) according to Regulation (EC) 1257/99. These had to be identified by 2008, and then be subjected to management modalities aimed at the conservation of biodiversity. Unfortunately, Italy is still today behind also in the determination of the HNV Farmland Areas. These HNV Farmland Areas were defined by Baldock et al. (1993, 1995) as systems of low input and with good levels of biodiversity that are characterised by the application of agricultural practices of low intensiy, a high proportion of semi-natural elements, and a high diversity in the soil coverage. Andersen et al. (2003) identified three typologies of HNV Farmland Areas:

- Type 1: Agricultural terrain with a high coverage of semi-natural vegetation;
- Type 2: Agricultural terrain dominated by low intensity agriculture or by a mosaic of semi-naural and cultivated territories;
- Type 3: Agricultural terrain where there are rare species or a high proportion of a population of a European or World animal or plant species.

For the identification of each typology, they indicated three types of approach:

- Approach 1: soil coverage (Corine Land Cover, not applicable to type 3).
- Approach 2: investigation of the farming system (system of agricultural accountability, RICA, not applicable to type 3).
- Approach 3: species and habitat (Natura 2000, IBA, PBA, IPA).

The method is based on the application of a series of indicators and indexes that have the function of assigning to every specific context a value as a function of characteristics relative to the coverage of the soil, cultivation practices, and the presence of rare species.

The limit of this method is determined by the fact that this is essentially based on large-scale cartographic data, while not making reference to the need for detailed territorial analyses, such as floristic-vegetational investigations. These last would be a valid instrument of support both for the definition phase of the areas and in the course of the successive stages of the planning and the implementation of the management measures.

4.1.3 Habitat Directive

Directive 92/43/EEC, known as the Habitat Directive, relates to the conservation of natural and semi-natural habitats and wild flora and fauna, and it was approved by the European

Commission on 21 May, 1992. Each Member State then had the job of accepting and implementing this Community Normative. The man aim of the Directive is to promote the protection of biodiversity through the conservation of the natural habitats and of the wild flora and the fauna of the European territory of the signing Member States, taking into account also the economic, social, cultural and regional needs.

In Article 1, the significance of the SCI and SAC were defined:

- A SCI (Site of Community Importance) is a site that contributes in a significant way to the maintenance or restoration of a type of natural habitat (Attached I) or a species (Attached II) in a satisfactory state of conservation.
- A SAC (Special Area of Conservation) is a site of Community importance designated by the Member States through an administrative Act where conservation measures are applied that are necessary for the maintenance or restoration in a satisfactory state of conservation of the natural habitats and/or populations of species for which the site is designated.

Each Member State identified the sites in their own territory that are fundamental for the conservation of the species and habitats of Community interest and proposed to the European Commission their own list of pSCIs (as stated in Article 4 of the Directive). On the basis of this list, the Commission drew up the list of SCIs. Within six years from the declaration of an SCI, a specific area is declared an SAC by the Member State.

Instead, the SPAs (Special Protection Areas) were created according to the Bird Directive (79/409/EEC), with the aim of protecting in a rigorous way the sites in which ornithological species live, as contained in attachment I of the same Directive.

The combination of the SCIs (which will anyway become a SACs according to the procedure indicated above) and the SPAs form the European ecological network known as Natura 2000 and defined in Article 3 of the Habitat Directive. The Natura 2000 Network is a combination of sites characterised by the presence of habitats and species of both animals and plants that are of Community interest (indicated in attachments I and II of the Directive), the function of which is to guarantee the long-term survival of the biodiversity, in all of its components, present on the European continent, recognising the interdependence of the biotic, abiotic and anthropic elements.

Article 6 of the Directive defines the management of the sites of Natura 2000 (including the ZSPs). The obligation derived from the Directive is that of adopting opportune measures to avoid the following:

1. The degradation of the habitats of Attachment I.
2. The degradation of the habitats of the species for which the zones were designated.
3. The disturbing of the species for which the zones were designated where this can have negative effects on their conservation.

The instrument of implementation of the Habitat Directive within the SCIs and SPAs is represented by the management plan. The Habitat Directive assumes great importance in the rural environment because with a large part of the sites identified by the Member States there are agricultural territories that are more or less extensive. These areas are often identifiable as HNV Farmland Areas, in which the presence of man and of agro-forestry-pastural activities have favoured the creation of a mosaic of environments, which are characterised by a high level of biological variety. Within these areas, the protection of the traditional cultural systems based on extensive agricultural techniques should be favoured. These considerations must be examined and evaluated carefully at the moment of the drawing up of the management plans.

4.2 Applications of the indicator system

As already illustrated, the analysis and evaluation system described in this study allows the acquisition of numerical data starting from qualitative information of the description of the ecological characteristics of every single phytocoenosis. The integration of these data with those obtained by the cartographic procedures then allows a measurement to be provided of the level of conservation of entire territories, which can also be operated on different scales of investigation. The model is well suited to be used as a means of decisional support in the politico-administrative and management fields, and it has numerous other potential applications, many of which can be applied to the legislative instruments illustrated above, which can be synthetically summarised as follows:

- Support for the definition of naturalistic values, of the state of conservation of the hydrographic network, and of management measures of the agroecosystems more appropriate to guarantee the conservation of the habitats within the SCIs and SACs. The measures should then be included in the specific management plans.
- Identification of the HNV Farmland Areas (agricultural areas of high naturalistic value) by means of the integration of the indexes in use (diversification in the use of the soil, proportion of semi-natural elements present, intensity of the impact of agricultural practices) and the data derived from studies of the plant landscape and the application of the floristic-vegetational indexes.
- Determination of the conservation level of the agricultural territories, from a hypothetical starting situation, to be used as a basis for subsequent monitoring.
- Evaluation of the respecting of the Cross Compliance regulations on the part of the single farms, with the aim of the reaching of the objective of increasing the verifiability and the efficacy of the Norms imposed by the same Cross Compliance.
- Supporting the determination of the economic value linked to the positive ecosystem benefits and measuring the ecosystem services (Finco et al., 2007) produced by specific management practices.
- Determination of the value linked to activities that can be potentially funded within the terms of the RDP.
- The possibility for single farms to autocertificate the quality of their own practices.
- Comparison between farms that adopt different management models (e.g. organic *versus* conventional agriculture; Lazzerini et al., 2004) and the choice of the culture to practice on the basis of the vegetational potentialities identified.
- Certification of the quality of the agricultural ecosystems in the districts of quality production.

In conclusion, it can be stated that the realisation of such a system of investigation on a large scale will represent an excellent tool for the acquisition of useful data for the planning and implementation of management practices that are functional for the protection of the residual habitats of greatest value present in the agroecosystem.

5. References

Allegrezza, M.; Ballelli, S. & Biondi E. (1987). *Su due associazioni di vegetazione nitrofila dei settori litoranei e collinari dell'Adriatico centrale italiano*. Studi sul Territorio, Ann. Bot. (Roma), Vol. 45, Suppl. 5, pp. 81-88. ISSN 0365-0812.

Allegrezza, M.; Biondi, E. & Felici S. (2006). *A phytosociological analysis of the vegetation of the Central Adriatic sector of the Italian peninsula.* Hacquetia, Vol. 5, No. 2, pp. 135-175. ISSN: 1581-4661

Andersen, E. et al. (eds.) (2004). *Developing a high nature value indicator. Internal report.* European Environment Agency, Copenhagen.

Baldock, D.; Beaufoy, G.; Bennet, G. & Clark, J. (1993). *Nature conservation and new directions in the common agricultural policy.* IEEP London.

Baldock, D.; Beaufoy, G. & Clark, J. (1995). *The nature of farming. Low intensity farming systems in nine European countries.* Report IEEP/WWF/JNRC, London/Gland/ Peterborough.

Baldoni, M. (1995). *Vegetazione infestante le colture erbacee delle Marche e dei piani carsici dell'Appennino umbro-marchigiano (Italia centrale) e serie di vegetazione.* Coll. Phytosoc., Vol. 24, pp. 787-812. ISBN: 3-443-70008-X.

Biondi, E. (1996). *La geobotanica nello studio ecologico del paesaggio.* Annali dell'Accademia Italiana di Scienze Forestali, Vol. 45, pp. 3-39.

Biondi, E. & Allegrezza, M. (1996). *Inquadramento fitosociologico di alcune formazioni prative del territorio collinare anconetano.* Giornale Botanico Italiano, Vol. 130, No. 1: 136-148. ISSN 0017-0070.

Biondi, E. & Allegrezza, M. (2004). *Lettura e modellizzazione sinfitosociologica del paesaggio vegetale del Bacino del Fosso della Selva.* I Quaderni della Selva, Vol. 2 (L'ambiente della Selva di Gallignano), pp. 36-57. ISSN: 1724-4064.

Biondi, E.; Allegrezza, M.; Casavecchia, S.; Pesaresi, S. & Vagge I. (2006). *Lineamenti vegetazionali e paesaggio vegetale dell'Appennino centrale e settentrionale.* Biogeographia Vol. 28, pp. 35-129. ISBN 88-7145-224-0.

Biondi, E.; Allegrezza, M. & Zuccarello, V. (2005). *Syntaxonomic revision f the Apennine grasslands belonging to* Brometalia erecti, *and an analysis of their relationships with the xerophilous vegetation of* Rosmarinetea officinalis (*Italy*). Phytocoenologia, Vol. 35, No. 1, pp. 129-163. ISSN 0340-269X.

Biondi, E. & Baldoni, M. (1991). *La vegetazione dei margini stradali dell'ordine* Brometalia rubenti-tectori *nell'Italia centrale.* Studi sul Territorio, Ann. Bot. (Roma), Vol. 49, Suppl. 8, pp. 213-218. ISSN 0365-0812.

Biondi E. & Baldoni M. (1993). *La vegetazione del Fiume Marecchia (Italia Centrale).* Lavori della Società Italiana di Biogeografia, Vol. 17, pp. 51-87.

Biondi, E.; Ballelli, S.; Allegrezza, M. & Zuccarello, V. (1995). *La vegetazione dell'ordine* Brometalia erecti *Br.-Bl. 1936 nell'Appennino (Italia).* Fitosociologia, Vol. 30, pp. 3-46. ISSN: 1125-9078.

Biondi, E.; Carni, A.; Vagge, I.; Taffetani, F. & Ballelli, S. (2001). *The vegetation of the* Trifolio medii-Geranietea sanguinei *Muller 1962 class in the central part of the Apennines.* Fitosociologia, Vol. 38, No. 1, pp. 55-65. ISSN: 1125-9078.

Biondi, E.; Casavecchia, S. & Radetic, Z. (2002). *La vegetazione dei "guazzi" e il paesaggio vegetale della pianura alluvionale del tratto terminale del Fiume Musone (Italia centrale).* Fitosociologia, Vol. 39 , No. 1, pp. 45-70. ISSN: 1125-9078.

Carni, A. (2005). *Vegetation of trampled habitat sin the Prekmurje region (NE Slovenia).* Hacquetia, Vol. 4, No. 2, pp. 151-159. ISSN: 1581-4661.

Carni, A.; Kostadinovski, M. & Matevski, V. (2002). *Vegetacija na Pohojenih Rastiscih v Republiki Makedoniji.* Hacquetia, Vol. 1, No. 2, pp. 209-221. ISSN: 1581-4661.

Conti, F.; Abbate, G.; Alessandrini, A. & Blasi, C. (2005). *An Annotated Checklist of the Italian Vascular Flora.* Palombi Editori, Roma. ISBN 88-7621-458-5.

Finco, A.; Pollonara, M. & Di Pronio, G. (2007). *Gestione sostenibile dell'agricoltura e tutela della biodiversità.* Fitosociologia, Vol. 44, No. 2, Suppl. 1, pp. 307-312. ISSN: 1125-9078.

Font, X.; Ninot, J.M.; Perdigò, M.T. & Vigo J. (1988). *L'ordre* Galio-Alliarietalia *a Catalunya.* Acta Botanica Barcinonensia, Vol. 37, pp. 201-222. ISSN 0210-7597.

Géhu, J.M. & Rivas-Martinez, S. (1981). *Notions fondamentales de phytosociologie.* Ber. Intern. Symposion. Syntaxonomie in Rinteln, pp. 1-33.

Guinochet, M. & Vilmorin, R. (1973). *Flore de France.* Editions du Céntre National de la Recherche Scientifique, Paris.

Hruska, K. (1983). *Ruderal xerotermic vegetation in the Marche (Central Italy).* Coll. Phytosoc., Vol. 12, pp. 149-154. ISBN: 3-443-70008-X.

Hruska, K. (1988). *Vegetazione nitrofila dei corsi d'acqua del versante adriatico dell'Appennino centrale.* Acta Botanica Barcinonensia, Vol. 37, pp. 253-256. ISSN 0210-7597.

Lazzerini, G.; Colom, M. R.; Camerà, A.; Sacchetti, P. & Vazzana, C. (2004). *Biodiversità aziendale e sua relazione con gli aspetti gestionali in aziende biologiche e convenzionali in Val d'Orcia in Toscana.* In: Atti XIV Congresso della Società Italiana di Ecologia (4-6 ottobre 2004-Siena).

Ninot, M.; Carreras, J.; Carrillo, E. & Vigo J. (2000). *Syntaxonomic conspectus of the vegetation of Catalonia and Andorra. I: Hygrophilous herbaceous communities.* Acta Botanica Barcinonensia, Vol. 46, pp. 191-237. ISSN 0210-7597.

Oberdorfer, E. (1990). *Pflanzensoziologische Excursionflora.* Eugen Ulmer GmbH & Co., Stuttgart.

Pignatti, S. (1982). *Flora d'Italia.* Vol. 1, 2, 3. Edagricole, Bologna.

Poldini, L. (1989). *La vegetazione del Carso isontino e triestino.* Edizioni Lint, Trieste. ISBN 88-95083-30-7.

Puppi, G. (2008). Monitoraggio della diversità vegetale negli ambienti agrari intensivi e semi-intensivi, In: *Monitoraggio della biodiversità selvatica negli agroecosistemi intensivi e semi-intensivi,* Genghini M., pp. 81-111. ISPRA, Ministero delle Politiche Agricole Alimentari e Forestali, Studi Ecologici Ricerca Natura Ambiente.

Rivas-Martinez, S.; Diaz, T.E.; Fernàndez-Gonzàlez, F.; Izco, J.; Loidi, J.; Lousa, M. & Penas, A. (2002). *Vascular plant communities of Spain and Portugal.* Itinera Geobotanica, Vol. 15, No. 2, pp. 433-922. ISSN: 0213-85.

Rameau, J.C.; Mansion, D.; Dumé, G.; Timbal, J.; Lecointe, A; Dupont, P. & Keller, R. (1989). *Flore Forestiere Francaise. Guide Ecologique Illustré.* Institut pour le Développement Forestier & Ministère de l'Agricolture at de la Foret.

Royer, J.M. (1991). *Synthèse eurosibérienne, phytosociologique et phytogéographique de la classe des* Festuco-Brometea. Dissertationes Botanicae, Band 178, J. Cramer, Berlin Stuttgart.

Silc, U. & Kosir, P. (2006). *Synanthropic vegetation of the city of Kranj (Central Slovenia).* Hacquetia, Vol. 5, No. 1, pp. 213-231. ISSN: 1581-4661.

Taffetani, F.; Giannangeli, A.; Micheletti, A.; Rismondo, M. & Zitti S. (2005). *Vegetazione forestale a* Quercus cerris *nel versante adriatico italiano.* Informatore Botanico Italiano, Vol. 37, No. 1, A, pp. 534-535. ISSN 0020-0697.

Taffetani, F.; Orlandini, S. & Zitti, S. (2009). *Paesaggio vegetale di un'area pre-appenninica dell'Italia centrale: il Bosco dei Monaci Bianchi nelle Marche (Italia).* Fitosociologia, Vol. 46, No. 1, pp. 27-47. ISSN: 1125-9078.

Taffetani, F. & Rismondo, M. (2009). *Bioindicators system for the evaluation of the environment quality of agro-ecosystems.* Fitosociologia, Vol. 46, no. 2, pp. 3-22. ISSN: 1125-9078.

Rismondo, M.; Taffetani, F. & Lancioni, A. *Integrated tools and methods for the analysis of agro-ecosystem's through vegetational investigations.* Fitosociologia, Vol. 48, no. 1, pp. 41-52. ISSN: 1125-9078.

Westhoff, V. & Van der Maarel, E. (1978). *The Braun-Blanquet approach.* Handbook of vegetation science. Ordination and classification of vegetation: 619-729.

16

Why Enhancement of Floral Resources in Agro-Ecosystems Benefit Honeybees and Beekeepers?

Decourtye Axel[1], Alaux Cédric[1], Odoux Jean-François[2], Henry Mickaël[1], Vaissière Bernard E.[1] and Le Conte Yves[1]
[1]*UMT PrADE, INRA – UMR Abeilles et environnement, Site AgroParc, AVIGNON Cedex 9*
[2]*INRA, Unité expérimentale d'entomologie Le Magneraud, Surgères*
France

1. Introduction

Honeybee (*Apis mellifera* L.) populations have experienced serious losses in Europe and North America (Neumann & Carreck, 2010; Oldroyd, 2007). These losses highlights the potential risks for our natural and agricultural landscapes through lack of pollination, and the repercussions these would have for human activities and nutrition (Aizen & Harder, 2009; Gallai et al., 2009; Klein et al., 2007; Rader et al., 2009; Ollerton et al., 2011). There is a great deal of concerns about the decline of honeybees across the world and what has been termed colony collapse disorder in the USA. Honeybee health is influenced by biotic factors such as availability of resources, competitors, pathogens, parasites, and predators, and abiotic factors such as climate and pollutants. Colony losses are likely due to multiple factors (Oldroyd, 2007), and most research to date has focused on epidemiological studies affecting honeybee health (Cox-Foster et al., 2007; Johnson et al., 2009; Le Conte et al., 2010), and on the negative effects of agricultural pesticides (Decourtye & Devillers, 2010; Desneux et al., 2007; Johnson et al., 2010; Kevan, 1977). Although the putative causes of this decline are still being investigated, it has long been recognized that a lack of food, and particularly a dearth of pollen, within intensively farmed agricultural landscapes has contributed to the loss of colonies (Mattila & Otis, 2006; Maurizio, 1950; Naug, 2009). Beekeepers have frequently cited starvation and poor foraging conditions as the principal causes of the recent bee losses (Allier et al., 2010; vanEngelsdorp et al., 2007). Naug (2009) suggested that nutritional stress due to habitat loss has played an important role in honeybee colony collapse, and thereby stressed the usefulness of protecting and enhancing the availability of flower resources by using rules and policies for efficient management of agricultural farmlands. Decourtye et al. (2010) reviewed potential approaches to provide and maintain diverse floral resources in a landscape, giving particular consideration to herbaceous plants, to sustain bee populations, and support honeybee health and beekeeping. These approaches include the cultivation and maintenance of large strips (6-12 m in width) of native or non native melliferous plants maintained between crop fields, fallow areas, field margins, and conservation buffer strips. To set up an favourable landscape for honeybees, it is essential to understand the relationship between this insect and the melliferous or polleniferous flora in their environment. This is the main goal of this review and we also describe some available data on the effects of diet on some biological functions of bees so as to build a scientific background for agro-environmental measures protecting floral resources and benefiting the beekeeping industry.

2. A brief presentation of the honeybee

2.1 A social insect

The honeybee is a eusocial insect, living in colonies that can be made of up to 50,000 individuals, including a fertile female, the queen, workers which are sterile females, males (drones), and brood (eggs, larvae and pupae). The workers do various tasks in succession during their lifetime depending on their age and the environment, such as maintaining and building combs, feeding larvae, feeding and grooming the queen and other members of the colony, ventilation, reception and storing of collected food, and defending the nest. From about 21-day old on to their death, the main part of their activity is outside the hive, gathering pollen, nectar, water and propolis. By visiting flowers, forager bees provide the colony with nectar and pollen, which are subsequently transformed into honey and "bee bread", respectively. Survival of the colony depends on the reserves made from these supplies in the combs, especially during the winter.

2.2 Nectar and pollen gathering

Nectar and pollen gathering often occur together as the flowers of most plant species are hermaphrodites and provide both resources. But foraging bee can also collect only one resource and be strict nectar or pollen collector. Indeed, some plants produce on one hand little or no nectar, but offer an abundance of pollen (e.g. kiwifruit *Actinidia deliciosa* A. Chev.; Vaissière et al., 1996), while others produce large echinate pollen grains that honeybees cannot pack on their hind legs (e.g. cotton *Gossypium hirsutum* L; Vaissière & Vinson, 1994). In addition it was demonstrated that a tendency to prefer pollen foraging is genetically determined (Page et al., 1995). Despite this, for reasons of adaptability to the environment and to the colony needs, honeybee worker can rapidly change their foraging activities based on mechanisms of communication and learning.

2.3 Foraging efficiency depends on communication and learning

At the colony level, honeybees are communicating through dances to exchange information about the location of food sources which have been located by scout workers (von Frisch, 1967). Also, the foraging activities vary according to the colony needs communicated by trophallaxis (exchange of food and secretions from mouth to mouth) and exchanges of pheromones inside the colony (Winston, 1987). At the individual level, the foraging worker can remember the configuration of the foraging area and the characteristics of the food sources (Menzel et al., 1993). This last point raises the question of the appreciation of flower attractiveness by the bees. Bees are first attracted by floral odours which can be perceived from far away (Loper & Waller, 1970). But overall it is the association between the reward (quantity and quality of pollen or nectar supplied) and the characteristics of the flower such as odour, colour, and shape, that enables the learning and subsequent recognition of a flower. A flower type providing little resources will be quickly abandoned to the expense of a more profitable type of flower, if the food supply in the environment allows it (Winston, 1987).

Under experimental conditions, honeybees prefer nectar containing saccharose (Waller, 1972) and are able to assess differences in sugar concentrations in the order of 5% (Jamieson & Austin, 1956). They prefer concentrated nectars that range in sugar content between 15% to 50% (Jamieson & Austin, 1956; Sigurdson, 1981a, 1981b; Waddington & Kirchner, 1992; Waller, 1972), but they avoid nectars which are too viscous and thus make their collection difficult. Nectar accessibility also influences floral choice by nectar foragers as nectar has to be within reach given their proboscis length. Honeybees have an average tongue length of 6 mm and can often gather little nectar from flowers with deep corolla such as those of red clover (*Trifolium pratense* L.) and several other Fabaceae (Jablonski, 2001).

For pollen, the protein content does not seem to be detected by the foragers (Maurizio, 1954). Pollens which have the highest protein content are not necessarily those which are the most sought after or the most appetizing. Thus, the fact that a pollen is collected by honeybee foragers does not provide information about its nutritive value. Indeed, pollen from female kiwifruit vines is readily collected by honeybees while its nutritive value is very low (Jay & Jay, 1993). It is not easy to establish a link between the protein content and the nutritive value of a given pollen, because pollen grains contain many other elements besides proteins, and also because their proteins are more or less digestible. Digestibility essentially depends on the thickness and ornementation of the pollen wall. Moreover, a lack of rich pollen in the environment can result result in workers collecting pollen completely deficient of any nutritive value (Louveaux, 1959; Maurizio, 1954; Wille et al., 1985). Also accessibility is an important factor for pollen collection since honeybees cannot harvest large pollen grains which are echinate with long spines such as those of cotton and several other Malvaceae (Vaissière & Vinson, 1994). And the overall appetence of pollen is also influenced by the presence or absence of phago-stimulants or repulsive compounds (Pernal & Currie, 2000).

2.4 Flower loyalty and foraging areas

Honeybees respond to groups of stimuli and rewards that are characteristic of each floral type. Both pollen and nectar gathering by a forager are made more efficient by the fact that the bee learns the handling of a floral type that it finds profitable and then remains loyal to this type for as long as it is available. Indeed, one can observe specialized foraging positions on most flower types (Robinson, 1989) and a strong area loyalty. Indeed, the individual foraging area of honeybee foragers is limited to an area of ca. 100 m^2 (Singh, 1950).

Yet one must not confuse the foraging area of an individual forager with that of the colony as a whole. Within the colony, different foraging groups target different sites. In an environment where nectar is abundant, the foraging area of a colony is on average ca. 2 km radius around the hive. And 90% of bees dancing gather pollen in a radius of less than 5 km from the hive (Beekman & Ratnieks, 2000; Steffan-Dewenter & Kuhn, 2003), that is a survey area of about 80 km^2. When needed, a colony can expand considerably its foraging area. Beekman & Ratnieks (2000) noted that 50% of the foraging bees foraged at > 6 km, and 10% at > 9 km. Based on these results, the flowering areas targeted at the honeybees could be safely created and protected within a 1 km radius around the hive. But in intensively farmed landscapes that are flower-poor, the foraging activity can take place readily over several kilometres around the hive (Decourtye et al., 2008; Steffan-Dewenter & Kuhn, 2003).

3. Characteristics and composition of food sources

3.1 Nectar

Nectar is the dominant source of carbohydrates and is therefore an energy source immediately available to fill colony needs. However, while the hive workers frequently visit honey cells, the food gathering foragers do not as they are fed by nurse honeybees through trophallaxis (Crailsheim et al., 1996). Thus nectar reserves, progressively transformed into honey by the workers, fuel the global activities of the colony, including food collection itself. Honey reserves also enable over-wintering survival of adults, as no extra food is gathered at this period.

Nectar is generally produced in the floral nectary, but some plants also have extra-floral nectaries (Elias, 1983). Studies of more than 900 plant species revealed that both floral and extra-floral nectaries contain 3 main sugars: glucose, fructose and saccharose, in proportions characteristic of the plant species (Baker & Baker, 1982). In addition, small quantities of proteins, lipids and amino acids are also found in nectar (Baker & Baker, 1982), as well as other substances that may help its spotting by bees (Thorp et al., 1975).

While the sugar composition of nectar is relatively stable, the volume and concentration of the nectar are quite variable. In entomophilous plants (pollinated by insects) growing in temperate regions, the volumes are most frequently in the order of 0.5 to 7 µL/flower, with a modal value of 2 µL (Cruden et al., 1983), while sugar concentrations vary from 5% to 50% depending on the species (Baker & Baker, 1982; Cruden et al., 1983). Nectar secretion is subject to a specific rhythm, and, for each species, not only is there a large difference between secretion rate between day and night, but also nectar production can vary considerably during the course of a day (Maurizio, 1975). According to Maurizio (1975) and Corbet et al. (1979), optimal nectar secretion, in volume and sugar concentration, is produced during a period of 5 hours on average, either in the morning, or at the end of the afternoon in the majority of plants that are pollinated by hymenopteran pollinators. More rarely, certain plants show two secretion peaks : one in the morning and one in the evening. It is also noteworthy that even within a given plant species, large differences can exist in the secretion pattern between populations and varieties (e.g., oilseed rape *Brassica napus* L.; Pierre et al., 1999).

Apart from genetic factors, nectar production (volume and concentration) depends on the developmental stage of the plant, as well as the time of day and the age and position of the flower on the plant. Even more, it depends largely on environmental factors such as soil type and moisture level, cultivation practices, and weather conditions (wind, temperature, relative humidity).

From this brief review, it appears that honeybees are clearly in a situation where nectar supply can not only be very diverse in that it is usually provided by an array of several plant species, but also extremely variable in amount and composition. The colony can thus go from a situation when nectar supplies are totally absent, to a situation where there is an abundance of nectar over just a few days (Crane, 1975). Honeybees have therefore to adapt rapidly their foraging depending upon the nectar and pollen availability in their environment.

3.2 Pollen

Pollen is produced and released by the anthers and it can be more or less accessible to the floral visitors depending on floral morphology. Honeybees rarely eat it in its natural state. Rather, it is first aggregated with nectar or diluted honey to form pellets (Vaissière & Vinson, 1994). Pellets are then placed in the cells, packed and covered with honey and it is transformed by lactic fermentation to make « bee bread ».

Fresh pollen contains proteins and amino acids, but also carbohydrates and lipids, including sterols. Each type of pollen can be characterised by its global caloric value, its protein content as % of dry matter, its nitrogen content, amino acid composition (classified as essential or non essential), and starch, sugar and lipid content as well as its vitamin and mineral elements. Specific techniques are needed for each of these measurements and results often differ between different authors. It is therefore difficult to obtain the complete biochemical profile for the pollen of a given plant species (Stanley & Linskens, 1974). For example, dandelion pollen (*Taraxacum campylodes* G.E. Haglund), has a protein content that ranges from 9.2 to 19.2% of its dry weight, depending on the authors, its main lipids are made up linoleic and palmitic acid, and its mineral element composition is known as well. We also know that it is deficient in arginine and is missing essential amino acids such as tryptophane and phenol-alanine (Loper & Cohen, 1987). A few other complete data exists on pollen of gymnosperms and corn *Zea mays* L., but generally such complete data sets are rare and does not allow a multi-criteria classification of pollen types.

3.3 Nectar and pollen budget of the colony

Large quantities of food are required by a honeybee colony. Food gathering at the height of the season must be enough to feed 50,000 workers and 9,000 larvae. Such colony may have an annual nectar budget of about 120 kg and a pollen budget of 20 kg (Seeley, 1995). It may stock from 60 to 80 kg of honey per year (Erber, 1992; Rosov, 1944; Seeley, 1995; Weipple, 1928. The larva of a worker honey bee consumes about 140 mg of honey during its development (Rosov, 1944; Seeley, 1985; Weipple, 1928; Winston, 1987).

An active foraging worker uses 0.5 mg of honey per km flown, and it can fly as much as 800 km during its lifetime (Gould & Gould, 1988). Others authors estimated that a forager consumed 11.5 mg of sugars per hour in flight, and only 0.7 mg per hour when inside the hive (Heinrich, 1979; Olaerts, 1956). During winter, bee activity is reduced, but the cluster must maintain a constant temperature in the centre of the nest of 34°C to 36°C, which requires a large energy expenditure. Thus, an average sized colony needs to stock ca. 25 kg of honey for winter consumption. Overall, the annual needs of a colony are estimated at about 80 kg of honey and 20 kg to 40 kg of pollen depending upon the authors (Crailsheim et al., 1992; Louveaux, 1954; Winston, 1987).

A colony gathers from 15 to 55 kg of pollen per year (Eckert, 1942; Hirschfelder, 1951; Louveaux, 1958; Ribbands, 1953; Seeley, 1985). Nursing bees are the ones that consume the most pollen as they eat ca. 60 mg of pollen over 10 days (Pain & Maugenet, 1966), to develop their hypopharyngeal glands which produce the 42 mg of food consumed by larvae during the first 5 days of larval development (Haydak, 1968). This food given by nursing bees constitutes a major part of the protein supply consumed by larvae since pollen is processed into brood food and only 5% of the protein derived from pollen are directly fed to larvae (Babendreier et al., 2004).

Clearly, the survival and development of honeybee colonies are influenced by the regularity, quality and quantity of the nectar and pollen supply. Within intensively farmed agricultural landscapes, nectar- and pollen-producing crops may provide a narrow window with mass flowering followed by a dearth of pollen and nectar resources. A typical example occurs with oilseed crops such as rape and sunflower (*Helianthus annuus* L.) for which nectar and pollen resources are usually abundant during the blooming, but only for a short period. The subsequent temporal dearth of resources could be partially filled simply by the creation and protection of additional non-cropped areas such as field margins (strips bordering crop fields), hedgerows (linear scrub along field boundaries), woodlands, ponds, ditches and fallow farm fields (Decourtye et al., 2010).
But, for both pollen or nectar , it is long and difficult to assess precisely their production in quantity and quality. Indeed, one must not only take into account the effects from the environment but also the effects of the genetic make-up of the plants at the species and variety levels. This last point is important to objectively assess the best flowering plant for bees, though it is rarely known.

4. Impact of pollen supply on the life of the colony

The majority of the protein intake of larvae comes from the hypopharyngeal gland secretions of nurse bees, and the development of these glands depends on their pollen diet. The development of hypopharyngeal glands is not dependant upon the essential amino acids present in the pollen consumed, but on the total quantity of protein ingested (Pernal & Currie, 2000). Therefore, a situation of pollen deficiency may have detrimental effects on the nurse bees' tasks and brood care, resulting in undernourished larvae (Blaschon et al., 1999). When such deficiencies occur, the nurse bees may also reduce the number of larvae to feed and cannibilize the eggs (Schmickl & Crailsheim, 2001). The young larvae are eliminated for the benefit of the older larvae which need more pollen input. In the case of extended pollen deficiency, another strategy is to reduce overall larval care (Blaschon & Crailsheim, 2001). Finally, pollen feeding during the larval stage has an important impact on the traits of the adults as it may determine the size of the future adult (Roulson & Cane, 2000) and its life expectancy (Schmidt et al., 1987).
Ovary development is another major process influenced by pollen intake. Experiments have been carried out on worker bees kept in isolation without queen, in order to evaluate their capacity to develop ovaries, a physiological process highly dependent protein utilization and pollen quality. This method has shown that pollen from apple (*Malus domestica* Borkh.) or sweet clover (*Melilotus* spp.) is favourable to ovary development whereas pollen from pine (*Pinus* spp.) is not and that of sunflower is below average (Pernal & Currie, 2000). Also, a diet consisting of several pollens of different nutritive value is not equivalent to their average nutritive value (Taséi & Aupinel, 2008; Alaux et al., 2011). We can also assume that a pollen of low nutritional value, like that of dandelion, can be compensated by mixing it with another pollen that has a higher nutritional value for reproductive needs (Genissel et al., 2002). This highlights the importance of avoiding putting bees in a situation where they do not have sufficient floral variety, as is often the case amidst large intensive agricultural landscapes.
Finally, it has recently been shown that a high concentration of pollen lipids (mainly linoleic, myristic and dodecanoic acids) protects the brood nest from certain bacteria such as *Paenibacillus larvae* and *Melissococcus pluton* which are the pathogenic agents responsible for

American foul-brood and European foul-brood, respectively (Manning, 2001). Likewise, a diet with reduced pollen quantity and diversity can induce not only a protein deficiency, but also a decreased synthesis of detoxification enzymes. As a result, bees fed with a mixed pollen diet can be less sensitive to some pesticides (Wahl & Ulm, 1983).

5. What are the benefits of diet diversity?

Within complex landscapes, honeybee colonies normally collect and consume pollen from a large array of plant species (Dimou & Thrasyvoulou, 2009; Louveaux, 1959; Severson & Parry, 1981). This generalist behavior for pollen collection (polylecty), as opposed to oligolecty or monolecty in other bee species, is further supported by experimental evidence that honeybees feed preferentially on plurispecific pollen mixtures rather than pollen from a single species (Schmidt, 1984). These results support the postulate that honeybees select a mixed diet when given a choice. However, today intensive agricultural landscapes most often provide colonies with a low variety of plant species. Indeed, honeybees used for the pollination service of large areas may even be forced to feed mainly on a single flowering crop, like sunflower that has pollen with a poor nutritional value and sesame that has pollen which contains only low levels of phagostimulants (Schmidt et al., 1995). Such practices may be stressful for colony health. In those cases, the presence of additional floral or food sources could to reduce these potential problems.

The protein content of pollen (2.5 to 61%), its amino acid composition, its lipid content (1 to 20%), and that of starch, sterols, vitamins and minerals vary tremendously among plant species (Roulston & Buchmann, 2000; Roulston & Cane, 2000; Stanley & Linskens, 1974). Therefore, regardless how much monospecific pollen is consumed, it might miss nutrients essential to the health of honeybee colonies. The deficiency of a given pollen in essential nutrients could be compensated for by a more diverse pollen diet. Indeed, a mixed pollen diet increases the lifespan of honeybees as compared to a monospecific pollen diet (Schmidt, 1984; Schmidt et al., 1987). In addition, a diversified pollen diet might help the bees to fight against pathogens. Alaux et al. (2010) found that a diverse pollen diet can actually enhance some immune functions in worker honeybees. For example, their production of glucose oxidase, an enzyme involved in the synthesis of the antiseptic hydrogenperoxide in honey and brood food, was 40% higher in honeybees fed with pollen from a variety of plants compared to those fed with pollen from a single plant species - even if this monofloral diet had a higher protein content. However, whether polyfloral diets might increase the actual resistance to diseases and pathogens remain to be tested. In bumblebees *Bombus terrestris*, larvae fed with a mixed pollen diet were heavier than larvae fed with monofloral pollen of equivalent or higher protein content (Taséi & Aupinel, 2008), and so we might also expect a similar beneficial effect of a diverse pollen diet on the health of honeybee larvae.

As stated previously, increasing the number of pollen species in the diet probably provides a buffer against deficiencies of specific nutrients (Schmidt, 1984) as well as the presence of toxic compounds in some pollen (Mesquita et al., 2010). Indeed, several studies have demonstrated the importance of some specific nutrients in bee health. For example, de Groot (1953) found that 10 essential amino acids in specific proportions are required for optimal honeybee health: arginine, histidine, lysine, tryptophan, phenylalanine, methionine, threonine, leucine, isoleucine, and valine. Those essential amino acids cannot be synthesized *de novo* by honeybees and therefore need to be supplied directly in the diet. Some fatty acids, found at different levels in pollens (Manning, 2001; Singh et al., 1999), can also be beneficial to honeybees due to their antimicrobial properties (Hornitzky, 2003). Pollen and nectar, beside being the primary food source for bees, also contain phytochemicals and are often rich in carotenoids, flavonoids, alkaloids and phenolic compounds that have antioxidant properties and antimicrobial activity (Adler, 2000; Balch & Balch, 1990; Basim et al., 2006; Campos et al., 2003; Leblanc et al., 2009; Morais et al., 2011). Sustaining the diversity of flower resources in the landscape might increase the chances for bees to find all those beneficial nutrients, and also to avoid or provide an alternative to toxic nectar and pollen. Indeed, some plant components are non-nutritive but toxic to honey bees (Barker, 1990), like the sugars galactose, arabinose, xylose, melibiose, raffinose, stachyose and lactose that can be found in pollen and nectars of some plants (Barker, 1977).

6. Availability of food resources in the environment: a question of temporal and spatial scales

Currently, there is no temporal continuity in floral resource availability within intensively farmed agricultural landscapes. In Europe, mass flowering crops providing bees with nectar or pollen are often limited to maize, sunflower and oilseed rape at large landscape scales in cereal farmland systems (Decourtye et al., 2010). Among these three, maize provides the greatest quantities of pollen collected by honeybees, due to its long-lasting availability and good accessibility to foragers (Charrière et al., 2010; Odoux et al., 2004; Vaissière & Vinson 1994). On the other hand, oilseed rape and sunflower can be important nectar sources and provide substantial honey crops. This abundance of resource often has a strong impact on colony dynamics as the high intake of food during the flowering of the oilseed rape induces a rapid demographic increase in the colonies. Unfortunately, these young and populous bee cohorts are likely to suffer from food resource scarcity after the mass flowering ends.

Beekeeping takes place in a great variety of agrosystems and is therefore exposed to a wide range of colony management issues. The foraging ecology of honeybees has to be considered within temporal and spatial scales as well. Cereal plains often include also semi-natural habitats, which are able to supply food resources for honeybees during periods of food shortage. Therefore, bees shift foraging habitats on a seasonal basis. The capacity of these agricultural environments to sustain honeybee colonies between crop mass-flowering periods depends on the presence of wild plants in hedgerows, grasslands and woody habitats. In other words, the carrying capacity of the landscapes is expected to increase with its structural complexity (Steffan-Dewenter & Kuhn, 2003). During periods of food shortage, bees can cover larger distances to fetch food, e.g. beyond 5 km away from the colony (Beekman & Ratnieks, 2000; Odoux et al., 2009). In addition, some weeds in cereal fields probably become keystone resources at this time. For instance, bees may collect on a daily basis more pollen on poppy flowers *(Papaver rhoeas* L.), than they would on maize (Odoux, 2010). The presence of this adventicious species is clearly dependant on agricultural practices such as crop rotation, soil preparation, and herbicide use.

The nutritional carrying capacity of farming landscapes for honeybees varies a lot, as food resources vary in quantity and quality. The cultivated lavender hybrid (*Lavandula x intermedia* Emeric ex Loisel) is an attractive melliferous crop for beekeeping, but it is male-sterile so that it does not provide any pollen at all. On the contrary, apple orchards produce an abundant pollen that is rich in proteins (Louveaux, 1959). Abundant intakes of sunflower and maize pollens, which are particularly poor in protein and lipid content (Feuillet et al., 2008; Roulston & Cane, 2000), are expected to exert stresses on colony demography. The pollen of some other grasses usually grown for fodder (e.g. *Setaria italica* L.P.Beauv.) are also actively foraged by honeybees during periods of pollen dearth, but this pollen does not appear satisfactory based on its nutrient content. In this regard, wild plant species such as wild cherries (*Prunus avium* L.) or wild poppies offer protein- and lipid-rich pollens, and enable a stable development of colonies in the spring, when the blooming of the oilseed rape is over.

In intensive agricultural landscapes where semi-natural habitats and weeds are sparse, floral schemes, i.e. management of flowering areas, might provide ecological compensation features for honeybees (Decourtye et al., 2010). They are intended foremost to ensure population sustainability during the periods of food shortage rather than to foster honey production. Such floral schemes might also contribute to the conservation of some wild bee species. Indeed Fabaceae, which are also visited extensively by wild bees, are often preferred in the agri-environmental schemes of farmers, since their management is well-known and their seeds are generally fairly cheap. The efficiency of floral schemes to provide pollen resources to colonies can be assessed by analyzing the pollen pellets obtained for example with pollen traps. And the benefits for beekeepers are clear as colonies that have had access to more regular resources during a given season produced more brood and larger populations in the following year (Decourtye et al., 2008).

7. Consideration of the landscape factors in agro-environmental actions promoting flowering areas

Some authors have questioned the effectiveness of agro-environmental schemes because their beneficial effects on target species do vary from one study to another (Kleijn et al., 2006). The same holds true for floral landscape enhancements dedicated specifically to bees (Dicks et al., 2010). Yet there is a growing body of evidence that the seemingly unpredictable effectiveness of floral schemes is actually dependent on the landscape context in which it is established (Decourtye et al., 2010). Landscape context typically refers to the degree of land use by humans. Intensive, simplified, agricultural landscapes are distinguished from complex landscapes with greater amounts of semi-natural habitats or habitat diversity. A handful of studies have specifically measured the influence of landscape context on the efficiency of experimental flower patches in attracting bees (Heard et al., 2007; Kohler et al. 2008; Steffan-Dewenter et al., 2002). However, to date, no consistent conclusion can be drawn on such floral schemes because different field protocols have been used with different study species and at different spatial scales.

Only one of these studies (Steffan-Dewenter et al., 2002) focused on honeybees and it reported a weak pattern of context-dependent floral scheme effectiveness. Authors have implanted experimental flower patches in a variety of landscape contexts, and have monitored honeybee foraging activity at those patches. Flower patches tended to attract fewer honeybees when implanted in landscapes characterized by a higher amount of semi-natural habitats (field and forest margins, hedgerows, fallows and extensive grasslands) within a 3-km radius. Conversely, foraging activity at flower patches located in more intensive agricultural landscapes was greater, indicating that honeybees compensated for the lack of natural resources by making a disproportionate use of the floral schemes.

Yet, theoretical evidence suggests that honeybees would rather respond positively to the presence of semi-natural habitats. In particular, hedgerows, forest margins and other linear landscape elements may be used as visual landmarks by honeybees to direct their flight path and locate food sources (Chittka & Geiger, 1995; Dyer, 1996; Dyer et al., 2008). As such, linear landscape elements are generally considered to promote landscape connectivity (Taylor et al., 1993), i.e. to facilitate movement of organisms among their resource patches by forming flight corridors (Townsend & Levey, 2005; Van Geert et al., 2010).

The mismatch between theoretical expectations of higher foraging activity in complex landscapes on one hand, and the opposite pattern actually observed by Steffan-Dewenter et al. (2002) on the other hand, most probably reveal an interplay of behavioral processes acting at different spatial scales. The study by Steffan-Dewenter et al. (2002) reports broad-scale patterns, with landscape context characterized within 250-m to 3-km radii around experimental plots, i.e. a spatial window scaled on the foraging range of honeybee colonies. However, the facilitative effect of visual landmarks for the orientation of foragers probably acts at a very local scale. Perception and orientation in honeybees imply short-scale behavioral processes. For instance, an experimental modification of landmark arrangements within a range of only several tens of meters altered the capacity of honeybees to retrieve food sources (Chittka & Geiger, 1995).

Recent foraging surveys conducted in an intensive cereal farming system support the previous hypothesis that semi-natural and/or linear landscape elements enhance the use of floral schemes by honeybees at a local scale (Henry et al., in press). Honeybee foraging activity was monitored in 170 plots (2×2m) of *Phacelia tanacetifolia* Benth. scattered among 17 field margins, all positioned within the foraging range of a single experimental apiary. The presence of hedgerows and forest edges within a 160-m radius around the phacelia plots had a strong positive effect on honeybee foraging activity. Conversely, the presence of mass flowering crops within the same distance had a negative effect, probably because they exerted a concurrent attraction. This critical range of 160 m was not covered by the study of Steffan-Dewenter et al. (2002).

Altogether, the above-mentioned studies suggest that the effectiveness of floral enhancement schemes varies with landscape context in a multi-scale fashion. On a broad scale, i.e. at the landscape level envisioned by the foraging range of a honeybee colony (3-km grain), floral schemes will be more successful in intensive agricultural landscapes (with <15% of land cover occupied by semi-natural habitat remnants). At a smaller scale, within intensive landscapes, floral schemes will be more successful if they are implanted nearby (< 160 m) existing networks of hedgerows, forest edges, or other salient semi-natural elements. Priority should be given to areas farther (> 160 m) from other existing mass-flowering crops. These tentative recommendations still need further validation, depending on the type of floral schemes.

8. Conclusions

Even if much research remains to be done to improve the assessment of the nutritive value of specific nectar, pollen and their mixes, it is clear that these resources are complementary and essential throughout the lifetime of honeybees. We have seen that, by a feedback loop system, a deficiency in quantity and quality of pollen and nectar can lead to a demographic decrease of the colony. A reduced pollen intake leads to low production of larvae and of course adults, and in turn less pollen and nectar foragers. The role of nectar and pollen is empirically so obvious that beekeepers now provide supplements in the form of syrup or pollen in case of deficiency. However, it must be highlighted that the suitability of pollen feeding depends on the storage method and duration of the pollen. The same holds true for the feeding of sugar syrup solutions. These cannot fully replace natural nectar in terms of nutritive value. Consequently, making sure that honeybees have access to pollen and nectar at the right time in their natural environment, remains the best way to guarantee colony survival. Moreover, recent results reported in literature (Alaux et al., 2010; DeGrandi-Hoffman et al., 2010) highlight a connection between diet, protein levels and immune response and suggest that colony losses might be reduced by alleviating diet stress through landscape enhancement of floral resources in agro-ecosystems.

The goal of agro-environmental measures aiming to favour honeybee protection should be to temporally and spatially increase heterogeneous habitats for these insects (Decourtye et al., 2010). As a first goal, it is urgent to preserve current semi-natural habitats in farmlands (hedgerows, woodlands, ponds, ditches). In addition, regulations advocating the management of uncropped farmlands (field margins, farmland set-aside) for increased floral availability to bees could contribute to the maintenance of colony viability (at risk today, especially in intensive cereal farming areas). The management of uncropped farmlands designed to introduce floral resources is likely to benefit other pollinators (wild species) and sustain beekeeping activities (domestic species). But the economic question has to be addressed in relation with farmers as well, depending on the possibility to derive any agronomic benefit from floral schemes. Land managers have to seek a trade-off between the costs of introducing floral schemes, and the benefits they might receive from it because of enhanced pollination service. Yet, enhancement of farmland set-aside may require disproportionate financial investments in some areas - generally where farming practices are intensive (Kleijn et al., 2009; Van Buskirk & Willi, 2004). Therefore, it appears critical to gain a prior knowledge about where enhancement schemes are likely to reach maximal efficiency in a given landscape. In that respect, we should study the efficiency of flower enhancement schemes according to local landscape context. This issue has been poorly explored to date (Heard et al., 2007). Reliable modeling tools are needed to delineate the landscape areas likely to be associated with greater effectiveness of floral enhancement schemes, specifically in intensive agricultural systems.

In these intensive farming areas, the conflict between the conservation of flora, to increase the availability of food to bees, and the management of pesticide treatments complicates the reliability of the positive impact of such flowering schemes. The protection of plants producing nectar and pollen in the farmland areas must actually come with a harmless management of pesticides. Thus, rather than any particular farming practice causing current population decline, such as insecticide application or reduction of flora, the multivariate effects of agricultural practices strongly interact and should be considered collectively, rather than individually.

9. Acknowledgements

This work is a product of UMT PrADE, a unit which associates different groups of stakeholders towards bee protection, from researchers to apicultural engineers. It was financially supported in part by a grant from the French Ministry of Agriculture (MIRES). The research on the melliferous and polleniferous flora and pollen nutrients was supported by grants from FEAGA, L'Institut National de la Recherche Agronomique (INRA "Département Santé des Plantes et Environnement"), and a French-Romanian bilateral program (PHC Brancusi).

10. References

Adler, L.S. (2000). The ecological significance of toxic nectar. *Oikos* Vol.91, pp. 409-420.

Aizen, M.A. & Harder, L.D. (2009). The global stock of domesticated honey bees is growing slower than agricultural demand for pollination. *Current Biology* Vol.19, pp. 915-918.

Alaux, C.; Ducloz, F.; Crauser, D. & Le Conte, Y. (2010). Diet effects on honeybee immunocompetence. *Biology Letter* Vol.6, pp. 562-565.

Allier, F.; Holzmann, C.; Britten, V.; Jourdan, P. & Vallon, J. (2010). Estimation of honeybee colony losses within professionnal beekeepers in France during winter 2008/2009, *Proceeding of Eurbee*, Ankara, Turkey, 2010.

Badendreier, D.; Kalberer; N.; Romeis; J.; Fluri, P. & Bigler, F. (2004). Pollen consumption in honey bee larvae: a step forward in the risk assessment of transgenic plants. *Apidologie*, Vol.35, pp. 293-300.

Baker, H.G. & Baker, I. (1982). Floral nectar constituents in relation to pollinator type. In: *Handbooks of experimental pollination biology*, C.E. Jones & R.J. Little (Eds.), 131-191, van Nostrand-Reinhold, New-York.

Barker, R.J. (1977). Some carbohydrates found in pollen and pollen substitutes are toxic to honey bees. *Journal of Nutition*, Vol.107, pp. 1859-1862.

Barker, R.J. (1990). Poisoning by plants, In: *Honey bee pests, predators, and diseases*, R.A. Morse & R. Nowogrodzki (Eds.), 306-328, Cornell University Press, New York and London.

Basim, E.; Basim, H. & Ozcan, M. (2006). Antibacterial activities of Turkish pollen and propolis extracts against plant bacterial pathogens. *Journal of Food Engineering*, Vol.77, pp. 992-996.

Beekman, M. & Ratnieks, F.L.W. (2000). Long-range foraging by the honey-bee, *Apis mellifera* L. *Functional Ecology*, Vol.14, PP. 490-496.

Blaschon, B.; Guttenberger, H.; Hrassnig, N. & Crailsheim, K. (1999). Impact of bad weather on the development of the brood nest and pollen stores in a honeybee colony (Hymenoptera: Apidae). *Entomology Genetic*, Vol.24, pp. 49-60.

Campos, M.G.; Webby, R.F.; Markham, K.R.; Mitchell, K.A. & Da Cunha, A.P. (2003). Aged induced diminution of free radicals scavenging capacity in bee-pollens and the contribution of constituents flavonoids. *Journal of Agricultural and Food Chemistry*, Vol.51, pp. 742-745.

Charrière, J.D.; Imdorf, A.; Koenig, C.; Gallmann, S. & Kuhn, R. (2010). Do sunflowers influence the development of honey bee, *Apis mellifera*, colonies in areas with diversified crop farming? *Journal of Apicultural Research*, Vol.49, pp.227-235.

Chittka, L. & Geiger, K. (1995). Can honey bees count landmarks? *Animal Behaviour*, Vol.49, pp. 159–164.

Corbet, S.A.; Unwin, D.M. & Prys-jones, O.E. (1979). Humidity, nectar and insect visits to flowers, with special reference to *Crataegus*, *Tilia* and *Echium*. *Ecological Entomology*, Vol.4, pp. 9-22.

Cox-Foster, D.L.; Conlan, S.; Holmes, E.C.; Palacios, G.; Evans, J.D.; Moran, N.A.; Quan, P.L.; Briese, T.; Hornig, M.; Geiser, D.M.; Martinson, V.; vanEngelsdorp, D.; Kalkstein, A.L.; Drysdale, A.; Hui, J.; Zhai, J.; Cui, L.; Hutchison, S.K.; Simons, J.F.; Egholm, M.; Pettis, J.S. & Lipkin, W.I. (2007). A metagenomic survey of microbes in honey bee colony collapse disorde. *Science*, Vol.318, pp. 283-287.

Crailsheim, K.; Schneider, L.H.W.; Hrassnigg, N.; Bühlmann, G.; Brosch, U.; Gmeinbauer, R. & Schôffmann, B. (1992). Pollen consumption and utilization in worker honeybees. *Journal of Insect Physiology*, Vol.38, pp. 409-419.

Crailsheim, K.; Hrassnigg, N. & Stabentheiner, A. (1996). Diurnal behavioural differences in forage and nurse honey bees (*Apis mellifera carnica*). *Apidologie*, Vol.27, pp. 235-244.

Crane, E. (1975). The flower honey comes from. In: *Honey, a comprehensive survey*, E. Crane (Ed), 3-105, Heinemann, London, UK.

Cruden, R.W.; Hermann, H.M. & Peterson, S. (1983). Patterns of nectar production and plant-pollinator coevolution. In: *The biology of nectaries*, B. Bentley & T. Elias, 80-125, Columbia Univ. Press, New-York.

Decourtye, A.; Odoux, J.F. & Cluzeau-Moulay, S. (2008). Influence des aménagements floristiques sur les abeilles. *Bulletin Technique Apicole*, Vol.35, No.3, pp. 114-123.

Decourtye, A. & Devillers, J. (2010). Ecotoxicology of neonicotinoids insecticides in the bees. In: *Insect Nicotinic Acetylcholine Receptors*, S. Thany (Ed.), 85-95, Landes Bioscience.

Decourtye, A; Mader, E. & Desneux, N. (2010). Landscape enhancement of floral resources for honey bees in agro-ecosystems. *Apidologie*, Vol.41, pp. 264-277.

Degrandi-Hoffman, G.Y.; Chen, E.; Huang, M. & Huang H. (2010). The effect of diet on protein concentration, hypopharyngeal gland development and virus load in worker honey bees (*Apis mellifera* L.). *Journal of Insect Physiology*, Vol.56, pp. 1184-1191.

de Groot, A.P. (1953). Protein and amino acid requirements of the honey bee (*Apis mellifica* L.). *Physiologia Comparata et Oecologa*, Vol.3, pp. 197-285.

Desneux, N.; Decourtye, A. & Delpuech, J.M. (2007). The sublethal effects of pesticides on beneficial arthropods. *Annual Review of Entomology*, Vol.52, pp. 81-106.

Dicks, L.V.; Showler, D.A. & Sutherland, W.J. (2010). *Bee conservation: evidence for the effects of interventions*. Pelagic Publishing (Ed.), Exeter, UK.

Dimou, M. & Thrasyvoulou, A. (2009). Pollen analysis of honeybee rectum as a method to record the bee pollen flora of an area. *Apidologie*, Vol.40, pp. 124-133.

Dyer, F.C. (1996). Spatial memory and navigation by honeybees on the scale of the foraging range. *Journal of Experimental Biology*, Vol.199, pp. 147–154.

Dyer, A.G., Rosa, M.G. & Reser, D.H. (2008). Honeybees can recognise images of complex natural scenes for use as potential landmarks. *Journal of Experimental Biology*, Vol.211, pp. 1180.

Elias T. (1983). Extrafloral nectaries: their structure and distribution. In: *The biology of nectaries*, B. Bentley & T. Elias (Eds.), 174–203, Columbia University Press, New-York, USA.

Feuillet, D.; Odoux, J.F.; Mateescu, C.; Aupinel, P.; Lamy, H.; Moreau, N.; Roucher, L. & Souché, T. (2008). Evolution floristique et physico-chimique des pollens récoltés au cours de l'année 2006 sur le site de l'INRA du Magneraud. *Bulletin Technique Apicole*, Vol.35, No.1, pp. 20-26.

Gallai, N.; Salles, J.M.; Settele, J. & Vaissière, B.E. (2009). Economic valuation of the vulnerability of world agriculture confronted with pollinator decline. *Ecological Economics*, Vol.68, pp. 810-821.

Genissel, A.; Aupinel, P.; Bressac, C.; Taséi, J-N. & Chevrier C. (2002). Influence of pollen origin on performance of *Bombus terrestris* micro colonies. *Entomologia Experimentalis et Applicata*, Vol.104, pp. 329-336.

Giersch, T.; Barchia, I. & Hornitzky M. (2010). Can fatty acids and oxytetracycline protect artificially raised larvae from developing European foulbrood? *Apidologie*, Vol.41, pp. 151–159.

Gould, J.L. & Gould, C.G. (1988). *The honey bee*. Scientific American Library, New-York, USA.

Haydak, M.H. (1970). Honey bee nutrition. *Annual Review of Entomology*, Vol.15, pp. 143–156.

Heard, M.S.; Carvell, C.; Carreck, N.L.; Rothery, P.; Osborne, J.L. & Bourke, A.F.G. (2007). Landscape context not patch size determines bumble-bee density on flower mixtures sown for agri-environment schemes. *Biology Letters*, Vol.3, pp. 638-641.

Henry, M.; Fröchen, M.; Maillet-Mezeray, J.; Breyne, E.; Allier, F.; Odoux, J.-F. & Decourtye, A. (2012). Spatial autocorrelation in honeybee foraging activity reveals optimal focus scale for predicting agro-environmental scheme efficiency. *Ecological modelling*, in press.

Jablonski, B. (2001). Agronomic and beekeeping value of short-tube populations of red clover (*Trifolium pratense* L.). *Journal of Apicultural Science*, Vol.45, pp. 37-50.

Jamieson, C.A. & Austin, G.M. (1956). Preferences of honeybees for sugar solutions, *Proceedings of International Congress of Entomology*, pp. 1059-1062, Montréal, Québec, 1956.

Jay, S.C. & Jay D.H. (1993). The effect of kiwifruit (*Actinidia deliciosa* A. Chev.) and yellow flowered broom (*Cytisus scoparius* Link) pollen on the ovary development of worker honey bees (*Apis mellifera* L). *Apidologie*, Vol.24, pp. 557-563.

Johnson, R.M.; Evans, J.D.; Robinson, G.E. & Berenbaum, M.R. (2009). Changes in transcript abundance relating to colony collapse disorder in honey bees (*Apis mellifera*). *Proceedings of the National Academy of Sciences* USA, Vol.106, pp. 14790-14795.

Johnson R.M., Ellis M.D., Mullin C.A., Frazier M. (2010) Pesticides and honey bee toxicity – USA, *Apidologie* vol.41, pp 312-331.

Kevan, P. (1977). Blueberry crops in Nova Scotia and New Brunswick - pesticides and crop reductions. *Canadian Journal of Agricultural Economics*, Vol.25, pp. 61-64.

Kleijn, D.; Baquero, R.A.; Clough, Y.; Díaz, M.; Esteban, J.; Fernández, F.; Gabriel, D.; Herzog, F.; Holzschuh, A.; Jöhl, R.; Knop, E.; Kruess, A.; Marshall, E.J.P.; Steffan-Dewenter, I.; Tscharntke, T.; Verhulst, J.; West, T.M. & Yela, J.L. (2006). Mixed biodiversity benefits of agri-environment schemes in five European countries. *Ecology Letters*, Vol.9, pp. 243-254.

Kleijn, D.; Kohler, F.; Báldi, A.; Batáry, P.; Concepción, E.D; Clough, Y.; Díaz, M.; Gabriel, D.; Holzschuh, A.; Knop, E.; Kovács, A.; Marshall, E.J.P.; Tscharntke, T. & Verhulst, J. (2009). On the relationship between farmland biodiversity and land-use intensity in Europe. *Proceedings of the Royal Society B: Biological Sciences*, Vol.276, pp. 903-909.

Klein, A. M., B. E. Vaissière, J. H. Cane, I. Steffan-Dewenter, S. A. Cunningham, C. Kremen, et T. Tscharntke. 2007. Importance of pollinators in changing landscapes for world crops. Proceedings of the Royal Society B 274:303-313.

Kohler, F.; Verhulst, J.; van Klink, R. & Kleijn, D. (2008). At what spatial scale do high-quality habitats enhance the diversity of forbs and pollinators in intensively farmed landscapes? *Journal of Applied Ecology*, Vol.45, pp. 753-762.

Leblanc, B.W.; Davis, O.K.; Boue, S.; DeLucca, A. & Deeby, T. (2009). Antioxidant activity of Sonoran Desert bee pollen. *Food Chemistry*, Vol.115, pp. 1299-1305.

Le Conte, Y.; Ellis, M. & Ritter, W. (2010). Varroa mites and honey bee health: can Varroa explain part of the colony losses? *Apidologie*, Vol.41, pp. 353–363.

Loper, G.M. & Waller, G.D. (1970). Alfalfa flower aroma and flower selection by honey bees. *Crop Science*, vol.10, pp. 66-68

Loper, G.M. & Cohen A.C. (1987). Amino-acid content of dandelion pollen, a honey bee (Hymenoptera, Apidae) nutritional evaluation. *Journal of Economic Entomology*, Vol.80, pp. 14-17.

Louveaux, J. (1959). Recherches sur la récolte du pollen par les abeilles (*Apis mellifera* L.). *Annales de l'abeille*, Vol.2, pp. 99.

Manning, R. (2001). Fatty acids in pollen: a review of their importance for honey bees. *Bee World*, Vol.82, pp. 60-75.

Mattila, H.R. & Otis, G.W. (2006). Influence of pollen diet in spring on development of honey bee (Hymenoptera: Apidae) colonies. *Journal of Economic Entomology*, Vol.99, pp. 604-613.

Maurizio, A. (1950). The influence of pollen feeding and brood rearing on the length of life and physiological condition of the honeybee. *Bee World*, Vol.31, pp. 9–12.

Maurizio, A. (1954). Pollen: its composition, collection, utilization, and identification. *Bee World*, Vol.35, pp. 49-50.

Maurizio, A. (1975). How bees make honey. In: *Honey, a comprehensive survey*, E. Crane (Ed), 77-105, Heinemann, London, UK.

Menzel, R.; Greggers, U. & Hammer, M. (1993). Functional organization of appetitive learning and memory in a generalist pollinator, the honey bee. In: *Insect learning*, D.R. Papaj & A.C. Lewis, 79-125, Chapman Hall, New-York, USA.

Mesquita, L.X.; Maracajà, P.B.; Sakamoto, S.M. & Soto-Blanco, B. (2010). Toxic evaluation in honey bees (*Apis mellifera*) of pollen from selected plants from the semi-arid region of Brazil. *Journal of Apicultural Research*, Vol.49, pp. 265-269.

Morais, M.; Moreira, L.; Feas, X. & Estevinho, L.M. (2011). Honeybee-collected pollen from five Portuguese Natural Parks: Palynological origin, phenolic content, antioxidant properties and antimicrobial activity. *Food Chemistry and Toxicology*, Vol.49, pp. 1096-1101.

Naug D. (2009). Nutritional stress due to habitat loss may explain recent honeybee colony collapses. *Biological Conservation*, Vol.142, pp. 2369–2372.

Neumann, P. & Carreck, N.L. (2010). Honey bee colony losses. *Journal of Apicultural Research*, Vol.49, No.1, pp. 1-6.

Odoux, J.F.; Lamy, H. & Aupinel, P. (2004). L'abeille récolte-t-elle du pollen de maïs et de tournesol? *La Santé de l'Abeille*, Vol.201, pp. 187-193.

Odoux, J.F.; Caro, G.; Touillet, C.; Peyra, E.; Derelle, D.; Aupinel, P. & Bretagnolle, V. (2009). Which landscape features influence population ecology of bee colonies in farmland intensive cereals systems? *Proceedings of Apimondia 41th Congress*, Montpellier, France, September 15-20, 2009.

Odoux, J.F.; Feuillet, D.; Aupinel, P.; Loublier, Y.; Taséi, J.N. & Mateescu C. (2010). Territorial biodiversity in cropping system and consequences on physico-chemical characteristics of collected pollen. *Proceedings of Eurbee 4th European Conference of Apidology*, Ankara, Turkey, September 7-9, 2010.

Oldroyd, B.P. (2007). What's killing American honey bees? *PLoS Biology*, Vol.5, pp. 1195-1199.

Ollerton, J.; Winfree, R. & Tarrant, S. (2011). How many flowering plants are pollinated by animals? *Oikos*, Vol.120, pp. 321-326.

Pain, J. & Maugenet, J. (1966). Recherches biochimiques et physiologiques sur le pollen emmagasiné par les abeilles. *Annales de l'abeille*, Vol.9, pp. 209-236.

Page, R.E.; Waddington, K.D.; Hunt, G.J. & Fondrk, M.K. (1995). Genetic determinants of honey bee foraging behaviour. *Animal Behaviour*, Vol.50, pp. 1617-1625.

Pernal S.F. and Currie R.W. (2000) Pollen quality of fresh and 1-year-old single pollen diets for worker honey bees (*Apis mellifera* L.). *Apidologie* vol.31, pp 387-409.

Pierre, J.; Mesquida, J.; Marilleau, R.; Pham-Delègue, M.H. & Renard, M. (1999). Nectar secretion in winter oilseed rape, *Brassica napus* - quantitative and qualitative variability among 71 genotypes. *Plant Breeding*, Vol.118, pp. 471-476.

Pierre, J. & Renard, M. (1999). Plant development mutants : incidence on honey bees behaviour and pollination. *Proceedings of 10th International Rapeseed Congress*, pp. 26-29 Canberra, Australia, September 26-29, 1999.

Rader, R.; Howlett, B.G.; Cunningham, S.A.; Westcott, D.A.; Newstrom-Lloyd, L.E.; Walker, M.K.; Teulon, D.A.J. & Edwards W. (2009). Alternative pollinator taxa are equally efficient but not as effective as the honeybee in a mass flowering crop. *Journal of Applied Ecology*, Vol.46, pp. 1080-1087.

Ribbands, C.R. (1953). *The behaviour and social life of honeybees.* London Bee Research Association, London, UK.

Robinson, W.S. (1981). Honey bees *Apis mellifera* development of foraging fidelity to Delicious apple flowers. *Journal of Economic Entomology*, Vol.74, pp. 127-130.

Roulston, T.H. & Buchmann, S.L. (2000). A phylogenetic reconsideration of the pollen starch-pollination correlation. *Evolutionary Ecology Research,* Vol.2, pp. 627-643.

Roulston, T.H. & Cane, J.H. (2000). Pollen nutritional content and digestibility for animals. *Plant Systematics and Evolution,* Vol.222, pp. 187-209.

Schmickl, T. & Crailsheim, K. (2001). Survival of honeybee larvae in times of pollen stress. *Apidologie*, Vol.32, pp. 496-498.

Schmidt, J.O. (1984). Feeding preferences of *Apis mellifera* L. (Hymenoptera: Apidae): individual versus mixed pollen species. *Journal of the Kansas Entomological Society*, Vol.57, pp. 323-327.

Schmidt, J.O.; Thoenes, S.C. & Levin, M.D. (1987). Survival of honey bees, *Apis mellifera* (Hymenoptera: Apidae), fed various pollen sources. *Journal of Economic Entomology,* Vol.80, pp. 176-183.

Schmidt, L.S.; Schmidt, J.O.; Rao, H.; Wang, W. & Xu, L. (1995). Feeding preference and survival of young worker honey bees (Hymenoptera:Apidae) fed rape, sesame, and sunflower pollen. *Journal of Economic Entomology,* Vol.88, pp. 1591-1595.

Seeley, T. (1985). *Honeybee ecology: a study of adaptation in social life.* Princeton University Press, Princeton, UK.

Severson, D.W. & Parry, J.E. (1981). A chronology of pollen collection by honey bees. *Journal of Apicultural Research,* Vol.20, pp. 97-103.

Sigurdson, J.E. (1981a). Automated discrete trials techniques of appetitive conditioning in honeybees. *Behaviour Research Methods and Instruments*, Vol.13, pp. 1-10.

Sigurdson, J.E. (1981b). Measurement of consummatory behaviour of honeybees. *Behaviour Research Methods and Instruments*, Vol.13, pp. 308-310.

Singh, S. (1950). Behavior studies of honeybees in gathering nectar and pollen. Annual Report of Cornell University Agricultural Experimental Station, Vol.288, 57 pp.

Singh, S.; Saini, K. & Jain, K.L. (1999). Quantitative comparison of lipids in some pollens and their phagostimulatory effects in honey bees. *Journal of Apicultural Research,* Vol.38, pp. 87-92.

Stanley, R.G. & Linskens, H.F. (1974). *Pollen: Biology, biochemistry, management.* Springer Heidelberg, Germany.

Steffan-Dewenter, I.; Münzenberg, U.; Bürger, C.; Thies, C. & Tscharntke, T. (2002). Scale-dependent effects of landscape context on three pollinator guilds. *Ecology,* Vol.83, pp. 1421–1432.

Steffan-Dewenter, I. & Kuhn, A. (2003). Honeybee foraging in differentially structured landscapes. *Proceedings of the Royal Society of London Series B-Biological Sciences,* Vol.270, pp. 569-575.

Taséi, J.N. & Aupinel, P. (2008). Nutritive value of 15 single pollens and pollen mixes tested on larvae produced by bumblebee workers (*Bombus terrestris,* Hymenoptera: Apidae). *Apidologie,* Vol.39, pp. 397-409.

Taylor, P.D.; Fahrig, L.; Henein, K. & Merriam, G. (1993). Connectivity is a vital element of landscape structure. *Oikos,* Vol.68, pp. 571–573.

Thorp, R.W.; Briggs, D.L.Jr.; Estes, R. & Erickson, E.H. (1975). Nectar fluorescence under ultraviolet irradiation. *Science,* Vol.189, pp. 476-478.

Townsend, P.A. & Levey, D.J. (2005). An experimental test of whether habitat corridors affect pollen transfer. *Ecology,* Vol.86, pp. 466-475.

Vaissière, B.E. &Vinson, S.B. (1994). Pollen morphology and its effect on pollen collection by Honey-bees, *Apis mellifera* (Hymenoptera, Apidae), with special reference to upland cotton, *Gossypium-Hirsutum* L. (Malvaceae). *Grana,* Vol.33, pp. 128-138.

van Buskirk, J. & Willi, Y. (2004). Enhancement of farmland biodiversity within set-aside land. *Conservation Biology,* Vol.18, pp. 987-994.

Vaissière, B.E.; Rodet, G.; Cousin, M.; Botella, L. & Torre Grossa, J.P. (1996). Pollination effectiveness of honey bees in a kiwifruit orchard. *Journal of Economic Entomology,* Vol.89, pp. 453-461.

vanEngelsdorp, D.; Underwood, R.; Caron, D. & Hayes, J. Jr. (2007). An estimate of managed colony losses in the winter of 2006-2007: a report commissioned by the Apiary Inspectors of America. *American Bee Journal,* Vol.147, pp. 599-603

vanEngelsdorp, D.; Evans, J.D.; Saegerman, C.; Mullin, C.; Haubruge, E.; Nguyen, B.K.; Frazier, M.; Frazier, J.; Cox-Foster, D.; Chen, Y.; Underwood, R.; Tarpy, D.R. & Pettis, J.S. (2009). Colony Collapse Disorder: A Descriptive Study. *PLoS ONE,* Vol.4, No.8.

vanEngelsdorp, D.; Hayes J. Jr.2; Underwood, R.M.; Caron, D. & Pettis J. (2011). A survey of managed honey bee colony losses in the USA, fall 2009 to winter 2010. *Journal of Apicultural Research,* Vol.50, No.1, pp. 1-10.

van Geert, A.; van Rossum, F. & Triest, L. (2010). Do linear landscape elements in farmland act as biological corridors for pollen dispersal? *Journal of Ecology,* Vol.98, pp. 178-187.

von Frisch, K. (1967). *The dance language and orientation of bees.* Cambridge, UK.

Waddington, K.D. & Kirchner, W.H. (1992). Acoustical and behavioral correlates of profitability of food sources in honey bee round dances. *Ethology,* Vol.92, pp. 1-6.

Wahl, O. & Ulm, K. (1983). Influence of pollen feeding and physiological condition on pesticide sensitivity of the honey bee *Apis mellifera carnica*. *Oecologia*, Vol.59, pp. 106-128.

Waller G.D. (1972). Evaluating responses of honeybees to sugar solutions using an artificial feeder. *Annals of Entomological Society of America*, Vol.65, pp. 857-861.

Wille, H.; Imdorf, A.; Buhlmann, G.; Kilchenmann, V. & Wille, M. (1985). Beziehung zwischen Polleneintrag Brutaufzucht und mittlerer Lebenserwartung der Arbeiterinnen in Bienenvölkern (*Apis mellifica* L.). *Bulletin de la Société Entomologique Suisse*, Vol.58, pp. 205-214.

Winston, M.L. (1987). *The Biology of the Honey Bee*. Harvard University Press, Cambridge, UK.

17

Biodiversity and Vegetable Breeding in the Light of Developments in Intellectual Property Rights

João Silva Dias
Technical University of Lisbon, Instituto Superior de Agronomia, Tapada da Ajuda, Lisboa
Portugal

1. Introduction

Farming and plant breeding have been closely associated since the early days when crops were first domesticated. Plant breeding was built on biodiversity and on the work of 10,000 years of farmers's selection and some generations of breeders. Without understanding the science behind it, early farmers saved the seed from the best portion of their crop each season. Over the years, they selected the traits that they liked the best, transforming and domesticating the vegetable crops they grew. Every vegetable product we see on the market has benefited from plant breeding in one way or another. New varieties were created by breeders by making use of the total genetic information that was present in a gene pool. Access to that genetic variation, the biodiversity, was required to achieve variety improvement.

In the 18th and 19th centuries the Vilmorin-Andrieux family, owner of the first commercial seed company, played an important role in a number of theoretical and technical advances in commercial vegetable breeding, such as producing the first vegetable seed catalog for horticulturists, developing the principles of genealogical breeding programs, improving seed quality through cross-breeding initiatives, and creating disease-resistant and hybrid varieties of vegetables (Gayon & Zallen, 1998). In 1856 Louis Vilmorin published "*Note on the creation of a new race of beetroot and considerations on heredity in plants*" establishing the theoretical groundwork for the modern vegetable breeding industry. The first suggestion to exploit hybrid vigor or heterosis in vegetables was made by Hayes & Jones (1916) for cucumber. Commercial hybridization of vegetable species began in the United States in the middle 1920s with sweet corn, followed by onions in the 1940s. Since that time, private breeding companies have been placing more and more emphasis on the development of vegetable hybrids, and many species of vegetables have been bred as hybrid varieties for the marketplace. Besides heterosis, hybrids also allow breeders to combine the best horticultural traits and multiple disease and stress resistances. Furthermore, if the parents are homozygous, the hybrids will be uniform, an increasingly important trait in commercial vegetable market production. The creation of vegetable hybrid varieties requires homozygous inbred parental lines, which provide a natural protection of plant breeders' rights without legal recourse and ensure a market for seed companies.

In the 1970's breeders' rights protection has been provided through International Union for the Protection of New Varieties of Plants (UPOV), which coordinates an international common legal regime for plant variety protection. Protection was granted for those who develop or discover varieties that are new, distinct, uniform, and stable. Varieties may be either sexually or asexually propagated. Coverage is for 20 years for herbaceous species. Protective ownership was extended by UPOV in 1991 to include essentially derived varieties. At the same time, the farmer's exemption (that permitted farmers to save seed for their own use) was restricted, giving member states the option to allow farmer's to save seed. In addition, after 1998 in Europe, and 2001 in the United States plant breeding companies can take advantages of patent laws to protect not only the variety itself but all of the plant's parts (pollen, seeds), the progeny of the variety, the genes or genetic sequences involved, and the method by which the variety was developed. The seed can be used only for research that does not include development of a commercial product i.e., another variety, unless licensed by the patent older. The patents are considered the ultimate protective device alowing neither a farmer's exemption nor a breeder's exemption (that permitted that the protected variety could be used by others in further breeding to create new varieties).

Research and development (R&D) for improved seed development is expensive. Such product protection has presented a business incentive to corporations to invest in the seed industry, which supported an enormous increase in private R&D leading to strong competition in the marketplace between the major seed companies. The majority of current vegetable varieties sold nowadays are proprietary products developed by private R&D. A significant consequence of this increase in R&D has been a reduction in public breeding programs. As a result, the cost for R&D to develop new varieties is shifting from the publicly supported research programs to the customers of the major seed companies.

One of the main factors that determine success in vegetable production is biodiversity and genetic capacity. No practical breeding program can succeed without large numbers of lines (genotypes) to evaluate, select, recombine and inbreed (fix genetically). This effort must be organized, so valid conclusions can be reached and decisions made. Scientists, breeders, support people and facilities, budgets, and good management are requirements to assure success in the vegetable seed business. Science must be state-of-the-art to maximize success in a competitive business environment. Since the continued need for fundamental breeding research is critical to support development of new technology and expansion of the knowledge base that supports variety development, competition among proprietary varieties results in owner-companies striving to do the best possible research to develop their own products and to compete on genetic and physiological quality of vegetable seed in the marketplace. Reasonable profit margins are necessary to pay back the R&D costs to the owner and to fund future research on developing even better vegetable varieties to stay competitive. There is considerable genetic variation within the various vegetable species, which can be exploited in the development of superior proprietary varieties. The consequences of this dynamic situation will mean relatively short-lived varieties replaced by either the owner of the variety or a competitor seed company. This intense competition means constantly improved and more sophisticated varieties for the vegetable industry. Seed companies are in the business of manipulating genes to improve plant variety performance for a profit. The success of the research is judged by the success of the product in making a reasonable profit. The research must improve economic performance starting with the seed production costs and include the grower-shipper/processor and the end user. If any link in this sequence of events is weak or broken, the new variety will likely fail.

Biotechnology is a new, and potentially powerful, tool that has been added by all the major seed corporations to their vegetable breeding research programs, and is part of ongoing public research for developing transgenic vegetable projects. It can augment and/or accelerate conventional variety development programs through time saved, better products, and more genetic uniformity, or achieve results not possible by conventional breeding.

In 2008 the global vegetable seed market was estimated at US$3.5 billion with the following shares of vegetables: solanaceous (30%), cucurbits (21%), roots and bulbs (16%), brassicas (13%), large seed (13%), leafy and other (7%) (Monsanto, 2009). In the last 8 years global commercial vegetable seed sales had an annual growth rate of 5.8% (Dias, 2010). This expansion of seed market is due to globalization of commercial seed market (more countries using commercial seed), more farmers within these countries purchasing seed and gradually increasing prices of seeds.

There are now over 6.8 billion human beings inhabiting this planet, and it has been projected that world population growth may exceed 70 million annualy over the next 40 years. It is expected to reach approximately 9.5 billion by 2050, when approximately 90% of the global population will reside in Asia, Africa and Latin America countries. With the increase in world population and consumption, and the advent of a high degree of added value through biotechnology, the global market of vegetable seeds is expected to expand in future years.

In recent times, there have been new challenges in the vegetable breeding and patenting domain. The objective of this paper is to discuss these challenges and to highlight the importance of biodiversity, plant breeding and improved vegetable varieties as key to modernize the vegetable production and to alleviate some protective measures that can create obstacles for innovation, and risks for biodiversity and food security.

2. World importance of vegetables

Vegetables make up a major portion of the diet of humans in many parts of the world and are considered essential for well-balanced diets since they supply vitamins, minerals, dietary fiber, and phytochemicals and have been associated with improvement of gastrointestinal health -good vision, and reduced risk of heart disease, stroke, chronic diseases such as diabetes, and some forms of cancer (Keatinge et al., 2010).

"Hidden hunger" or micronutrient deficiency is a pernicious problem that is caused by a lak of vitamins and minerals such as vitamin A, iodine and iron in the human diet and affects the health of between 2 and 3.5 billion people in the developing world (Pfeiffer & McClafferty, 2007). The consequences of micronutrient deficiency are: higher mortality, higher morbidity, lower cognitive ability and work productivity, and impaired growth and reproduction. Vegetables, due to their biodiversity, with increased available iron, zinc, and caroten, and enhanced protein quality could greatly improve the nutrition, health, and quality of life of these people. Diversifying diets with vegetables is a potent weapon in the current global battle against malnutrition. Food security and nutrition is not only about solving the urgency in the short-term; it must also address the long-term issue of poverty alleviation and economic growth. Greater investment in agriculture, including breeding and variety development, more effective development aid, and reforms to trade and domestic policies are all part of the solution.

Vegetables are grown worldwide, on large and small farms, on good and marginal land, and by large commercial growers and small subsistence farmers. According to FAO statistics, the production of vegetables in the world in 2007 was almost 900 million tons (FAO, 2009). Asia produced 74.7% of the world's vegetables (671 million t) on 72.8% of the world's vegetable production area (52.7 million ha). China has always been a large contributor to world vegetable production, and currently produces over 50% of the world's vegetables, which translates to 313 kg per capita. India is the second largest producer of vegetables in the world but at almost a six-fold lower level than China. Worldwide the area of arable land devoted to vegetables is expanding at 2.8% annually, higher than fruits (1.75%), oil crops (1.47%), root crops (0.44%) and pulses (0.39%), and at the expense of cereals (-0.45%) and fiber crops (-1.82%) (FAO, 2009).

The worldwide consumption and importance of vegetables in the diet is difficult to estimate owing to scant production statistics. Even where crop reporting services are an integral part of the agricultural infrastructure, information is available on only a small percentage of the vegetable crops grown. The consumption and caloric contribution of vegetables to the diet varies widely with geographical region, nationality, local customs, and cuisine. China is the largest consumer of vegetables in the world. Vegetables make up about 35% of per capita food consumption in China, a much higher share than the world average (Dias, 2011). Besides India, other southern Asian countries such as Bangladesh, Cambodia, North and South Vietnam, Laos and Philippines are also high producers and great consumer of vegetables. For example, vegetables comprise 40% of the Bangladeshi diet (Dias, 2011). Many vegetables are consumed near where they are produced, especially in Asia. The per capita consumption of vegetables in Asia has increased from 41 kg to 141 kg between 1975 and 2003 (FAO, 2009). Particularly in China the per capita consumption has increased from 43 kg (1975) to 154 kg (2003).

Rapid growth in mean per capita incomes in developed countries during the 1990s enabled consumers to purchase a broader range of relatively expensive vegetable commodities such as off-season produce, relatively new or renewed vegetables, and organic produce. Higher incomes of consumers in developed countries have also raised the demand for other attributes such as better quality vegetables and more variation in the daily menu. In developing countries consumption and domestic vegetable markets are also expanding because of an emerging educated middle class with increasing incomes. As worldwide health awareness increases and household income grows, an increasing global demand for vegetables is expected. At the same time, available arable land and a suitable water supply are lessening, so energies should be directed to enhance vegetable productivity and quality. Increasingly, consumers in developed and developing countries are also concerned about the quality and safety of their food, as well as the social and the environmental conditions where it is produced. It is expected that the assurance of safe vegetable products will become increasingly important. Food safety legislation in the European Union, in the United States, and in Japan is introducing increasingly stricter standards.

Desire for year-round availability and increased diversity, and growing health awareness, have also been important reasons for increased consumption of vegetables in developed countries. For example, the dietary benefit of fresh produce is the major reason for the 25% increase in fresh vegetable consumption in the United States during the 1977–1999 period (Regmi & Gehlar, 2001). On the other hand, factors such as increased participation by women in the labor market have created demand for processed, ready-to-eat convenience vegetable products.

A world vegetable survey showed that 392 vegetable crops are cultivated worldwide, representing 70 families and 225 genera (Kays & Dias, 1995, 1996); non-cultivated species, lower organisms (e.g. fungi), most trees and woody shrubs, and plants grown in or gathered from salt water were excluded. Vegetable crops, of which the leaves or young leafy shoots are consumed, were the most common group of vegetables utilized (53% of the total), followed by vegetable fruits (15%). Below ground crop vegetable organs ranked as follows in frequency of use: roots>tubers>rhizomes>corms>stolons, and together comprised 17% of the total number. Many vegetable crops have more than one part used. Most of the vegetables are marketed fresh with only a small proportion processed. Of these marketed vegetables, only 67 (17%) have attracted great breeding attention by international seed companies, due to their large area of production and substantial consumption, 52 (13%) were considered minor, and 85 (22%) were considered rare.

3. Vegetable marketing

Vegetables typically are perishable products that are of specific high value and that usually are sold through specialized markets. Currently more than 60% of the vegetables produced in the world are sold by vegetable growers to wholesale dealers or huge supermarket chains. Relatively few growers sell their product at retail prices to consumers in farm markets. Globally the horticultural product markets are still dominated by a large number of wholesalers or middlemen, which means not only that the producers have a lower profit but also the consumer often does not have access to lower-priced vegetables. Globally growers receive only 30% or less of the retail price. This situation is a serious problem for growers.

Domestic and international markets for vegetables are changing rapidly all over the world, partially fueled by the spread of supermarkets (Dias, 2011). Consumers increasingly purchase their vegetables and other foods in large convenience stores such as supermarkets and hypermarkets. The proliferation of supermarkets in developed and developing countries creates both challenges and opportunities for vegetable producers (Shepherd, 2005; Dias, 2011). Indeed, supermarkets may contribute to a higher demand for horticultural products and increase expectations for quality, safety, and presentation while simultaneously excluding small growers from participating in procurement and contracts. The growing importance of supermarket outlets has implications of its own regarding methods of procurement and quality standards. Supermarkets in the cities bring quality to the shelves. Vegetables are well packed and presented, providing scope for premium quality as well as novelty items. The difficulties that growers can experience is reflected in fairly rapid declines in the numbers of growers involved, as companies tend to delist suppliers who do not meet expectations in terms of volume, quality, and timely delivery. The standards of quality, safety, and presentation make it difficult for the small producers to compete. The market is getting more refined in terms of quality and yield expectations, and there is a clear demand for excellent hybrid vegetable varieties. Success for vegetable growers will depend on their ability to access diverse markets and respond promptly to changes in market conditions. Growers grow vegetables for immediate marketing, and their produce is subject to competition to decide the prices. Hence they cannot compromise the quality of the seeds they use for fear of rejection of their produce. They buy the best seeds, mainly hybrids, and frequently try new products to remain successful. The seed companies in turn get instant response and success if they develop new promising hybrids. Price of seeds is a more critical factor in marginal vegetable areas, where capital and spending input regimes are low, but is less important where high yields can be obtained and the grower's produce can be sold profitably. Providing that the benefits of the hybrid seed are understood, the price of seeds is less important than other factors, such as availability of capital, confidence in the produce market, and ability to buy other inputs, such as fertilizers and pesticides.

The increase in total volume of vegetables traded worldwide has been dramatic. Still, compared to overall exports of agricultural products, the importance of vegetable exports remains minor, comprising less than 10%of the total value. However, in recent years, the share of vegetables traded worldwide has been rising and is projected to continue to rise faster than other agricultural products. During the 1990-2010 period, the value of fresh and processed vegetables imported by developed countries surpassed all other categories. Growth in these commodities is also linked to changing trends in consumer preference and food retailing. In this situation, many vegetable growers are eager to produce value-added horticultural crops as compared to field crops and to obtain higher yields of high-quality products. International supermarket chains and large processors are becoming the main buyers of exported fresh vegetables, and small-scale growers worldwide need to be trained and organized to meet the challenge of supplying these international players. The major constraints against the participation of small-scale growers in international vegetable exports are the increasing attention that food quality and safety are receiving in food trade and an expansion in the number of nontariff measures that developed countries apply to horticultural products (Henson & Loader, 2001; Dinham, 2003; Henson et al., 2005).

Vegetables belong to the class of food items most frequently affected by sanitary and phytosanitary measures. Sanitary issues refer to ensuring a safe food supply for consumers, while phytosanitary issues concern the protection of domestic crops from imported pests and diseases. The Sanitary and Phytosanitary Agreement (SPS) of the World Trade Organization specifies that countries can pursue their own levels of food safety standards. However, SPS issues sometimes are used as a protectionist tool against imports since multilateral trade agreements have reduced the ability to protect domestic production with tariffs and quotas (Henson & Loader, 2001). SPS regulations may be the most important barrier to international trade in fresh vegetables. Thus exporters from less developed countries must be provided training opportunities and information access on how to produce and supply safe products to developed countries. Traceability, phytosanitary, infrastructure, and productivity issues will continue to be a barrier for participation in the vegetable trade for most of the developing world. Application of agricultural chemicals is often poorly regulated, and industrial pollutants are common hazards in the soil, water, and air of developing countries. In the future, the inability of these countries to meet increasingly strict phytosanitary and traceability requirements for food products will constrict exports to developed countries. Small-scale growers and processors in developing countries thus will have to learn to supply safe products with traceability labels, if their participation in global trade is to continue and expand. Technologies for safe and environmentally friendly vegetable production as well as capacity building should therefore gain particular attention for training to enable small-scale growers to participate in vegetable production for international markets.

Horticultural production, particularly in Mediterranean, subtropical, and hot-wet tropical environments, is severely constrained by postharvest losses, which have been estimated as 15%, depending on the crop and season (Kader, 2003). Vegetables often are highly perishable, restricting the ability of producers to store them to cope with price fluctuations. Reducing postharvest losses would make diversification into vegetable production less risky and more attractive. Postharvest-related quality losses also reduce opportunities for export and export revenues. Improved vegetable varieties subject to fewer postharvest losses can help improve this situation. Competitive participation in international markets requires relatively sophisticated marketing, information, and transportation networks as well as improved varieties, quality control, product standardization, and, for some future markets, traceability.

4. Biodiversity, vegetable migration and vegetable breeding

4.1 Introduction

Breeders play a key role in determining what we eat, since the plant varieties (=cultivars) they develop begin the dietary food chain. Vegetable breeding is the development of new vegetable varieties with new proprieties. Innovation in vegetable breeding is dependent on biodiversity and access to genetic resources, on specific knowledge, on the development and application of new technologies, and capital to utilise those factors. Access to genetic biodiversity as well as to technology is essential for the development of new vegetable varieties. The impact of vegetable breeding on vegetable production is dependent upon the complex relationships involving the growers, the varieties available to them, and the developers of those varieties. Vegetable growers consist of commercial producers with varying size land holdings ranging from moderately small farms to very large ones, and poor growers many of them subsistence farmers with small farms often on marginal lands. The subsistence farmers are usually also poor. Several types of varieties are available. The least sophisticated in terms of the method of development are landraces, also known as local varieties. Modern varieties consist of those developed by crossing and selection alone, those developed by crossing and selection but with specific important improvements often obtained from crosses with wild species or by transgenic methods, and F1 hybrids between desirable inbred lines. The developers of landraces are usually the farmers themselves, and are obtained by repeated simple selection procedures generation after generation. Improved varieties and hybrids are created either by public sector breeders or seed companies.

Farmers in some cases can plant and save their own vegetable seeds, but there are real problems in this system in commercial production, where typically many different species may be grown. In farmer-grown seeds, viability may be low, due to poor seed storage environment, pollination is often uncontrolled, genetic improvement is lacking and seed born diseases including virus may be a problem. Thus, in modern vegetable production the seed business is most efficiently conducted as a distinct industry dominated by international private seed companies.

The consequences of these relationships may be quite profound for the farmers at each level, the seed producers, the availability of food worldwide, and the future of crop biodiversity and sustainability. Therefore, it is worthwhile to examine international policy with regard to genetic resources genetic resources, vegetable breeding and its connections with commercial breeding industry to assess our future expectations.

4.2 Biodiversity

As stated biodiversity is the basis for plant breeding. Selection is impossible without diversity and new varieties for farmers and growers cannot be developed without it. This makes access to this variation essential for breeders.

Biodiversity of crops is characterised by a number of historical bottlenecks, i.e. critical moments diminishing this diversity. The first bottleneck was the result of domestication of crops in which only a subset of the diversity of the wild species remained after repeated selection for desired traits, e.g., non-shatering, plump seeds, etc. This was followed by a dispersal bottleneck which arises when only a subset of the crop was exported to another region, in which diversity was further reduced through adaptive selection to the new conditions (Zeder et al., 2006). This last case led for example to the famine in Ireland caused by potato blight exerting its disastrous effect as result of the narrow genetic base of the cultivated potato in comparison with those in the Andes areas of origin. The last bottleneck is the result of the modern scientific plant breeding that replaced genetically diverse landraces by uniform varieties, mainly F1 hybrids in vegetables. Besides recently "trait breeding", i.e. introducing a new trait in an existing variety through genetic modification or repeated backcrossing in combination with the use of markers, genetic biodiversity could also decrease. Where in the past conventional breeding introduced a much wider load when farmers' landraces or wild relatives were used to introduce such traits, a much more precise introduction of the desired trait alone is now possible. This may lead to a narrower genetic base of crops. Other modern bottlenecks such as "molecular bottleneck" and "cooporation companies bottleneck" are a risk and are discussed later.
The modern selection of uniform varieties from a wide biodiversity and the resulting fear of global genetic erosion led to the establishmemt of international, national and corporate genebanks. International policy with regard to genetic resources started in 1983 with the adoption of the "International Undertaking of Plant Genetic Resources for Food and Agriculture" of the UN Food and Agriculture Organization (FAO). This agreement primarily treated genetic resources as a "Heritage of Mankind" that should be freely available for all. The Convention on Biological Diversity (CBD) which came into force in 1993 put genetic resources under the sovereignty of the nations where such resources have obtained their distinctive character. Countries can, since then, make access to their genetic resources subject to mutually agreed terms. Countries differ in their implementation of the CBD and in their policy regarding access to genetic material. In some countries it is very difficult to gain access, e.g. because consent of the farmer, the land owner, the local community leader, local administrators and national authorities is required (e.g. Philippines). Such problems led to the development by FAO in 2001 of the "International Treaty for Plant Genetic Resources for Food and Agriculture" (see www.planttreaty.org). An important novelty of this International Treaty is the "Multilateral System" that should facilitate both the collecting and the sharing of benefits originated from the use of the genetic resources. Although most of the vegetables are not included in this International Treaty that applies to almost all crops and forages important for global food security. Materials of crops to which the Multilateral System apply that are under the control of the signatory governments are available under a single Standard Material Transfer Agreement (SMTA). The terms of the SMTA include a mandatory payment of 1.1% of the value of the seed sales in the case that a crop variety is produced using genetic resources from the Multilateral System, and if that variety is not freely available for further research and development. Alternatively, breeders can contribute with a flat rate of 0.5% of their gross sales for use of all genetic resources of the crop.

4.3 Vegetable migration

The introduction and trade of crops and seeds from one region to another has been continuous throughout history due to the migration of people, conquests, discoveries, and development of commercial trade routes. The "Silk Route," for example, was responsible for many plant exchanges and trades between the West and China. As people immigrated to new countries, they carried not only languages, religions, and traditional customs but foods and seeds. Vegetables, in particular, are attractive candidates for introduction into a new environment, as most vegetables tend to be fairly short-season crops and, as a consequence, lend themselves for cultivation in many diverse areas. Thus, the tropical tomato has been transformed to a temperate annual. Superior vegetable crops and varieties often are assimilated into the cuisine of the indigenous populations.

Worldwide there has been, and continues to be, substantial emigration of peoples from a diverse range of countries. In western Europe and in the United States, immigration has had a pronounced effect on the vegetables consumed. For example, in the 1960s, southern Europeans immigrated to work in the northwestern European industrial zone, bringing with them their distinct consumer behavior patterns and vegetable preferences. They were responsible for the introduction of broccoli, eggplant, pepper, and fennel. Portuguese immigrants carried tree kales and tronchuda cabbages to Brazil in the 16th century and to France and Germany in the 1960s. The introduction and popularity of a small number of new vegetable crops in some countries have resulted in a steady increase in their utilization. For example, Chinese cabbage and pak-choy were novelties in much of the United States in the 1970s but are now widely consumed due to the immigration of Asiatic people. Likewise, since the 1970s, sweet corn consumption has increased in several European countries. Nowadays many cities in the United States are excellent examples of increasing ethnic diversity and its impact on the vegetables available. Ethnic markets sprang up followed by shopping malls catering to Chinese, Korean, Vietnamese, or Hispanic populations. Vegetables such as Thai eggplant, tindora, parval, cactus leaves, and others are now readily found in ethnic American markets. New vegetables also have moved into traditional markets, as superior crops are accepted by the general public, greatly enriching the biodiversity of vegetables grown and consumed.

As a result of the increased frequency of world travel, a substantially greater variety of vegetable crops is available worldwide in many local markets. However, the number of new crops that become mainstream vegetables remains relatively small. Examples are rocket (*Eruca vesicaria* syn. *E. sativa*), lamb's lettuce (*Valeriana olitoria*), physalis (*Physalis peruviana, P. pruinosa, P. ixocarpa*), and pepino (*Solanum muricatum*). The number of crops that can be utilized within a local area will depend on the ethnic diversity of the location, affluence of the population, production and marketing constraints, and other factors.

Consumers also want more vegetable diversification and a continuous supply. Vegetables are purchased based partly on eye appeal, which means that the development of desire to consume increases market demand. Diversification also tends to increase consumption. Product differentiation, including new or renewed product introductions, is still a key strategy for expanding sales in vegetable markets. For example, the fresh tomato category has been differentiated to more than 10 offerings (beefsteak tomato, Roma-type tomato, vine-ripe tomato, cocktail tomatoes on vine, tiny-plum tomatoes, mini-plum tomatoes, red cherry tomatoes on vine, attractive yellow and orange cherry tomatoes, mini San Marzano-type tomatoes, teardrop or pear-shaped tomatoes, super or premium taste tomatoes). The introduction of specialty fresh baby leaf vegetable salads and fresh-cut products has opened new opportunities for domestic producers. The increased production of baby leaf vegetables in the world is intended to increase desire among elite consumers and is also an excellent way of supplying micronutrients. For example, baby leaf curly kale, as well as other dark green leafy vegetables, is a rich source of lutein and zeaxanthin carotenoids and, when cooked, contains seven times as much vitamin A as cooked broccoli and provides more calcium per 100 g than milk, yogurt, cooked broccoli, or cooked spinach. Until recently, kale was not a particularly popular vegetable in Europe (except Portugal), but as a baby leaf vegetable with 5 leaves, it is now accepted by many European consumers. An example of diversification in peas is the pea shoot. It is a nutritious leaf vegetable with high levels of vitamin C, folic acid, and vitamin A. To exploit such opportunities, it is important to continue research in biodiversity and to disseminate information regarding the benefits of vegetables, develop new improved vegetable varieties and processed products, evaluate the economic opportunities and the market scope of these new products, and identify marketing trends and alternatives.

Increasingly more wealthy and healthy people will demand greater vegetable dietary diversity in a global bio-based economy, which means that biodiversity will be crucial for the future. Besides biodiversity remains the main raw material for vegetable agricultural systems to cope with climate change because it can provide traits for plant breeders and farmers to select resilient climate-ready crop germplasm and release new varieties. Thus collecting samples of endangered vegetables to be preserved in genebanks is the first step, but also protecting the agricultural systems where those vegetables are produced is also important to ensure the *in situ* evolutionary processes.

4.4 Objectives of vegetable breeding

The genetic improvement of vegetables through breeding has to address and satisfy the needs of both consumers and growers. The general objectives for farmers are good yield, disease and pest resistance, uniformity and abiotic stress resistance. Objectives for consumers are quality, appearance, shelf life, taste, and nutritional value. Vegetable product innovation is necessary for maintaining the interest of today's consumer. Quality in vegetable crops, in contrast to field crops, is often more important than yield. For farmers to survive, varieties must be accepted by the market. Thus, color, appearance, taste, shape, are usually more important than productivity. For example, tomatoes to be used either fresh or in processing must have distinct quality characteristics. Fresh tomatoes must have acceptable flavor, color, texture, and other taste parameters to satisfy consumer demands and handling requirements. Processing tomatoes, on the other hand, must have intrinsic rheological characteristics that make them suitable for various processing applications, such as juice, ketchup, or sauce production. Traditional breeding requires the selection of a tomato genotype or a related wild species that has a desirable trait, such as early ripening or disease resistance, and crossing it with another tomato variety that has a good genetic background. The desired result is an earlier ripening tomato that makes it to the market sooner, or varieties that resist pathogen attack. In this way, several thousands of tomato varieties have been developed over the years. The final goal of vegetable breeding programs is then to release new varieties having elite combinations of many desirable horticultural characteristics. Plant breeding for improved taste, convenience, and consumer appeal has already contributed to increased per capita vegetable consumption with the development of products such as baby carrots, yellow and orange peppers, cherry and pear tomatoes, non-bitter cucumbers, mild tasting eggplants, seedless watermelons, and lettuces with different colors, textures and flavors for baby leaf and precut salads.

Other important objectives of vegetable breeding are disease and pest resistances. Since the early days of the 20th century, traditional breeding for disease resistance in vegetables has been a major method for controlling plant diseases. Varieties that are resistant or tolerant to one or a few specific pathogens are already available for many vegetable crops. Resistant hybrids with multiple resistances to several pathogens exist and are currently used in vegetable production. For example, in tomato, the genetic control of pathogens is a very useful practice and most resistances are monogenic and dominant. So far, tomato breeding has resulted in varieties with resistance to at least 15 pathogens, although with varying stability and level of expression (Grube et al., 2000). Tomato varieties with some resistance to fungi or oomycetes (*Alternaria alternata* f. sp. *lycopersici*, *Cladosporium fulvum*, *Fusarium oxysporum* f. sp. *lycopersici*, *Fusarium oxysporum* f. sp. *radicis-lycopersici*, *Phytophthora infestans*, *Pyrenochaeta lycopersici*, *Verticillium dahliae*), bacteria (*Corynebacterium michiganense*, *Pseudomonas solanacearum*, *Pseudomonas syringae* pv. *tomato*), virus (beet curly top hybrigeminivirus, tomato mottle bigeminivirus, tomato spotted wilt tospovirus, and several variants of the tomato yellow leaf curl bigeminivirus), and nematodes (*Meloidogyne* spp.) are available (Lacerrot, 1996). Many open-pollinated varieties of tomato presently cultivated possess genetic resistance to three or four pathogens. With the increasing use of F1 hybrids it is possible to use varieties combining from four up to six resistances (Grube et al., 2000).

Pest resistance is essential in vegetable production but is marginal in vegetable breeding research. There are few vegetable varieties resistant to insects. Resistance may be unstable due to genetic variants of the insect that are able to overcome that source of resistance. Depending on the complexity of the interaction between the pest and the vegetable plant, plant resistance may break down rapidly or be long-lived. Insects, including aphids, whiteflies, thrips and leafhoppers, are also very important in vegetables because they vector many viruses. Viruses can substantially reduce production and quality and are becoming increasingly problematic worldwide due to the absence of virus resistant germplasm for many important vegetable crops. Aphid vectored viruses are particularly problematic because many are transmitted in a non-circulative and non-persistent manner (Zitter et al., 1996). This means that a very short time, i.e. a few seconds or minutes, is sufficient for aphids to acquire virus particles when probing on infected plants. A similarly short time period is enough for aphids to release virus particles when probing on healthy plants. The primary injury caused by aphid-vectored viruses arises not from direct feeding damage by the aphids, but from their ability to allow the virus to enter the plant and initiate the disease. The economic return of investment in breeding for disease and pest resistance may be low because it is dispersed among many different vegetable crop types. Also resistant varieties compete directly with non-resistant ones that may still be used by growers with minimum problems. Therefore vegetable disease resistance is most important when the disease is a limiting factor in production, and is especially important for many virus diseases. The high interest in, and the increasing present demand for breeding for disease and pest resistance is related to a generalized interest in releasing "environmentally friendly" vegetable varieties requiring sparse or no use of pesticides.

Breeding for postharvest traits, mainly transport quality, shelf life and cosmetic problems, is of increasing importance in vegetables. For example, in tomato, textural properties of fruits are important contributors to the overall quality for the fresh market and to the properties of products processed from tomatoes. Because cell wall disassembly in ripening fruit contributes to fruit texture, modification of cell wall proteins and enzymatic activity during ripening can impact cell wall polysaccharide metabolism and influence texture. Lettuce and other leafy vegetables used for salads deteriorate rapidly following harvest, requiring a considerable investment of effort to maintain quality and shelf life of cut material.

Harvesting increases respiration, stimulating deterioration, with increase in the synthesis of phenylalanine ammonia lyase and phenolic compounds, such as chlorogenic acid, which cause tissue browning. Consequently, delaying leaf senescence is an important target for breeding of leafy vegetables. Also in lettuce, breeding efforts have targeted tipburn, marginal browning, and rib discoloration, which detract from overall appearance. Vegetable products with good transport quality, better shelf life and good appearance will be preferred by traders and also by consumers.

Since vegetables are rich in vitamins, minerals and other micronutrients, and therefore vital for health, breeding objectives should include improving their nutritional value. Historically vegetable breeders have applied selection pressure to traits related to agronomic performance, particularly yield and quality, because these are the traits important to the producer. Rarely have growers been paid for nutritional factors, so there have not been economic incentives to pay much attention to these traits. However, consumers are becoming more aware of these traits.

Vegetable breeding for nutritional quality was not mentioned as a primary goal in plant breeding text books through the mid-20th century. However vegetable breeding efforts targeting improved micronutrient content and composition had begun in the 1940s and 1950s with research describing the inheritance and development of tomato breeding stocks and lines high in provitamin A carotenoids and vitamin C (Lincoln et al., 1943; Lincoln & Porter, 1950; Tomes et al., 1953). Lincoln et al. (1943) noted a fourfold variation in vitamin C among commercial varieties and up to 1194 ppm in red-fruited tomato interspecific crosses with *Solanum pimpinellifolium*. Similar research leading to the development of darker orange, and consequently high provitamin A, carrots began in the 1970s (Gabelman & Peters, 1979). Yellow core color occurs only in older open-pollinated carrot varieties since uniform orange storage root color has been a trait of interest in carrot for over a century. Similar studies were made in squash where rapid gains in carotenoid content have been made with phenotypic selection for orange color versus green and cream (Sudhakar et al., 2002). Genetic improvement to increase levels of specific micronutrients has been pursued in several other vegetables such as melon, spinach, sweet potato, potato, lettuce, broccoli, pepper, watermelon, collard, kale, peas, and bean. This field of study is relatively new, and also complex because of mineral interactions with each other, and numerous other compounds in the soil and in the plant. There is usually a large environmental effect, when the component is present in tiny amounts, such as for some micronutrients and phytochemicals. Success in vegetable breeding for higher vitamin and mineral content must consider not only substance concentration but also organic components in plants that can be abundant and either reduce or increase bioavailability. With these numerous considerations, breeding vegetable plants for improved nutritional value is a complicated goal that needs expertise in many disciplines such as plant breeding, nutrition, and soil science. When a vegetable compound (micronutrient or phytochemical) is found to be important for human health, and growers, vegetable markets and seed companies can capitalize on the value of the compound, there may be an opportunity for vegetable breeders to increase the amount of this compound. Breeders can be successful in reaching this goal, if the vegetable crop contains genetic variability for the compound, if selection is effective without detrimental pleiotropic effects, and if there is an easy method to measure the compound.

Enhanced nutritional content would add value for poor, malnourished populations. Breeding for provitamin A carotenoids, iron, and zinc is of keen interest as a strategy to alleviate nutrient deficiencies in developing countries (Graham et al., 2007; Hotz &

McClafferty, 2007; Pfeiffer & McClafferty, 2007). An example is the "golden tomato" which contains three to six times more provitamin A carotenoids than standard tomatoes. Developed at World Vegetable Center (AVRDC) with conventional breeding techniques after evaluation of tomato biodiversity, these improved nutritionally-rich tomato lines could help prevent many children of developing countries from going blind, since vegetarians and populations with limited access to animal products depend on provitamin A carotenoids for vitamin A. One "golden tomato" can provide a person's full daily vitamin A requirements. Tomato fruit and its processed products are the principal dietary sources of carotenoids such as lycopene. Lycopene is a potent antioxidant with the potential to prevent epithelial cancers and improve general human health. Therefore, there is considerable interest in elevating the levels of carotenoids in tomato fruit and thereby improve the nutritional quality of the crop. The B gene from *Solanum hirsutum* shifts tomato carotenoid accumulation from lycopene almost entirely to β-carotene and results in orange fruit color (Premachandra, 1986). This consequently dramatically increases the provitamin A carotenoid content. β-carotene content of commercial varieties, as mentioned, is of interest and several high β-carotene orange cherry tomato breeding lines have been bred (Stommel et al., 2005). Rainbow carrots, which are super sweet and crunchy, have multi-pigmented roots that naturally contain several antioxidants, such as lycopene, lutein, and anthocyanin. Similarly, yellow sweet potatoes are much more nutritious than white ones since they are high in provitamin A carotenoids. Unfortunately, the popularity of white fleshed sweet potato varieties in many tropical regions may complicate the acceptance of more nutritious orange ones. Recent studies across a range of Andean potato varieties show wide variation in calcium, iron and zinc content (Andre et al., 2007) as well as anthocyanins (Brown et al., 2005; Reyes et al., 2005) due to the existence of red-, blue- and purple-fleshed potatoes.

A vegetable, in order to have impact for its nutrient content, must be appealing to consumers. Sensory appeal, including color, is an attribute important to consumers when selecting many vegetables. Enhanced pigmentation of carrot, potato, tomato, and pepper, for example, is considered a quality factor. In peppers carotenoid content of green, yellow, orange and particularly red peppers can be relatively high. Selection for high pigment is an important goal because these carotenoids are important for visual appeal in many markets but until now didn't have any specific nutritional impact (Biacs et al., 1993; Hanson et al., 2004). In the past yellow or orange tomatoes could not compete with red tomatoes because they were unfamiliar to consumers, but now they are commercialized, and are challenging the market. This fact and recent development of orange colored cauliflower and orange flesh cucumbers reflect a new direction in vegetable breeding: nutritional quality. The white versions of these two vegetables lack carotenoids present in high enough concentration to alter their appearance. As nutritional quality becomes a more common breeding goal, and the novelty of unusual colors brings added value to seed companies and vegetable growers, unusual colors will quite certainly become more available and perhaps more widely consumed. But consumer requirements for quality: appearance, shelf life, and taste, must be met. Breeding to increase consumer appeal by improving convenience and the quality factors of a moderately nutritious crop often can be a more effective approach to increase intake of shortfall nutrients (Simon et al., 2009). Nutritional quality identifiable by the consumer and available at a moderate price might induce increased consumption and thus confer an important marketing incentive for breeding activity.

4.5 The commercial breeding industry

In the case of most vegetable crops, biodiversity and genetics are delivered in a marvelous package known as the seed. The special techniques of seed production, such as seed treatment for the control of planting diseases and viruses, and the combination of breeding improvement program such as development of hybrids and the incorporation of biotechnology, cannot be efficiently carried by individual growers. By these reasons in modern vegetable production the seed business is usually conducted as a distinct industry. High tech seed industry is a key part of modern horticulture that combines, seed production, genetic improvement, seed production, storage, and distribution.

Private breeding companies are placing more and more emphasis on the development of hybrids to exploit heterosis, and to combine multiple disease and stress resistance, but also for economic purposes to ensure growers must purchase seed for each planting. Control of the parents prevents other seed companies from reproducing the hybrid.

Farmers pay all the breeding work and seed marketing costs when purchasing improved or hybrid vegetable seed. International seed companies are mainly interested in the breeding and production of vegetable seeds with a high commercial value. Traditional vegetable landraces have largely been neglected by seed companies, policymakers, and researchers. But while their production often takes place under low-input conditions, they contribute substantially to household food and livelihood security, particularly for small resource-poor farmers (Weinberger & Msuya, 2004). For example, in Africa traditional landraces constitute an important source of micronutrients, contributing between 30% and 50% of iron and vitamin A consumed, respectively, in poor households (Gockowski et al., 2003; Weinberger & Msuya, 2004).

Breeding companies strived more and more to bring hybrid seeds onto the market. Worldwide the share of hybrid seed is increasing at a fast pace of 8 to 10% annually in most of the vegetables. More than two-thirds of the 5000 non-hybrid vegetable varieties available in 1984 seed catalogs from North America were dropped during last 25 years. A hybrid is produced by crossing two carefully selected parent lines to produce seed combining the best characteristics of each. Hybrids often exhibit higher productivity and vigor than open pollinated varieties due to heterosis. The superior characters of hybrid plants, unlike that of open pollinated varieties, cannot be maintained by farmers by saving their seeds for growing in the next growing season since the uniformity, vigor and overall performance of the hybrid is lost during seed multiplication. Therefore farmers are obliged to buy seeds from the seed companies every growing season if they want to compete in the marketplace. Many growers have been skceptical about the cost of hybrid seeds but found that they gave excellent returns. The main reasons which influenced the growers' decision to adopt improved or hybrid vegetable seeds are the ability of the product to meet the market demand, for high productivity, uniformity, resistance or tolerance to diseases or pests, better response to costly inputs, high quality, and storability. Hybrid vegetable varieties are used increasingly in large-scale intensive production because they provide increased marketable product for commercial growers and thus add to commercial incentives for the seed companies.

Open-pollinated varieties are derived by repeated selection of superior plants from within the same line and are genetically uniform for appearance traits. In naturally self-pollinated plants such as tomato, lettuce, and legumes, inbreeding leads to homozygosity. It has generally been found that self-pollinating vegetables such as tomato, after they have become stabilized, do not substantially change their genetic constitution. Thus it is easy to maintain the purity of self-pollinated vegetable crops by removing occasional variants. In the case of cross-pollinated plants such as cucurbits, cole crops, sweet corn, beet, and spinach, uniformity can be achieved by rigorous selection and confining natural crossing within the selected population. Isolation distances of at least 1,500 m between different varieties of the same species are recommended for all cross-pollinated crops that are to be harvested for seed.

Hybrid vegetable technology has made significant impact on most vegetable crops in developed countries, but a major limitation to vegetable production in many developing countries is the unavailability of high quality seeds. Hybrid seed production is a high level technology and cost intensive venture. Only well organized seed companies with good scientific manpower and well-equipped research facilities can afford hybrid seed production. The public sector in developing countries frequently does not have sufficient capacity to supply adequate quantities of good quality vegetable seed to poor growers and at present, there are few private sector seed companies adapting varieties to local environments, especially in the poorer countries (Rohrbach et al., 2003). Farmers themselves often produce seeds of locally preferred or traditional landraces, as the individual markets are too small and private companies have little interest in producing open pollinated varieties (Weinberger & Msuya, 2004). Without proper seed production, processing technology, quality assurance, and management supervision, locally produced seeds are often contaminated by seed transmitted viruses and other disease organisms, and are genetically diverse. Lack of proper storage facilities and an effective monitoring mechanism often leads to low or uncertain seed viability and vigor. Moreover, low capital resources and poor market information discourage the development of seed-related agribusinesses. Seed quality and treatment are keys to product quality, and there is a need for upgrading quality control laboratories to meet international standards.

By those reasons in modern vegetable production the seed business is usually conducted as a distinct industry. The global seed trade is now dominated by international corporations whose vast economic power has effectively marginalized the roles of public sector plant breeding and local, small scale seed companies. Thirty years ago there were thousands of seed companies in the world, most of which were small and family owned. Today, the top six global seed companies control almost 50% of the commercial seed trade. Some of these companies belong to worldwide corporations that also have other business areas besides seed like pesticides and biotechnology. A large number of acquisitions of small and big seed companies happened between 1996 and 2008 and these companies have increased their turnover both in conventional and in organic vegetable production.

The analysis of the companies involved in vegetable breeding reveals mainly five business models: i) vegetable breeding companies traditionally integrating variety development, production and marketing of seed; ii) vegetable breeding companies that breed and produce seed in their home country but licence their varieties to companies in other countries; iii) vegetable breeding companies that have meanwhile developed their own capacity in applied biotechnology; iv) vegetable breeding companies specialized in plant breeding biotechnology only, without being active in practical breeding, variety development, and seed production; and v) globally operating companies that have a strategic research capacity between fundamental and applied. Some of these companies belong to worldwide corporations that are also involved with pesticides and biotechnology. In the traditional vegetable breeding companies (i and ii) their income is primarily the selling of seeds. Although even these traditional companies are now also increasingly using biotechnology in their breeding programmes. In the companies that have still developed their own capacities in applied biotechnology (iii) their income remains by selling seed and not by generating income via licences on patents. Although this group of companies also comprises some companies originated from the agrochemical sector and that later became breeding companies via acquisitions and mergers. These last companies are combining two business: selling seeds and acquiring market positions via licences on their patents. Biotechnology companies (iv) are focusing on income from contract research for seed companies and on licence income from their biotechnological findings based on patent rights. This in particular concerns patents on molecular breeding techniques, marker platforms (e.g. Keygene) and on properties of the plants ("traits"), marketing of traits (e.g. BASF-CropLife). The value of such patents will in the end have to be paid at the level of the market for the seeds and planting materials by the end users (farmers and growers). Globally operating companies (v) combine a large biotechnological capacity with the production and marketing of seed while at the same time licensing technologies to other breeding companies. This category comprises most multinationals in the seed sector that are also active in agrochemicals and/or pharmacy, but also larger traditional breeding companies with a significant biotechnology capacity (e.g. Rijk Zwaan). For these companies the income from seed sales is the most important but some also generate income from licences.

A greater desire for year-round availability of vegetables has had a significant impact on seed companies, requiring full year-round production and consequently a global presence. Unlike for instance agrochemicals "where one size fits all", seed varieties need to be adapted and differentiated to suit the agronomic needs of the respective region where the vegetable is grown. Active international trade and overseas vegetable seed production by contract is common in many countries. Each multinational company vies to provide better vegetable seeds to compete with domestic seed producers. In China, whose seed market is estimated to be valued at more than US$1.4 billion, the increased recognition of new and high-yielding hybrid varieties has encouraged the local development of a large number of vegetable seed producers and distributors. Four types of vegetable seed producers were established: public seed companies, research institutes, foreign seed companies, and local seed companies. Private seed companies have been expanding rapidly in recent years and there are now thousands of small firms. Some companies have started to breed their own varieties and establish marketing networks. They play a strong role in the Chinese vegetable seed industry. About 60 foreign seed companies have opened branch companies or stations in China. Most of them not only sell their vegetable seeds but also have established breeding stations. In other southeastern Asian countries such as India, Indonesia, Vietnam and Malaysia, the percentage of hybrid vegetable varieties is lower than in China, and so a large expansion of seed companies has not yet occurred. While there has been rapid growth in the seed markets of developing countries due to a shift away from farm saved seed, the seed markets in developed countries, particularly those of Europe and Japan are stagnant. In Europe and the United States, the seed industry has been concentrated and is largely in the hands of large corporations and many small firms are closing.

Commercial vegetable breeding has brought a paradigm shift in the agricultural cropping system by developing superior and productive vegetable crop varieties in a short span of time. The vegetables attracting the most breeding attention vary considerably between small and huge seed companies/corporations. Small seed companies have a tendency to specialize in a few vegetable crops. In large international companies the breeding activity is more diverse, but is concentrated on the more economically important crops. In these companies, the application of modern biotechnologies such as molecular marker technology has become an integral component of many commercial vegetable breeding programs (Dias, 1989). The access to modern tools of plant breeding such as genomic information to develop markers for important traits and genetic resources are the key drivers of successful modern vegetable plant breeding. In an era of continuous change, vegetable plant breeding is contributing towards fulfilling requirements of producers and consumers as well as in assessing climate or growing conditions, through continuous innovations to develop new and better varieties. Vegetable breeding strategy and targets are dependent on market trends. Successful breeders anticipate changes in the market by developing new varieties that are ready to be released to the growers when their demand increases. Therefore it will be interesting to see how breeding companies react to changes in vegetable consumption and to evaluate the potential influence that the vegetable market and growing systems may have on breeding targets and priorities.

The commercial vegetable breeding sector produces a continuous flow of innovative new varieties for a number of vegetables. Breeding focuses on the following most important properties: resistances against pests and diseases, increasing yields, quality improvement (such as shelf life, taste), and increasing production efficiency. Companies that are introducing a new variety with a new trait usually have a lead of about four years, after which the competitors can introduce their own new varieties with the same trait. In such cases they make use of the "breeder´s exemption". This is how this "open innovation" system leads to a wide availability of such an innovation. Investiments in R&D by the top companies in this sector are very high, between 15 to 25% of their turnover, and this level keeps track with the annual increase in turnover. Most of the top companies show an annual growth of 5-7% with net profits exceeding 10%. Such growth can be realised in two different ways: by mergers and acquisitions or by autonomous growth. Companies with autonomous growth have to spend more on innovative R&D since they have to creat new varieties and new technology themselves.

Plant breeding is a long-term and therefore costly activity. Until the 1980s breeding was merely an empirical activity where breeders, on the basis of much knowledge and experience about traits of the reproductive material made crosses and select the most suitable plants. This process was strongly affected by growing season, lenght of the generation cycle, growing conditions, and available space. This meant that the developement of a new variety (e.g. a new hybrid) took 10-24 years, depending on the species. This developement period decreased to 4-11 years over the last 30 years by application of a wide range of biotechnological methods, such as *in vitro* tissue culture, *in vitro* haploidization, mutation breeding, DNA technologies, molecular breeding, etc. The application of modern technology has made plant breeding less time and space-depend and breeding processes have become much more efficient. This resulted in a reduction of the developement period of a new variety by a factor 2.5. Even though the R&D costs increase strongly (by about 10% annually) the return of such investiments in ensured by the faster production of new varieties.

A breeding company tries to maintain, or preferably expand, its market share by developing good varieties. A company can only continue the developement of new varieties if a good "return on investment" is ensured. The long time needed for the development of a new variety entails high risks and costs. This requires an adequate protection against the misuse of varieties developed by the breeder with a lot of creativity and professionalism. In Europe, Plant Breeder´s Rights (see 5.2) provide, depending on the vegetable crop, a protection of 25 or 30 years; this is long enough because the success period of a variety is usually 3 to 7 years. Seed companies can recover their investments by increasing the price of innovative seeds. This is possible in view of the usually fairly low price elasticity of vegetable seeds caused by the seed price being only marginal in comparison to the total production costs of a plant, by seeds being essential as basic material for production, and by innovations giving the seed a worthwhile added value. Currently, its also possible to protect a new trait in a variety via patent rights (see 5.3), provide that the new trait does at least meet the criteria of novelty, inventiviness and industrial applicability, and if the inventation is not restricted to one variety. The exclusivity for the patent holder means that these innovative traits cannot be used in breeding without permission (licence) of the patent holder.

4.6 Techniques of vegetable breeding

4.6.1 Introduction

From the 1980's onwards, major changes took place in plant breeding research as result of the application of modern biotechnologies. Commercial breeding had until then been based on traditional/classical breeding methods and plant biotechnology was limited to rapid *in vitro* multiplication of propagation material, the production of *in vitro* haploid plants for the rapid development of homozigous lines, etc. Since that time molecular technologies are now part of the plant breeder's tool box. The knowledge of molecular genetic is developing at high speed. Before long we will have access to genetic information of the complete genome of all major crops including some vegetables. Application of these techniques is called "molecular breeding", which uses enormous amounts of genetic data (bioinformatics), and which enables the combination of genetic information with information on gene expression (transcriptomics and proteomics), physiological data (metabolomics) and phenotypic data. The main breakthrough techniques are briefly described below.

One of the largest leaps forward in molecular breeding was the development of marker technologies (e.g. AFLP DNA fingerprint technology) to visualize the genetic make-up of plants. Breeders can use DNA images based on these technologies to analyse the germplasm of their vegetable crops. Marker assisted selection (MAS) has become an integral component of many commercial vegetable breeding programs (Dias, 1989). The aim of molecular breeding is to supplement conventional methods with faster and more efficient breeding through MAS and/or marker-assisted backcrossing (MAB). Molecular markers that are closely linked to the trait of interest may be identified and applied in gene pyramiding, facilitating introgression of desirable traits into varieties, early selection, etc. For more complex traits conferred by polygenes, quantitative trait loci (QTL) analysis is carried out. The decision to implement molecular markers as a selection tool is based on the fact that the cost of molecular marker analyses has steadily decreased due to the development of new, less expensive marker technologies, and by the implementation of automation for marker detection. Thus, for phenotypes that are simply inherited, screening with markers is often less expensive, and may be more precise, than screening for the phenotype in the field. QTL analysis is more complex and therefore may be more expensive. Markers bring additional value when they can be used to accelerate the development of new improved vegetable varieties. This may occur when phenotypes can be assessed only in specific seasons (since markers can be done anytime), or by improving the efficiency of selecting a recurrent parent for backcrossing or for selecting multi-locus genotypes. For example, in southern India, molecular-based tools were used to identify strains of the tomato leaf curl virus (ToLCV) and to select genes of resistance from wild tomatoes germplasm that were then bred into cultivated lines. ToLCV-resistant lines were developed that produce twice the yield of the most popular varieties in the region after only a few years (Hanson et al., 2000). Tomato resistant varieties to geminivirus were also developed with the help of the molecular markers. SCAR and CAPS markers linked to the beta gene in tomato have been developed for MAS (Zhang & Stommel, 2001). MAS for higher provitamin A carotenoids content is under way in muskmelons. Recently developed markers for watermelon lycopene β-cyclase allows selection differentiating lutein yellow carotenoid plants from lycopene red fruited plants early in development (Bang et al., 2007).

In vitro haploidization techniques like anther and microspore culture and *in vitro* gynogenesis, are also used in many breeding seed companies to accelerate the production of homozygous inbred parental lines and new hybrid vegetable varieties. For example successful *in vitro* microspore culture of broccoli and an improvement in the existing cole protocols, to make this technique available for the purpose of routine breeding, was described by Dias (2001, 2003) in 10 different broccoli genotypes. This protocol is now used by several seed companies worldwide to produce doubled haploid lines for pure inbred progenitors to obtain and ensure uniformity in generated hybrid varieties. Developmental cycles of commercial hybrid varieties range between 5 and 12 years. Besides breeding lines necessary for creating hybrid seed need to be refined for specific markets.

Another important tool for improving vegetable varieties is through genetically modified (GM) technology. Dias & Ortiz (2012) made recently a review of the status of transgenic vegetables to improve vegetable production. These authors analised the advances and potentials in transgenic research on tomato, eggplant, potato, cucurbits, brassicas, lettuce, alliums, sweet corn, cowpea, cassava, sweet potato, and carrots. Highlighted was host plant resistance to pathogens and pests, tolerance to herbicide, quality (both fresh and processed), and vaccine delivery in transgenic vegetables. They suggested, by their review, that the most promising traits seem to be host plant resistances to insects and pathogens, especially for vegetables such as tomato, eggplant, potato, summer squah and sweet corn. Traits such as disease and pest resistance or product quality creat value in the vegetable chain for farmers, traders and consumers.

Many vegetable crops have been genetically modified to include resistance to insects, plant pathogens (including viruses) and herbicides, and for improved features such as slow ripening, higher nutritional status, seedless fruit, and increased sweetness (Dias & Ortiz, 2012). Transgenic crops enable breeders to bring favorable genes, often previously inaccessible, into already elite varieties, improving their value considerably and offer unique opportunities for controlling insects and pathogens. The first commercially grown genetically modified crop was Flavr Savr™ tomato by Calgene in 1994, where the tomato fruit was made more resistant to rotting by adding an antisense gene that interferes with the production of the enzyme polygalacturonase (Kramer & Redenbaugh, 1994). This tomato was deemed to have a long shelf-life. However, it proved later to have a very short "market shelf-life" since the variety was considered inferior by growers, and was rapidly withdrawn from the market. An important lesson was learned by plant genetic engineers: the importance of cooperation with breeders. New vegetable varieties must be tested for performance in all markets before sales.

Two successful commercial examples of GM vegetables are GM squash and sweet corn varieties. In the United States squash yield losses due to viruses often range from 20% to 80% in summer squash (*Cucurbita pepo*) with a reported US$2.6 million economic loss in the state of Georgia in 1997 (Gianessi et al., 2002). Three of the most important viruses affecting squash production are *zucchini yellow mosaic virus* (ZYMV), *watermelon mosaic virus* (WMV), and *cucumber mosaic virus* (CMV) (Zitter et al., 1996). No summer squash variety with satisfactory resistance to CMV, ZYMV, and WMV has yet been developed by conventional breeding (Gaba et al., 2004). Control of squash viruses has focused on cultural practices, including delayed transplanting relative to aphid flights, use of reflective film mulch to repel aphids, and application of stylet oil to reduce virus transmission, in combination with insecticides to reduce aphid vector populations (Perring et al., 1999). In the state of Georgia, it is estimated that ten applications of stylet oils and insecticides are made routinely to control aphids and, hence, limit virus incidence and transmission (Gianessi et al., 2002). Two lines of squash expressing the CP gene of ZYMV, WMV, and CMV were deregulated and commercialized in 1996 (Medley, 1994). Subsequently, many squash types and varieties have been developed, using crosses and backcrosses with the two initially deregulated lines. This material is highly resistant to infection by one, two or all three of the target viruses. (Ochoa et al., 1995; Tricoli et al., 1995; Fuchs et al., 1998; Schultheis & Walters, 1998). The adoption of virus resistant squash varieties has steadily increased in the United States since 1996. In 2005, the adoption rate was estimated at 12% (approximately 3,100 ha) across the country with the highest rates in New Jersey (25%), Florida (22%), Georgia (20%), South Carolina (20%) and Tennessee (20%) (Shankula, 2006). Virus-resistant transgenic squash has allowed growers to achieve yields comparable to those obtained in the absence of viruses with a net benefit of US$22 million in 2005 (Shankula, 2006). Engineered resistance was the only practical approach to development of varieties with multiple sources of resistance to CMV, ZYMV, and WMV. US vegetable grower benefited drom having GM squash varieties resistant to those virus in markets where GM squash is allowed.

Sweet corn, expressing Cry1Ab endotoxin, was introduced commercially in the United States in 1998 into an industry that is highly sensitive to damage to corn ears from lepidopteran pests (Lynch et al., 1999). Research showed that this endotoxin was very effective against the European corn borer (*Ostrinia nubilalis*) in the state of New York, providing 100% clean ears when no other lepidopteran species were present and in excess of 97% when the two noctuids, corn earworm *(Helicoverpa zea)* and fall armyworm *(Spodoptera frugiperda)*, were also present (Musser and Shelton, 2003). Studies in other states in the USA have shown that *Bt*-sweet corn provided consistently excellent control of the lepidopteran pest complex and the potential for 70 to 90% reductions in insecticide requirement (Lynch et al., 1999; Burkness et al., 2001; Hassell & Shepard, 2002; Musser & Shelton, 2003; Speese et al., 2005; Rose & Dively, 2007). Although it was estimated that of the 262,196 ha of sweet corn (fresh and processing) grown in the United States less than 5% was *Bt*-sweet corn in 2006 (NASS, 2007), because processors have avoided growing *Bt*-sweet corn due to concerns about export markets. Since then it has been therefore grown only as a fresh market vegetable crop. By using appropriately timed insecticide applications with *Bt*-sweet corn varieties, fresh market sweet corn growers in South and North Carolina have been able to extend their production later into the season when populations of *H. zea* and *S. frugiperda* are generally too high to control satisfactorily with insecticide applications alone (Hassell & Shepard, 2002). Even when two insecticide sprays are required on *Bt*-sweet corn (e.g., for late season control of *H. zea*), an economic assessment in Virginia found a gain of US$ 1,777 ha^{-1} for fresh-market sweet corn vs. non-*Bt*-sweet corn sprayed up to six times with pyrethroid insecticides (Speese et al., 2005). *Bt*-sweet corn has also proven to be soft on the major predators of *O. nubilalis*, including the lady beetles *Coleomegilla maculata* and *Harmonia axyridis*, the hemipteran *Orius insidiosus* (Musser & Shelton, 2003; Hoheisel & Fleischer,

2007), and a complex of epigeal coleopterans (Leslie et al., 2007). Overall, *Bt*-sweet corn was much better at preserving these predators while controlling *O. nubilalis* than were the commonly used insecticides lambda-cyhalothrin, indoxacarb and spinosad. *Bt*-sweet corn can replace the traditional method of controlling *Lepidoptera* with broad-spectrum insecticides. It may however allow secondary pests to arise. Results from these studies led to the development of a decision guide for sweet corn growers that uses information on biological control and can advise them on the economic return of using various options, including *Bt*-sweet corn (Musser et al., 2006). These results demonstrate also that some of the new *Bt*-sweet corn hybrids allow a truly integrated biological and chemical pest control program in sweet corn, making future advances in conservation, augmentation and classical biological control more feasible. The use of *Bt*-sweet corn has proven to be very effective against the targeted lepidopteran key pests and plantings of *Bt*-sweet corn continue to rise in the United States, with new *Bt*-fresh-market hybrids being released each year.

The introduction of a GM plant on the market requires a safety analysis, both in terms of consumer health and ecological effects. According to a recent COGEM publication the total costs associated with the market introduction of one transgenic plant with one transgenic event is between 6 and 10 million euro. Biotechnology products will be successful if clear advantages and safety are demonstrated to consumers. However, countries vary in their market standards of acceptance of GM products.

Pest and virus-resistant transgenic plants are particularly valuable if no genetic sources of resistance have been identified or if host resistance is difficult to transfer due to genetic incompatibility or links to undesired traits. In such cases, engineered resistance may be the only viable option to develop pest and virus-resistant varieties. However, vegetables are considered minor crops and traditionally have had fewer resources channeled to them compared to field crops. While it is becoming less expensive to create GM crops for pest management, developing a marketable product and a regulatory package remains costly. Development and regulatory costs can be more readily recouped if the product is grown on an extensive area, as would be done with field crops, but which is not generally the case for individual vegetable crops. For example, the large agricultural biotechnology companies have for the most part abandoned the development of GM vegetable crops because of the high costs associated with product development and deregulation. For vegetables, there are many varieties of the same crop and the expected life of a particular variety can be quite limited. Introducing a GM trait into a breeding program can be complicated and cost prohibitive, especially in crops where backcrossing is difficult or impossible (e.g. potatoes). In most countries, deregulation of a GM trait is event specific. For many vegetable crops, it is not possible to develop a single GM event that can be converted into many different varieties of a single or closely related group of vegetable species via conventional breeding. For example, *Brassica* contains about 40 closely related commercialized crops, including cabbage, cauliflower, broccoli, Brussels sprouts, turnip, Chinese cabbage, pak-choy, various mustards, swede, vegetable rapes, etc. (Kays & Dias, 1995, 1996). No single parent exists that can be used to backcross the transgene into the many different types of *Brassica* botanical varieties. Individual events would have to be developed for many of the crop types and deregulation of more than one event for a single protein is problematic for most business models. For the few transgenic vegetable crops that are being developed, novel or unconventional strategies have been employed to bring the crops to market based generally on private and public partnerships, in which the private sector would focus on selling hybrids to higher end producers while the public sector would focus on low resource growers. Consumers concerns on GM vegetables, in particular in Europe and Japan, is also important. Globally it is widely accepted that the introduction of GM varieties of vegetables is some way off, but it remains a future possibility mainly in developing countries where more than 80% of the world's 6.8 billion population lives, the majority of the vegetables are produced and pest problems are most acute.

5. Intelectual property in vegetable breeding

5.1 Introduction

As stated the development of a new vegetable variety, or a new breeding technique, requires much time, effort and money. Making new vegetable varieties requires high investments that can only be recouped if the breeding companies can comercialise the variety for a certain period of time. To encourage innovations, compensate and reward innovators, and to protect the rights of the breeder the legislator has developd systems to be used by the breeder and/or discoverer to protect himself against the risk that others can without permission simply copy, imitate and commercialise his result, the new variety or the new finding.

The first half of the 20thcentury saw the development of a specific type of property right for the breeding sector: "plant breeder´s right". The advent of modern biotechnology in plant breeding in the 1980s brought along another form of intellectual property right (IPR): "patent right´s". The plant breeder's right under plant variety protection laws protects breeder's innovation and authorizes him to exclude others from commercializing his newly developed variety without his permission. On the other hand, "breeder's exemption" ensures that the other breeders in the field can use the protected variety to develop new varieties by making use of the commercially accepted properties of the protected variety.

The technological intervention in plant breeding brought about rapid development in crop improvement and the availability of genomic, proteomic and metabolomics information, in turn, enhanced strategic interests in patent rights in plant breeding. The overall system of intellectual property rights (IPR) applicable to a broader range of protections made plant breeding and the commercial seed sector more efficient and effective. The values created by patents or by other IP protection laws have resulted in the consolidation of the plant breeding industry. Both IPR systems are described in this chapter.

5.2 Plant breeder's rights

Plant breeder´s rights (PBR) represent the oldest form of protection available to plant breeder´s. PBR is a protection system specifically designed for the breeding of new plant varieties. After the approval of "Plant Patent Act" in 1930 in United States, that permitted the patenting of asexually propagated plants, seed companies compaigned for international protection for sexually derived varieties as well. Their efforts culminated in 1961 with the establishement of the "International Union for the Protection of New Varieties of Plants" (UPOV) which purpose was the legal protection of plant varieties through defined plant breeder's rights. UPOV was first enacted for countries of the European Union, and the original law was the result of a revision of the "Netherlands Seeds and Planting Materials Act" that was introduced in the Netherlands in 1941 for the protection of plant breeder's rights. This original law was revised in 1972, 1978 and 1991 (UPOV, 1994). A drastic revision was made in 1991 where protective ownership was extended to include "essentialy derived varieties", which was defined as a variety that is more than 85 per cent similar to the originally protected variety. This implies that breeder's rights on a new, distinguishable variety that strongly resemble the parent variety as a result of the application of particular method of plant breeding or biotechnology (such as repeated backcrossing, genetic transformation or mutation) may be dependent on the rights over the parent variety. At the same time, the "farmer's exemption" was restricted, giving member states the option to allow farmer's to save seed. The spirit of the original attempts of UPOV to protect plant breeder's rights was that the granting of a certificate of protection should not inhibit the flow of information and products through continued research by the entire plant breeding community. Unfortunately, those collective revisions detract from that spirit. Membership was expanded, first to other developed countries and later to 16 developing countries. By 2004, a total of 54 countries subscribed to UPOV.

PBR protection available to breeders allows them to protect their new plant varieties and restrict others from using these varieties without the permission of title holder. PBRs only provide protection to the variety and offer no protection for the method necessary to obtain the variety. This means that PBR is only meant to protect the product in the market. The varieties in order to qualify for protection must fulfil the conditions of distinctness, uniformity, and stability (DUS), and novelty. Distinctness: the variety is deemed to be distinct if it is clearly distinguishable from any other variety of which the existence at the time of submission of the application is a matter of common knowledge. A variety can be distinguishable from an existing variety by a difference in morphological (e.g. flower colour) or physiological (e.g. salt tolerance) properties. Uniformity (Homogeniety): the variety is deemed to be homogeneous if it is, having regard to the variation that may be expected from the particular features of its own reproduction, sufficiently homogeneous as regards its relevant characteristic. Stability: the variety is deemed to be stable if in its essential characteristics it remains true to its description after repeated reproduction or propagation. Novelty: the variety is deemed to be new if propagating or harvested material has not been sold or otherwise disposed of, for the purpose of exploiting the variety. The novelty concept in PBRs does not refer to a certain variety not having existed before but to a variety not having been sold before. This plant breeder´s rights approach of the novelty concept made, many define this condition as the condition of commercial novelty. PBRs grant the holder the authority to forbid others to reproduce, handle, offer for sale, sell, import and export or store material of the protected variety. PBR does not extend to private actions, experimental actions ("research exemptions"), or actions for breeding activities ("breeder's exemption"). In breeder´s exemption, the breeder can neither act against third parties that use the protected variety as basic material for the development of new breeding products. In other words, the breeder cannot act against actions carried out in order to breed other varieties. This restriction holds a fundamental confirmation of the exploitation thereof. This means that PBR do not protect or restrict genetic material for further use and that existing, already successful varieties can be used as basis for new varieties. This results in the genetic potencial of the varieties showing an increasing line year after year. PBR do thus not prevent buiding on existing varieties already protected under plant breeder´s rights. "Research exemption" is provided to other researchers to use the variety for experimental purposes, which means that the breeder cannot act against third parties that use he protected variety for experimental purposes. "Farmer's privilege" is another exemption where the farmer who has lawfully acquired the protected variety can keep a small amount of seed from his own harvest to be re-sown without asking the breeder for permision.

5.3 Patent rights

Patent rights offer a second protection system that has for some years been available to the plant breeder. From the end of the 1980's with the biotechnological intervention becoming widely prevalent, patent protection for plant related inventions were awarded in many countries, mostly from the developed world. In the United States patent systems became important for the breeding sector following subsequent court decisions: i) Diamond vs Chakrabarty in 1980, which involved the first patent on a man-made microorganism; ii) in 1985, plants were considered patentable following the ruling in *Ex* parte Hibberd; and iii) in J.E.M. AG Supply, Inc. v. Pioneer HiBred International, Inc. (Stoll, 2001), where the US Supreme Court allow the use of the general patent law to grant true patents for plants, plant materials and methodologies. In Europe patent rights alow patents on plant properties, plants *per se* and numerous molecular plant techniques since 1998, the year in which the European Union Biotechnology Directive was enated (EU, 1998). In United States the purpose of the utility patent law was extended to plants in 2001. These facts mean that new varieties resulting from modern molecular breeding are legally protected under patent laws offering wider protection. Unlike PBRs where only plant variety is protected, patents are granted for inventions in general, not just for plants. However, in many countries, the patents for plants are not granted. Where patent rights are granted, they include plant genes and to some extent the plants or plant varieties. Though there is no specific rule of patentability of plant genes or the gene sequence, it is necessary that they meet the requirement of patentability i.e. novelty, inventiveness and commercial utility. As per the EU Biotechnology Directive (Article 5), the commercial use of gene sequence must be stated in the patent application. Initially, there plant patents were not available in the European Union. Due to the fact that some biotechnological inventions were not protectable using existing protection like PBR, a new biotechnology directive was implemented in 1998 in all member states of European Union. Although, plant varieties *per se* are not patentable, the additional provision introduced in the EU Biotechnology Directive allowed invention concerning plants to be patentable if the technical feasibility of the invention is not confined to a particular plant variety. Inspite of this restriction, there are a number of cases where patented genes are inserted into the plant material which leads to indirect patent protection of the plant variety. The patented genetic material can then only be used further after obtaining the permission of the patent holder. The plant covered by a patent, for instance cannot be used as a crossing parent in breeding without permission.

In patent laws, only pure scientific research comes under exemption whereas research aimed at the development of a new commercial product is not exempted. In United States, the exemptions under patent laws are strongly restricted as a result of a court ruling in the case of *Maley* v. *Duke University* (Ludwig & Chumney, 2003). This led to further complex situation where GM plants cannot be used for scientific research, a restriction that many researchers are critical about (Waltz, 2009). This lack of a research exemption in the United States has created the unusual situation where a university invention, if licensed exclusively, may be unavailable for ongoing research even in the very laboratory where the invention itself was made. To address this situation, many universities in their exclusive license agreements now reserve rights for the use of inventions within their own institution or, even more broadly, within all academic or nonprofit research institutions. Now, the question that arises is that since plants are self-reproducing, up to which generation does one grant patent protection?

It must also be clear that the patent system does not allow for any reproduction of seeds by farmers. The only exception is the "farmer's privilege" that has been explicity introduced in the European Biotechnology Directive. This experience may be an interesting example for developing countries.

5.4 Plant breeder's rights *versus* patent rights

In the pre-protection era, most of the innovators were compensated in terms of their professional growth. In private breeding, the 'first mover advantage' and 'trade secrets' inbuilt in hybrids gave sufficient compensation to innovators, but after the enactment of IPR laws related to agriculture, in most countries, private research increased and research companies rushed to gain as much IPR protection or patents as possible to gain commercial benefits. The rapid development in technologies, particularly, molecular biology has led to "breeding by design". The knowledge of molecular genetics is making rapid advances and soon access to genetic information of the complete genome of all major crops shall be gained. These approaches in plant breeding are anticipated to produce lot of alternative processes like "breeding by chromosomes" resulting in patentable products. Presently, big corporation companies not only earn money by selling products but also from royalty on their patents. A scenario study by the OECD (2009) predicts widespread use of the technologies based on high-throughput sequencing, proteomics, metabolomics and phenotyping, new types of genetic markers and new genetic modification system by 2030. In GM vegetable plants the genes will be inserted in plants for production of pharmaceuticals and other valuable products. Whether these technologies will become commercial successes depends upon costs related to research, market introduction and regulations, public acceptance and balanced intelectual property policies that stimulate innovations and competition.

The descriptions of both forms of intellectual property protection reveal fundamental differences between plant breeder´s rights and patent rights as regards the subject of protection (PBR is granted for one, physically existing, variety; a patent is granted for products or processes as formulated in the claims), the condition for protection, content of the right, and the exemptions. It particularly applies to the exemptions that when both systems apply at the same time (as in case of a plant variety coming under the patent on a property or method) only those exemptions apply that are applicable in both systems. All other exemptions, such as the breeder´s exemption, are subsidiary to the right of the other system (the patent). This means that when a variety protected under PBR is part of a patent claim, the variety may under PBR be used for further breeding whereas this may not under patent rights.

The coexistence of patent rights and plant breeder´s rights is recognise in the European Biotechnology Directive (EU, 1998). The legal instrument provided in the Directive to enable coexistence between patent rights and plant breeder´s is the compulsory licence. Art. 12(1) aproaches the problems from the position of the breeder and stipulates that when a breeder can neither obtain nor exploit a PBR without infringing a patent of an earlier date, he may request a compulsory licence for non-exclusive exploitation of the inventation protected by such a patent. Art. 12(2) then deals with the problems in a similar way from the point of view of the patent holder and stipulates that when the holder of a patent on a biotechnological invention cannot exploit such an invention without infringing on a PBR of an earlier date, the patent holder may request a compulsory licence for non-exclusive exploitation of this plant variety that is protected under PBR. Art. 12(3) lists the conditions that must be met by breeder and patent older to obtain a compulsory licence. Breeder and patent older must demonstrate that they have unsuccessfully approached the patent older or PBR holder, respectively, to obtain a contractual licence and that the plant variety or the invention represent a "significant technical progress or significant economic interest" in relation to the invention for which a patent is requested or for the protected plant variety.

The PBR have a few weaknesses but they were written specifically for plants, and thereby implicitly recognize the differences between plants and inanimate objects. This is a saving grace. Much more egregious is the application to plants of the patent laws, which do not recognize these differences and therefore creates serious problems. The patent laws were written and amended over the years to protect a process, a machine, a manufacture, etc., but not for living things. Therefore, it became necessary to apply the criteria of the patent laws to living entities, for which they were not intended. This has had some interesting consequences. Consider the bases for granting a patent. Under the patent law, an invention must be novel, non-obvious, and useful. The use of the term novel in IPR laws may be confusing. For plant breeder's rights protection, as explained before, it means new in a commercially available sense. Under patent law, it means: "of a remarkably new and different kind". As stated by Ryder (2005), this criterion is badly abused in plant patents. For example in 12 lettuce patents involving lettuce found in an internet search, eight are for new varieties. All are unequivocally obvious. Hundreds of lettuce varieties have been developed and released over the years; these eight are not remarkable in any way. The concept, breeding methods, and characteristics claimed are all ordinary. Most plant varieties are developed by shuffling known genes in various combinations; the genes code for obvious, known traits. The other four patents were for characteristics or procedures. One was for aphid resistance transferred by traditional breeding crosses from a related wild species. The resistance was closely linked with a deleterious character; they were separated through crossing-over, and the recombinants were identified by molecular methods. The overall process was clever but obvious: breeders often find it necessary to break undesirable linkages. The second patent was for a trait called "multileaf characteristic" and refers to lettuce plants subject to fasciation, a flattening of the stem due to a wide meristematic apex. The trait was selected to occur very early in the life of the plant and resulted in the production of many leaves within a relatively narrow size range. This trait would be advantageous in producing cut leaves for packaging. This innovation may be considered non-obvious. The third trait is an elongated iceberg type lettuce produced by crossing iceberg lettuce x romaine lettuce. Iceberg lettuce is normally spherical. The head leaves are closely appressed and cup-shaped and are therefore hard to separate. Romaine lettuce has elongated leaves that remain separated. The claimed trait specifies iceberg type leaves (characteristic texture and taste) in an elongated head where the leaves also separate easily. This combination of traits is non-obvious. The fourth patent is for a chemical treatment that inhibits head formation of iceberg or butterhead lettuce, so that the leaves remain upright and open. Interior leaves are exposed to light and therefore are green instead of white. This presumably increases the content of certain nutrients, for example, beta-carotene, of these leaves. This may qualify as a non-obvious invention, although the idea of producing all green head leaves has been proposed before. The last criterion for protection is utility. The meaning of this is straightforward: the invention is marketable and therefore has potential economical use. This criterion is particularly important to the inventor, because the driving purpose of the invention is to sell it and make money. The difficulties noted above stem from a failure to properly apply two of the three basic requirements that qualify an invention as patentable under the patent law. Much of the above discussion leads to the inevitable conclusion that the patent laws are inadequate for plants and should be replaced. IPR protection for plants must be framed in different terms than for inanimate objects. Many patent applications are granted broad claims on traits and processes which are essential in nature. So, in patents, the essential processes like crosses, segregations and recombinant selections which are used for developing new varieties should be excluded. However the term "essentially biological" processes are not well defined. In the European Biotechnology Directive, these are defined as entirely natural phenomenon of crossing and selection. A technology step in breeding seems sufficient to make whole process not entirely a natural phenomenon, thus patentable.

Patent for plant varieties is considered the ultimate protector of breeder's rights, affording the opportunity to control as many aspects of the invention as possible, thereby strangling the innovative capacity of the competition. The patent is a means to slow the flow of progress of plant breeding research, except within the company holding the patent. While obviously benefiting that company, it is a big step backwards for the plant breeding community and by extension, for agriculture itself. Theoretically, if each seed company could obtain a patent on a new variety with certain favorable traits, each would do further breeding only with its own protected variety. So there would be parallel lines of research without the enrichment to each program that comes from crossing those lines with varieties in other programs. The owner of a patented variety can share it by licensing its use in breeding to other companies. The cost of the license, in outright payment or in royalty fees, may be quite steep. This would certainly limit the interest in using that variety, since the cost may negate any profit from a new variety.

Patents allow elevation of the profit motive far above the good-of-society and biodiversity requirements. There are two major products of plant biotechnology: traits and methods. Traits such as a disease resistance or product quality (e.g. increase antioxidant content) create value in the process of vegetable breeding. For vegetable breeding companies, specialized in plant breeding biotechnology that have based their business model on the development and marketing of traits or marker platforms (*cf.* 4.5) the protection through patents is essential. For them patent system is the only way to create freedom to operate for further innovation. Patents are also necessary to enter into public-private partnerships, to maintain freedom to operate for scientists, assist in the downstream utilisation of public inventions, and to obtain cash benefits for the institute facing increasing difficulties to secure public financing.

It is recognized that these IPR have provided an essential contribution to the innovation and the success of plant breeding until now but breeder's exemption which allows them to benefit from the availability of the competitor's genetic resources and to use protected varieties for further breeding seems crucial for the future of biodiversity and food security. Breeder's exemption plays an essential role in innovation in practical plant breeding which motivation is to find creative solutions for problems in vegetable farming and in the value chain that can capture a market segment. It should also be noted that nowadays no breeder's rights are requested for many vegetable crops in view of the fact that the economic life of a new variety is no more than few years and that most income can be generated during the time required to register such varieties (1-2 years). Another reason is that most vegetable varieties are hybrids than cannot be reproduced.

6. Trends in biodiversity and vegetable breeding

As referred above, about one half (52%) of the total number of vegetables cultivated in the world receive commercial breeding attention by seed companies and, of those, only 17% are in large scale breeding programs, fostering a need for serious attention to maintenance of vegetable crop biodiversity. There has been a severe decline in the vegetable variety genetic base, as evidenced by the significant reduction, especially within the last 50 years, in the number and range of vegetable varieties grown. During this period vegetable genetic biodiversity has been eroding all over the world and vegetable genetic resources are disappearing, on a global scale, at an unprecedented rate of 1.5-2% per annum. Widespread adoption of simplified vegetable systems with low genetic biodiversity carries a variety of risks including food insecurity. In the short term, such systems risk potential crop failure. In the longer term, they encourage the reduction of the broad genetic base that contributes to high yields, quality traits, disease and pest resistance, etc. This compromises the future genetic health of vegetables. Especially prominent among the "enemies" of genetic biodiversity are the commercial markets and economic social pressures that have practiced breeding methods that promote uniformity, encouraging extensive cultivation of preferred improved and hybrid vegetable varieties with insufficient biodiversity. In addition, globalization has stimulated the consolidation of vegetable seed companies into huge corporations and the decline of small seed companies that serve local and regional markets. In consequence some vegetable breeding programs have been merged or eliminated to reduce costs. Thus fewer and fewer companies/corporations are making critical decisions about the vegetable research agenda, and the future of vegetables worldwide. Inevitably, two things will happen. There will be fewer vegetable breeders in the future and growers will be dependent on a narrower genetic background that could contribute in the near future to food insecurity for poor growers and consumers. Also, with the advent of genetic engineering, these huge seed corporations are also assuming ownership of a vast array of living organisms and biological processes. Of equal concern are expanded uses of legal mechanisms, such as patents and plant breeder's rights that are removing vegetable plant germplasm from general public use (Ryder, 2005). Intellectual property rights for plants were intended as a defensive mechanism to prevent the loss of invented varieties to competitors. However, with the more stringent enforcement of plant breeding rights, and particularly with the application of the utility patent law in the United States to protect all forms of an innovation, this has become an offensive weapon to stifle competition and inhibit the flow of germplasm and information. This can have serious implications for the future conservation of vegetable genetic resources and for world food security.

Some landraces and old open-pollinated varieties of vegetables have existed for long periods outside the commercial and professional plant breeding circles because they have been kept alive within communities by succeeding generations of seed savers. Unfortunately, there are fewer and fewer active seed savers among the millions of vegetable growers, due to the demand of commercial markets and the professionalization of the sector. This is an additional threat to genetic biodiversity. So continued survival of landraces and open-pollinated varieties of vegetables depends largely on popular interest and initiative as well as preservation in gene banks. We should be alerted and concerned about the loss of biodiversity in vegetables and about this impact on food security.

Vegetable growers have an important role in conserving and using vegetable biodiversity. The future of world food security depends not just on stored vegetable genes, but also on the people who use and maintain crop genetic biodiversity on a daily basis. In the long run, the conservation of plant genetic biodiversity depends not only on a small number of institutional plant breeders and seed banks, but also on the vast number of growers who select, improve, and use vegetable biodiversity, especially in marginal farming environments. That is why we should be also alerted and particularly alarmed by the current trend to use improved and hybrid vegetable varieties exclusively. Growers do not just save seeds, they are plant breeders who are constantly adapting their vegetable crops to specific farming conditions and needs. For many generations, vegetable growers have been selecting seeds and adapting their plants for local use. This genetic biodiversity is the key to maintaining and improving the world's food security and nutrition. No plant breeder or genetic engineer starts from scratch when developing a new variety of tomato, pepper, cabbage or lettuce. They build on the accumulated success of generations of growers, who have selected and improved vegetable seeds for thousands of years. If poor small-scale growers in marginal areas stop saving seeds, we will lose genetic biodiversity. Growers will lose the means to select and adapt vegetable crops to their unique farming conditions, which are characterized by low external inputs. Hybrid seed technology is designed to prevent growers from saving seed from their harvest, thus forcing them to return to the commercial seed market every year. Hybrid vegetable seeds alone, and used globally, can be a dead-end to biodiversity. If growers abandon completely their traditional vegetable landraces in the process of adopting only hybrids, crop genetic biodiversity achieved over centuries will be lost forever. Many agronomic benefits will be lost to worldwide growers and thus to consumers.

The exclusive adoption of hybrid varieties in marginal areas may restrict the vegetable producing capacity of growers. It will also destroy biodiversity, and it may contribute in the long-term to food insecurity. For example, a study by Daunay et al. (1997) points out that the release of F_1 hybrids (in Europe and some Asian countries like China and Japan) displaying higher productivity, but with poor phenotypic variability, has contributed to the losses of eggplant landraces, thus inevitably leading to genetic erosion of *S. melongena.* Moreover, some African cultivated eggplants have been lost following social, economic, and political changes (Lester et al., 1990). Therefore, the cultivated eggplant has been considered a priority vegetable species for the preservation of genetic resources since 1977. Several studies have been carried out in Asia and Africa (Lester et al., 1990; Gousset et al., 2005), and collections built up (Bettencourt & Konopka, 1990), particularly in China (Mao et al., 2008).

Fortunately, in some developed countries new independent seed companies, offering unique collections of regionally adapted landrace vegetable varieties, have recently emerged. Furthermore vegetable hobbyist groups, mainly from organic horticulture, are thriving and maintaining old vegetable landraces, in organizations known as "Seed Savers." In this way traditional landraces are being restored to native growers and urban and peri-urban growers. Some of these traditional landraces display combinations of traits that make them especially responsive to local or regional conditions, or are well-suited to particular growing methods, such as those used in organic horticulture or low-external-input systems, or are tolerant to local pests and diseases or other stresses and constraints. Organic growers who seek to grow "full-cycle" or seed-to-seed, are also working to ensure the continued availability of organically grown seeds. There are also considerable ongoing efforts by national governments and international organizations to preserve plant vegetable germplasm in gene banks. This is a valuable but static approach, as further evolutionary changes and improvements will not occur until the seeds are planted, and selection takes place. It is also an activity that relies heavily on continued political stability and support, including sustained governmental funding. Active and positive connections between the private breeding sector and large-scale gene banks are required to avoid possible conflict involving breeders' rights and gene preservation.

The biodiversity of vegetable crop species will be promoted by the maintenance of crop gene banks by governmental and non-governmental organizations, the continued use of diverse sources by plant breeders, especially in the public sector, and by the use of local varieties and landraces by farmers.

7. Prospects for developing countries and poor vegetable farmers

Breeding of vegetables in developing countries is reduced and focused on a very limited number of crops. It is strongly dependent on public investments in the centres of the Consultative Group on International Agricultural Research (CGIAR). One interesting exception it is the recent partnership for new vegetable varieties between Rijk Zwaan international breeding company and Tanzania Government which aim is jointly to develop new vegetable varieties with hight quality standards for the African farmers and consumers. This program will be strongly supported by technology, facilities and know-how of Rijk Zwaan in colaboration with local Tanzania partners. The general lack of private investment in developing countries can be explained by the dominance of the public sector on the one hand and the low purchasing power of the majority of the farmers. Besides in some of these developing countries the market is too small to generate the interest of the international breeding companies for specific programmes. Those CGIAR centres, e.g. AVRDC, provide varieties or half-bred materials to national public institutes, universities and to the private seed sector particularly in Asia.

A major concern in developing countries is that under UPOV, farmers are not allowed to exchange seed of protected varieties, and that only for specific crops farmers may be allowed to reuse their own seed. This is opposed to traditional seed handling practices by farmers in these countries, and exchange is an important tool to maintain biodiversity and in preventing seed shortages among poor farmers. During the Green Revolution, the local exchange of seed was stimulated in order to increase access to better varieties. The level of implementation of breeder's rights (UPOV) legislation differs widely within developing countries. Most Latin American countries responded by joining UPOV under its 1978 or 1991 Acts. Most Asian countries developed systems that are closed to UPOV (but didn't join) or combine breeder's rights with aspects of politically important farmer's rights because these are considered insuficiently protected by UPOV. In Africa, few countries are members of UPOV (South Africa, Kenya – and Tanzania has applied). Developing countries are sometimes asked to become a member of UPOV in exchange for trade agreements and sometimes they are asked to introduce patent rights for plants and even plant varieties. This puts developing country policy makers, who are aware of the importance of local seed exchange among farmers for basic vegetable crops in a difficult position. UPOV recognises an exemption for private and non-commercial use, but this is interpreted by many as to be valid only for farmers who consumed all of their crop within the family. Since almost all farmers take some surplus vegetables to the local market, this strict interpretation does not help much and is not likely to lead to UPOV membership least developed countries.

The legal systems of most of these developing countries is relatively new and not mature enough to tackle the growing patent complexities and they expect a rational approach and freedom to operate in using biodiversity.

Nearly half of the world's vegetable farmers are poor and cannot afford to buy hybrid seed every growing season. What are the prospects for these growers since they produce 15-20% of the world's vegetables and they directly feed almost one billion people in Asia, Latin America, and Africa?. Capital and risk factors are the key constraints that limit the adoption of improved vegetable varieties by small and poor farmers, because these vegetables generally are much more costly to produce per hectare than traditional landrace varieties (Key & Runsten, 1999; Ali & Hau, 2001; Ali, 2002), and most farmers require credit to finance their production. While landraces are usually cultivated using a level of input intensity appropriate to the financial resources available within a household, improved vegetable varieties often require an intensive input regime, including large labor inputs for planting and harvest that cannot be met with family labor alone (Weinberger & Genova, 2005). For small and poor farmers improved vegetable varieties also tend to be riskier than landraces, since the higher costs associated with seeds and production impose a greater income risk. Small farmers may have lower production costs with landraces, because they achieve adequate yields with fewer inputs. In addition, the profits from improved varieties or hybrids tend to vary because yields are often higher but prices fluctuate. From another perspective variable prices and yields increase the variability in market supply (Key & Runsten, 1999).

The lack of capital available to small and poor farmers denies them the opportunity to invest in vegetable production inputs. Without collateral help these farmers are usually unable to secure a loan from a bank or money lender. For those who can get a loan, rates are often unmanageably high, with strict penalties for late repayments. Similarly, a lack of education, resources, skill training, and support prevent these farmers from using improved varieties and then to generate a stable income from their production. In addition, governments usually do not regulate the price of vegetable crops or even provide market information, unlike for field crops. Improving market information systems for vegetable crops and facilitating farmers' access to credit are then essential components of a strategy to enable poor farmers to grow improved vegetable varieties and to overcome the insecurity of their food supplies. The problem of food insecurity in this situation, like that of poverty, is thus frequently traceable to macroeconomic conditions and market failures due to actions of exploitative intermediaries, including landowners, moneylenders, and traders. A major obstacle to success in vegetable production is the shortage of affordable credit. In some cases vegetable farmers must pay high interest rates of 15 to 25 % per 100 days. Desperate for cash, subsistence farmers are forced to sell their crops immediately after the harvest to middlemen or their creditors at unfavorable prices. As pointed by HKI (2010) low cost quality seeds are essential for these farmers. Credit facilities and other inputs must be also part of these vegetable production systems, so that the use of improved vegetable varieties can help subsistence vegetable growers to overcome their poverty and food insecurity.

The benefits from the use of improved varieties are shown by a project supported by the United States Agency for International Development (USAID) and conducted by AVRDC from 1991–2000 at Bangladesh, in two districts (Jessore and Savar), with the aim of overcoming constraints in vegetable production (Weinberger & Genova, 2005). The two districts were selected because they have large vegetable production areas. In Savar three-quarters of all agricultural land is in vegetable production, while in Jessore, the share is 50%. Technological interventions included germplasm evaluation and variety development for many vegetables, off-season production technologies, and grafting technologies for tomato and watermelon to control soil-borne diseases. Between 1996 and 2000, after variety development and the introduction of some new facilities, rural infrastructures and extension services, vegetable production grew at an average annual rate of 5.4%. In a survey conducted with 300 growers, the adoption of improved varieties (42%) and hybrid seed (30%) were the most responsible for this increase in vegetable production of all the technologies used. Among 27 vegetable crops propagated by seed there were 5 (19%) where the only change in vegetable production was the adoption of improved varieties or hybrid variety seeds; in 14 (52%) there was the additional adoption of simple cultural practices like row sowing and fertilization. Tomato grafting and other more sophisticated practices were not implemented. Of the hybrid varieties used, 92% were of cross-pollinated species in which hybrids are important for uniformity. Vegetatively propagated vegetables were not included in the activities. Eager to increase their production, the majority of the growers (91%), regardless of farmer type, invested in some new vegetable technology over the last five years of the project. The average proportion of growers who adopted an improved technology was 43%, and the average adoption rate among all technologies was 31%. Improved or hybrid vegetable varieties were 72% responsible for the increase of vegetable production, since it was easier and cheaper for the growers to buy improved variety seeds than to adopt other technologies. This fact per se highlights the importance of improved varieties and the need to invest in varietal improvement research, since it requires fewer behavioral changes compared to adopting new crop management practices. In terms of farmer's receptiveness to these improved technologies, small-scale farmers, particularly small landowners, tend to be late adopters due to skepticism about the cost of improved and hybrid seeds and capital and risk constraints compared with larger-scale farmers with large cultivated areas (Collins, 1995).

Increased vegetable production has also resulted in important employment benefits for the community such as: new employment opportunities, substitution of family labor by hired labor, and increased wages. Local support industries have also benefited from the expansion of vegetable cultivation both on the input and output side. A higher degree of input commercialization was observed for vegetables as compared to cereals and included all inputs such as seed, inorganic fertilizers, pesticides, farm manure, plastic, mesh netting and bamboo poles. In general, a higher share of vegetable output was sold on markets as compared to the production of cereals. Vegetable growers were highly integrated into markets, selling a large share of their products and retaining a small portion for home consumption. This was true for both small-scale and large-scale farmers. Since supermarkets continue to play a minor role in Bangladesh, most of the vegetable produce was sold either in the local markets or to wholesalers.

In general, the survey project found that vegetable production has contributed to widespread welfare improvement and poverty alleviation in Bangladesh. While nearly all communities agreed that they were benefiting from increased vegetable production (either in terms of enhanced consumption, enhanced investment, saving opportunities, or increased welfare), the grower level data also showed that larger-scale farmers have been able to capitalize more. On average, 90.3% of households experienced an improvement in their lives over the past five years, but large-scale growers reported greater increases in well-being as compared to smaller-scale farmers. The study has also shown that more impact can still be expected, particularly if agro-technology industries develop further. However, the availability of cheap, high quality vegetable seed may be restricted and a major impediment to progress.

Similar projects were implemented by AVRDC in other Asia-Pacific countries and in Africa and the results were similar showing that farmers receive more income from vegetables per hectare than grain crops, and that efficient vegetable production using cheap improved seeds contributes to poverty reduction and less food insecurity. These projects show that improved vegetable varieties could benefit poor farmers and landless laborers by increasing both production and employment. It could benefit the rural and urban poor through growth in the rural and urban non-farm economies and by making food available that is high in nutrients.

8. Concluding remarks

Vegetable breeding is the development of new vegetable varieties with new proprieties. In this era of changes, vegetables will play a major role in well-balanced diets and in the current global battle against malnutrition. There will be continuous need of biodiversity and new and performing varieties for sustainability of vegetable production. Biodiversity is the basis for vegetable breeding and for the introduction of new varieties and hybrids to improve quality and productivity.

Creation of vegetable hybrids is a key means towards the development of varieties for modern vegetable production. Hybrid seed production is high technology and a cost intensive venture. Only well organized seed companies with good scientific manpower and well equipped research facilities can afford seed production. Due to globalization, most vegetable breeding research and variety development in the world is presently conducted and funded in the private sector, mainly by huge multinational seed companies. Few companies are controlling a large part of the world market. Public vegetable breeders and public sector variety development are disappearing worldwide. This means in general that there will be fewer decision-making centers for vegetable breeding and variety development. This has also resulted in the focus on relatively few major vegetables produced worldwide, to the detriment of all other cultivated vegetables. It is imperative that national governments and policymakers, as part of a social duty, invest in breeding research and variety development of traditional open-pollinated varieties and in the minor and so-called "forgotten" vegetables. Smaller seed companies, which are usually specialized in few vegetable crops, must be supported, possibly through autonomous affiliation with the larger companies. More investments in this area will mean less expensive seed for growers to choose from, and increased preservation of vegetable biodiversity. The accomplishment of this goal may require new approaches to vegetable breeding research and development by both the public and private sector.

Domestic and international vegetable markets are changing rapidly, and a variety of factors such as supermarkets and improvements in transportation and refrigeration have largely contributed to this development. Trade liberalization has impacted on the increasing importance of exports, which are increasing for high value vegetable crops. Increasing urbanization, with increasing incomes mainly of growing middle classes in most parts of the world, requires large quantities of vegetables. These may be produced locally or at great distances from where they are consumed, with effects on vegetable post-harvest processing and value-added activities. The standards for participation in high value vegetable markets have increased, both in developed and developing countries and supply chains are increasingly complex, undergoing rapid changes, and often based on strong vertical integration. The participation of small-scale growers in dynamic vegetable markets for higher value vegetables is a major challenge. Participation requires, particularly in developing countries, a set of institutional changes, training, and credit facilities to allow them to compete in increasingly competitive global markets that demand safe, uniform and high-quality produce. Credit facilities and other inputs must be part of subsistence vegetable production systems, so that the use of improved vegetable varieties can help subsistence vegetable farmers to overcome their poverty and food insecurity.

Since the introduction of modern biotechnology in the 1980's many new (in particular molecular) technologies have been devepoded that are important for plant breeding, which enables, e.g., speeding up of the breeding process and the discovery of genetic information. These technological breakthoughs have led to major changes in plant breeding and the development of molecular breeding. Molecular tools will be useful for selecting resistance genes, and increasing quality, nutritional value, and yields. These traits plus food safety will be important aspects of future breeding efforts. Overall, there is great genetic and phenotypic diversity for types and amounts of micronutrients in the various vegetables. Consequently there is a good potential for increasing micronutrient content and thus enrich the diet of the average consumer. More research is needed with the goals of providing benefit to poor and malnourished populations.

Biotechnology provides the ability to produce a broad array of insect-resistant and pathogen-resistant varieties that also express a variety of other value-added traits such as nutritional and post-harvest traits. As the number of value-added, GM traits increases, the number of potential combinations of traits that could be stacked within individual varieties increases geometrically, as to the cost associated with maintaining inventories of geographically adapted varieties expressing different combinations of traits. Consequently, we can expect that commercially available, GM vegetable varieties of the future will express multiple, unrelated, transgenic traits, and farmers in many cases likely will not have the option of planting varieties expressing only single traits. The availability of transgenic vegetable crops does not however ensure that they will be adopted by growers. The benefits of their adoption must exceed their costs for a large proportion of vegetable growers from one season to the next to be widely adopted.

The objective of plant breeding is to produce better varieties for farmers and growers. Investor's interests usually are not the breeder's first priority. Development of new technologies and its use in plant breeding have led to escalation of costs for the breeding companies. Further, protection and regulatory costs add to the high risk investments for the smaller breeding companies. The increased complexity of markets and the higher demands force modern plant breeding to reduce the time for new varieties development, thus, further escalating the cost. The short span of life of a variety in the environment is resulting in shorter earn back period. This double impact is bound to put more pressure on investors to recover their investment through protection of intellectual property, and consequently royalties.

Protection of intellectual property in plant breeding is not the primary driver to develop new, innovative varieties but it is an adequte tool to protect the new varieties in the market against illegal reproduction and sales. Plant breeder's rights as well as patent rights play a major role in supporting plant breeding and innovation. Despite the large differences between both systems, plant breeder's rights and patent rights may have two fundamental, identical objectives: i) on one hand, both rights systems ensure that the developer/inventor is recognised for his creation/invention by granting an exclusive right. For the proprietor this serves in practice a business-economic purpose that may provide; ii) on the other hand, plant breeder's rights as well as patent rights include an important socio-economic objective, by disclosing information on the patentable invention and by making a plant variety under

PBR available for further breeding ("breeder's exemption") . This offers possibilities to built on such inventions and may stimulate further innovation by others, including competitors, which serves the public objective of economic development, food security and preservation of biodiversity. As regards patent rights, however a clear distinction needs to be made between patents on technologies for plant breeding and patents on genetic properties of plants. Granting patent rights for genetic traits is conflicting with plant breeder's rights, the breeder's exemption in particular. The access to genetic variation/biodiversity is so crucial to further innovation in breeding that a form of breeder's exemption within patent rights is required. Amendement of regulations is necessary to increase room for innovation in vegetable breeding. This can be reached by restricting the scope of patents in plant breeding, and more specifically by reinstating the exemption of patents on varieties or by introducing full breeder's exemption in patent rights.

Until recently vegetable breeding research and development that targets small-scale and poor vegetable growers has largely been undertaken by public sector institutions and national agricultural research institutes. Public plant breeding remains a key component of vegetable research systems worldwide, especially in developing countries. However the increasing presence of private sector breeding and a decrease in national and international support makes it difficult for the public sector to continue operating in the traditional manner. Declining funding for public vegetable breeding coupled with the rapid increase of vegetable production and consumption and an urbanizing population, has created a difficult situation. More public sector vegetable breeders are needed worldwide to select and to produce non-hybrid varieties of the minor and "forgotten" vegetables. Breeding of vegetables and other minor crops must continue as a viable endeavor. This will benefit small-scale growers, and will safeguard biodiversity and food security in developing countries. A good example is China where there are four types of vegetable breeders and seed producers: public seed companies, research institutes, foreign seed companies, and local seed companies.

While the maintenance of vigorous public sector breeding programs in areas where private companies are not interested in providing low cost varieties is highly desirable, an additional approach to maximize vegetable and horticultural research input would be the development of global programs with public–private partnerships. The public sector may support portions of vegetable and horticultural R&D that are not attractive to the private sector, and feed improved breeding lines and systems to the private sector for exploitation in regions where the private sector is active, and nurture private sector development in regions where it is lacking. Many in the public and private sectors support such a complementary approach to overcome poverty and malnutrition in developing countries.

In summary, we must ensure that society will continue to benefit from biodiversity and from the vital contribution that plant breeding offers, using both conventional and biotechnological tools, because improved and hybrid vegetable varieties are, and will continue to be, the most effective, environmentally safe, and sustainable way to ensure global food security in the future.

9. References

Ali, M. (2002). The vegetable sector in Indochina: a synthesis. The vegetable sector in Indochina countries: farm and household perspectives on poverty alleviation. AVRDC-ARC. Bangkok.

Ali, M. & Hau, V.T.B. (2001). Vegetables in Bangladesh. Technical Bull. 25. AVRDC, Shanhua, Tainan.

Andre, C.M.; Ghislain, M.; Berlin, P.; Oufir, M.; Herrera, M.R.; Hoffman, L.; Hausman, J.F.; Larondelle, Y. & Evers, D. (2007). Andean potato cultivars (*Solanum tuberosum* L.) as a source of antioxidant and mineral micronutrients. J. Agr. Food Chem. 55:366–378.

Bang, H.J.; Kim, S.G.; Leskovar, D. & King, S. (2007). Development of a codominant CAPS marker for allelic selection between canary yellow and red watermelon based on SNP in lycopene beta-cyclase (LCYB) gene. Mol. Breed. 20:63-72.

Bettencourt, E. & Konopka, J. (1990). Vegetables. pp. 204–220. In: Directory of germplasm collection. IBPGR, Rome.

Biacs, P.A.; Daood, H.G.; Huszka, T.T. & Biacs, P.K. (1993). Carotenoids and carotenoide esters from new cross-cultivars of paprika. J. Agr. Food Chem. 41:1864–1867.

Brown, C.R.; Culley, D.; Yang, C.P.; Durst, R. & Wrolstad, R. (2005).Variation of anthocyanin and carotenoid contents and associated antioxidant values in potato breeding lines. J. Am. Soc. Hort. Sci. 130:174–180.

Burkness, E.C.; Hutchison, W.D.; Bolin, P.C.; Bartels, D.W.; Warnock, D.F. & Davis, D.W. (2001). Field efficacy of sweet corn hybrids expressing a *Bacillus thuringiensis* toxin for management of *Ostrinia nubilalis* (Lepidoptera: Crambidae) and *Helicoverpa zea* (Lepidoptera: Noctuidae). J. Econ. Entom. 94:197-203.

Collins, J. (1995). Farm size and non traditional exports: Determinants of participation in world markets. World Development 23:1103-1114.

Daunay, M.C.; Lester, R.N. & Ano, G. (1997). Les Aubergines. In: *L'amélioration des Plantes Tropicales. Repéres.* Charrier A., Jacquot M., Hamon S. & Nicolas D. (eds.) pp. 83–107 CIRAD-ORSTOM, Paris.

Dias, J.S. (1989). The use of molecular markers in selection of vegetables. SECH, Actas Hort. 3:175-181.

Dias, J.S. (2001). Effect of incubation temperature regimes and culture medium on broccoli microspore culture embryogenesis. Euphytica 119:1-6.

Dias, J.S. (2003). Protocol for broccoli microspore culture. In: *Doubled haploid production in crop plants. A manual.* Maluszynski, M., Kasha K.J., Forster B.P. & Szarejko I. (eds.). pp. 195-204. Kluwer Academic Publishers, Dordrecht.

Dias, J.S. (2010). Impact of improved vegetable cultivars in overcoming food insecurity. Euphytica 176:125-136.

Dias, J.S. (2011). World importance, marketing and trading of vegetables. ICH 2010, Lisbon. Acta Horticulturae (in press).

Dias, J.S. & Ortiz, R. (2012). Transgenic vegetable crops: progress, potentials and prospects. Plant Breeding Reviews 35:151-246.35:151-246.

Dinham, B. (2003). Growing vegetables in developing countries for local urban populations and export markets: Problems confronting small-scale producers. Pest Manag. Sci. 59: 575–582.

EU (1998). Directive 98/44/EC of the European Parliament and of the Council of 6 July 1998 on the legal protection of biotechnological inventions. PBL no. 213. EU Brussels, pp.13.

FAO (2009). FAOSTAT data. Available from http://www.fao.org.

Fuchs, M.; Tricoli, D.M.; McMaster, J.R.; Carney, K.J.; Schesser, M.; McFerson, J.R. & Gonsalves D. (1998). Comparative virus resistance and fruit yield of transgenic squash with single and multiple coat protein genes. Plant Disease 82:1350–1356.

Gaba, V.; Zelcer, A. & Gal-On, A. (2004). Cucurbit biotechnology - the importance of virus resistance. In vitro Cell. & Develop. Biol. - Plant 40:346–358.

Gabelman, W.H. & Peters, S. (1979). Genetical and plant breeding possibilities for improving the quality of vegetables. Acta Hort. 93:243–270.

Gayon, J. & Zallen, D.T. (1998). The role of the Vilmorin Company in the promotion and diffusion of the experimental science of heredity in France, 1840–1920. J. History Biol. 31:241–262.

Gianessi, L.P.; Silvers, C.S.; Sankula, S. & Carpenter, J.E. 2002. Virus resistant squash. In: *Plant Biotechnology: Current and Potential Impact for Improving Pest Management in U.S. Agriculture. An Analysis of 40 Case Studies*. National Center for Food and Agricultural Policy, Washington, DC, USA.

Gockowski, J.; Mbazo'o, J.; Mbah, G. & Moulende, T. (2003). African traditional leafy vegetables and the urban and peri-urban poor. Food Policy 28:221–235.

Gousset, C.; Collonnier, C., Mulya, K.; Mariska, I.; Rotino, G.L.; Besse, P.; Servaes, A. & Sihachakr, D. (2005). *Solanum torvum,* as a useful source of resistance against bacterial and fungal diseases for improvement of eggplant (*S. melongena* L.). Plant Science 168:319–327.

Graham, R.D.; Welch, R.M.; Saunders, D.A.; Ortiz-Monasterio, I.; Bouis, H.E.; Bonierbale, M.; Haan, S.; Burgos, G.; Thiele, G.; Dominguez, M.R.L.; Meisner, C.A.; Beebe, S.E.; Potts, M.J.; Kadian, M.; Hobbs, P.R.; Grupta, R.K. & Twomlow, S. (2007). Nutritious subsistence food systems. Adv. Agron. 92:1–74.

Grube, R.C.; Radwanski, E.R. & Jahn, M. (2000). Comparative genetics of disease resistance within the *Solanaceae*. Genetics 155:873–887.

Hanson, P.M.; Bernacchi, D.; Green, S.K.; Tanksley, S.D.; Muniyappa, V. & Padmaja, A. (2000). Mapping a wild tomato introgression associated with tomato leaf curl virus resistance in a cultivated tomato line. J. Am. Soc. Hort. Sci. 125:15–20.

Hanson, P.M.;Yu, Y.R.; Lin, S.; Tsou, S.C.S.; Lee, T.C.; Wu, J.; Jin, S.; Gniffke, P. & Ledesma, D. (2004). Variation for antioxidant activity and antioxidants in subset of AVRDC—the World Vegetable Center *Capsicum* core Collection. Plant Gen. Res.: Characterization and Utilization 2:153.

Hassel, R. & Shepard, B.M. (2002). Insect population on *Bacillus thuringiensis* transgenic sweet corn. J. Entom. Sci. 37:285-292.

Hayes, H.K. & Jones, D.F. (1916). First generation crosses in cucumber. Republic Conf. Agr. Exp. Stat. 5:319-322.

Henson, S. & Loader, R. (2001). Barriers to agricultural exports from developing countries: The role of sanitary and phytosanitary requirements. World Devel. 29:85–102.

Henson, S.; Masakure, O. & Boselie, D. (2005). Private food safety and quality standards for fresh produce exporters: The case of Hortico Agrisystems, Zimbabwe. Food Policy 30:371–384.

HKI (Helen Keller International) (2010). Homestead food production model contributes to improved household food security, nutrition and female empowerment-experience from scaling-up programs in Asia (Bangladesh, Cambodia, Nepal and Philipines). Nutr Bull 8:1-8.

Hoheisel, G.A. & Fleischer, S.J. (2007). Coccinelids, aphids, and pollen in diversified vegetable fields with transgenic and isoline cultivars. J. Insect Sci. 7:64.

Hotz, C. & McClafferty, B. (2007). From harvest to health: Challenges for developing biofortified staple foods and determining their impact on micronutrient status. Food Nutr. Bul. 28:S271–S279.

Kȧder, A. (2003). A perspective on postharvest horticulture (1978–2003). HortScience 38:1004–1008.

Kays, S.J. & Dias, J.S. (1995). Common names of commercially cultivated vegetables of the world in 15 languages. Economic Botany 49:115-152.

Kays, S.J. & Dias, J.S. (1996). *Cultivated Vegetables of the World. Latin Binomial, Common Names in 15 Languages, Edible Part, and Method of Preparation*. Exon Press, Athens, Georgia.

Keatinge, J.D.H.; Waliyar, F.; Jammadass, R.H.; Moustafa, A.; Andrade, M.; Drechsel, P.; Hughes, J.d'A.; Kardivel, P. & Luther, K. (2010). Re-learning old lessons for the future of food: By bread alone no longer – diversifying diets with fruit and vegetables. Crop Sci. 60:S51-S62.

Kerr, E.A. (1969). Do tomato cultivars deteriorate after they are introduced? Reports of the Horticultural Research Institute of Ontario, Vineland Station, Ontario, Canada. pp. 75–80.

Key, N. & Runsten, D. (1999). Contract farming, small-holders, and rural development in Latin America: The organization of agroprocessing firms and the scale of outgrower production. World Devel. 27:381–401.

Kramer, M.G. & Redenbaugh, K. (1994). Commercialization of a tomato with an antisense polygalacturonase gene: the FLAVR SAVR™ tomato story. Euphytica 79:293–297.

Lacerrot, H. (1996). Breeding strategies for disease resistance in tomato with emphasis to the tropics: Current status and research challenges. 1st Intl. Symp. on Tropical Tomato Disease. Recife, Brazil. pp. 126–132.

Leslie, T.W., Hoheisel, G.A.; Biddinger, D.J.; Rohr, J.R. & Fleisher, S.J. (2007). Transgenes sustain epigeal insect biodiversity in diversified vegetable farm systems. Env. Entom. 36:234-244.

Lester, R.N.; Jaeger, P.M.L.; Spierings, B.H.B.; Bleijendaal, H.P.O. & Hooloway, H.L.O. (1990). African eggplant: a review of collecting in West Africa. Plant Gen. Resources Newsl. 81–82:17–26.

Lincoln, R.E. & Porter, J.W. (1950). Inheritance of beta-carotene in tomatoes. Genetics 35:206–211.

Lincoln, R.E.; Zscheile, F.P.; Porter, J.W.; Kohler, G.W. & Caldwell, R.M. (1943). Provitamin A and vitamin C in the genus *Lycopersicon*. Bot. Gaz. 105:113–115.

Lynch, R.; Wiseman, B.; Sumner, H.; Plaisted, D. & Warnick, D. (1999). Management of corn earworm and fall armyworm (Lepidoptera: Noctuidae) injury on a sweet corn hybrid expressing a cryIA (b) gene. J. Econ. Entom. 92:1217-1222.

Ludwig, S.P. & Chumney, J.C. (2003). No room for experiment: the federal circuit's narrow construction of the experimental use defence. Nature Biotechnology 21:453.

Mao,W.; Yi, J. & Sihachakr, D. (2008). Development of core subset for the collection of Chinese cultivated eggplant using morphological-basal passport data. Plant Gen. Res. 6:33–40.

Medley, T.L. (1994). Availability of determination of non-regulated status for virus resistant squash. Federal Register 59:64187–64189.

Monsanto (2009). Supplemental toolkit for investors. Updated June 2009. Monsanto.

Musser, F.R. & Shelton, A.M. (2003). Bt sweet corn and selective insecticides: their impacts on sweet corn pests and predators. J. Econ. Entom. 96:71-80.

Musser, F.R.; Nyrop, J.P. & Shelton, A.M. (2006). Integrated biological and chemical controls in decision-making: European corn borer control in sweet corn as an example. J. Econ. Entom. 99:1538-1549.

NASS (2007). Vegetables: 2006 annual summary. National Agricultural Statistics Service, Washington DC.

OECD (2009). The Bioeconomy to 2030. Designing a policy agenda. OECD, Paris.

Ochoa, J.P.A.; Dainello, F.; Pike, L.M. & Drews, D. (1995). Field performance comparison of two transgenic summer squash hybrids to their parental hybrid lineage. HortScience 30:492–493.

Perring, T.M.; Gruenhagen, N.M. & Farrar, C.A. (1999). Management of plant viral diseases through chemical control of insect vectors. An. Rev. Entom. 44:457–481.

Pfeiffer, W. & McClafferty, B. (2007). HarvestPlus: breeding crops for better nutrition. Crop Sci. 57:S88-S105.

Premanchandra, B.R. (1986). Genetic regulation of carotene biosynthesis in selected tomato strains: Aspects of beta-carotene biosynthesis and B gene specificity. Intl. J. Vitam. Nutr. Res. 56:35–43.

Regmi, A. & Gehlar, M. (2001). Consumer preferences and concerns shape global food trade. Food Rev.: Global Food Trade 24:2–8.

Reyes, L.F.; MillerJr, J.C. & Zevallos, L.C. (2005). Antioxidant capacity, anthocyanins and total phenolics in purple and red-fleshed potato (*Solanum tuberosum* L.) genotypes. Am. J. Pot. Res. 82:271–277.

Rohrbach, D.D.; Minde, I.J. & Howard, J. (2003). Looking beyond national boundaries: Regional harmonization of seed policies, laws and regulations. Food Policy, 28: 317–333.

Rose, R. & Dively, G.P. (2007). Effects of insecticide-treated and lepidopteran-active *Bt* transgenic sweet corn on the abundance and diversity of arthropods. Env. Entom. 36:1254-1268.

Ryder, E.J. (2005). Intellectual property rights for plants. The case for a new law. Chronica Horticulturae 45(2):5-11.

Schultheis, J.R. & Walters, S.A. (1998). Yield and virus resistance of summer squash cultivars and breeding lines in North Carolina. HortScience 8:31–39.

Shankula, S. (2006). Quantification of the Impacts on US Agriculture of Biotechnology-Derived Crops Planted in 2005. Available from http://www.ncfap.org/

Shepherd, A. (2005). The implications of supermarket development for horticultural farmers and traditional marketing systems in Asia. FAO, Rome.

Simon, P.W.; Pollak, L.M.; Clevidence, B.A.; Holden, J.M. & Haytowitz, D.B. (2009). Plant breeding for human nutritional quality. Plant Breed. Rev. 31:325–392.

Speese, I.J.; Kuhar, T.P.; Bratsch, A.D.; Nault, B.A.; Barlow, V.M.; Cordero, R.J. & Shen, Z. (2005). Efficacy and economics of fresh-market Bt transgenic sweet corn in Virginia. Crop Protection 24:57-64.

Stommel, J.R. & Haynes, K.G. (1994). Inheritance of beta carotene content in the wild tomato species *Lycopersicon cheesmanii*. J. Hered. 85:401–404.

Stoll, R.L. (2001). J.M.E. Ag Supply v. Pioneer Hi-Bred international, Inc. - a commentary. UPOV Gazette 92.

Sudhakar, P.; Singh, J.; Upadhyay, A.K. & Ram, D. (2002). Genetic variability for antioxidants and yield components in pumpkin (*Cucurbita moschata* Duch. Ex Poir.). Veg. Sci. 29:123–126.

Tomes, M.L.; Quackenbush, F.W.; Nelson Jr., O.E. & North, B. (1953). The inheritance of carotenoid pigment systems in the tomato. Genetics 38:117–127.

Tricoli, D.M.; Carney, K.J.; Russell, P.F.; McMaster, J.R.; Groff, D.W.; Hadden, K.C.; Himmel, P.T.; Hubbard, J.P.; Boeshore, M.L. & Quemada, H.D. (1995). Field evaluation of transgenic squash containing single or multiple virus coat protein gene constructs for resistance to cucumber mosaic virus, watermelon mosaic virus 2, and zucchini yellow mosaic virus. BioTechnology 13:1458–1465.

UPOV (1994). International Convention for the Protection of New Varieties of Plants of 2 December 1961, as revised at Geneva on 10 November 1972, 23 October 1978, and March 1991. Publication 221, UPOV, Geneva.

Zeder, M.A.; Emschwiller, E.; Smith, B.D. & Bradley, D.G. (2006). Documenting domestication; the intersection of genetics and archeology. Trends in Genetics 22:139-155.

Zhang, Y.P. & Stommel, J.R. (2001). Development of SCAR and CAPS markers linked to beta gene in tomato. Crop Sci. 41:1602-1608.

Zitter, T.A.; Hopkins, D.L. & Thomas, C.E. (1996). *Compendium of Cucurbit Diseases*. APS Press, St. Paul, MN, USA.

Waltz, E. (2009). Under wraps. Nature Biotechnology 27:880-882.

Weinberger, K. & Genova, C. (2005). Vegetable production in Bangladesh: Commercialization and rural livelihoods. Technical Bull. 33. AVRDC, Shanhua, Tainan.

Weinberger, K. & Msuya, J. (2004). Indigenous vegetables in Tanzania: Significance and prospects. Technical Bull. 31. AVRDC, Shanhua, Tainan.

18

How Past Vicariant Events Can Explain the Atlantic Forest Biodiversity?

Gisele Pires Mendonça Dantas[1],
Gustavo Sebastián Cabanne[3] and Fabrício Rodrigues Santos[2]
[1]*Instituto de Biociências – Universidade de São Paulo- Rua do Matão, Cidade Universitária, São Paulo, SP*
[2]*Instituto de Ciências Biológicas- Universidade Federal de Minas Gerais- Av. Antônio Carlos, Belo Horizonte, MG*
[3]*Museo Argentino de Ciencias Naturales, "Bernadino Rivadavia", Av. Angel Gallardo 470, Buenos Aires*
[1,2]*Brazil*
[3]*Argentina*

1. Introduction

Biodiversity is a wide term that includes all the hierarchy of life in the Earth. However, this word refers to the whole biological diversity: ecosystem diversity, species diversity and genetic diversity. Those three levels of diversity are melt one in another. The basal level involves genetic diversity that includes variation within and among individuals that are grouped in populations. In the next level, populations may differentiate due to mutations, genetic drift and different environmental pressures into distinct species. Finally, ecosystems are characterized by different assemblages of species (Hunter, 1996).

The biological communities observed today were formed along millions of years, although most of those biomes have been already affected by human activity, including many severally endangered regions of the world (Primack & Rodrigues, 2001). Some human activities that affect natural environments are as deforestation, coast occupation, overhunting and introduction of exotic species. Thus, nowadays, the great challenge for conservation of natural systems is to conciliate human activities and conservation. The discipline of conservation biology emerge as answer to this crisis, with multidisciplinary approaches that aim to investigate the human impacts on natural populations, biological communities and ecosystems; to developed practice to prevent the environmental degradation and species extinction, restoration of ecosystems and reintroduction of populations, to establish sustainable relationship between human communities and ecosystems (Rozzi et al., 1998). However, all remaining ecosystems have been previously affected by multiple natural impacts such as climatic changes during the Pleistocene. Then, conservation biology also aims to discriminate between impacts due to natural events from those due to anthropogenic causes affecting current biodiversity distribution.

Biogeography, community ecology and population genetics attempt to describe how biological diversity is spatially distributed at different geographic scales (Miller et al.

2010, Diniz-Filho et al. 2008). Into this context, the molecular biology provides the tools to further investigate phylogenetic relationships among organisms, which can be associated with geographical distribution. With technological advances, the molecular markers have been increasingly applied to access genetic partitions among geographically isolated populations. The relationship between gene genealogies and geography can be used to estimate historical processes that can be responsible for contemporary geographic distributions of individuals and species. This new discipline, the phylogeography, is enabling us to understand processes of diversification, and to reconstruct the historical relationships considering explicit biogeographic hypotheses (Smith & Patton 1993, Patton et al. 1994).

One of the oldest and likely most recognized biodiversity patterns is the latitudinal gradient of species richness (Rosenzweig 1995, Miller et al. 2010). The marked difference in biodiversity richness from regions of high and low latitudes is well documented across distinct taxonomic levels and constitutes a widely recognized biogeographical pattern (D´Horta et al. 2011, Willig et al. 2003). The description of geologic, biogeographic and genetic patterns along tropical ecosystems helps us to better understand the differential effects of evolutionary history of low latitudes in the biodiversity dynamics. In this context, the objective of this chapter was to review the hypotheses of diversification proposed to explain the current biodiversity distribution observed in Brazilian Atlantic Forest. We present here each hypothesis and the different studies supporting or rejecting them.

The Atlantic forest is distributed along eastern and southwestern Brazil, eastern Paraguay, and north-eastern Argentina (Gusmão Câmara, 2003). The Brazilian Atlantic Rain Forest originally presented an area of 1.1 million km^2 and covered a large extension of the coast. Given this large geographic extent, the Atlantic Forest is floristically diverse with severe regional forms of rainforest (ombrophilous) and semi-deciduos forest, depending on rainfall regimes (Oliveira-Filho & Fontes, 2000) (Figure 1). Nowadays, this biome is considered one of the most important conservation hotspots of the World, because of its high levels of endemism and degradation. For example, although near 200 endemic species of birds are reported there, only 5% of its original area remain (Myers et al., 2000). The last estimates account for approximately 20.000 vascular plant species and over 2.300 vertebrate species, half of them being endemic and about 150 with threatened status (Conservation International do Brasil *et al.*, 2000).Furthermore, most of the remaining forested areas are located in regions of steep topography, where agriculture and cattle ranching are not economically viable.

The Atlantic forest biota is probably the result of a complex evolutionary history; however, the processes that shaped it are not well known (Mustrangi & Patton, 1997; Costa *et al.*, 2000; Smith & Patton, 2001; Pellegrino *et al.*, 2005). The knowledge of these evolutionary processes is extremely important for conservation purposes (Moritz, 2002). Among the hypothesis for diversification in Atlantic Forest, the models of Pleistocene refuges, gradient hypothesis; rivers as barriers and orogeny changes have been well discussed. All hypotheses are based in some provisional reductions gene flow among populations, which promoted divergence in allopatry, when the populations became different because they were somehow geographically isolated.

In the following section we discuss each hypothesis.

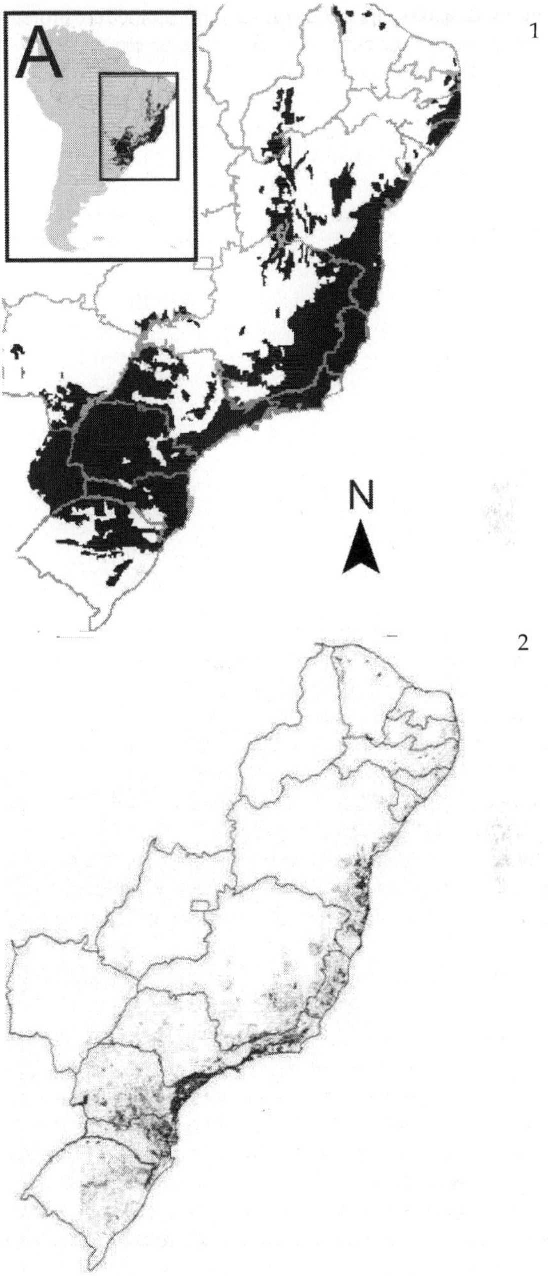

Fig. 1. Original (1) and remain (2) spatial distribution of Brazilian Atlantic Forest .

2. Pleistocene refuges

The refuges theory is one of the most discussed models of diversification to explain the origin of the diversity of the Atlantic forest. In the Neotropics, the refuge theory was originally proposed to explain speciation during the Pleistocene mainly in the Amazon basin (Haffer, 1969; Vanzolini & Williams, 1970; Brown & Ab'Saber, 1979; Haffer & Prance, 2001). This theory proposes that during the glaciations the rainforests were reduced to refuges isolated by open areas, and that organisms isolated in these refuges could have diverged and originated new lineages. Then, in the next interglacial period, the forest expanded and the new clades would be in secondary contact. Brown and Ab'Saber (1979) proposed that open areas dominated the Atlantic forest's landscape during the maximum of Late Pleistocene glaciations, suggesting that the refuge theory can be important to understand the biological diversification of the biome. Taxa may have evolved in allopatry within refuges (rainforest relicts) due to evolutionary factors as genetic drift and divergent selection.

The refuges hypothesis predicts to find evidence of high species endemism and high genetic diversity in the areas with high stability or forest in the past (refugial zones) and, in contrast, lower diversity, lower endemism and molecular signatures of recent range expansion within species in unstable, recently recolonized regions (non-refugial areas) (Carnaval & Moritz 2008). Moritz et al. (2000) and Thomé et al. (2010) affirmed the refuges hypothesis still need to consider additional predictions: the presence of sister taxa in adjacent refugia, secondary contact zones between refugia and range expansion out of refugia area refuges areas.

Carnaval & Moritz (2008) used climatic and forest distribution models and predicted the existence of a large and stable forest refuge in the state of Bahia, in the northeast of Brazil, and smaller refuges located along the Brazilian coast, one area north of the Paraiba river, called Pernambuco refuge, and possibly many small patches south of the Doce River and severe forest contraction south of the São Paulo state (Figure 2). Thomé et al. (2010) also used paleoclimatic modeling to suggest five stable areas in Atlantic Forest to *Rhinella crucifer* (toad) (1) the coastal region of north eastern Brazil, ranging from Alagoas to Rio Grande do Norte, called Pernambuco region; 2) southeastern Brazil, ranging from Rio de Janeiro to Espírito Santo and eastern Minas Gerais; 3) coastal south-southeastern Brazil, ranging from north Santa Catarina to São Paulo (called Serra do Mar); 4) the interior of the Paraná state; and 5) central-north Rio Grande do Sul state and western Santa Catarina state. Thomé et al. (2010) and Carnaval & Moritz (2008) showed many concordant refuges, with a difference that Tomé et al. (2010) found more areas in south Brazil due likely to specific habitat conditions of *Rhinella crucifer* (Figure 2).

Many studies have also found the phylogeographic patterns along Atlantic Forest that are compatible with predictions of the refuge hypothesis. For example, D´Horta et al. (2011) observed that in the study of intrapopulation genetic variation of *Sclerurus scansor* (Rufous-breasted Leaftosser) is compatible with that proposed by refuges hypothesis. They found three groups well defined, one in north of Atlantic Forest (Ceará state), another in central (Bahia, Minas Gerais and north São Paulo State), and a last one the south (Southern São Paulo, Santa Catarina and Rio Grande do Sul State). The estimate of the divergence time between lineages point to events during the middle and late Pleistocene, a period for which there are extensive records documenting change in forest distribution associated with climatic cycles.

Cabanne et al. (2007) also demonstrated demographic changes in *Xyphorhynchus fuscus* (Lesser Woodcreeper) consistent with responses to Pleistocene forest contractions and subsequent advances into southern areas of the Atlantic biome in responses to late Quaternary climate change. The same pattern was found to *Conopophoga lineata* (Rufous Gnateater), which showed data consistent with differentiation in the Pleistocene period (Pessoa, 2008). In some cases, those lineages showed also a secondary contact due to recent expansion in the Holocene period, as it has been found between south Minas Gerais State and North São Paulo, for *Xynphorhynchus fuscus* (Lesser Woodcreeper)(Cabanne et al. 2007) and *Conopophoga lineata* (Rufous Gnateater) (Pessoa 2008) and *Sclerurus scansor* (Rufous-breasted Leaftosser) (D´Horta et al. 2011).

Martins et al. (2009) also found two phylogroups in *Desmodus rotundus* (common vampire bat) whose estimate divergence times fall within the Pleistocene epoch, suggesting this bat is susceptible to forest fragmentation into refuges. Pavan et al. (2010) studied other species of bat *Carollia perspicilatta* (Short-tailed fruit bat), also found two clades whose dating corroborated the vicariant event occurring in the Pleistocene, following by recent population expansion. Moraes-Barros et al. (2006) inferred two main phylogeographic groups exist in the Atlantic forest for *Bradipus torquatus* and *Bradipus variegatus* (Sloth) representing north (Southeastern region of Bahia State north of Minas Gerais) and south (Espiírito Santo and São Paulo)

The difference between clades north and south observed in several Atlantic Forest species, led to the discussion about the influence of latitudinal gradient. The Atlantic Forest covers the 2° to 30 °S alongside the Brazilian coast, consequently presents significant differences in temperature and humidity, which in the past could have affected the number of refuges. The influence of the latitudinal gradient affecting the biodiversity is one of the oldest and most recognized patterns associated to species richness (Rosenzweig 1995). Because of the strong historical effect that Pleistocene era glaciers had on the biogeography of higher latitudes, it is perhaps not surprising that post-glacial expansion is usually considered primarily responsible for the observed genetic diversity patterns (Hewitt 1996, Miller et al. 2010). Vellend (2003) and Vellend and Geber (2005) noted that the same biogeographic conditions favorable to high species richness within community (i.e. high immigration rates and low extinction rates) should promote high genetic diversity within the species comprising that community (Miller et al. 2010). Many studies focused on temperate zone organisms have suggested that latitudinal patterns of within population genetic diversity are most likely due to a history of post-glacial poleward habitat expansion (Miller et al. 2010). The latitudinal biodiversity gradient may reflect the distinct influence of Pleistocene glacial and interglacial cycles in the geographic landscape (Hewitt 2004). Because of the strong historical effect that Pleistocene glaciers had on the biogeography of higher latitudes, it is perhaps not surprising that post-glacial expansion is usually considered primarily responsible for the observed genetic diversity patterns (Hewitt 1996, Miller et al. 2010). In accordance with D´Horta et al. (2011) the latitudinal gradient hypothesis makes some explicit predictions: 1) populations form higher latitudes experienced more pronounced change in their effective population sizes and therefore exhibit signatures of recent demographic expansion and a lower genetic structure; 2) populations from lower latitudes experienced smaller or no changes in effective sizes, thus presenting higher diversity and genetic structure. Carnaval et al. (2009) observed that amphibians from Atlantic Forest showed higher levels of genetic diversity and structure of population in lower than higher latitudes. Some studies of mammals, birds and reptiles have found latitudinal differentiation along the Brazilian Atlantic Forest, and showed an expansion signal in lower latitudes (Pavan et al. 2011, Grazziotion et al. 2006, Martins et al. 2009). However, these studies did not report higher genetic diversity in northern population (lower latitudes), as it would be expected under gradient hypothesis.

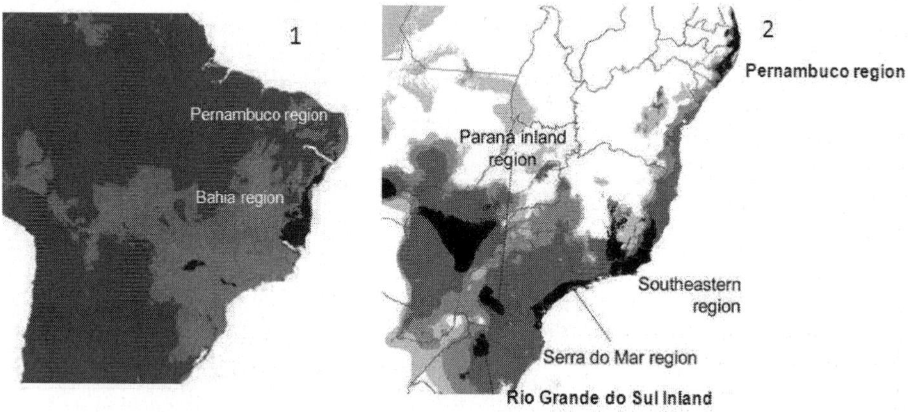

Fig. 2. Summary maps of historically stable areas for the Atlantic forest definitions, obtained by(1) Carnaval and Moritz (2008) summing across BIOCLIM and MAXENT output grids for forest absence/presence under current and (2) Thomé *et al.* (2010) models of habitat distribution for current time, last interglacial period (LIG), last glacial maximum period (LGM),.

3. Neotectonic hypothesis

The Atlantic margin of the South American plate is tectonically passive (see Thomé et al. 2010), although little changes occur, causing faults and fractures and consequently affect dated sedimentary deposits, regional uplifts consequently remodeling the landscape (Ricommini & Assumpção 1999). In the Brazilian Atlantic Forest many changes may have been caused by the uplift of the coastal Brazilian mountains (Serra do Mar). Those events possibly interrupted precipitation in southeastern Brazil by the early Pliocene at about 5.6 Ma and therefore altered the distribution of humid and dry habitats. This period coincides with the transition from tropical humid to semiarid or arid conditions described by some authors (Simpson 1979; Vasconcelos *et al.* 1992). This orogenic process deeply changed the geomorphologic and climatic conditions of south and southeast areas of Brazil, and consequently fragmented Brazilian Atlantic Forest with drier areas (Grazziotini et al. 2006). The palynological record of the Quartenary showed that between 33,000 and 25,000 years ago, the central Brazilian region was moister than today and was covered by rainforest (Ledru 1993), and during the last glaciation (18,000-12,000 years ago) the present day corridor of xeric vegetation was covered by extensive woodland (Prado & Gibbs 1993, Costa et al. 2003). It is believed that during drier periods, forest formations were more likely to occur in mountain areas, because of the higher pluviometric level resulting from orographic effect. Such phenomenon is currently observed in north-eastern region of Brazil, where the occurrence of humid forests is strictly associated with areas of mountain ridges (D´Horta et al. 2011).

Mountain chains often delimit Atlantic Forest distribution, but few studies have established geomorphological events as promoter of allopatric diversification in this biome (Thomé et al. 2010). Neotectonic activity has significantly remodeled the landscape of eastern Brazil during the Quaternary, confounding the signatures of isolation mechanisms along this Tertiary-Quaternary time scale. Thomé et al. (2010) found that the distinct phylogroups concordant with neotectonic barriers in Guapiara lineament and the Cubatão Shear zone in the São Paulo State, both including recent superficies ruptures (Ricommni and Assumpção, 1999). Although, the tectonic events in the region occupied by Brazilian Atlantic Forest are still poorly understood, they may be an alternative explanation to observed patterns.

4. Riverine barriers

The rivers can play an important role in biological diversification as they may act as primary or secondary barriers to gene flow and may have been important to model the current biota distribution. Siedchlang et al. (2010) suggest that the São Francisco River was an important barrier to *Calyptommatus* (lizards), allowing speciation on opposite margins of the river, being responsible to present distribution of *C. sinebrachiatus* and *C. leiolepis*, as well as that of *C. nicterus* and *C. leiolepis*, which occurred in adjacent banks on opposite margins. Thomé et al. (2010) observed that *Rhinella crucifier* group presents divergent lineages spatially concordant with Doce River systems and refute the refuges model to diversification this group. Also, Lacerda *et al.* (2007) presented genetic data that suggested a role of the Jequitinhonha river and Doce river for separating populations of passeriformes *Thamnophilus ambiguous* (Sooretama). Pellegrino *et al.* (2005) show also that the genetic structure of lizards of the *Gymnodactylus darwinii* complex coincides with the river system in the northern regions of the Brazilian Atlantic Forest and that major coastal rivers in this region may have played a key role in its diversification

On the other hand, D´Horta et al. (2011) suggested for *Sclerurus scansor* that tectonic activity associated with the Paraiba Valley can be congruent with the scenario that the river was important for the secondary contact of lineages of the south and central of Atlantic Forest, but not for the origin of these lineages due to phylogeography rupture, because the divergence time is much more recent (middle/late Pleistocene). This hypothesis of secondary contact among lineages is corroborated by Cabanne et al. (2007) and Pessoa (2008), who also suggested Paraíba do Sul Valley as contact region of divergent mitochondrial lineages from *Xyphorhynchus fuscus* and *Conopophoga lineata*. Furthermore, in both margins of the Paranapanema river were also found two phylogroups of *Bothrops jararaca* (Grazziotin et al. 2006).

In summary, the riverine systems seem important to differentiation between lineages and species, thus, are relevant to consider in the evolutionary processes related to the Atlantic Forest diversification, mainly the São Francisco, Jequitinhonha, Doce and Paranapanema (Fig 3).

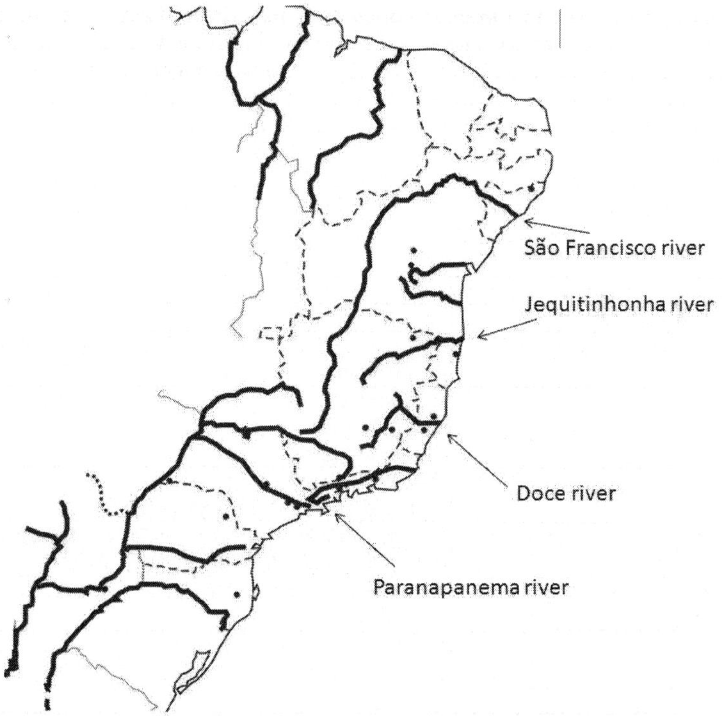

Fig. 3. Localization of mainly rivers that influence the distribution of species at Brazilian Atlantic Forest.

5. Gradient hypothesis

The Atlantic Forest is surrounded by dry forests and forested savannas (Cerrado). Because of the existence of a gradual transition from humid forest to those drier biomes, many organisms associated to humid forests are also found intermingled in the open biomes. Each region, the Atlantic forest and the neighboring regions, present different characteristics, and therefore it is expected to find differential selective regimes that could make organisms to diverge between regions. This hypothesis is known as the ecological gradient hypothesis. Also, there are different types of forests within the Atlantic Forest that could imply differential selective regimes. Even though this scenario is very plausible, few studies addressed the problem of divergence across ecological gradients in this biome. For example, Lara et al. (2005) mentions that the occurence of species of tree rat *Phyllomys* and of spiny rat *Trinomys* is associated with vegetation types and with humidity gradients indicate that evolution across gradients may be important. Bird species show distributed in different zone of humidity and temperature across forest types in the Atlantic forest with, which could suggest and important role of environmental gradients in their evolution. So far only one study addressed the problem of evolution across gradients in the Atlantic forest. Cabanne et al (in press) studied whether the plumage color in *Dendrocolaptes platyrostris* was related to change by drift in different populations historically isolated, or by selective change in different forest types. They found that the plumage variation was related to different forest types and not to historically isolated lineages, suggesting an important role of selection. *D. platyrostris* at the open vegetation corridor was lighter and less streaked than at the forest habitat, a morph which is suggested to be an adaptation of woodcreepers for habitats with high luminosity levels, as are forests at the open vegetation corridor (Marantz, 1997; Willis, 1992). On the other hand, rainforest individuals are darker and more streaked, what is considered to be an adaptation to live in low luminosity and very humid conditions (Marantz, 1997; Willis, 1992; Zink & Remsen, 1986).

Era	Period	Epoch	MYA	Event	Reference
CENOZOIC	Quartenary	Holocene	0.01		
		Pleistocene	1.8	Divergence lineage from *Sclerurus scansor* (Passeriformes)	D´Horta et al. 2011
				Divergence of lineages of *Xyphorhynchus fuscus* (Passeriformes)	Cabanne et al. 2007
				Divergence of lineages of *Conopophoga lineata* (Passeriformes)	Pessoa 2008
				Divergence of lineages from South the *Gymnodectylus darwinii* (lizards)	Pellegrino et al. 2005
				Divergence between lineages of *Carollia perspicillata* (bat)	Pavan et al. 2011
				Divergence of lineages of *Rhinella crucifier* center and north Atlantic Forest (toad)	Thomé et a. 2010
				Divergence of lineages of *Desmodus rotundus* (bat)	Martins et al. 2009
				Divergence of lineages of *Bradypus torquatus* (Xenarthra)	Moraes-Barros et al. 2006
	Tertiary	Pliocene	5.3	Divergence of phylogrops *of Bothrops jararaca* (Serpentes)	Grazziotini et al. 2006.
				Divergence of lineages of North and south of *Rhinella crucifier* (toad)	Thomé et a. 2010
				Uplift Brazilian coast mountain	
		Miocene	23.0	Drainage of Parana river	Grazziotin et al. 2006
				Neogene sediments of the Barreiras Formation Doce River	
				Divergence of lineages North and South Gymnodactylus darwinii (Lizards)	Pellegrino et al. 2005

Table 1. The geological time scale and the resume of principal studies of Phylogeography in Atlantic Forest

The open vegetation corridor and its network of gallery forests and dry forests are contiguous with the Atlantic and Amazon forests. The results of Cabanne et al. (in press) supported the idea that the two plumages types of *D. platyrostris* may have evolved by divergent selection regimes between habitats. There are several other species that occur in both habitats and might present a similar evolutionary story. Puorto et al. (2001) found one clinal morphological variation to *Bothrops atrox* group, although did not show association with genetic variation, which revealed two clades concordant with division North and South of Forest Atlantic.

6. Conclusions

In conclusion, the separation of the northern and southern phylogroups observed at Atlantic Forest endemic species is a pattern found for several taxa, however the discontinuities were observed in distinct zones of the Atlantic Forest. Some discrepancies can be explained by sampling bias, but others can be due to real differences in the dynamics of the species or the associated ecosystem. Anyway, distinct mechanisms have been invoked to explain the breaks, sometimes they were attributed differences are attributed to refuges hypothesis and another times to riverine barriers or tectonic activities. Silva et al. (in press) studied the panbiogeographic nodes in Atlantic Forest found six of the seven nodes found one node (Pernambuco) in North AF, two nodes (Bahia and Espírito Santo) are in Central AF, and three nodes (São Paulo, Paraná and Santa Catarina) correspond to South AF and concluded that the pattern of latitudinal subdivision of *taxa* distributions has originated at least since the Miocene and is more complex than previously thought. The endemism areas are concordant with the stability areas proposal for Atlantic Forest, the great part of studies show differentiation between lineages in Pleistocene Epoch (Table 1). Due to the complexity of the geomorphological and ecological features of the Atlantic Forest, and the intrinsic complexity of the ecophysiology of the Atlantic Forest organisms, it seems too simplistic to imagine that one single diversification mechanism can explain the origin of the current biogeographical patterns exhibit by Atlantic Forest species (D´Horta et al. 2011). A realist scrutiny of the Atlantic Forest diversity and past ecosystem dynamics should consider multiple mechanisms operating at different spatial and temporal scales (Thomé et al. 2010).

The elucidating the process that acted in Atlantic forest is essential to understand the biodiversity present in this biome, and to conservation of lineages and species. Understanding the speciation process, the effects of climate oscilations will be important to estimate the consequence of global warming in this ecosystem. Predictions of ecological niche modeling to Brazil has indicated a decline of 80% of current distribution for half of the birds of the family Pipridae in Amazon and Atlantic Florest (Anciães & Peterson, 2006). Niches paleoclimate modeling combined with molecular analysis has pointed to cases of recent population expansion from refuges or ecologically stable areas with high diversity and population structure in the Atlantic in response to environmental changes in the Quaternary Period: amphibians, lizards (Carnaval et al. 2009, Carnaval and Moritz 2008) and birds (Cabanne *et al.*, 2008). Thus, the importance of the characterization of the distribution of the genetic diversity of threatened and non-threatened species is important to future conservation plans and politics efforts.

7. Acknowledgement

This chapter aim to understand the principal evolutionary processes that are related to the diversification and maintenance of the Atlantic forest biodiversity.

The author thank to Fundação de Amparo a Pesquisa do Estado de São Paulo (FAPESP), Conselho Nacional de Pesquisa do Governo Brasileiro (CNPq), Coordenação de Aperfeiçoamento de Pessoal de Nível Superior (CAPES), Fundação de Amparo a Pesquisa do Estado de Minas Gerais (FAPEMIG) for given funding to develop these studies.

8. References

Brown, K.S. & Ab'Saber, A.N. (1979). Ice-age forest refuges and evolution in Neotropics: correlation of paleoclimatoligical, geomorphological and pedological data with biological endemism. *Paleoclimas,* 5, 1–30.

Cabanne, G.S., D´Horta, F., Meyer, D., Myiaki, C. & Silva, J.C.. (in press) Plumage and genetic variation in Dendrocolaptes platyrostris (Aves:Dendrocolaptidae): recent discontinuity between the south American open vegetation corridor and Atlantic Forest. *Biological journal of Linnean Society*

Cabanne, G.S., d'Horta, F.M., Sari, E.H., Santos, F.R. & Miyaki, C.Y. 2008. Nuclear and mitochondrial phylogeography of the Atlantic forest endemic *Xiphorhynchus fuscus* (Aves: Dendrocolaptidae): biogeography and systematics implications. *Molecular Phylogenetic and Evolution,* 49, 760-773.

Cabanne, G.S., Santos, F. & Miyaki, C.Y. (2007) Phylogeography and demographic history of Xiphorhynchus fuscus (Passeriformes: Dendrocolaptidae) in the southern Atlantic forest of Brazil and Argentina. *Biological Journal of the Linnean Society,* 91, 73– 84.

Carnaval, A.C. & Moritz, C. (2008) Historical climate modeling predicts patterns of current biodiversity in the Brazilian Atlantic forest. *Journal of Biogeography,* 35, 1187–1201.

Carnaval, A.C., Hickerson, M.J., Haddad, C.F.B., Rodrigues, M.T., Moritz, C. (2009) Stability predicts genetic diversity in the Brazilian Atlantic Forest hotspot. Science 323, 785-789.

Conservation International do Brasil, Fundação SOS Mata Atlântica, Fundação Biodiversitas, Instituto de Pesquisas Ecológicas, Secretaria do Meio Ambiente do Estado de São Paulo, SEMAD/Instituto Estadual de Florestas-MG (2000) Avaliação e acções prioritárias para a conservação da biodiversidade da Mata Atlântica e Campos Sulinos (ed. by Ministério do Meio Ambiente / Secretaria Nacional de Biodiversidade e Florestas). pp. 40. Brasília.

Costa, L.P. (2003) The historical bridge between the Amazon and the Atlantic forest of Brazil: a study of molecular phylogeography with small mammals. *Journal of Biogeography,* 30, 71–86

Costa, L.P., Leite, Y.L.R., da Fonseca, G.A.B. & da Fonseca, M.T. (2000) Biogeography of South American forest mammals: endemism and diversity in the Atlantic forest. *Biotropica,* 32, 872–881.

D'Horta, F.M., Cabanne, G.S., Meyer, D. & Miyaki,.C.Y. (2011). The genetic effects of Late Quaternary climatic changes over a tropical latitudinal gradient: diversification of an Atlantic Forest passerine. *Molecular Ecology* xxxx

Diniz-Filho, J. A. F., De Campos Telles, M. P., Bonatto, S. L., Eizirik, E., De Freitas, T. R. O., De Marco, P., Santos, F. R., Sole-Cava, A. & Soares, T. N. (2008) Mapping the evolutionary twilight zone: molecular markers, populations and geography. *Journal of Biogeography,* 35, 753–763.

Grazziotin, F.G., Monzel, M., Echeverrigarauy, S. & Bonato, S.L. (2006) Phylogeography of the Bothrops jararaca complex (Serpentes: Viperidae): past fragmentation and island colonization in the Brazilian Atlantic Forest. *Molecular Ecology*, 15, 3969–3982.

Gusmão Câmara, I. (2003). Brief history of conservation in the Atlantic forest. In: Galindo-Leal C, Gusmão Câmara I, eds. *The state of the hotspots: the Atlantic forest.* Washington, DC: Island Press, 31–42.

Haffer, J. & Prance, G.T. (2001) Climatic forcing in Amazonia during the Cenozoic: on the refuge theory of biotic differentiation. *Amazoniana,* 16, 579–607.

Haffer, J. (1969). Speciation in Amazonian birds. *Science,* 165, 131–131.

Hewitt, G.M. (1996). *Some genetic consequences of ice ages, and their role in divergence and speciation.* Biol. J. Linn. Soc. Lond., 58, 247–276.

Hunter, M.L. (1996) *Fundamentals of Conservation Biology*. Blackwell Science, Massachusetts.

Lacerda, D.R., Marini, M.A. & Santos, F.R., (2007). Mitochondrial DNA corroborates the species distinctiveness of the Planalto (Thamnophilus pelzelni Hellmayr, 1924), the Sooretama (T. ambiguus Swainson, 1825) Slaty-antshrikes (Passeriformes: Thamnophilidae). *Braz. J. Biol.* 67, 873–882.

Lara, M.C., Geise, L. & Schneider, C.J. (2005) Diversification of small mammals in the Atlantic forest of Brazil: testing the alternatives. In: Mammalian diversification: From Chromosomes to Phylogeography (A Celebration of the Career of James L. Patto) (eds Lacey EA, Myers P), pp. 311-335. University of California Press Berkeley, Los Angeles, London

Ledru, M.P. (1993) Late quaternary environmental and climatic changes in central Brazil. *Quaternary Research*, 39, 90–98.

Martins, F., Templeton, A R., Pavan, A. C., Kohlbach, B. C. & MORGANTE, J. S. (2009) Phylogeography of the common vampire bat (Desmodus rotundus): Marked population structure, Neotropical Pleistocene vicariance and incongruence between nuclear and mtDNA markers. *BMC Evolutionary Biology* 9, 294

Miller, M.J., Bermingham, E., Klicka, J., Escalante, P. & Winker, K.(2010) Neotropical birds show a humped distribution of within-population genetic diversity along a latitudinal transect. *Ecology Letters*, 13, 576–586

Moraes-Barros, N., Silva, J.A.B., Miyaki, C.Y. & Morgante, J.S. (2006) Comparative phylogeography of the Atlantic forest endemic sloth (Bradypus torquatus) and the widespread three-toed sloth (Bradypus variegatus) (Bradypodidae, Xenarthra). *Genetica*, 126, 189–198.

Moritz, C., Patton, J.L., Schneider, C.J. & Smith, T.B. (2000) Diversification of rainforest faunas: an integrated molecular approach. *Annual Review of Ecology and Systematics*, 31, 533–563.

Mustrangi, M.A. & Patton, J.L. (1997). Phylogeography and systematics of the slender mouse opossum *Marmosops* (Marsupialia,Didelphidae). *University of California Publications in Zoology* 130: 1–86

Myers, N., Mittermeier, R.A. & Mittermeier, C.G., Fonseca, G.A.B., Kent, J. (2000). Biodiversity hotpots for conservation priorities. *Nature*, 403, 853-858.

Oliveira-Filho, A.T. & Fontes, M.A. (2000). Patterns of floristic differentiation among Atlantic Forests in SOutheartern Brazil and the influence of the climate. *Biotropica,* 32, 793-810.

Patton, J.L., Silva, M.N.F. & Malcolm, J.R. (1994) Gene genealogy and differentiation among arboreal spiny rats (Rodentia: Echimyidae) of the Amazon Basin: a test of the riverine barrier hypothesis. *Evolution,* 48, 1314–1323.

Pavan, A.C., Martins, F., Santos, F.R, Ditchfield, A. & Redondo, R.A.F. (2011). Patterns of diversification in two species of short-tailed bats (*Carollia* Gray, 1838): the effects of historical fragmentation of Brazilian rainforests.*Biological Journal of the Linnean Society,* 102, 527–539

Pellegrino, K.C.M., Rodrigues, M.T., Waite, A.N., Morando, M., Yassuda, Y.Y. & Sites, J.W. (2005). Phylogeography and specieslimits in the *Gymnodactylus darwinii* complex (Gekkonidae, Squamata): genetic structure coincides with river system in the Brazilian Atlantic Forest. *Biological Journal of the Linnean Society* 85, 13–26.

Pessoa, R.O. (2007) Sistemática e Biogeografia Histórica da Família Conopophagidae (Aves: Passeriformes): Especiação nas Florestas da América do Sul. Tese de Doutorado, Instituto de Biociências da Universidade de São Paulo, São Paulo.

Prado, D.E. & Gibbs, P.E. (1993) Patterns of species distributions in the dry seasonal forests of South America. Annals of the Missouri Botanical Garden, 80, 902–927

Primack, R.B. & Rodrigues, E. (2001) *Biologia da Conseração.* Midiograf, Londrina, Paraná.

Puorto G., Salomão M. G., Theakston, R.D.G., Thorpe, R.S., Warrell D.A. & wuester W. (2001) Combining mitochondrial DNA sequences and morphological data to infer species boundaries: phylogeography of lanceheaded pitvipers in the Brazilian Atlantic forest, and the status of *Bothrops pradoi* (Squamata: Serpentes: Viperidae). *Journal of. Evolutionary* Biology. 14, 527-538.

Ricommini, C. & Assumpção, M., (1999). Quaternary tectonics in Brazil. *Episodes* 22, 221–225.

Rosenzweig, M.L. (1995). *Species Diversity in Space and Time.* Cambridge University Press, Cambridge.

Rozzi R., Primack, R., Feinsinger, P., Dirzo, R. & Massardo. F. (1998) qué es la biología de la conservación? Pg 35-58 In: Primack, R., Rozzi, R., Feinsinger, P., Dirzo, R. & Massardo, F. *Fundamentos de Conservación Biológica: Perspectivas latinoamericans.* Fondo de cultura Económica México.

Siedchlag, A.C., Benozzati, M.L., Passoni, J.C. & Rodrigues, M. T.(2010) Genetic tructure, phylogeny, and biogeography of Brazilian eyelid-less lizards of genera Calyptommatus and Nothobachia (Squamata, Gymnophthalmidae) as inferred from mitochondrial DNA sequences. *Molecular Phylogenetics and Evolution,* 56, 622–630

Silva, S. M., Moraes-Barros, N., Ribas, C. C., Ferrand, N. & Morgante, J. S.(in press). Brazilian Atlantic Forest Biogeography - complexity revealed by current knowledge. *Biogeography*

Simpson, B.B. (1979) Quaternary biogeography of the high montane regions of South America. In: The South American Herpetofauna: its Origin, Evolution, and Dispersal (ed. Duellman WE), pp. 157–188. Monograph of the Museum of Natural History, University of Kansas, Lawrence, Kansas.

Smith, M.F. & Patton, J.L. (1999) Phylogenetic relationships and the radiation of sigmodontine rodents in South America: evidence from cytochrome b. *Journal of Mammalian Evolution*, 6, 89–128.

Thomé, M.T.C, Zamudio, K.R., Giovanelli, J.G.R., Haddad, C.F.B., Baldissera, F.A. Jr & Alexandrino, J. (2010) Phylogeography of endemic toads and post-Pliocene persistence of the Brazilian Atlantic Forest. *Molecular Phylogenetics and Evolution* ,55, 1018–1031

Vanzolini, P.E. & Williams, E.E. (1970) South american anoles: the geographic differentiation and evolution of the *Anolis chrysolepis* species group (Sauria: Iguanidae). *Arquivos de Zoologia,* 19, 1–298.

Vellend, M. & Geber, M.A. (2005). Connections between species diversity and genetic diversity. Ecol. Lett., 8, 767–781.

Vellend, M. (2003). Island Biogeography of genes and species. Am. Nat., 162, 358–365.

Willig, M.R., Kaufman, D.M. & Stevens, R.D. (2003) Latitudinal gradients of biodiversity: patterns, process, scale and synthesis. Annual Review of Ecology. *Evolution and Systematics*, 34, 273–309.

19

SPAR Profiles for the Assessment of Genetic Diversity Between Male and Female Landraces of the Dioecious Betelvine Plant (*Piper betle* L.)

Shirish A. Ranade[1], Anjali Soni[1,2] and Nikhil Kumar[3,4]
[1]*Plant Molecular Biology (Genomics), CSIR - National Botanical Research Institute, Rana Pratap Marg, Lucknow U.P. State*
[2]*(Present Address) Department of Biotechnology, Veer Narmad South Gujarat University, Surat, Gujarat State*
[3]*Betelvine Biotechnology, CSIR - National Botanical Research Institute, Rana Pratap Marg, Lucknow. U.P. State*
[4]*(Present Adress) B2/M91 SBI Colony, Sector B, Janakipuram, Lucknow, U.P. State India*

1. Introduction

Betelvine (*Piper betle* L., family Piperaceae) is an important, traditional and an ancient crop of India and is a shade loving, perennial evergreen climber of tropical origin. It is generally known as "Paan" in Hindi in the Indian subcontinent and by different names in the Asiatic region and is a plant of considerable antiquity. It is distributed in several countries in including India as well as other countries of Indochina region - Indonesia, Malaysia, Vietnam, Laos, Kampuchea, Thailand, Myanmar, Singapore and the Far-East (Figure 1), where its cultivation or ethnomedicinal properties are well known. In India, betelvine is widely cultivated in the states of Uttar Pradesh, Bihar, Madhya Pradesh, Northeastern India, Maharashtra, Karnataka, West Bengal, Orissa, Andhra Pradesh, Tamil Nadu, Kerala and Andamans and almost the entire production of betel leaves is consumed fresh as a masticatory.

Betelvines are dioecious and therefore, under controlled hybridization, attempts have been made to cross different landraces and in some of these experiments, viable seed set has also been reported. However, as a crop, propagation is obligately only through vegetative means. Its cultivation in northern India under sub-tropical conditions (Figure 2) has been shown to be a unique case of plant establishment under anthropogenically regulated microclimatic conditions (Kumar 1999). Cultivated betelvine is grown in traditional farming systems many of which are managed exclusively by families or communities. The betelvine growers invariably named their cultivars with local or vernacular names. These cultivated betelvines are therefore, nothing but landraces and it is this description that will be used consistently throughout the manuscript. A survey over several years indicated between 125 to 150 local cultivars (landraces) of betelvines in India.

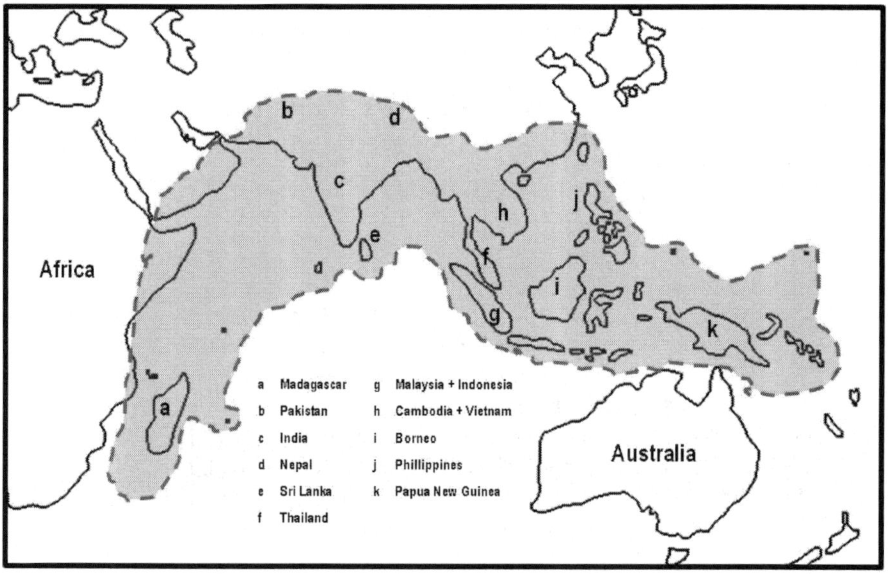

Fig. 1. Schematic map showing the regions of the world where betelvine is cultivated or consumed.

Currently more than 200 landraces are cultivated in India and are often named after the localities, villages or towns where they are grown. As a result of this, though the geographical distribution of the betelvines under cultivation is vast, it is possible that the genetic variation may not be so well distributed. However, no systematic study of betelvine genetic diversity had been carried out to our knowledge. Thus the assessment of genetic variation amongst the different landraces became an important objective for which, PCR-based parallel approaches were used. We have shown, in an earlier study with different landraces using RAPD method, a clear distinction between male and female betelvines or between types of landrace groups (Ranade et al. 2002; Verma et al. 2004). A further substantiation of those results as well as an in-depth analysis was carried out using four different primer sets including RAPD primers on a specific subset of landraces clearly distinguished from each other by flowering behavior as either male or female betelvines (Figure 3). On such a set of known male and female betelvines, hereafter referred to as the "gender set" we have assessed genetic diversity using four primer types revealing polymorphic profiles from discrete but widely distributed genomic regions. These primers include (i) arbitrary sequence decamer primers amplifying from several anonymous regions; (ii) SSR primers amplifying several inter-SSR regions; (iii) minisatellite core sequence primers amplifying from miniatellite rich regions and (iv) primers derived from X and Y chromosomes of a known dioecious plant with heteromorphic sex chromosomes, *Silene latifolia* that were expected to specifically amplify from genomic regions homologous to the X or Y chromosomes in case of the male and female betelvines. Such a PCR-based study was planned because there were no prior reports about chromosomal basis of gender distinction in case of betelvines. A better understanding of how gender discrimination takes place in those plants where chromosomal basis of inheritance of sex is undocumented or negligible or non-existent too is important as the study also assumes significance in understanding the resource allocation and metabolism costs attributed to gender discrimination in those plants where the economic importance of plant is defined by a metabolite or a phytochemical that is recovered from a somatic tissue. Such a study in case of a dioecious plant is expected to throw light on whether or not there is a difference in diversity of the two genders and if so whether such differences are significant in terms of the economic values of the plant.

Fig. 2. The stages during the construction of the unique man-made structures created to regulate the micro-climate for cultivation of the betelvines in the sub-Tropical parts of India. These structures are called as "bareja" and are almost exclusively made out of natural raw materials such as palm thatch and bamboo. The panels labeled A through H depict the various stages. The small black arrows in panels A through C point to the bamboo sticks that are used to support the betelvines. The white larger white arrows in panels B and C point to the thatch walls of the bareja sides.

Fig. 3. Photos of the female (top panel) and male (bottom panel) betelvines. The solid white arrows (top panel) and the solid black arrow (bottom panel) point to the female and male inflorescences respectively.

2. Materials and methods

Plant material: Betelvine landraces were collected from three of the centers of the All India Co-coordinated Research Project (AICRP) on Betelvine at Chinthalapudi (Bapatla) in Andhra Pradesh, Sirugamani (Tiruchirappalli) in Tamil Nadu and Digraj (Sangli) in Maharastra. The landraces of betelvine for which leaf tissue were collected are listed in Table 1. Young leaf tissue was harvested from the vines, washed free of dirt and dust and then quickly mopped dry on blotting sheets. The leaves were de-ribbed and powdered rapidly in liquid nitrogen, and then either the DNA isolation procedures were immediately followed or the powdered tissue was stored at -70°C till further use. The leaves of, the *Piper* species outgroup (*Piper hamiltonii*) as well as non-*Piper* outgroup, Mulberry variety 'MI-0129' were collected from NBRI, Lucknow. As in the case of betelvines the young leaf tissue was harvested, washed free of dirt and dust and then quickly mopped dry on blotting sheets. Then the leaves were de-ribbed and powdered rapidly in liquid nitrogen, and then either the DNA isolation procedures were immediately followed or the powdered tissue was stored at -70°C till further use.
Isolation of DNA: Total plant DNA was isolated from the frozen tissue powder according to the method of de Kochko and Hamon (1990) with some modifications as described earlier (Bhattacharya and Ranade, 2001). At least three to five independent DNA preparations were made from leaf tissues collected from each plant. The quantity and quality of DNA samples were estimated by comparing band intensities on agarose gel as well as by fluorometry (DyNA Quant 200, Pharmacia) using Hoechst 33258 as the fluorochrome.
Minisatellite, SSR, SLXY and RAPD Primers: Four minisatellite core sequence primers and five SSR primers and three SLXY primers were custom synthesized from Bangalore Genei, Bangalore, India. The fifteen RAPD primers were procured from Qiagen Operon Technology Inc., Alameda, CA, USA. The sequences of all these primers as well as the annealing temperatures used in PCR with these primers are given in Table 2.

Sample #	Landrace	Collection	
		Source	Time
207F	Bangla Nagaram	Chinthalapudi (A.P.)[a]	Jan. 2000
211M	Kapoori Chillumurru	-- Ditto --	Jan. 2000
213M	Kapoori Tuni	-- Ditto --	Jan. 2000
214M	Kapoori Peddachapelli	-- Ditto --	Jan. 2000
218M	Kapoori Doddipatla	-- Ditto --	Jan. 2000
219M	Kapoori Chinnachapelli	-- Ditto --	Jan. 2000
223F	Bangla (U.P.)	-- Ditto --	Jan. 2000
226M	Kapoori Vuyyur (A.P.)	-- Ditto --	Jan. 2000
234F	Godi Bangla (Orrisa)	-- Ditto --	Jan. 2000
235F	Bangla (M.P.)	-- Ditto --	Jan. 2000
239F	Bangla Ramtek	-- Ditto --	Jan. 2000
244F	Kali Bangla (Assam)	-- Ditto --	Jan. 2000
301M	Kapoori Mhaisal	Digraj (Maharashtra)[b]	Jul. 2000
304M	Kapoori Bolvad	-- Ditto --	Jul. 2000
305M	Kapoori Viddi	-- Ditto --	Jul. 2000
306M	Kapoori Karve	-- Ditto --	Jul. 2000

307M	Kapoori Indapur	-- Ditto --	Jul. 2000
310M	Kapoori Arvi	-- Ditto --	Jul. 2000
319F	Calcutta Bangla	-- Ditto --	Jul. 2000
323M	Kapoori Maharashtra	-- Ditto --	Jul. 2000
326M	Kapoori Solapur	-- Ditto --	Jul. 2000
409F	Bangla Jal	Sirigamani (T.N.)[c]	Aug. 2000
410F	Bangla Ramtek	-- Ditto --	Aug. 2000
423F	Bangla Jabalpur	-- Ditto --	Aug. 2000
426F	Bangla Desi	-- Ditto --	Aug. 2000
427F	Nav Bangla	-- Ditto --	Aug. 2000
601F	Kakair	Sirugamani (T.N.)[c]	Aug. 2003
602F	Banchi Kodi	-- Ditto --	Aug. 2003
603F	Bangla Jal	-- Ditto --	Aug. 2003
604F	Kuljedu	-- Ditto --	Aug. 2003
605M	Kapoori Vasani	-- Ditto --	Aug. 2003
606M	Kapoori Bihar	-- Ditto --	Aug. 2003
607F	Deshawari	-- Ditto --	Aug. 2003
608F	SGM-1	-- Ditto --	Aug. 2003
609F	Sreenivasa Nallur	-- Ditto --	Aug. 2003
610F	Bangla Ramtek	-- Ditto --	Aug. 2003
612F	Bangla	-- Ditto --	Aug. 2003
613M	Kapoori Chittikavata	-- Ditto --	Aug. 2003
614M	Tellaku Ponnur	-- Ditto --	Aug. 2003
615F	Shirpurkata	-- Ditto --	Aug. 2003
616M	Tellaku	-- Ditto --	Aug. 2003
617M	Tellaku Uttukuru	-- Ditto --	Aug. 2003
618M	Karpuri (T.N.)	-- Ditto --	Aug. 2003
619M	Sangli Kapoori	-- Ditto --	Aug. 2003
620F	Kalipatti	-- Ditto --	Aug. 2003
621F	Gach Pan	-- Ditto --	Aug. 2003
PH	Piper hamiltonii	NBRI (Lucknow)[d]	Aug. 2000
MO	Mulberry	NBRI (Lucknow)[e]	Sep. 1999

[a] AICRP on Betelvine Center at Acharya NG Ranga Agricultural University, Bapatla, A. P. State.
[b] AICRP on Betelvine Center at MPKV ARS at Kasbe Digraj, Sangli, Maharashtra State.
[c] AICRP on Betelvine Center at TNAU Sugarcane Research Station at Sirugamani, Tiruchirapalli, T. N. State.
[d] Betelvine Conservatory, at Betelvine Biotechnology Laboratory, NBRI, Lucknow, U. P. State.
[e] Mulberry accession number MI-0129 from an earlier NBRI collection of tissue for R&D work on mulberry.
Samples in shaded cells were not included in data analysis since their PCR profiles were not always consistent amongst replicate experiments.

Table 1. Betelvine landraces (gender set), *Piper hamiltonii* and mulberry included for the studies on PCR-based profiling are given below. In all cases, leaf tissue was harvested, washed and stored frozen as described. The source locations have been detailed as footnotes to the table. The sample number refers to the numbers assigned to the tissue samples while **M** or **F** suffixes to sample numbers indicate the appropriate gender as Male or Female respectively.

Primer Name	Primer Sequence (5′- 3′) / (Length in numbers of bases)	Annealing Temperature (°C)
RAPD Method primers		
OPG-02	GGCACTGAGG / (10)	**35**
OPG-03	GAGCCCTCCA / (10)	**35**
OPG-08	TCACGTCCAC / (10)	**35**
OPG-10	AGGGCCGTCT / (10)	**35**
OPG-11	TGCCCGTCGT / (10)	**35**
OPG-13	CTCTCCGCCA / (10)	**35**
OPG-16	AGCGTCCTCC / (10)	**35**
OPG-17	ACGACCGACA / (10)	**35**
OPG-18	GGCTCATGTG / (10)	**35**
OPG-19	GTCAGGGCAA / (10)	**35**
OPH-04	GGAAGTCGCC / (10)	**35**
OPH-06	ACGCATCGCA / (10)	**35**
OPH-12	ACGCGCATGT / (10)	**35**
OPH-18	GAATCGGCCA / (10)	**35**
OPH-19	CTGACCAGCC / (10)	**35**
ISSR method primers		
$(GACA)_4$	GACAGACAGACAGACA / (16)	37, **40**, 45
$(GATA)_4$	GATAGATAGATAGATA / (16)	**30**, 33, 37
$(GAA)_6G$	GAAGAAGAAGAAGAAGAAG / (19)	41, 45, **48**
$(CA)_8GC$	CACACACACACACACAGC / (18)	**45**,48,53
$(ACTG)_4$	ACTGACTGACTGACTG / (16)	**38**, 41, 45
DAMD method primers		
M-13	GAGGGTGGCGGTTCT / (15)	41, 45, **48**
HVR(-)	GCTCCTCCCTCCT / (13)	**50**, 55
HBV	GGTGTAGAGAGGGGT / (15)	41, 45, **48**
33.6	GGAGGTGGGCA / (11)	**50**, 55
SLXY primers		
SLXY6	TGGACTTCCACTGGAATTCGAT / (21)	45, **50**, 55
SLXY7	ACTTGCAACGACTTCACTTTGAG / (25)	45, **50**, 55
SLXY8	ATCGAATTCCAGTGGAAGTCC / (22)	45, **50**, 55

Table 2. Various primers used for PCRs in present study. Annealing temperatures given in boldface are the optimum for that primer.

SPAR with arbitrary sequence decamers (RAPD): Decamers from kits G and H (Operon Technologies Inc., Alameda, California, USA) were used as primers. DNA was amplified essentially following Williams *et al.* (1990). Initially a pilot experiment was carried out varying primer, template DNA and Mg^{++} ion concentrations. The final amplification reactions contained 1x buffer, 1.5 mM $MgCl_2$, 200 μM each dNTP, 10 pmoles primer, 0.6 U *Taq* DNA polymerase (Bangalore Genei, Bangalore, India) and 50 ng betelvine DNA template in a 10μl reaction volume. The reaction was predenatured at 94°C for 1 min and thereafter cycled 44 times at 94°C for 15 s, 35°C for 20 s and 72°C for 45 s in the Air Thermal Cycler. Additionally a final extension cycle allowed incubation for 240 s at 72°C. The PCR products were separated by electrophoresis (at constant current of 5A) through 1% agarose gels in 0.5X TBE buffer, visualized and imaged using Nighthawk™ gel documentation system (pdi Inc., USA) after staining with ethidium bromide. Only distinct and well-separated bands repeated in replicated experiments were included in the analysis.

SPAR with SSR primers (ISSR-PCR): The PCR conditions were according to Gupta *et al.* (1994). A pilot experiment was carried out to determine the optimum annealing temperature for each primer in the range 3-10°C lower than the denaturation temperature. The denaturation temperature was calculated according to Berger and Kimmel (1987), by adding 2°C for each A or T and 4°C for each G or C in the oligomer. The final reaction was carried out in 10 μl volumes and contained 50 ng of template DNA, 10 pmoles of SSR primer, 200 μM each dNTP, 1.5 mM Mg^{2+} ion concentration in suitable 1X assay buffer supplied along with the enzyme and 0.6 Units of the thermostable *Taq* DNA polymerase (Bangalore Genei, Bangalore, India). The tubes were placed in the Air Thermal Cycler (Idaho Technology, USA) for the PCR. The Air Thermal Cycler was programmed to include pre-denaturation at 94°C for 60 s. This was followed by 30 cycles of denaturation at 94°C for 20 s, annealing at the optimized temperature for 30 s, and extension at 72°C for 45 s. The final cycle allowed an additional 240 s period of extension at 72°C. The reaction products obtained after PCR were analyzed on 1.2 or 1.8% agarose gels. The gel was stained in ethidium bromide and visualized and imaged on Nighthawk™ gel documentation system (pdi Inc., USA). Only distinct and well-separated bands repeated in replicated experiments were included in the analysis.

SPAR with minisatellite primers (DAMD): The reaction was carried out essentially according to Zhou *et al.* (1997) for the primers 33.6, HBV and HVR(-) while for primer M13 the reactions were essentially according to Lorenz *et al.* (1995). The optimum annealing temperature was determined by carrying out DAMD at different annealing temperatures in the range 40°C to 55°C. The PCR parameters were as follows: First cycle of incubation at 94°C for 60 s followed by 30 cycles of incubation at 94°C for 20 s, at annealing temperature (41°C in case of M13 primer and 50°C in case of the other three primers) for 30 s and at 72°C for 45 s. The final cycle allowed an additional incubation at 72°C for 240 s. The reactions contained 100 ng template DNA, 40 pmoles of primer, 1.5 mM Mg^{2+} ions, 200 μM each dNTP, 0.3 U *Taq* DNA polymerase in 1x assay buffer in a final volume of 15 μl in case of the primers 33.6, HBV and HVR(-). In case of the M13 primer, the template, primer and enzyme and the reaction volumes were 60 ng, 10 pmoles, 0.6 Units and 10 μl respectively. All amplification reactions were carried out in Air Thermal Cycler (Model ATC1605, Idaho Technology, Inc.) and the products were separated by electrophoresis (at constant current of

5A) through 1.2% agarose gels in 0.5X TBE buffer, visualized and imaged using Nighthawk™ gel documentation system (pdi Inc., USA) after staining with ethidium bromide. Only distinct and well-separated bands repeated in replicated experiments were included in the analysis.

SPAR with SLXY primer (SLXY-PCR): The primers used for profiling, were synthesized from the prior known sequences specific to the X- and Y-chromosomes of the dioecious perennial plant *Silene latifolia* (Filatov *et al.*, 2000). The reaction conditions were optimized for these primers also in the same way as for the RAPD primers. The final reactions included 100 ng of template, 20 pmoles of primer, 1.5 mM Mg^{2+} ions, 200 μM each dNTP, 0.9 U *Taq* DNA polymerase in 1x assay buffer in a final volume of 15 μl. All PCR were carried out in Air Thermal Cycler (Model ATC1605, Idaho Technology, Inc.). The Air Thermal Cycler was programmed to include pre-denaturation at 94°C for 60 s. This was followed by 30 cycles of denaturation at 94°C for 20 s, annealing at 50°C for 30 s, and extension at 72°C for 45 s. The final cycle allowed an additional 240 s period of extension at 72°C. The PCR products were separated by electrophoresis (at constant current of 5A) through 1.4% agarose gels in 0.5X TBE buffer, visualized and imaged using Nighthawk™ gel documentation system (pdi Inc., USA) after staining with ethidium bromide. Only distinct and well-separated bands repeated in replicated experiments were included in the analysis.

Data Analysis: Data (fragment sizes of all amplification products, estimated from the gel by comparison with standard molecular weight marker, λ DNA double digested with *Hin*d III and *Eco*R I) were scored as discrete variables, using "1" to indicate presence and "0" to indicate absence of a band. A pair wise matrix of distances between genotypes was determined for the band data from each method using Jaccard coefficient (Jaccard, 1908) in the **FreeTree** program (ver. 0.9.1.5; Pavlicek *et al.* 1999). These pairwise distance data were used to compare the average distances estimated within and between the male and female betelvine landraces for each method. Additionally, a cumulative distance matrix, for the band data of all four methods considered together, was also computed separately to generate a single NJ tree after allowing a 1000 replicate bootstrap test using the same program. The tree was viewed, annotated and printed using **TreeView** (ver. 1.6.5; Page 2001). The robustness of the SPAR methods was tested in each case with a suitable non-*Piper* outgroup DNA included in the analysis, and in all cases outgroup was resolved as distinct from the betelvine DNAs (data not shown).

Comparison of the different SPAR methods: To determine the utility of each of the marker systems used, diversity index (**DI**), effective multiplex ratio (**E**) and marker index (**MI**) were calculated according to Powell *et al.* (1996).

DI for genetic markers was calculated from the sum of the squares of the allele frequency:

$$\mathbf{DI_n = 1\text{-}\sum p_i^2}$$

where 'p_i' is the allele/band frequency of the 'i^{th}' allele and '**n**' is the total number of loci. **DI** for polymorphic markers ($\mathbf{DI_p}$) was calculated from:

$$\mathbf{DI_p = \sum DI_n / np}$$

where '**np**' is the number of polymorphic loci analyzed. Effective Multiplex Ratio (**E**) is the product of the fraction of polymorphic loci and the number of polymorphic loci for an individual assay and was calculated from:

$$E = np(np / n)$$

MI is defined as the product of the average diversity index for polymorphic bands in any assay and the effective multiplex ratio (**E**) for that assay. It was calculated as:

$$MI = DI_p \times E$$

The Mantel matrix correspondence test (Mantel, 1967) was used to compare distance matrices for each SPAR method with the help of the program **Mantel 2** (Liedloff, 1999).

3. Results

Optimization of the different PCRs: Screening of the genotypes and various optimization experiments was carried out so as to identify the reaction parameters and conditions including concentrations of template, Mg^{2+} ions, primers and dNTPs used as well as the annealing temperatures in the PCRs that enabled the most repeatable results. Likewise, consistency of the profiles was judged by using duplicate or triplicate template samples for the same primers such that all the prominent bands were consistently produced with each of the template replicates. In all 22 male and 24 female betelvine variety DNAs were used in the present study along with *Piper hamiltonii* and mulberry DNAs as outgroups. The last named mulberry DNA outgroup was used just to determine the robustness of the PCR reactions and data from this was not scored for the analysis. On the basis of the preliminary optimizations (data not shown) template DNAs of male varieties, 211, 301 and 310 (Table 1) were excluded from data analysis since profiles with these were not consistent with all primers and all replicates tested.

RAPD analysis of the gender set of betelvine landraces: A total of 15 primers (Table 2) were used for RAPD analysis and all the primers resulted in discrete amplification products. Three landraces 211, 301 and 310 failed to give discrete profiles with some of the primers so data for these landraces was not scored from the profiles for final calculations and analysis. The RAPD data for all 15 primers were considered cumulatively and included a total of 219 bands (17.13 bands per primer) for the analysis of the relationship amongst the betelvine landraces. Of these some 98% bands were polymorphic and only four bands were present in all betelvine and *Piper hamiltonii* DNAs. Jaccard coefficients were computed from the band data and the highest (0.86) in female group was between two Bangla Ramtek accessions 239 (from Chinthalapudi) and 610 (from Sirugamani) while the least (0.22) was between 604 (Kuljedu) and 620 (Kallipati) and 427 (Nav bangla). The corresponding values in case of the male landraces were 0.96 between 213 (Kapoori Tuni) and 214 (Kapoori Peddachapelli) and 0.23 between 307 (Kapoori Indapur) and 218 (Kapoori Doddipatla). Between male and female, the highest (0.40) was between 326 (Kapoori Solapur) and 612 (Bangla) whereas least (0.18) was between both 305 (Kapoori Viddi) and 326 (Kapoori Solapur) and 427 (Nav Bangla). The distance data computed from the Jaccard coefficients were analyzed further by NJ method, using the FreeTree program, as described earlier in the materials and methods, to describe the relationship between betelvine landraces. The NJ tree after 1000 replicate bootstrapping revealed two major clusters respectively for the female and male landraces and the out group 501 (*Piper Hamiltonii*) was separated from these two clusters (NJ tree not shown).

ISSR-PCR profiles of gender set of betelvine landraces: Here five primers resulted in discrete patterns of 73 closely spaced bands (16.20 bands per primer). Of these some 93%

bands were polymorphic across the betelvine landraces and only 5 bands were present in all landraces as well as *P. hamiltonii*. The Jaccard coefficients were computed and the highest (0.94) in female group was between 223 (Bangla U.P.) and 235 (Bangla M.P.) while the least (0.31) was between 409 (Bangla Jal) and 612 (Bangla). In male group the corresponding values were (0.97) between 219 (Kapoori Chinnachapelli) and 226 (Kapoori Vuyyur) and (0.47) between 307 (Kapoori Indapur) and 618 (Karpuri) and between 616 (Tellaku) and 304 (Kapoori Bolvad). Among female and male groups the highest was 0.50 between 618 (Karpuri), 410 (Bangla Ramtek), 223 (Bangla U.P.), 609 (Sreenivasa Nallur) and 619 (Sangli Kapoori), 608 (SGM-1) and 304 (Kapoori Bolvad) while the least (0.28) was between 616 (Tellaku) and 427 (Kapoori Bihar). The NJ tree after 1000 replicate bootstrapping revealed the betelvine landraces grouped together in two major clusters of the male and female landraces and here also as in the case of RAPD data 501 (*Piper Hamiltonii*) was separated out distinctly (NJ tree not shown).

DAMD profiles of the gender set of betelvine landraces: The DAMD profiles consisted of distinctly polymorphic banding patterns and the four DAMD primers resulted in a total of 68 polymorphic bands (19.00 bands per primer). Jaccard coefficients in the female group here revealed highest (0.83) was between 410 (Bangla Ramtek) and 610 (Bangla Ramtek). In male group the highest (0.95) was between 305 (Kapoori Viddi) and 326 (Kapoori Solapur). Between female and male, the highest value is 0.44 between 606 (Kapoori Bihar) and 604 (Kuljedu) and 617 (Tellaku Uttukuru). The distance data analyzed by NJ method using the FreeTree program, as described earlier revealed the NJ tree after 1000 replicate bootstrapping was divided into two major clusters here also.

SLXY primer PCR profiles of the gender set of betelvine landraces: PCR with SLXY primers used singly in the amplification reactions resulted in distinctly polymorphic banding patterns. A total of 60 bands (all polymorphic) were scored with the three SLXY primers with the average number of 17.25 bands per primer. The highest (0.91) between 207 (Bangla Nagaram) and 235 (Bangla M.P.) and least (0.22) between 234 (Godi Bangla), 426 (Bangla Desi) and 427 (Nav Bangla) coefficients were computed in the female group. In male group the corresponding values are 1.0 between 218 (Kapoori Doddipatla) and 213 (Kapoori Tuni); 0.26 between 619 (Sangli Kapoori) and 614 (Tellaku Ponnur). Between female and male, the highest value is 0.45 between 617 (Tellaku Uttukuru) and 606 (Kapoori Bihar) and least 0.95 was between 427 (Nav Bangla) and 619 (Sangli Kapoori). The NJ tree after 1000 replicate bootstrapping reveals *Piper Hamiltonii* (501) were clearly separated from the rest of the betelvine landraces that were in turn divided into two major clusters of male and female landraces respectively.

Comparison of the different PCR methods used: Four different PCR-based methods were used to assess diversity and to distinguish between the genders. In order to determine the utility of each of these PCR-based methods used, a comparative statistical assessment was done according to Powell *et al.*, (1996). Diversity Index (**DI**) or Heterozygosity Index, Marker Index (**MI**) and Effective Multiplex Ratio (**E**) were calculated in case of each of the methods used as described in the materials and methods. By RAPD method **DI** value was 0.29, **E** value was 17.13 and **MI** value was 4.91. For the same set of landraces with ISSR-PCR, **DI** was 0.33, **E** was 16.2 and **MI** was 5.3. When DAMD primers were used for the analysis, **DI** value was 0.23, **E** was 19.0 and **MI** was 4.4. The corresponding values for the SLXY primer PCR were 0.25, 17.25 and 4.37 respectively (Table 3).

Comparison	Average similarity index (SI)	Method	Average # of bands per primer	Diversity Index (DI)	Effective Multiplex Ratio (E)	Marker Index (MI)
Within Male	0.544	RAPD	14.60	0.29	17.13	4.914
Within Female	0.492					
Between Male and Female	0.282					
Within Male	0.671	**ISSR**	16.20	0.33	16.20	5.289
Within Female	0.606					
Between Male and Female	0.400					
Within Male	0.389	**DAMD**	19.00	0.23	19.00	4.423
Within Female	0.245					
Between Male and Female	0.183					
Within Male	0.545	**SLXY**	17.25	0.25	17.25	4.366
Within Female	0.489					
Between Male and Female	0.266					

Table 3. The comparison of different PCR methods for the assessment of genetic diversity in the gender set of betelvine landraces.

Methods compared Matrix 1 / Matrix 2	Standard Normal Variate (g)	Mantel Coefficient for 100 random iterations (Z)	Correlation coefficient (r)
RAPD / ISSR	21.179**	673.781	0.8432+++
RAPD / DAMD	9.968**	970.869	0.5009+
RAPD / SLXY	18.021**	828.749	0.7853+++
ISSR / DAMD	8.895**	790.384	0.4619+
ISSR / SLXY	17.562**	680.878	0.7854+++
DAMD / SLXY	6.399**	978.962	0.3850+

** Highly significant values (Critical value $\mathbf{p}_{0.005}$ = 2.575)
+++: High correlation
+: Moderate or low correlation

Table 4. Correlations among the distance matrices for the RAPD, ISSR, DAMD and SLXY primer PCR band data in case of the gender based set of the betelvine landraces. A significant, (p = 0.005), standard normal variate **(g)** was obtained among the matrices generated by all the four methods in all possible pairs of matrix comparison.

Another parameter for comparing the different profiling methods is to correlate the respective distance matrices (computed from the Jaccard coefficients) by a Mantel test. Such a correlation analysis was carried out (Table 4) and resulted in highly significant correlation (standard variate **g > p** $_{0.005}$ value) in all cases. However, for the four methods the highest correlation was between RAPD and ISSR distance matrices (**r** = 0.8432, Table 4). This clearly indicates that relative distances of the landraces to each other were almost similarly estimated by both RAPD and ISSR methods. On the other hand, in case of the DAMD method, it was less strongly correlated (**r** = 0.5009, 0.4619 and 0.3850) to the other three methods (RAPD, ISSR and SLXY respectively). The correlation between RAPD and ISSR with SLXY primer data was almost equally strong in both cases (**r** = 0.7853, 0.7854 respectively).

4. Discussion

Dioecy is a widespread condition in flowering plants. Despite their recent evolutionary origin, 6% of the 240,000-angiosperm species are dioecious and 7% of 13,000 genera of angiosperms include dioecious species, suggesting that it has arisen many times during flowering plant evolution (Renner and Ricklefs, 1995). Dioecy is correlated with perennial climbing growth, wind or water pollination and has a preponderence in tropical flora. Natural selection, development of complex physiological and morphological traits, male fitness, ecological context, sex ratio, phylogenetic perspective and nuclear-cytoplasmic gynodioecy are some of the factors affecting distribution of dioecy. The plant taxa can offer insights into level of dimorphism that exists prior to the evolution of complete dioecy and the tradeoffs or constraints faced by hermaphrodites. The betelvine is one such interesting dioecious plant with an almost obligate vegetative propagation, lianaceous habit and perennial growth that provides a good system for studying molecular aspects of dioecy in general and functional dioecy in particular.

The study with PCR profiles was the first step towards resolving gender differences, if any, in betelvines. In an earlier study with 53 landraces that included only a few landraces with known genders, the RAPD method had provided a distinction between male and female betelvines (Verma *et al.*, 2004). In the present study with a larger sampling specifically amongst the betelvines with known genders, the bootstrap NJ tree for the RAPD data clearly differentiated the male and female landraces into two separate broad clusters, thereby supporting our original grouping of the landraces in terms of male and female vines based on known or actual flowering data. Banerjee *et al.* (1999) have reported a similar study of RAPD profile variation in another dioecious species, *Piper longum* L. and have further shown that at least two RAPD fragments were consistently associated with male plants. The clear separation of the betelvines on the basis of gender provides important leads for the identification and development of gender specific primers and probes. This work, however, has an important caveat that the economically most important product of the betelvines are the leaves and leaves of both male and female vines have a market value. Therefore the gender specific detection and or diagnosis will have only an academic value in case of betelvines unlike that in plants such as Papaya (Deputy *et al.*, 2002; Ma *et al.*, 2004). ISSR-PCR has been successfully utilized to distinguish gender or gender-specific markers in dioecious plants like hops and datepalm (Jakse et al. 2008; Younis et al. 2008). In the present study one ISSR primer each with dinucleotide and trinucleotide motifs and three primers with tetranucleotide motifs were tried in PCR and revealed distinct profiles that were however broadly similar across all the landraces. This result indicated that the ISSR regions were apparently conserved at least in length if not in sequence. The polymorphic bands helped to resolve the NJ tree into two major clusters with two groups of male and female betelvine clearly separating out. DAMD method has not been specifically used for any gender distinction studies in plants. For betelvine landraces, the NJ tree separated the male and female genders of the landraces and at the same time the DAMD method also revealed a relatively greater diversity amongst the male betelvines. The primers based on sequences specific to known X and Y chromosome of a dioecious plant *Silene latifolia* were tested with the dioecious betelvines in the gender set of landraces. Here the expectation was that the X and Y specific primers would reveal clear differences in the PCR profiles of the female and male betelvines respectively. These primers when used singly as well as in combination, however, did not reveal any such discrete profile differences. Surprisingly, these primers actually resulted in RAPD like multibanded profile and hence the band data for this was also scored in the same way as for the other three methods. This kind of study has never been done for any other plant to our knowledge. The conclusion about the most similar or most dissimilar landraces was based on cumulative data for all the SLXY primers. The result is actually interesting. Primarily the results indicate that multiple primer binding sites were present in all the DNAs. If these primers were generating multiple products it would appear that they have no co-relation with gender determining sequences in betelvines. This however is not entirely true since the NJ tree from the cumulative data can be clearly resolved into two major clusters, for the male and female betelvine respectively. Thus the PCR products resulting from these primers seemingly differs between the two genders and since several products are formed. It would appear that these primers are amplifying sequences from more than one gender determinant. Further since primers from both X and Y specific sequences gave multiple amplification products from the both male and female betelvines, our results indicate that gender determination in betelvine may either not be dependent on specific sex chromosomes or if such chromosome do exist male and female betelvines have similar sets of sequences on these chromosomes. Considering that all dioecious plants do not always have heteromorphic and distinct sex chromosomes, on the basis of our results with SLXY primers we infer that betelvine is one of those dioecious plants that lack distinct sex chromosomes.

In a novel approach, the fourth group of primers used was actually a heterologous set of primers were derived from X and Y chromosome specific sequences of the dioecious plant *Siliene latifolia*. The use of such primers for the dioecious betelvines was expected to reveal specific information about chromosomal basis for dioecy, if any. Interestingly, the SLXY primers resulted in multibanded profiles of several distinct bands even under stringent PCR conditions clearly indicating that several dispersed sequences homologous to the primers used were present. Though the primers collectively generated data that segregated the male and female vines, no single primer gave a sharply defined dimorphic profile for the two genders. This observation leads us to the conclusion that dioecy in betelvine may not follow the same chromosomal basis as in the case of *Silene latifolia*. Of course it is possible that such a lack of dimorphic profile could also be attributed to lack of strong homology of the primers to the appropriate regions of the betelvine. In this situation however, it would be very interesting to determine the identity and sequence of genomic regions of betelvines that have resulted in the discrete multibanded profiles even under high PCR stringency.

Comparison of the different methods used to assess the genetic diversity in betelvines: The gender based set of betelvine landraces was systematically analyzed with four different types of primers. For each set of primers used, the profile data were used for the calculation of the Diversity Index (DI), Effective Multiplex Ratio (E) and Marker Index (MI) and Mean probability (p). These calculations allowed a comparison of the four PCR-methods. The values calculated in each case, are given in Table 3. Parson *et al.* (1997) suggested that differences in the chromosomal location of the three types of markers could influence the diversity assessment. Kojima *et al.* (1998), indicated that in wheat RAPDs were more representative of chromosomal regions enriched in repeated sequences, while ISSRs were related, as RFLPs, to coding sequences. A similar case was seen in case of lentil (Sonnate and Pignone, 2001) where the authors could not find any congruence between the RAPD and ISSR method. Carvalho and Schaal (2001) also obtained different levels of polymorphism in cassava, where the SSR-primed markers showed less polymorphism than the RAPD markers. In their case also grouping of varieties from different geographical habitats varied between the RAPD and ISSR techniques. In absence of any pedigree information about the varieties, we could not address the issue of concordance between the molecular profiling based estimates of genetic similarity and pedigree but, we may expect greater genetic information about genetic similarity from the molecular profiling based estimates in accordance with Russell *et al.* (1997). From their result also it was apparent that the different techniques reveal information about distinct regions of the genome. Moreover, the rate of evolution of the primer target site sequence is most likely different for the two types of markers. So depending on this aspect, the divergence shown amongst the genotypes by the different techniques would also differ. Powell *et al.* (1996) suggested, that any estimation of genetic relationships between individual genotypes was affected by, the number of markers, the distribution of markers in the genome (genome coverage) and the nature of evolutionary mechanisms underlying the variation measured.

The four methods do not reveal polymorphism within gender set to the same extent. Under our experimental conditions ISSR method was found to have the higher Marker index as well as PIC. This is clearly reflected in the bootstrapped Neighbour Joining tree for the ISSR data where the entire female vines clustered together in one broad group while the male vines were separated in at least three distinct sub-groups. In general during the course of the present study we have observed that the male betelvines are invariably more heterogenous than the female vines. Such a result is of significance for the application of breeding methods for the improvement of betelvines. Unfortunately betelvine as a crop is cultivated by vegetative means ever since it was first domesticated. There have been only sporadic attempts at the improvement of betelvines through controlled process. The present study has resulted in the assessment of range of diversity in the betelvines for a breeder who can now exploit the diversity judiciously.

In our study we found that ISSR method, which showed the highest Diversity and Marker Index, can be the method of choice for diversity analysis type of studies, in so far as polymorphism or Marker Index is the criterion. In the present study actually all four methods were almost equally useful for the analysis of the betelvine landraces (MI values in the range 4.366 to 5.289, Table 3). In fact the Mantel Test (distance matrix correlation) analysis (Table 4) also resulted in highly significant normal standard variate g > $p_{0.005}$ = 2.575 in all cases. However, here RAPD and ISSR have resolved the landraces almost equally similarly since the correlation (r = 0.8432, Table 4) was strong between the distance matrices of the two methods. This is an interesting observation and it is suggestive of the possibility that the landraces have more differences amongst themselves in their minisatellite and related tandem repeat sequences. This is actually supported by the observation that all DAMD bands in betelvine were polymorphic.

Sex detemination systems based on heteromorphic X and Y sex chromosomes are particularly interesting to study from both a developmental and evolutionary perspective. There are many paralleles between the sex chromosomes, in different species even between animals and plants. The evolution of heteromorphic sex chromosome systems in widely differing species suggests that similar forces may have been involved in each case (Charlesworh, 1992; Ellis, 1998; Charlesworth and Guttman, 1999). The *Silene* genus is an example of how the evolution of an XY system contributes to morphological change and speciation. The Y chromosome differs from all other chromosomes not only in that it is the only chromosome that does not recombine along majority of its length, but also is being present only in the male sex in a permanent haploid condition (Y genetic isolation), in having a common ancestry and persistent meiotic relationship with the X, and the tendency of its genes to degenerate during evolution (Y genetic erosion). The Y becomes a specialized male chromosome, which essentially behaves like a single recombination unit. The functional coherence of the Y can be achieved relatively early during Y evolution, which might be an essential condition for the maintenance of an XY system. Filatov *et al.* (2001) found several differences in polymorphism of the regions of X and Y chromosomes. In another study done by Lebel-Hardenack *et al.* (2002) for genetically mapping of the sex-determination loci on the male-specific Y chromosome, it was found that *S.latifolia* has three dispersed male-determining loci on the Y chrosome. The sorrel *Rumex acetosa* (Polygonaceae) is a perennial dioecious weed, which possesses sex chromosomes (XX in females, XY1 Y2 in males). These studies indicated that it is only in the recent years that some details are emerging about molecular mechanisms and profiles vis-à-vis gender discrimination in plants.

The four methods tested with the gender set of betelvines have resulted in separation of the male and female betelvines. Though the individual sub-clusters in the four methods are not congruent however the primary separation of the male and female betelvines is clear and unambiguous. Thus the gender distinction of the betelvine landraces in terms of flowering is strongly supported by molecular profiling with four different PCR methods. This is an important and an interesting result. The four methods result in discrete profiles that reflect different genomic regions and in spite of that, the methods allow the landraces to be segregated on the basis of gender. Thus, it appears that there are several levels of genomic differences between the male and female betelvines. In other words, it appears that gender distinction in betelvine may not be confirmed to a few chromosomes or chromosomal regions. This inference is actually supported by lack of male specific and female specific PCR profiles when Y and X chromosome specific primers were respectively used. This further strengthens our inference that dioecy in betelvines is not apparently determined by distinct heteromorphic sex chromosomes. In order to have a comprehensive distribution of landraces, the data generated by all four methods combined and considered cumulatively.

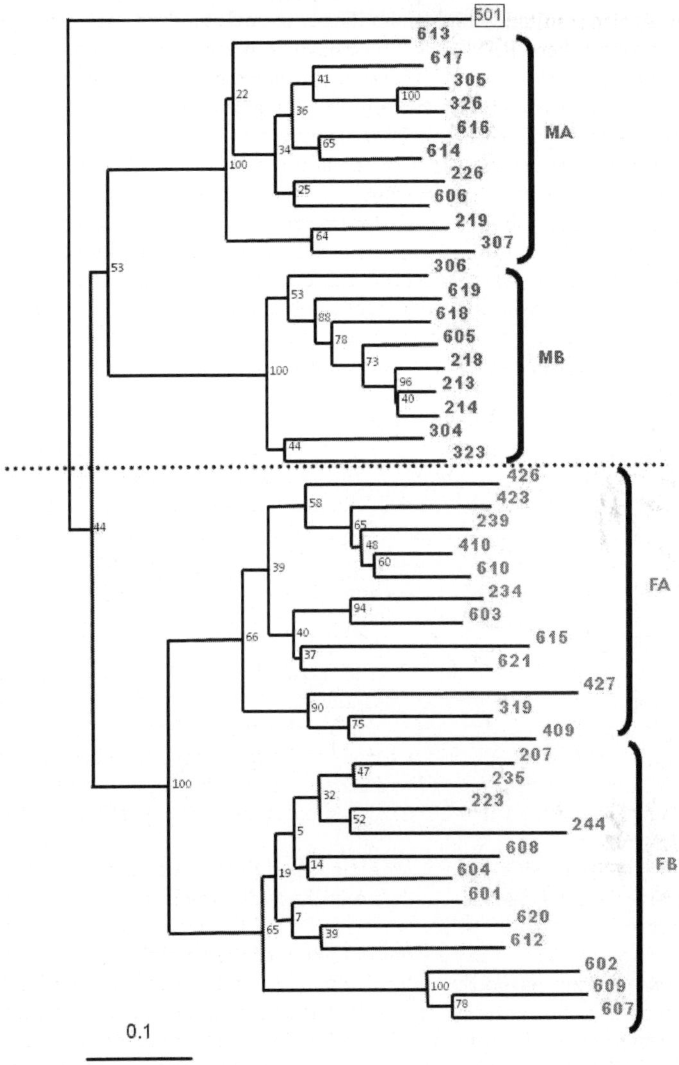

Fig. 4. The NJ tree for the cumulative band data for all the four PCR methods after 1000 replicate bootstrap analysis clearly depicts the separation of the out group taxon, *Piper hamiltonii* (501) from the betelvines which in turn are clearly distributed into separate clusters of male and female landraces. Interestingly each cluster of male or female landraces is further resolved into two sub clusters each, demarked in the figure with smaller labeled parenthesis. The scale at the bottom of the figure is for the Jaccard coefficients. Each landrace is identified by its number and color code as in Table 1. The dashed line through the figure separates the NJ tree into the respective gender halves.

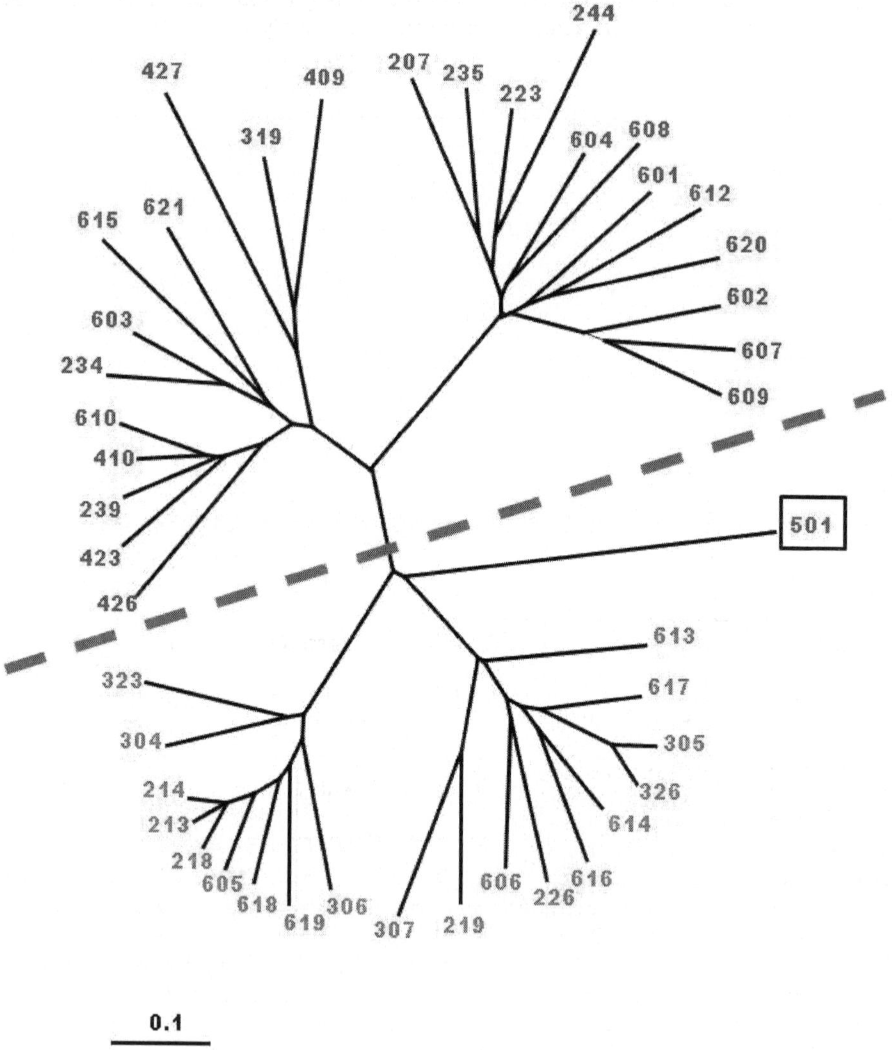

Fig. 5. The radial NJ tree for the cumulative band data for all the four PCR methods after 1000 replicate bootstrap analysis clearly depicts the separation of the out group taxon, *Piper hamiltonii* (501) from the betelvines which in turn are clearly distributed into separate clusters of male and female landraces. Each cluster of male or female landraces is very distinctly resolved into two sub clusters each. The scale at the bottom of the figure is for the Jaccard coefficients. Each landrace is identified by its number and color code as in Table 1. The dashed line through the figure separates the NJ tree into the respective gender halves.

The NJ tree from such a combined data for all methods is given in Figure 4. This tree also clearly separates the male and female betelvines. Interestingly the tree can be resolved into two subclusters within both male and female clusters as in the figure. MA, MB, FA and FB are the subclusters for males and females betelvines respectively (illustrated best by a radial tree form of the NJ tree as shown in Figure 5). We infer possibility of having at least two distinct ancient lineages for the male and female betelvines. In the absence of historical data and chronology of cultivation of these landraces it is not possible to confirm that there were actually a few discrete lineages of the cultivated betelvines and that the present day betelvine landraces are descendants of these lineages. The primary interest in all these studies is fuelled by the fact that dioecy as a proportion of accounts for only a small fraction of the numbers of flowering plants, yet in distribution across the plant families it is wide. Further, at the applied level, for plants of economic importance, detection and diagnosis of plant sex as an early event is desirable when the economically viable and important plant part is gender associated such as for example the fruits. In this context, betelvine is an exception in that for both male and female plants, the economically important plant part is the leaf.

5. References

Abe J., Xu D. H., Suzuki Y., Kanazawa A. and Shimamoto Y. (2003). Soybean germplasm pools in Asia revealed by nuclear SSRs. *Theor. Appl. Genet.* 106: 445-453.

Ajibade S. R., Weeden N. F. and Chite S. M. (2000). Inter simple sequence repeat analysis of genetic relationships in the genus *Vigna*. *Euphytica* 111: 47-55.

Banerjee N. S., Manoj P. and Das M. R. (1999) Male sex-associated RAPD markers in *Piper longum* L. *Current Science* 77(5): 693-695.

Bhattacharya E. and Ranade S. A. (2001). Molecular distinction amongst varieties of mulberry using RAPD and DAMD profiles. *BMC Plant Biology* 1: 3.

Carvalho L. J. C. B. and Schaal B. A. (2001). Assessing genetic diversity in the cassava (*Manihot esculenta* Crantz) germplasm collection in Brazil using PCR-based markers. *Euphytica* 120: 133-142.

Charlesworth D. (1992). The evolution of sex chromosomes. *Curr. Biol.* 2: 515-516.

Charlesworth D. and Guttman D. S. (1999). The evolution of dioecy and plant sex chromosome systems in *Sex determination in plants* edited by C. C. Ainsworth. BIOS Scientific Publishers, Oxford pp. 25-50.

deKochko A. and Hamon S. (1990). A rapid and efficient method for the isolation of restrictable total DNA from plants of genus *Abelmoschus*. *Plant Mol. Bio. Rep.* 8: 3-7.

Deputy J. C., Ming R., Ma H., Liu Z., Fitch M. M., Wang M., Manshardt R., Stiles J.I. (2002) Molecular markers for sex determination in papaya (*Carica papaya* L.). *Theor Appl Genet.* 106: 107-11.

Ellis N. A. (1998). The war of the sex chromosomes. *Nature Genet.* 20: 9-10.

Filatov D. A., Moneger F., Negrutiu I. and Charlesworth D. (2000). Low variability in a Y-linked plant gene and its implications for Y-chromosome evolution. *Nature* 404: 388-390.

Filatov D. A., Laporte C., Vitte C. and Charlesworth D. (2001) DNA diversity in sex-linked and autosomal genes of the plant species *Silene latifolia* and *Silene dioica*. *Mol. Biol. Evol.* 18: 1442-1454.

Gupta M., Chyi Y. S., Romero-Severson J. and Owen J. L. (1994). Amplification of DNA markers from evolutionarily diverse genomes using single primers of SSRs. *Theor. Appl. Genet.* 89: 998-1006.

Heath D. D., Iwama. G. K. and Devlin. R. H. (1993). PCR primed with VNTR core sequences yield species specific patterns with hypervariable probes. *Nucl. Acids Res.* 21: 5782 - 5785.

Jaccard P. (1908). Etude comparative dela distribution florale dans une portion des Alpes et des jura. *Bull. Soc. Vaudoise Sci. Nat.* 37: 547-579.

Jakse J., Stajner N., Kozjak P., Cerenak A. and Javornik B. (2008) Trinucleotide microsatellite repeat is tightly linked to male sex in hop (*Humulus lupulus* L.). *Mol. Breeding* 21: 139–148

Kojima T., Nagaoka T., Noda K. and Ogihara K. (1998). Genetic linkage map of ISSR and RAPD markers in einkorn wheat in relation to that of RFLP markers. *Theor. Appl. Genet.* 96: 37-45.

Kumar N. (1999). Betelvine (*Piper betle* L.) cultivation: A unique case of plant establishment under anthropogenically regulated microclimatic conditions. *Indian J. Hist. Sci.* 34: 19-32.

Lebel-Hardenack S., Hauser E., Teresa L. F., Schmid J. and Grant S. R. (2001). Mapping of sex determination loci on the White Campion (*Silene Latifolia*) Y chromosome using Amplified Fragment Length Polymorphism. *Genetics* 160: 717-725.

Li A. and Ge S. (2001). Genetic variation and clonal diversity of *Psammochloa villosa* (Poaceae) detected by ISSR markers. *Ann. Bot.* 87: 585-590.

Liedloff A. (1999) Mantel Nonparametric Test Calculator for Windows, Version 2.00 Distributed by author.

Lorenz M., Partensky F., Borner T. and Hess W. R. (1995). Sequencing of RAPD fragments amplified from the genome of the prokaryote *Prochlorococcus marinus* (Prochlorophyta). *Biochem. Mol. Bio. Int.* 36: 3705-3713.

Ma H., Moore P. H., Liu Z., Kim M. S., Yu Q., Fitch M. M., Sekioka T., Paterson A. H., Ming R. (2004) High-density linkage mapping revealed suppression of recombination at the sex determination locus in papaya. *Genetics* 166: 419-36.

Metais I., Aubry C., Hamon B., Jalouzot R. and Peltier D. (2000). Description and analysis of genetic diversity between commercial bean lines (*Phaseolus vulgaris* L.). *Theor. Appl. Genet.* 101: 1207-1214.

Nybom H. and Schaal B. A. (1990). DNA fingerprintings applied to paternity analysis in apple. *Theor. Appl. Genet.* 79: 763-768.

Nybom H., Rogstad S. H. and Schaal B. A. (1990). Genetic variation detected by use of M13 "DNA Fingerprints" probe in *Malus, Prunus* and *Rubus. Theor. Appl. Genet.* 79: 153-156.

Page R. D. M. (2001). *Tree View (Win 32) ver. 1.6.5.* Distributed by author.

Parsons B. J., Newbury H. J., Jackson M. T. and Ford-Llyod B. V. (1997). Contrasting genetic diversity relationships are revealed in rice (*Oryza sativa* L.) using different marker types. *Mol. Breed.* 3: 115-125.

Pavlicek A., Hrda S. and Flegr J. (1999). FreeTree – Freeware program for construction of phylogenetic trees on the basis of distance data and bootstrapping / jackknife analysis of the tree robustness. Application in the RAPD analysis of the genus *Frenkelia. Folia Biologica (Praha.).* 45: 97-99.

Perera L., Russell J. R., Provan J. and Powell W. (2001). Levels and distribution of genetic diversity of coconut (*Cocos nucifera* L., var. Typica form typical) from Sri Lanka assessed by microsatellite markers. *Euphytica* 122: 381-389

Perera L., Russell J. R., Provan J. and Powell W. (2003). Studying genetic relationships among Coconut varieties/populations using microsatellite markers. *Euphytica* 132: 121-128.

Powell W., Morgante M., Andre C., Hanafey M., Vogel J., Tingey S. and Rafalski A. (1996). The utility of RFLP, RAPD, AFLP and SSRP (microsatellite) markers for germplasm analysis. *Mol. Breed.* 2: 225-238.

Ranade S. A. and Farooqui N. (2002). Assessment of profile variations amongst provenance's of neem using single-primer amplification reaction (SPAR) techniques. *Mol. Biol.Today* 3: 1 - 10.

Ranade S. A., Verma A., Gupta M. and Kumar N. (2002) RAPD profile analysis of betelvine cultivars. *Biol. Plant.* 45: 523-527.

Reddy M. P., Sarla N. and Siddiq E.A. (2002). Inter simple sequence repeat (ISSR) polymorphism and its application in plant breeding. *Euphytica* 128: 9-17.

Renner S. S. and Ricklefs R. E. (1995). Dioecy and its correlates in the flowering plants. *American Journal of Botany* 82 (5): 596-606.

Russell J. R., Fuller J. D., Macaulay M., Hatz B. G., Jahoor A., Powell W. and Waugh R. (1997). Direct comparison of levels of genetic variation among barley accessions detected by RFLPs, AFLPs, SSRs, RAPDs. *Theor. Appl. Genet.* 95: 714-722.

Sonnate G. and Pignone D. (2001). Assessment of genetic variation in a collection of lentil using molecular tools. *Euphytica* 120: 301-307.

Verma A., Kumar N. and Ranade S. A. (2004) Genetic diversity amongst landraces of a dioecious vegetatively propagated plant, betelvine (*Piper betle* L.). *Journal of Biosciences* 29: 319-328.

Williams J. G. K., Kubelik A. R., Livak K. J., Rafalski J. A. and Tingey S.V. (1990). DNA polymorphisms amplified by arbitrary primers are useful as genetic markers. *Nucl. Acids Res.* 18: 6231-6235.

Younis R. A. A., Ismail O. M. and Soliman S. S. (2008) Identification of Sex-specific DNA Markers for Date Palm (*Phoenix dactylifera* L.) Using RAPD and ISSR Techniques. *Research Journal of Agriculture and Biological Sciences* 4: 278-284.

Zhebentyayeva T. N., Reighard G. L., Gorina V. M. and Abbott A. G. (2003). Simple sequence repeat (SSR) analysis for the assessment of genetic variability in apricot germplasm. *Theor. Appl. Genet* 106: 435-444.

Zhou Z., Bebeli P. J., Somers D. J. and Gustafson J. P. (1997). Direct amplification of minisatellite- region DNA with VNTR core sequences in the genus *Oryza*. *Theor. Appl. Genet.* 95: 942-949.